JOINT FORCES

Operations&Doctrine

Guide to Joint, Multinational & Interorganizational Operations

The Lightning Press

2227 Arrowhead Blvd.
Lakeland, FL 33813
24-hour Voicemail/Fax/Order: 1-800-997-8827
E-mail: SMARTbooks@TheLightningPress.com

www.TheLightningPress.com

Fifth Edition with Change 1

(JFODS5-1) The Joint Forces Operations & Doctrine SMARTbook

Guide to Joint, Multinational & Interorganizational Operations

JFODS5-1 is Change 1 to our fifth revised edition of The Joint Forces Operations & Doctrine SMARTbook. In addition to new/updated material from the latest editions of JP 3-0 Joint Operations (w/Change 1, Oct '18), JP 4-0 Joint Logistics (Feb '19), JP 3-33 Joint Task Force Headquarters (Jan '18), and JP 3-16 Multinational Operations (Mar '19), JFODS5-1 features a completely new chapter on Joint Air, Land, Maritime and Special Operations (JPs 3-30, 3-31, 3-32 & 3-05). Additional topics and references include JP 1 Doctrine for the Armed Forces of the United States (w/Change 1, Jul '17), JP 5-0 Joint Planning (Jun '17), and JP 3-08 Interorganizational Cooperation (val. Oct '17).

ISBN: 978-1-935886-80-8

Notice of Liability

The information in this SMARTbook and quick reference guide is distributed on an "As Is" basis, without warranty. While every precaution has been taken to ensure the reliability and accuracy of all data and contents, neither the author nor The Lightning Press shall have any liability to any person or entity with respect to liability, loss, or damage caused directly or indirectly by the contents of this book. If there is a discrepancy, refer to the source document. This SMARTbook does not contain classified or sensitive information restricted from public release.

"The views presented in this publication are those of the author and do not necessarily represent the views of the Department of Defense or its components."

Photo Credits. Cover photo: U.S. Army Rangers from Alpha Company, 3rd Battalion, 75th Ranger Regiment, search through documents found during their annual Task Force Training on Fort Knox KY., April 20, 2014 (U.S. Army photo by Spc. Philip Diab/RELEASED). Other photos courtesy of the Department of Defense and the Services.

Printed and bound in the United States of America.

(JFODS5-1) Notes to Reader

The nature of the challenges to the United States and its interests demand that the Armed Forces operate as a fully integrated **joint team** across the conflict continuum.

Joint operations are military actions conducted by joint forces and those Service forces employed in specified command relationships with each other, which of themselves do not establish joint forces. The potential range of **military activities and operations** extends from military engagement, security cooperation, and deterrence in times of relative peace up through major operations and campaigns that typically involve large-scale combat.

Joint planning is the deliberate process of determining how (the ways) to use military capabilities (the means) in time and space to achieve objectives (the ends) while considering the associated risks.

Joint logistics is the coordinated use, synchronization, and sharing of two or more Military Departments' logistics resources to support the joint force. Sustainment provides the joint force commanders freedom of action, endurance, and the ability to extend operational reach.

A joint task force (JTF) is established when the scope, complexity, or other factors of the contingency or crisis require capabilities of Services from at least two Military Departments operating under a single joint force commander.

Achieving national strategic objectives requires effective **unified action** resulting in unity of effort -- to include interagency, intergovernmental, nongovernmental and **multinational partners**. This is accomplished by **interorganizational cooperation**, synchronization, and coordination in the use of the diplomatic, informational, military, and economic **instruments of national power**.

JFODS5-1 is Change 1 to our fifth revised edition of The Joint Forces Operations & Doctrine SMARTbook. In addition to new/updated material from the latest editions of JP 3-0 Joint Operations (w/Change 1, Oct '18), JP 4-0 Joint Logistics (Feb '19), JP 3-33 Joint Task Force Headquarters (Jan '18), and JP 3-16 Multinational Operations (Mar '19), JFODS5-1 features a completely new chapter on Joint Air, Land, Maritime and Special Operations (JPs 3-30, 3-31, 3-32 & 3-05). Additional topics and references include JP 1 Doctrine for the Armed Forces of the United States (w/Change 1, Jul '17), JP 5-0 Joint Planning (Jun '17), & JP 3-08 Interorganizational Cooperation (val. Oct '17).

SMARTbooks - DIME is our DOMAIN!

SMARTbooks: Reference Essentials for the Instruments of National Power (D-I-M-E: Diplomatic, Informational, Military, Economic)! Recognized as a "whole of government" doctrinal reference standard by military, national security and government professionals around the world, SMARTbooks comprise a comprehensive professional library.

SMARTbooks can be used as quick reference guides during actual operations, as study guides at education and professional development courses, and as lesson plans and checklists in support of training. Visit **www.TheLightningPress.com**!

The following references were used to compile *The Joint Forces Operations & Doctrine SMARTbook*. All references are considered public domain, available to the general public, and designated as "approved for public release; distribution is unlimited." *The Joint Forces Operations & Doctrine SMARTbook* does not contain classified or sensitive material restricted from public release.

JFODS5-1 is Change 1 to our fifth revised edition of The Joint Forces Operations & Doctrine SMARTbook. In addition to new/updated material from the latest editions of JP 3-0 Joint Operations (w/Change 1, Oct '18), JP 4-0 Joint Logistics (Feb '19), JP 3-33 Joint Task Force Headquarters (Jan '18), and JP 3-16 Multinational Operations (Mar '19), JFODS5-1 features a completely new chapter on Joint Air, Land, Maritime and Special Operations (JPs 3-30, 3-31, 3-32 & 3-05). Additional topics and references include JP 1 Doctrine for the Armed Forces of the United States (w/Change 1, Jul '17), JP 5-0 Joint Planning (Jun '17), and JP 3-08 Interorganizational Cooperation (val. Oct '17).

Joint Publications (JPs)

JP 1	Jul 2017	Doctrine for the Armed Forces of the United States (with Change 1)
JP 3-0*	Oct 2018	Joint Operations (with Change 1)
JP 3-02*	Jan 2019	Amphibious Operations
JP 3-05*	Jul 2014	Special Operations
JP 3-08*	Oct 2017	Interorganizational Cooperation (validated w/no changes)
JP 3-13	Nov 2014	Information Operations (with Change 1)
JP 3-16*	Mar 2019	Multinational Operations
JP 3-30*	Jul 2019	Joint Air Operations
JP 3-31*	Feb 2019	Joint Land Operations
JP 3-32*	Jun 2018	Joint Maritime Operations
JP 3-33*	Jan 2018	Joint Task Force Headquarters
JP 4-0*	Feb 2019	Joint Logistics
JP 5-0	Jun 2017	Joint Planning

Other Publications and Manuals

CJCSM 3122.05	Dec 2011	Operating Procedures for Joint Operation Planning and Execution System (JOPES) (Current as of 18 Nov 2014)
CJCSM 3130.03A	Feb 2019	Planning and Execution Planning Formats and Guidance

* *New/updated references in this edition.*

(JFODS5-1) Table of Contents

Chap 1 Joint Doctrine Fundamentals

Chap 2

Joint Operations

Chap 3 Joint Planning

II(b). Application of Guidance

Chap 4

Joint Logistics

Chap 5

Joint Task Forces (JTFs)

Chap 6 Joint Force Operations (Air, Land, Maritime, SOF)

Chap 8

Interorganizational Cooperation

Chap 1

I. Joint Doctrine Theory & Foundations

Ref: JP 1 (w/Chg 1), Doctrine for the Armed Forces of the U.S. (Jul '17), chap. I.

Joint Publication 1 provides overarching guidance and fundamental principles for the employment of the Armed Forces of the United States. It is the capstone publication of the US joint doctrine hierarchy and it provides an overview for the development of other joint service doctrine publications. It is a bridge between policy and doctrine and describes authorized command relationships and authority that military commanders use and other operational matters derived from Title 10, United States Code (USC).

I. Fundamentals

The purpose of joint doctrine is to enhance the operational effectiveness of joint forces by providing fundamental principles that guide the employment of US military forces toward a common objective. With the exception of Joint Publication (JP) 1, joint doctrine will not establish policy. However, the use of joint doctrine standardizes terminology, training, relationships, responsibilities, and processes among all US forces to free joint force commanders (JFCs) and their staffs to focus their efforts on solving strategic, operational, and tactical problems. Using historical analysis of the employment of the military instrument of national power in operations and contemporary lessons, these fundamental principles represent what is taught, believed, and advocated as what works best to achieve national objectives.

As a nation, the US wages war employing all instruments of national power— diplomatic, informational, military, and economic. The President employs the Armed Forces of the United States to achieve national strategic objectives. The Armed Forces of the United States conduct military operations as a joint force. "Joint" connotes activities in which elements of two or more Military Departments participate. Joint matters relate to the integrated employment of US military forces in joint operations, including matters relating to:

- National military strategy (NMS)
- Deliberate and crisis action planning
- Command and control (C2) of joint operations
- Unified action with Department of Defense and interagency partners

The capacity of the Armed Forces of the United States to operate as a cohesive joint team is a key advantage in any operational environment. Unity of effort facilitates decisive unified action focused on national objectives and leads to common solutions to national security challenges.

Jointness and the Joint Force

The Armed Forces of the United States have embraced "jointness" as their fundamental organizing construct at all echelons. Jointness implies cross-Service combination wherein the capability of the joint force is understood to be synergistic, with the sum greater than its parts (the capability of individual components). Some shared military activities are less joint than are "common;" in this usage "common" simply means mutual, shared, or overlapping capabilities or activities between two or more Services.

II. War

War can result from failure of states to resolve their disputes by diplomatic means. Some philosophers see it as an extension of human nature. Thomas Hobbes stated that man's nature leads him to fight for personal gain, safety, or reputation.

Thucydides said nearly the same thing in a different order, citing fear, honor, and interest as the common causes for interstate conflict.

Individuals, groups, organizations, cultures, and nations all have interests. Inevitably, some of those interests conflict with the interests of other individuals, groups, organizations, cultures, and nations. Nearly all international and interpersonal relationships are based on power and self-interests manifested through politics. Nations exercise their power through diplomatic, informational, military, and economic means. All forms of statecraft are important, but as the conflicts approach the requirement for the use of force to achieve that nation's interests, military means become predominant and war can result. The emergence of non-state actors has not changed this concept. Non-state actors may not use statecraft as established; however, they do coerce and threaten the diplomatic power of other nations and have used force, terrorism, or support to insurgency to compel a government to act or refrain from acting in a particular situation or manner or to change the government's policies or organization.

See facing page for discussion of forms of warfare.

The Principles of War

War is socially sanctioned violence to achieve a political purpose. War historically involves nine principles -- objective, offensive, mass, economy of force, maneuver, unity of command, security, surprise, and simplicity -- collectively and classically known as the principles of war. The basic nature of war is immutable, although warfare evolves constantly.

The application of these classic "Principles of War" in the conduct of joint operations is amplified and expanded in JP 3-0, Joint Operations. See pp. 2-2 to 2-3.

Strategy in War

The two fundamental strategies in the use of military force are strategy of annihilation and strategy of erosion. The first is to make the enemy helpless to resist us, by physically destroying his military capabilities. This has historically been characterized as annihilation or attrition. It requires the enemy's incapacitation as a viable military force. The second approach is to convince the enemy that accepting our terms will be less painful than continuing to aggress or resist. This can be characterized as erosion, using military force to erode the enemy leadership's or the enemy society's political will. In such an approach, we use military force to raise the costs of resistance higher than the enemy is willing to pay. We use force in this manner in pursuit of limited political goals that we believe the enemy leadership will ultimately be willing to accept.

Particularly at the higher levels, waging war should involve the use of all instruments of national power that one group can bring to bear against another (diplomatic, informational, military, and economic). While the military focuses on the use of military force, we must not consider it in isolation from the other instruments of national power.

III. Strategic Security Environment and National Security Challenges

The strategic security environment is characterized by uncertainty, complexity, rapid change, and persistent conflict. This environment is fluid, with continually changing alliances, partnerships, and new national and transnational threats constantly appearing and disappearing. While it is impossible to predict precisely how challenges will emerge and what form they might take, we can expect that uncertainty, ambiguity, and surprise will dominate the course of regional and global events. In addition to traditional conflicts to include emerging peer competitors, significant and emerging challenges continue to include irregular threats, adversary propaganda, and other information activities directly targeting our civilian leadership and population, catastrophic terrorism employing weapons of mass destruction (WMD), and other threats to disrupt our ability to project power and maintain its qualitative edge.

Forms of Warfare

Ref: JP 1 (w/Chg 1), Doctrine for the Armed Forces of the U.S. (Jul '17), pp. I-4 to I-7.

Warfare is the mechanism, method, or modality of armed conflict against an enemy. It is "the how" of waging war. Warfare continues to change and be transformed by society, diplomacy, politics, and technology.

A useful dichotomy for thinking about warfare is the distinction between traditional and irregular warfare (IW). Each serves a fundamentally different strategic purpose that drives different approaches to its conduct; this said, one should not lose sight of the fact that the conduct of actual warfare is seldom divided neatly into these subjective categories. Warfare is a unified whole, incorporating all of its aspects together, traditional and irregular. It is, in fact, the creative, dynamic, and synergistic combination of both that is usually most effective.

Traditional Warfare

This form of warfare is characterized as a violent struggle for domination between nation-states or coalitions and alliances of nation-states. This form is labeled as traditional because it has been the preeminent form of warfare in the West since the Peace of Westphalia (1648) that reserved for the nation-state alone a monopoly on the legitimate use of force. The strategic purpose of traditional warfare is the imposition of a nation's will on its adversary nation-state(s) and the avoidance of its will being imposed upon us.

In the traditional warfare model, nation-states fight each other for reasons as varied as the full array of their national interests. Military operations in traditional warfare normally focus on an adversary's armed forces to ultimately influence the adversary's government. With the increasingly rare case of formally declared war, traditional warfare typically involves force-on-force military operations in which adversaries employ a variety of conventional forces and special operations forces (SOF) against each other in all physical domains as well as the information environment (which includes cyberspace).

Typical mechanisms for victory in traditional warfare include the defeat of an adversary's armed forces, the destruction of an adversary's war-making capacity, and/or the seizure or retention of territory. Traditional warfare is characterized by a series of offensive, defensive, and stability operations normally conducted against enemy centers of gravity. Traditional warfare focuses on maneuver and firepower to achieve operational and ultimately strategic objectives.

Irregular Warfare (IW)

This form of warfare is characterized as a violent struggle among state and non-state actors for legitimacy and influence over the relevant population(s). This form is labeled as irregular in order to highlight its non-Westphalian context. The strategic point of IW is to gain or maintain control or influence over, and the support of, a relevant population.

IW emerged as a major and pervasive form of warfare although it is not a historical form of warfare. In IW, a less powerful adversary seeks to disrupt or negate the military capabilities and advantages of a more powerful military force, which usually serves that nation's established government. The less powerful adversaries, who can be state or non-state actors, often favor indirect and asymmetric approaches, though they may employ the full range of military and other capabilities in order to erode their opponent's power, influence, and will. Diplomatic, informational, and economic methods may also be employed.

Military operations alone rarely resolve IW conflicts. For the US, which will always wage IW from the perspective of a nation-state, whole-of-nation approaches where the military instrument of power sets conditions for victory are essential. Adversaries waging IW have critical vulnerabilities to be exploited within their interconnected political, military, economic, social, information, and infrastructure systems.

IV. Instruments of National Power (DIME)

Ref: JP 1 (w/Chg 1), Doctrine for the Armed Forces of the U.S. (Jul '17), pp. I-11 to I-14.

The ability of the United States to achieve its national strategic objectives is dependent on the effectiveness of the US Government (USG) in employing the instruments of national power. The appropriate governmental officials, often with National Security Council (NSC) direction, normally coordinate these instruments of national power (diplomatic, informational, military, and economic). They are the tools the United States uses to apply its sources of power, including its culture, human potential, industry, science and technology, academic institutions, geography, and national will.

At the President's direction through the interagency process, military power is integrated with the other instruments of national power to advance and defend US values, interests, and objectives. To accomplish this integration, the armed forces interact with the other responsible agencies to ensure mutual understanding of the capabilities, limitations, and consequences of military and civilian actions. They also identify the ways in which military and nonmilitary capabilities best complement each other. The NSC plays key roles in the integration of all instruments of national power facilitating mutual understanding, cooperation, and integration of effort. This process of different USG agencies and organizations coordinating and working together is called "interagency coordination." The use of the military to conduct combat operations should be a last resort when the other instruments of national power have failed to achieve our nation's objectives.

Instruments of National Power

D Diplomatic	I Informational	M Military	E Economic
▪ Embassies/ Ambassadors ▪ Recognition ▪ Negotiations ▪ Treaties ▪ Policies ▪ International forums	▪ Military information ▪ Public diplomacy ▪ Public affairs ▪ Communications resources ▪ International forums ▪ Spokespersons, timing, media and venues for announcements	▪ Military operations ▪ Engagement, Security Coop, Deterrence ▪ Show of force ▪ Military technology ▪ Size, composition of force	▪ Trade policies ▪ Fiscal and monetary policies ▪ Embargoes ▪ Tariffs ▪ Assistance

D - Diplomacy

Diplomacy is the principal instrument for engaging with other states and foreign groups to advance US values, interests, and objectives. The Department of State (DOS) is the lead agency for the USG for foreign affairs. The credible threat of force reinforces, and in some cases, enables the diplomatic process. Leaders of the Armed Forces of the United States have a responsibility to understand US foreign policy and to assure that those responsible for US diplomacy have a clear understanding of the capabilities, limitations, and consequences of military action. Geographic combatant commanders (GCCs) are responsible for integrating military activities with diplomatic activities in their areas of responsibility (AORs). The US ambassador and the corresponding country team are normally in charge of diplomatic-military activities in countries abroad. When directed by the President or Secretary of Defense (SecDef), the GCC employs military forces in concert with the other instruments of national power. In these circumstances, the US ambassador and the country team or another diplomatic mission team may have complementary activities (employing the diplomatic instrument) that do not entail control of military forces, which remain under command authority of the GCC. Since diplomatic efforts are often complementary with military objectives, planning should be complementary and coincidental.

I - Information

In a broad sense, the informational instrument of national power has a diffuse and complex set of components with no single center of control. The United States believes in the free market place of ideas. Therefore, information is freely exchanged with minimal government controls. Constraints on public access to USG information normally may be imposed only for national security and individual privacy reasons. Information readily available from multiple sources influences domestic and foreign audiences including citizens, adversaries, and governments. It is important for the official agencies of government, including the armed forces, to recognize the fundamental role of the media as a conduit of information.

The USG uses strategic communication (SC) to provide top-down guidance relative to using the informational instrument of national power in specific situations. SC is focused USG processes and efforts to understand and engage key audiences to create, strengthen, or preserve conditions favorable to advancing national interests and objectives through the use of coordinated information, themes, messages, and products synchronized with the actions of all instruments of national power. SC's primary communication capabilities are coupled with defense support to public diplomacy (DSPD) and military diplomacy activities to implement a holistic SC effort.

The predominant military activities that support SC themes and messages are information operations (IO), public affairs (PA), and DSPD. IO are those military actions to attack an adversary's information and related systems while defending our own. PA are those public information, command information, and community relations activities directed toward both the external and internal publics with interest in the Department of Defense. DSPD comprises those activities and measures taken by DOD components to support and facilitate USG public diplomacy efforts.

M - Military

The purpose of the Armed Forces is to fight and win the Nation's wars. As the military instrument of national power, the Armed Forces must ensure their adherence to US values, constitutional principles, and standards for the profession of arms. The United States wields the military instrument of national power at home and abroad in support of its national security goals in a variety of military operations.

E - Economy

The United States free market economy is only partially controlled by governmental agencies. In keeping with US values and constitutional imperatives, individuals and entities have broad freedom of action worldwide. The responsibility of the USG lies with facilitating the production, distribution, and consumption of goods and services worldwide. A strong US economy with free access to global markets and resources is a fundamental engine of the general welfare, the enabler of a strong national defense, and an influence for economic expansion by US trade partners worldwide.

The USG's financial management ways and means support the economic instrument of national power. The Department of the Treasury, as the steward of US economic and financial systems, is an influential participant in the international economy. In the international arena, the Department of the Treasury works with other federal agencies, the governments of other nations, and the international financial institutions to encourage economic growth, raise standards of living, and predict and prevent, to the extent possible, economic and financial crises.

The strategic security environment presents broad national security challenges likely to require the employment of joint forces in the future. They are the natural products of the enduring human condition, but they will exhibit new features in the future. All of these challenges are national problems calling for the application of all the instruments of national power. The US military will undertake the following activities to deal with these challenges:

- Secure the Homeland
- Win the Nation's Wars
- Deter Our Adversaries
- Security Cooperation
- Support to Civil Authorities
- Adapt to Changing Environment

V. Campaigns and Operations

Tactics, techniques, and procedures are the fundamental building blocks of concrete military activity. Broadly, actions generate effects; they change in the environment or situation. Tactical actions are the component pieces of operations.

An **operation** is a sequence of tactical actions with a common purpose or unifying theme. An operation may entail the process of carrying on combat, including movement, supply, attack, defense, and maneuvers needed to achieve the objective of any battle or campaign. However, an operation need not involve combat. A major operation is a series of tactical actions, such as battles, engagements, and strikes, conducted by combat forces coordinated in time and place, to achieve strategic or operational objectives in an operational area.

A **campaign** is a series of related major operations aimed at achieving strategic and operational objectives within a given time and space. Planning for a campaign is appropriate when contemplated military operations exceed the scope of a single major operation. Thus, campaigns are often the most extensive joint operations in terms of time and other resources.

See p. 2-43 for related discussion of the range of military operations (ROMO) and p. 2-59 for discussion of the conflict continuum.

VI. Task, Function, and Mission

A **task** is a clearly defined action or activity assigned to an individual or organization. It is a specific assignment that must be done as it is imposed by an appropriate authority. Function and mission implicitly involve things to be done, or tasks. It is, however, important to delineate between an organization's function and its mission.

A **function** is the broad, general, and enduring role for which an organization is designed, equipped, and trained. Organizationally, functions may be expressed as a task, a series of tasks, or in more general terms. Broadly, a function is the purpose for which an organization is formed. In the context of employing a joint force, joint functions are six basic groups of related capabilities and activities—C2, intelligence, fires, movement and maneuver, protection, and sustainment—that help JFCs integrate, synchronize, and direct joint operations.

Mission entails the task, together with the purpose, that clearly indicates the action to be taken and the reason therefore. A mission always consists of five parts: the who (organization to act), what (the task to be accomplished and actions to be taken), when (time to accomplish the task), where (the location to accomplish the task), and why (the purpose the task is to support). Higher headquarters commanders typically assign a mission or tasks to their subordinate commanders, who convert these to a specific mission statement through mission analysis. A mission is what an organization is directed to do. Functions are the purposes for which an organization is formed. The two are symbiotic. Tasks are relevant to both functions and missions.

Chap 1

II. Unified Direction of Armed Forces

Ref: JP 1 (w/Chg 1), Doctrine for the Armed Forces of the U.S. (Jul '17), chap. II.

I. National Strategic Direction

National strategic direction is governed by the Constitution, federal law, USG policy regarding internationally-recognized law and the national interest. This direction leads to unified action. The result of effective unified action is unity of effort to achieve national goals. At the strategic level, unity of effort requires coordination among government departments and agencies within the executive branch, between the executive and legislative branches, with NGOs, IGOs, the private sector, and among nations in any alliance or coalition, , and during bilateral or multilateral engagement. The responsibilities for national strategic direction as established by the Constitution and federal law and practice are as follows:

The President of the United States

The President of the United States, advised by the NSC, is responsible to the American people for national strategic direction.

When the United States undertakes military operations, the Armed Forces of the United States are only one component of a national-level effort involving all instruments of national power. Instilling unity of effort at the national level is necessarily a cooperative endeavor involving a number of Federal departments and agencies. In certain operations, agencies of states, localities, or foreign countries may also be involved.

Complex operations may require a high order of civil-military integration. Presidential directives guide participation by all US civilian and military agencies in such operations. Military leaders should work with the members of the national security team in the most skilled, tactful, and persistent ways to promote unity of effort. Operations of departments or agencies representing the diplomatic, economic, and informational instruments of national power are not under command of the Armed Forces of the United States or of any specific JFC. In US domestic situations, another department such as the Department of Homeland Security (DHS) may assume overall control of interagency coordination including military elements. Abroad, the chief of mission, supported by the country team, is normally in control.

The Secretary of Defense (SecDef)

The SecDef is responsible to the President for creating, supporting, and employing military capabilities. The SecDef provides authoritative direction, and control over the Services through the Secretaries of the Military Departments. SecDef exercises control of and authority over those forces not specifically assigned to the combatant commands and administers this authority through the Military Departments, the Service Chiefs, and applicable chains of command. The Secretaries of the Military Departments organize, train, and equip forces to operate across the range of military operations and provide for the administration and support of all those forces within their department, including those assigned or attached to the CCDRs.

See pp. 1-11 to 1-12 for further discussion.

The Chairman of the Joint Chiefs of Staff (CJCS)

The CJCS is the principal military adviser to the President, the NSC, and SecDef and functions under the authority of the President and the direction and control of the President and SecDef, and oversees the activities of the CCDRs as directed by SecDef. Communications between the President or the SecDef and the CCDRs are normally transmitted through the CJCS.

See pp. 1-13 to 1-14 for further discussion.

The Commanders of Combatant Commands (CCDRs)

Commanders of combatant commands exercise combatant command (command authority) (COCOM) over assigned forces and are responsible to the President and SecDef for the performance of assigned missions and the preparedness of their commands to perform assigned missions.

See pp. 1-17 to 1-22 for further discussion.

The US Chief of Mission (COM)

In a foreign country, the US chief of mission is responsible to the President for directing, coordinating, and supervising all USG elements in the HN, except those under the command of a CCDR. GCCs are responsible for coordinating with chiefs of mission in their geographic AOR (as necessary) and for negotiating memoranda of agreement (MOAs) with the chiefs of mission in designated countries to support military operations.

See pp. 8-19 to 8-21 for further discussion.

II. Unified Action

The term "unified action" in military usage is a broad term referring to the synchronization, coordination, and/or integration of the activities of governmental and nongovernmental entities with military operations to achieve unity of effort. Within this general category of operations, subordinate CDRs of assigned or attached forces conduct either single-Service or joint operations to support the overall operation. Unified action synchronizes, coordinates, and/or integrates joint, single-Service, and multinational operations with the operations of other USG agencies, NGOs, and IGOs (e.g., UN), and the private sector to achieve unity of effort. Unity of command within the military instrument of national power supports the national strategic direction through close coordination with the other instruments of national power.

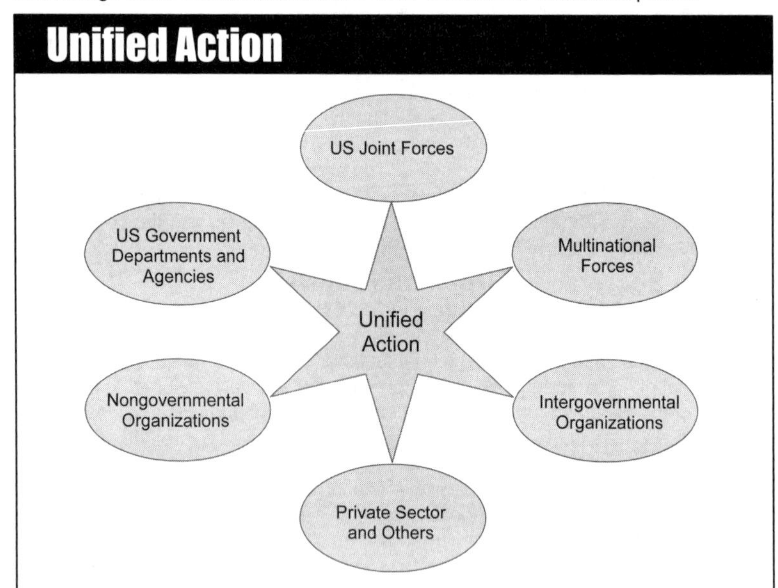

Ref: JP 1 (w/Chg 1), fig. II-2, p. II-8.

III. Strategy, Planning, and Resourcing

Ref: JP 1 (w/Chg 1), Doctrine for the Armed Forces of the U.S. (Jul '17), p. II-3 to II-6.

Military planning consists of joint strategic planning with its three subsets: security cooperation planning, force planning, and joint operation planning. Regarding force planning for the future, DOD conducts capabilities-based planning (CBP). The essence of CBP is to identify capabilities that adversaries could employ against the US or a multinational opponent and to defend themselves; identify capabilities, US and multinational, that could be available to the joint or combined force to counter/defeat the adversary; and then identify and evaluate possible outcomes (voids or opportunities), rather than forecasting (allocating) forces against specific threat scenarios. Integral to a capabilities-based approach are joint capability areas (JCAs), DOD's capability management language and framework. (National planning documents, fig. II-1, below.)

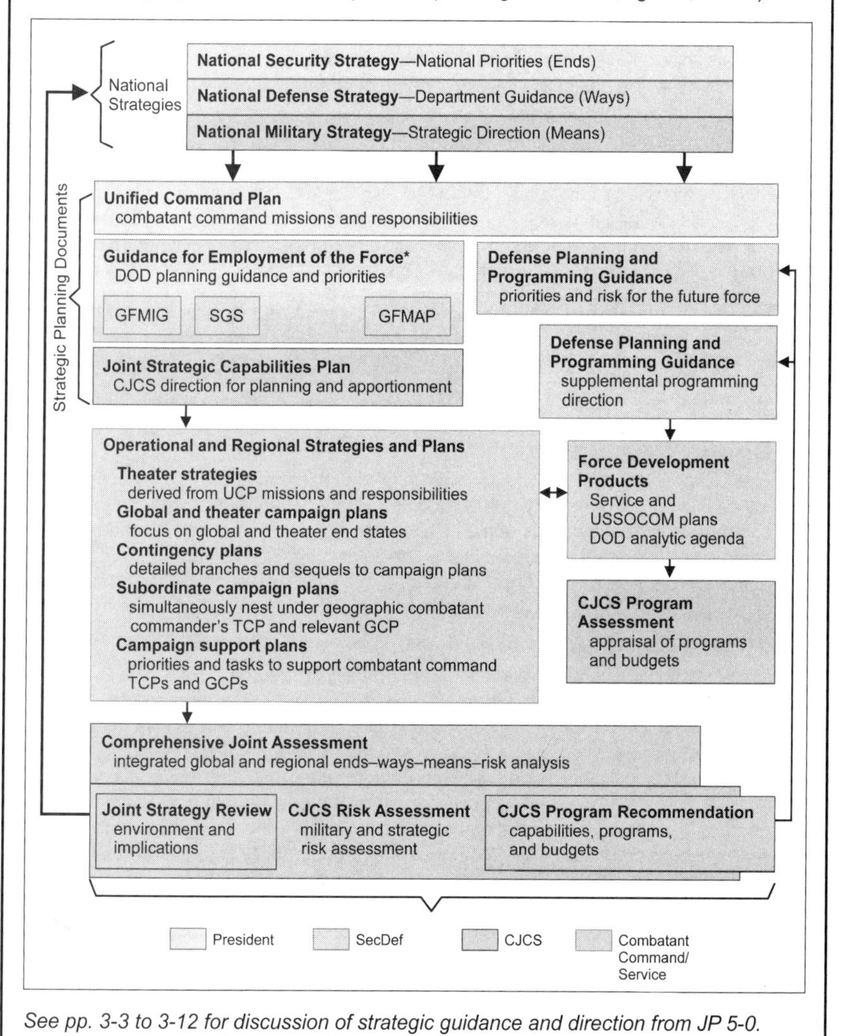

See pp. 3-3 to 3-12 for discussion of strategic guidance and direction from JP 5-0.

IV. Chain of Command

The President and SecDef exercise authority, direction, and control of the Armed Forces through two distinct branches of the chain of C2. One branch runs from the President, through SecDef, to the CCDRs for missions and forces assigned to their commands. For purposes other than the operational direction of the CCMDs, the chain of command runs from the President to SecDef to the Secretaries of the Military Departments and, as prescribed by the Secretaries, to the commanders of Military Service forces. The Military Departments, organized separately, operate under the authority, direction, and control of the Secretary of that Military Department. The Secretaries of the Military Departments may exercise administrative control (ADCON) over Service forces through their respective Service Chiefs and Service commanders. The Service Chiefs, except as otherwise prescribed by law, perform their duties under the authority, direction, and control of the Secretaries of the respective Military Departments to whom they are directly responsible.

The CCDRs exercise COCOM over assigned forces and are directly responsible to the President and SecDef for the performance of assigned missions and the preparedness of their commands. CCDRs prescribe the chain of command within their CCMDs and designate the appropriate command authority to be exercised by subordinate commanders.

The Secretaries of the Military Departments operate under the authority, direction, and control of SecDef. This branch of the chain of command is responsible for ADCON over all military forces within the respective Service not assigned to CCDRs (i.e., those defined in the Global Force Management Implementation Guidance [GFMIG] as "unassigned forces"). This branch is separate and distinct from the branch of the chain of command that exists within a CCMD.

V. Relationship Between Combatant Commanders, Military Department Secretaries, Service Chiefs, and Forces

Continuous Coordination

The Military Services and USSOCOM (in areas unique to special operations) share the division of responsibility for developing military capabilities for the combatant commands. All components of the DOD are charged to coordinate on matters of common or overlapping responsibility. The Joint Staff, Service, and USSOCOM headquarters play a critical role in ensuring that CCDRs' concerns and comments are included or advocated during the coordination.

Interoperability

Unified action demands maximum interoperability. The forces, units, and systems of all Services must operate together effectively. This effectiveness is achieved in part through interoperability. This includes the development and use of joint doctrine, the development and use of joint OPLANs; and the development and use of joint and/or interoperable communications and information systems. It also includes conducting joint training and exercises. It concludes with a materiel development and fielding process that provides materiel that is fully compatible with and complementary to systems of all Services. A key to successful interoperability is to ensure that planning processes are joint from their inception. Those responsible for systems and programs intended for joint use will establish working groups that fully represent the services and functions that will be affected and interoperability must be considered in all joint program reviews. CCDRs will ensure maximum interoperability and identify interoperability issues to the CJCS, who has overall responsibility for the joint interoperability program.

Chap 1

III(a). The Department of Defense (DOD)

Ref: JP 1 (w/Chg 1), Doctrine for the Armed Forces of the U.S. (Jul '17), chap. III.

DOD is composed of OSD, the Military Departments, the Joint Chiefs of Staff (JCS), the Joint Staff, the CCMDs, the Inspector General, agencies/bureaus, field activities, and such other offices, and commands established or designated by law, by the President, or by SecDef. The functions of the heads of these offices shall be as assigned by SecDef according to existing law.

See pp. 1-13 to 1-14 for further discussion of the Joint Chiefs of Staff, pp. 1-15 to 1-16 for the military departments, and pp. 1-17 to 1-22 for the combatant commands.

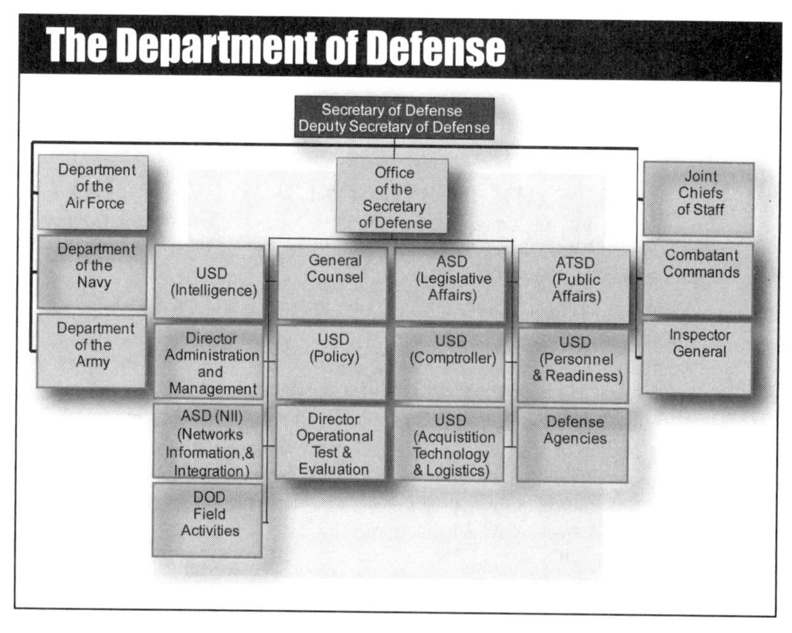

As prescribed by higher authority, the DOD will maintain and employ Armed Forces to fulfill the following aims.

- Support and defend the Constitution of the United States against all enemies, foreign and domestic
- Ensure, by timely and effective military action, the security of the US, its territories, and areas vital to its interest.
- Uphold and advance the national policies and interests of the United States

I. The Secretary of Defense (SecDef)

The SecDef is the principal assistant to the President in all matters relating to the DOD. All functions in the DOD and its component agencies are performed under the authority, direction, and control of the SecDef.

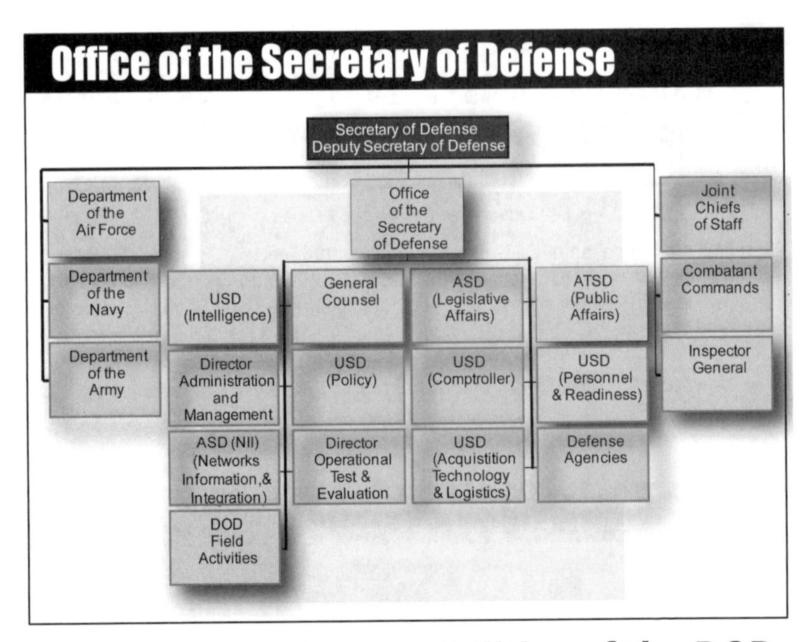

II. Functions and Responsibilities of the DOD

The functions and responsibilities assigned to the Secretaries of the Military Departments, the Services, the JCS, the Joint Staff, and the CCMDs are carried out in such a manner as to achieve the following:

- Provide the best military advice to the President and the SecDef
- Effective strategic direction of the Armed Forces
- Employment of the Armed Forces in joint force commands whenever such arrangement is in the best interest of national security
- Integration of the Armed Forces into an effective and efficient team
- Prevention of unnecessary duplication or overlapping capabilities among the Services by using personnel, intelligence, facilities, equipment, supplies, and services of any or all Services such that military effectiveness and economy of resources will thereby be increased
- Coordination of Armed Forces operations to promote efficiency and economy and to prevent gaps in responsibility
- Effective multinational operations and interagency, IGO, and NGO coordination

III. Executive Agents (EA)

SecDef or Deputy Secretary of Defense may designate a DOD executive agent (EA) and assign associated responsibilities, functions, and authorities within DOD. The head of a DOD component may be designated as a DOD EA. The DOD EA may delegate to a subordinate designee within that official's component the authority to act on that official's behalf for those DOD EA responsibilities, functions, and authorities assigned by SecDef or Deputy Secretary of Defense. Designation as EA confers no authority. The exact nature and scope of the DOD EA responsibilities, functions, and authorities shall be prescribed in the EA appointing document at the time of assignment and remain in effect until SecDef or Deputy Secretary of Defense revokes or supersedes them. Responsibilities of an EA are established in DODD 5101.1, Department of Defense Executive Agent, and specific DODDs on specific EAs.

See p. 4-14 for further discussion of executive agents from JP 4-0.

III(b). The Joint Chiefs of Staff (JCS)

Ref: JP 1 (w/Chg 1), Doctrine for the Armed Forces of the U.S. (Jul '17), chap. III.

The Joint Chiefs of Staff (JCS) consists of the Chairman of the Joint Chiefs of Staff (CJCS); the Vice Chairman of the Joint Chiefs of Staff (VCJCS); the Chief of Staff, US Army; the Chief of Naval Operations; the Chief of Staff, US Air Force; and the Commandant of the Marine Corps. The Joint Staff supports the JCS and constitutes the immediate military staff of the SecDef. The CJCS is the principal military advisor to the President, NSC, HSC, and SecDef.

The Joint Staff

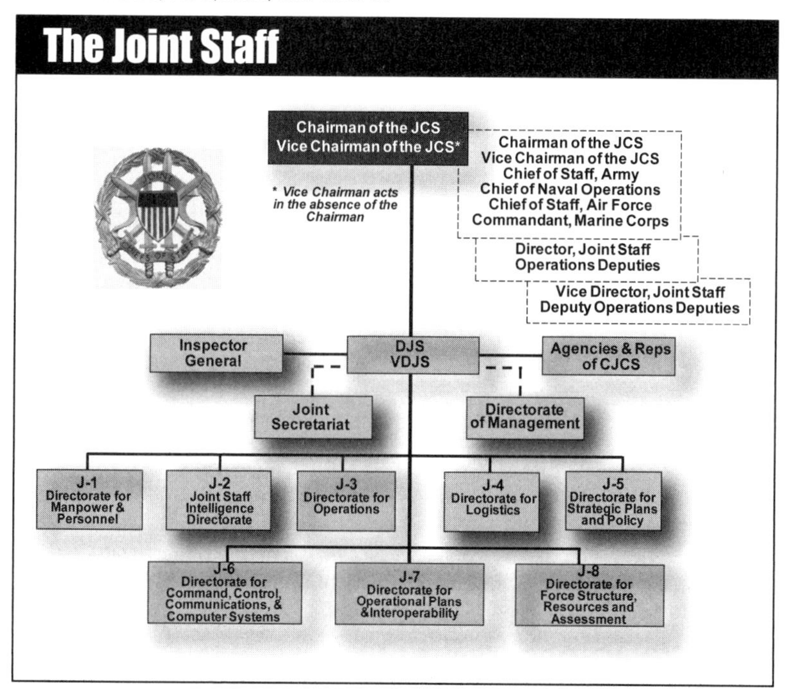

The other members of the JCS are military advisors to the President, NSC, HSC, and SecDef as specified below.

- A member of the JCS may submit to the CJCS advice or an opinion in disagreement with, or in addition to, the advice or opinion presented by the CJCS. If a member submits such advice or opinion, the CJCS shall present that advice or opinion to the President, NSC, or SecDef at the same time that he presents his own advice. The CJCS shall also, as he considers appropriate, inform the President, the NSC, or SecDef of the range of military advice and opinion with respect to any matter.
- The members of the JCS, individually or collectively, in their capacity as military advisors, shall provide advice on a particular matter when the President, NSC, HSC, or SecDef request such advice.

I. Chairman of the Joint Chiefs of Staff (CJCS)

The CJCS is appointed by the President, with the advice and consent of the Senate, from the officers of the regular component of the United States Armed Forces. The CJCS is the principal military advisor to the President, the NSC, and the SecDef. The CJCS arranges for military advice, as appropriate, to be provided to all offices of the SecDef.

While holding office, the CJCS outranks all other officers of the Armed Forces. The CJCS, however, may not exercise military command over the CCDRs, JCS, or any of the Armed Forces.

Subject to the authority, direction, and control of the SecDef, the CJCS serves as the spokesperson for the CCDRs, especially on the operational requirements of their commands. CCDRs will send their reports to the CJCS, who will review and forward the reports as appropriate to the SecDef, subject to the direction of the SecDef, so that the CJCS may better incorporate the views of CCDRs in advice to the President, the NSC, and the SecDef. The CJCS also communicates, as appropriate, the CCDRs' requirements to other elements of DOD.

The CJCS assists the President and the SecDef in providing for the strategic direction of the Armed Forces. The CJCS transmits orders to the CCDRs as directed by the President or SecDef and coordinates all communications in matters of joint interest addressed to the CCDRs.

II. Vice Chairman of the Joint Chiefs of Staff

The VCJCS is appointed by the President, by and with the advice and consent of the Senate, from the officers of the regular components of the United States Armed Forces.

The VCJCS holds the grade of general or admiral and outranks all other officers of the Armed Forces except the CJCS. The VCJCS may not exercise military command over the JCS, the CCDRs, or any of the Armed Forces.

The VCJCS performs the duties prescribed as a member of the JCS and such other duties and functions as may be prescribed by the CJCS with the approval of the SecDef. When there is a vacancy in the office of the CJCS, or in the absence or disability of the CJCS, the VCJCS acts as and performs the duties of the CJCS until a successor is appointed or the absence or disability ceases.

The VCJCS is a member of the Joint Nuclear Weapons Council, is the Vice Chairman of the Defense Acquisition Board, and may be designated by the CJCS to act as the Chairman of the Joint Requirements Oversight Council (JROC).

III. The Joint Staff

The Joint Staff is under the exclusive authority, direction, and control of the CJCS. The Joint Staff will perform duties using procedures that the CJCS prescribes to assist the CJCS and the other members of the JCS in carrying out their responsibilities.

The Joint Staff includes officers selected in proportional numbers from the Army, Marine Corps, Navy, and Air Force. Coast Guard officers may also serve on the Joint Staff. Selection of officers to serve on the Joint Staff is made by the CJCS from a list of officers submitted by the Services. Each officer whose name is submitted must be among those officers considered to be the most outstanding officers of that Service. The CJCS may specify the number of officers to be included on such a list.

After coordination with the other members of the JCS and with the approval of SecDef, the CJCS may select a Director, Joint Staff. The CJCS manages the Joint Staff and its Director. Per Title 10, USC, Section 155, the Joint Staff will not operate or be organized as an overall Armed Forces general staff and will have no executive authority.

III(c). The Military Depts and Services

Ref: JP 1 (w/Chg 1), Doctrine for the Armed Forces of the U.S. (Jul '17), chap. III.

Subject to the authority, direction, and control of SecDef and subject to the provisions of Title 10, USC, the Army, Marine Corps, Navy, and Air Force, under their respective Secretaries, are responsible for the functions prescribed in detail in DODD 5100.01, Functions of the Department of Defense and Its Major Components. Specific Service functions also are delineated in that directive.

See p. 1-16 for discussion of common functions of the military depts and Services.

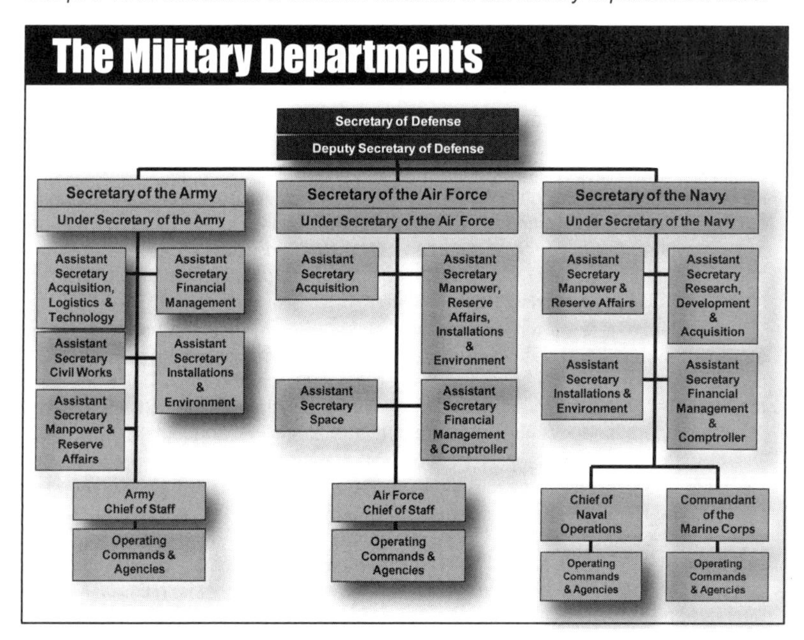

USSOCOM is unique among the CCMDs in that it performs certain Service-like functions (in areas unique to SO) (Title 10, USC, Sections 161 and 167), including the following:

- Organize, train, equip, and provide combat-ready SOF to the other CCMDs and, when directed by the President or SecDef, conduct selected SO, usually in coordination with the GCC in whose AOR the SO will be conducted. USSOCOM's role in equipping and supplying SOF is generally limited to SO-peculiar equipment, materiel, supplies, and services.
- Develop strategy, doctrine, and tactics, techniques, and procedures for the conduct of SO, to include military information support operations (MISO) and CA forces. (Note: Joint doctrine is developed under the procedures approved by the CJCS.)
- Prepare and submit to SecDef program recommendations and budget proposals for SOF and other forces assigned to USSOCOM.

Common Functions of the Military Services

Ref: JP 1, Doctrine for the Armed Forces of the U.S. (Mar '13), p. II-11 to II-12.

The authority vested in the Secretaries of the Military Departments in the performance of their role to organize, train, equip, and provide forces runs from the President through SecDef to the Secretaries. Then, to the degree established by the Secretaries or specified in law, this authority runs through the Service Chiefs to the Service component commanders assigned to the CCDRs and to the commanders of forces not assigned to the CCDRs. ADCON provides for the preparation of military forces and their administration and support, unless such responsibilities are specifically assigned by SecDef to another DOD component.

The Secretaries of the Military Departments are responsible for the administration and support of Service forces. They fulfill their responsibilities by exercising ADCON through the Service Chiefs. Service Chiefs have ADCON for all forces of their Service. The responsibilities and authority exercised by the Secretaries of the Military Departments are subject by law to the authority provided to the CCDRs in their exercise of COCOM.

Each of the Secretaries of the Military Departments, coordinating as appropriate with the other Military Department Secretaries and with the CCDRs, has the responsibility for organizing, training, equipping, and providing forces to fulfill specific roles and for administering and supporting these forces. The Secretaries also perform a role as a force provider of Service retained forces until they are deployed to CCMDs. When addressing similar issues regarding National Guard forces, coordination with the National Guard Bureau (NGB) is essential.

Commanders of Service forces are responsible to Secretaries of the Military Departments through their respective Service Chiefs for the administration, training, and readiness of their unit(s). Commanders of forces assigned to the CCMDs are under the authority, direction, and control of (and are responsible to) their CCDR to carry out assigned operational missions, joint training and exercises, and logistics.

The USCG is a military Service and a branch of the US Armed Forces at all times. However, it is established separately by law as a Service in DHS, except when transferred to the Department of the Navy (DON) during time of war or when the President so directs. Authorities vested in the USCG under Title 10, USC, as an armed Service and Title 14, USC, as a federal maritime safety and law enforcement agency remain in effect at all times, including when USCG forces are operating within DOD/DON chain of command. USCG commanders and forces may be attached to JFCs in performance of any activity for which they are qualified. Coast Guard units routinely serve alongside Navy counterparts operating within a naval task organization in support of a maritime component commander.

The NGB is a joint activity of DOD. The NGB performs certain military Service-specific functions and unique functions on matters involving non-federalized National Guard forces. The NGB is responsible for ensuring that units and members of the Army National Guard and the Air National Guard are trained by the states to provide trained and equipped units to fulfill assigned missions in federal and non-federal statuses.

In addition to the Services above, a number of DOD agencies provide combat support or combat service support to joint forces and are designated as CSAs. CSAs, established under SecDef authority under Title 10, USC, Section 193, and Department of Defense Directive (DODD) 3000.06, Combat Support Agencies, are the DIA, National Geospatial-Intelligence Agency (NGA), Defense Information Systems Agency (DISA), DLA, Defense Contract Management Agency (DCMA), DTRA, and National Security Agency (NSA). These CSAs provide CCDRs specialized support and operate in a supporting role. The CSA directors are accountable to SecDef.

III(d). The Combatant Commands (CCMDs)

Ref: JP 1 (w/Chg 1), Doctrine for the Armed Forces of the U.S. (Jul '17), chap. III and https://www.defense.gov/Our-Story/Combatant-Commands.

The Unified Command Plan (UCP)

The Unified Command Plan (UCP) is the document that sets forth basic guidance to all combatant commanders. The UCP establishes combatant command missions, responsibilities, and force structure; delineates geographic areas of responsibility for geographic combatant commanders; and specifies functional responsibilities for functional combatant commanders. Every two years, the chairman of the Joint Chiefs of Staff is required to review the missions, responsibilities, and geographical boundaries of each combatant command and recommend to the President, through the secretary of defense, any changes that may be necessary.

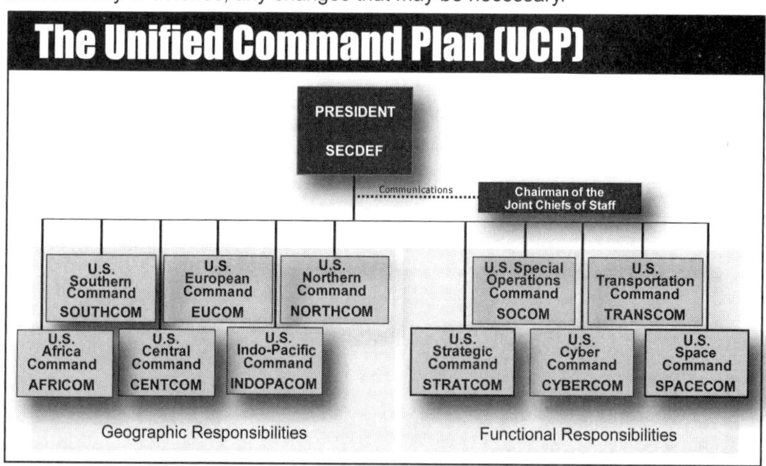

Geographic Combatant Commands (GCCs) are assigned a geographic AOR by the President with the advice of SecDef as specified in the UCP. Geographic AORs provide a basis for coordination by CCDRs. GCCs are responsible for the missions in their AOR, unless otherwise directed.

Functional Combatant Commands (FCCs) have transregional responsibilities and are normally supporting CCDRs to the GCC's activities in their AOR. FCCs may conduct operations as directed by the President or SecDef, in coordination with the GCC in whose AOR the operation will be conducted. The FCC may be designated by SecDef as the supported CCDR for an operation. The implementing order directing an FCC to conduct operations within a GCC's AOR will specify the CCDR responsible for mission planning and execution and appropriate command relationships.

Global Synchronizer. SecDef or Deputy Secretary of Defense may assign a CCDR global synchronizer responsibilities. A global synchronizer is the CCDR responsible for the alignment of specified planning and related activities of other CCMDs, Services, DOD agencies and activities, and as directed, appropriate USG departments and agencies within an established, common framework to facilitate coordinated and decentralized execution across geographic and other boundaries. The global synchronizer's role is to align and harmonize plans and recommend sequencing of actions to achieve the strategic end states and objectives of a GCP.

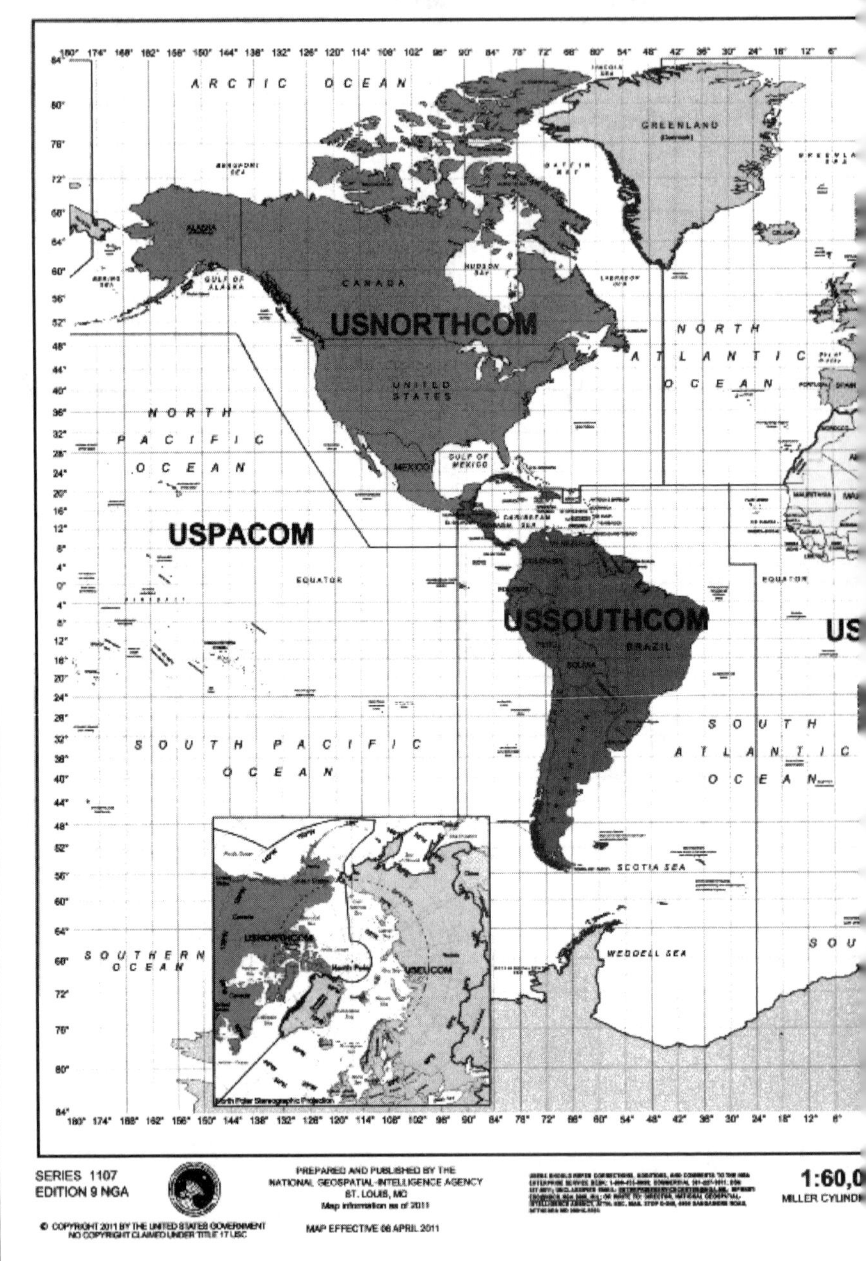
THE WORLD 1:60,000,000
THE WORLD WITH COMMANDERS
ARCTIC OCEAN
GREENLAND
CANADA
USNORTHCOM
UNITED STATES
NORTH ATLANTIC OCEAN
NORTH PACIFIC OCEAN
MEXICO
GULF OF MEXICO
USPACOM
EQUATOR
USSOUTHCOM
BRAZIL
SOUTH PACIFIC OCEAN
SOUTH ATLANTIC OCEAN
SCOTIA SEA
WEDDELL SEA
SOUTHERN OCEAN
USNORTHCOM
North Pole
USEUCOM
SERIES 1107
EDITION 9 NGA
PREPARED AND PUBLISHED BY THE
NATIONAL GEOSPATIAL-INTELLIGENCE AGENCY
ST. LOUIS, MO
MAP EFFECTIVE 08 APRIL 2011
© COPYRIGHT 2011 BY THE UNITED STATES GOVERNMENT
NO COPYRIGHT CLAIMED UNDER TITLE 17 USC
MILLER CYLINDR

' AREAS OF RESPONSIBILITY

EDITION 9 NGA
BASED ON
UNIFIED COMMAND PLAN
06 APRIL 2011

SERIES 1107

ARCTIC OCEAN
BARENTS SEA
KARA SEA
LAPTEV SEA
EAST SIBERIAN SEA
RUSSIA
USEUCOM
SEA OF OKHOTSK
POLAND
BELARUS
UKRAINE
KAZAKHSTAN
MONGOLIA
TURKEY
NORTH PACIFIC OCEAN
USCENTCOM
USPACOM
CHINA
LIBYA
EGYPT
SAUDI ARABIA
INDIA
NIGER
CHAD
SUDAN
ETHIOPIA
KENYA
TANZANIA
EQUATOR
AFRICOM
ANGOLA
ZAMBIA
NAMIBIA
SOUTH AFRICA
INDIAN OCEAN
CORAL SEA
AUSTRALIA
TASMAN SEA
NEW ZEALAND
SOUTH PACIFIC OCEAN
SOUTHERN OCEAN
USPACOM
ANTARCTICA
USSOUTHCOM
USAFRICOM
INDIAN OCEAN
SOUTH ATLANTIC OCEAN
South Pole Stereographic Projection
SOUTHERN OCEAN
ANTARCTICA

0,000
L PROJECTION

* West Bank and Gaza Strip - "Israeli-occupied with current status subject to the Israeli - Palestinian Interim Agreement - - permanent status to be determined through further negotiation"

THE REPRESENTATION OF BOUNDARIES IS NOT NECESSARILY AUTHORITATIVE

Combatant Commands (CCMDs)

Ref: JP 1 (w/Chg 1), Doctrine for the Armed Forces of the U.S. (Jul '17), pp. I-10, I-14, III-12 to III-15, and https://www.defense.gov/Our-Story/Combatant-Commands.

In accordance with the UCP, combatant commands are established by the President, through the SecDef, with the advice and assistance of the CJCS. Commanders of unified commands may establish subordinate unified commands when so authorized by the SecDef through the CJCS. JTFs can be established by the SecDef, a CCDR, subordinate unified commander, or an existing JTF commander.

Geographic Combatant Commanders are assigned a geographic AOR by the President with the advice of the SecDef as specified in the UCP. Functional CCDRs support GCCs, conduct operations in direct support of the President or the SecDef normally in coordination with the GCC in whose AOR the operation will be conducted, and may be designated by the SecDef as the supported CCDR for an operation.

See pp. 1-23 to 1-30 for further discussion of joint forces organization, roles and responsibilities, to include unified commands.

Geographic Combatant Commanders (GCCs)

GCCs are the vital link between those who determine national security policy and strategy and the military forces or subordinate JFCs that conduct military operations within their geographical AORs. GCCs are responsible for a large geographical area requiring single responsibility for effective coordination of the operations within that area. Directives flow from the President and SecDef through CJCS to the GCCs, who plan and conduct the operations that achieve national, alliance, or coalition strategic objectives. GCCs provide guidance and direction through strategic estimates, command strategies, and plans and orders for the employment of military force. As military force may not achieve national objectives, it must be coordinated, synchronized, and if appropriate, integrated with OGAs, IGOs, NGOs, MNFs, and elements of the private sector.

Six combatant commanders have geographic area responsibilities. These combatant commanders are each assigned an area of responsibility (AOR) by the Unified Command Plan (UCP) and are responsible for all operations within their designated areas: U.S. Northern Command, U.S. Central Command, U.S. European Command, U.S. Indo-Pacific Command, U.S. Southern Command and U.S. Africa Command.

U.S. Northern Command

northcom.mil

USNORTHCOM was established Oct. 1, 2002 to provide command and control of Department of Defense (DoD) homeland defense efforts and to coordinate defense support of civil authorities. USNORTHCOM defends America's homeland — protecting our people, national power, and freedom of action.

USNORTHCOM plans, organizes and executes homeland defense and civil support missions, but has few permanently assigned forces. The command is assigned forces whenever necessary to execute missions, as ordered by the president and secretary of defense.

U.S. Indo-Pacific Command

pacom.mil

USINDOPACOM encompasses about half the earth's surface, stretching from the west coast of the U.S. to the western border of India, and from Antarctica to the North Pole. The 36 nations that comprise the Asia-Pacific region are home to more than fifty percent of the world's population, three thousand different languages, several of the world's largest militaries, and five nations allied with the U.S. through mutual defense treaties.

USINDOPACOM protects and defends, in concert with other U.S. Government agencies, the territory of the United States, its people, and its interests. With allies and partners, USINDOPACOM is committed to enhancing stability in the Asia-Pacific region by promoting security cooperation, encouraging peaceful development, responding to contingencies, deterring aggression, and, when necessary, fighting to win. This approach is based on partnership, presence, and military readiness.

U.S. Southern Command

southcom.mil

USSOUTHCOM is responsible for providing contingency planning, operations, and security cooperation for Central and South America, the Caribbean (except U.S. commonwealths, territories, and possessions), Cuba; as well as for the force protection of U.S. military resources at these locations. SOUTHCOM is also responsible for ensuring the defense of the Panama Canal and canal area.

The U.S. Southern Command Area of Focus encompasses 31 countries and 10 territories. The region represents about one-sixth of the landmass of the world assigned to regional unified commands.

The services provide SOUTHCOM with component commands which, along with our Joint Special Operations component, two Joint Task Forces, one Joint Interagency Task Force, and Security Assistance Offices, perform SOUTHCOM missions and security cooperation activities.

U.S. Central Command

centcom.mil

USCENTCOM's area of responsibility covers the "central" area of the globe and consists of 20 countries -- Afghanistan, Bahrain, Egypt, Iran, Iraq, Jordan, Kazakhstan, Kuwait, Kyrgyzstan, Lebanon, Oman, Pakistan, Qatar, Saudi Arabia, Syria, Tajikistan, Turkmenistan, United Arab Emirates, Uzbekistan, and Yemen. There are also 62 coalition countries contributing to the war against terrorism.

With national and international partners, U.S. Central Command promotes cooperation among nations, responds to crises, and deters or defeats state and nonstate aggression, and supports development and, when necessary, reconstruction in order to establish the conditions for regional security, stability, and prosperity.

U.S. European Command

eucom.mil

USEUCOM is a Unified Combatant Command of the United States military, headquartered in Stuttgart, Germany. USEUCOM's mission is to maintain ready forces to conduct the full range of operations unilaterally or in concert with coalition partners; enhance transatlantic security through support of NATO; promote regional stability; counter terrorism; and advance U.S. interests in the AOR.

The area of responsibility (AOR) of the United States European Command includes 51 countries and territories. This territory extends from the North Cape of Norway, through the waters of the Baltic and Mediterranean seas, most of Europe, and parts of the Middle East. Several other countries and territories are considered to be part of EUCOM's area of interest (AOI).

U.S. Africa Command

africom.mil

Based in Germany, with select personnel assigned to U.S. Embassies and diplomatic missions in Africa, USAFRICOM is responsible for coordinating military-to-military relationships between the United States and African nations and military organizations.

Africa Command's military and civilian staff is dedicated to working closely with African nations and organizations, U.S. agencies and the international community to promote security and prevent conflict in support of U.S. foreign policy in Africa. U.S. Africa Command is prepared to respond to requests for support from African nations and from other U.S. government agencies by providing humanitarian or crisis response options.

Established in October 2007, U.S. Africa Command is focused on synchronizing hundreds of activities inherited from three regional commands that previously coordinated U.S. military relations in Africa – U.S. European Command, U.S. Pacific Command and U.S. Central Command.

Continued on next page

Continued on next page

Combatant Commands (Continued)

Functional Combatant Commanders (FCCs)

Continued from previous page

FCCs provide support to and may be supported by GCCs and other FCCs as directed by higher authority. FCCs are responsible for a large functional area requiring single responsibility for effective coordination of the operations therein. These responsibilities are normally global in nature. The President and SecDef direct what specific support and to whom such support will be provided. When an FCC is the supported CDR and operating within GCC's AORs, close coordination and communication between them is paramount.

Five combatant commanders assigned worldwide functional responsibilities not bounded by geography: U.S. Special Operations Command, U.S. Strategic Command, and U.S. Transportation Command, U.S. Cyber Command and U.S. Space Command:

U.S. Special Operations Command

socom.mil

USSOCOM is unique among the combatant commands in that it performs certain Service-like functions (Title 10, USC, Chapter 6). CDRUSSOCOM is a functional CCDR who exercises COCOM of all Active and RC special operations forces (SOF) minus US Army Reserve civil affairs and PSYOP forces stationed in CONUS.

U.S. Transportation Command

transcom.mil

USTRANSCOM mission is to develop and direct the Joint Deployment and Distribution Enterprise to globally project strategic national security capabilities; accurately sense the operating environment; provide end-to-end distribution process visibility; and responsive support of joint, U.S. government and Secretary of Defense-approved multinational and non-governmental logistical requirements.

U.S. Space Command

spacecom.mil

The USSPACECOM mission is to deter aggression and conflict, defend U.S. and allied freedom of action, deliver space combat power for the Joint/Combined force, and develop joint warfighters to advance U.S. and allied interests in, from, and through the space domain.

U.S. Strategic Command

stratcom.mil

Located at Offutt Air Force Base near Omaha, Neb., USSTRATCOM combines the synergy of the U.S. legacy nuclear command and control mission with responsibility for space operations; global strike; Defense Department information operations; global missile defense; and global command, control, communications, computers, intelligence, surveillance and reconnaissance (C4ISR), and combating weapons of mass destruction. This dynamic command gives National Leadership a unified resource for greater understanding of specific threats around the world and the means to respond to those threats rapidly.

U.S. Cyber Command

cybercom.mil

Commander, USCYBERCOM, has the mission to direct, synchronize, and coordinate cyberspace planning and operations to defend and advance national interests in collaboration with domestic and international partners. The Command unifies the direction of cyberspace operations, strengthens DoD cyberspace capabilities, and integrates and bolsters DoD's cyber expertise. The Command has three main focus areas: Defending the DoDIN, providing support to combatant commanders for execution of their missions around the world, and strengthening our nation's ability to withstand and respond to cyber attack.

Continued from previous page

IV. Joint Command Organizations

Ref: JP 1 (w/Chg 1), Doctrine for the Armed Forces of the U.S. (Jul '17), chap. IV.

Joint forces are established at three levels: unified commands, subordinate unified commands, and JTFs. In accordance with the National Security Act of 1947 and Title 10, USC, and as described in the UCP, CCMDs are established by the President, through SecDef, with the advice and assistance of the CJCS. Commanders of unified CCMDs may establish subordinate unified commands when so authorized by SecDef through the CJCS. JTFs can be established by SecDef, a CCDR, subordinate unified commander, or an existing JTF commander.

See pp. 1-19 to 1-24 for discussion of the UCP, the six geographic combatant commands, and the three functional combatant commands. See chap. 5 for further discussion of joint task forces.

Joint Commands

- I — **Unified Combatant Command**
- * — **Specified Combatant Command** *(There are currently no specified CCMDs designated.)*
- II — **Subordinate Unified Command**
- III — **Joint Task Force (JTF)**

Basis for Establishing Joint Forces

Joint forces can be established on either a geographic area or functional basis:

Geographic Area

Establishing a joint force on a geographic area basis is the most common method to assign responsibility for continuing operations. The title of the areas and their delineation are prescribed in the establishing directive. Note: Only GCCs are assigned AORs. GCCs normally assign subordinate commanders an operational area from within their assigned AOR.

- The UCP contains descriptions of the geographic boundaries assigned to GCCs. These geographic AORs do not restrict accomplishment of assigned missions; CCDRs may operate forces wherever required to accomplish their missions. The UCP provides that, unless otherwise directed by SecDef, when significant operations overlap the boundaries of two GCCs' AORs, a JTF will be formed. Command of this JTF will be determined by SecDef and forces transferred to the JTF commander through a CCDR, including delegation of appropriate command authority over those forces.

Organizing Joint Forces

Ref: JP 1 (w/Chg 1), Doctrine for the Armed Forces of the U.S. (Jul '17), p. IV-2 to IV-5.

A JFC has the authority to organize assigned or attached forces with specification of OPCON to best accomplish the assigned mission based on the CONOPS. The organization should be sufficiently flexible to meet the planned phases of the contemplated operations and any development that may necessitate a change in plan. The JFC will establish subordinate commands, assign responsibilities, establish or delegate appropriate command relationships, and establish coordinating instructions for the component commanders. Sound organization should provide for unity of command, centralized planning and direction, and decentralized execution. Unity of effort is necessary for effectiveness and efficiency. Centralized planning and direction is essential for controlling and coordinating the efforts of the forces. Decentralized execution is essential because no one commander can control the detailed actions of a large number of units or individuals. When organizing joint forces with MNFs, simplicity and clarity are critical. Complex or unclear command relationships or organization are counterproductive to developing synergy among MNFs.

Possible Components in a Joint Force

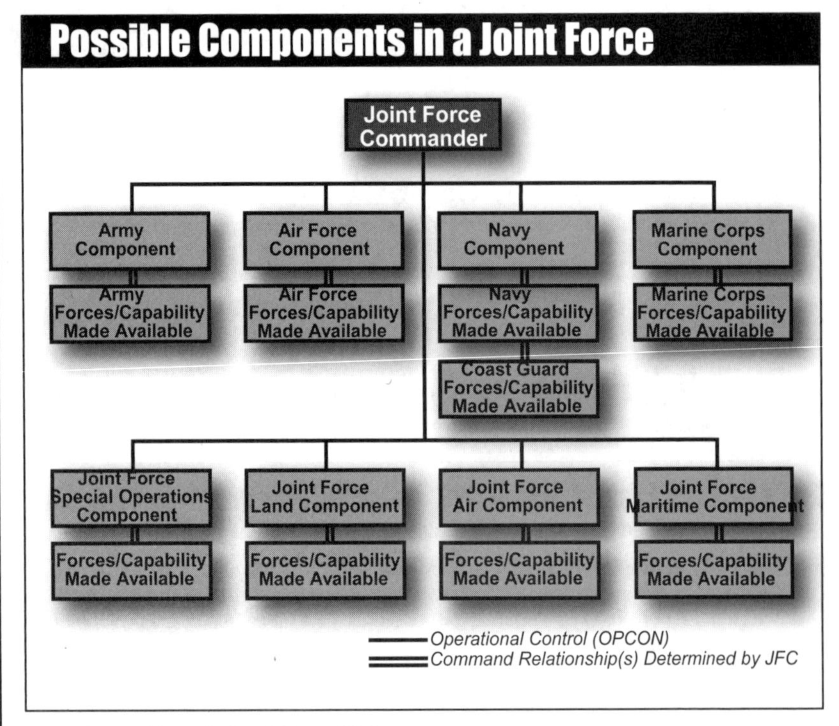

Ref: JP 1 (w/Chg 1), fig. IV-1, p. IV-3.

The composition of the JFC's staff will reflect the composition of the joint force to ensure that those responsible for employing joint forces have a thorough knowledge of the capabilities and limitations of assigned or attached forces.

See chap. 5 for further discussion of joint task forces, including functional component commands and service component commands.

Service Components

All joint forces include Service components, because administrative and logistic support for joint forces are provided through Service components. Service forces may be assigned or attached to subordinate joint forces without the formal creation of a respective Service component command of that joint force. The JFC also may conduct operations through the Service component commanders or, at lower echelons, Service force commanders. This relationship is appropriate when stability, continuity, economy, ease of long-range planning, and the scope of operations dictate organizational integrity of Service forces for conducting operations.

The JFC can establish functional component commands to conduct operations. Functional component commands are appropriate when forces from two or more Military Departments must operate within the same mission area or geographic domain or there is a need to accomplish a distinct aspect of the assigned mission. Joint force land, air, maritime, and SO component commanders are examples of functional components. (NOTE: Functional component commanders are component commanders of a joint force and do not constitute a "joint force command" with the authorities and responsibilities of a JFC, even when employing of forces from two or more Military Departments.) When a functional component command employs forces from more than one Service, the functional component commander's staff should include Service representatives from each of the employed Service forces to aid in understanding those Service capabilities and maximizing the effective employment of Service forces. Joint staff billets for needed expertise and individuals to fill those billets should be identified. Those individuals should be used when the functional component command is formed for exercises, contingency planning, or actual operations.

Service and Functional Component Commands

Normally, joint forces are organized with a combination of Service and functional component commands with operational responsibilities. Joint forces organized with Army, Navy, Air Force, and Marine Corps components may have SOF (if assigned) organized as a functional component. The JFC defines the authority, command relationships, and responsibilities of the Service and functional component commanders; however, the Service responsibilities (i.e., administrative and logistics) of the components must be given due consideration by the JFC.

The JFC has full authority to assign missions, redirect efforts, and direct coordination among subordinate commanders. JFCs should allow Service tactical and operational assets and groupings to function generally as they were designed. The intent is to meet the needs of the JFC while maintaining the tactical and operational integrity of the Service organizations.

See chap. 5 for further discussion of joint task forces, including functional component commands and service component commands.

Marine Air-Ground Task Force (MAGTF)

The following policy for C2 of United States Marine Corps tactical air (TACAIR) recognizes this and deals with Marine air-ground task force (MAGTF) aviation during sustained operations ashore. The MAGTF commander will retain OPCON of organic air assets. The primary mission of the MAGTF aviation combat element is the support of the MAGTF ground combat element. During joint operations, the MAGTF air assets normally will be in support of the MAGTF mission. The MAGTF commander will make sorties available to the JFC, for tasking through the joint force air component commander (JFACC), for air defense, long-range interdiction, and long-range reconnaissance. Sorties in excess of MAGTF direct support requirements will be provided to the JFC for tasking through the JFACC for the support of other components of the joint force or the joint force as a whole.

See p. 6-15 for an overview and further discussion of a MAGTF.

- Each GCC and subordinate JFC will be kept apprised of the presence, mission, movement, and duration of stay of transient forces within the operational area. The subordinate JFC also will be apprised of the command channels under which these transient forces will function. The authority directing movement or permanent location of transient forces is responsible for providing this information.
- Forces not assigned or attached to a GCC or attached to a subordinate JFC often are assigned missions that require them to cross boundaries. In such cases, it is the duty of the JFC to assist the operations of these transient forces to the extent of existing capabilities and consistent with other assigned missions. The JFC may be assigned specific responsibilities with respect to installations or activities exempted from their control, such as logistic support or area defense, particularly if adversary forces should traverse the operational area to attack the exempted installation or activity. GCC force protection policies take precedence over all force protection policies or programs of any other DOD component deployed in that GCC's AOR and not under the security responsibility of DOS. The GCC or a designated representative (e.g., a JTF or component commander) shall delineate the force protection measures for all DOD personnel not under the responsibility of DOS.
- Transient forces within the assigned AOR of a CCDR are subject to that CCDR's orders in some instances (e.g., for coordination of emergency defense, force protection, or allocation of local facilities).

Functional

Sometimes a joint force based solely on military functions without respect to a specific geographic region is more suitable to fix responsibility for certain types of continuing operations (e.g., the unified CCMDs for transportation, SO, and strategic operations). The commander of a joint force established on a functional basis is assigned a functional responsibility by the establishing authority.

- When defining functional responsibilities, the focus should be on the effect desired or service provided. The title of the functional responsibility and its delineation are prescribed in the establishing directive.
- The missions or tasks assigned to the commander of a functional command may require that certain installations and activities of that commander be exempt, partially or wholly, from the command authority of a GCC in whose area they are located or within which they operate. Such exemptions must be specified by the authority that establishes the functional command. Such exemptions do not relieve the commanders of functional commands of the responsibility to coordinate with the affected GCC.

I. Unified Combatant Command

A unified CCMD is a command with broad continuing missions under a single commander and composed of significant assigned components of two or more Military Departments that is established and so designated by the President through SecDef and with the advice and assistance of the CJCS. When either or both of the following criteria apply generally to a situation, a unified CCMD normally is required to ensure unity of effort.

- A broad continuing mission exists requiring execution by significant forces of two or more Military Departments and necessitating a single strategic direction.
- Any combination of the following exists and significant forces of two or more Military Departments are involved:
  - A large-scale operation requiring positive control of tactical execution by a large and complex force;
  - A large geographic or functional area requiring single responsibility for effective coordination of the operations therein; and/or
  - Necessity for common use of limited logistic means.

Authority of the Commander of a Unified Command in an Emergency

Ref: JP 1 (w/Chg 1), Doctrine for the Armed Forces of the U.S. (Jul '17), p. IV-7 to IV-8.

In the event of a major emergency in the Geographic Combatant Command's (GCC's) AOR requiring the use of all available forces, the GCC (except for CDRUSNORTHCOM) may temporarily assume OPCON of all forces in the assigned AOR, including those of another command, but excluding those forces scheduled for or actually engaged in the execution of specific operational missions under joint OPLANs approved by the SecDef that would be interfered with by the contemplated use of such forces. CDRUSNORTH-COM's authority to assume OPCON during an emergency is limited to the portion of USNORTHCOM's AOR outside the United States. CDRUSNORTHCOM must obtain SecDef approval before assuming OPCON of forces not assigned to USNORTHCOM within the U.S. The commander determines when such an emergency exists and, on assuming OPCON over forces of another command, immediately advises:

- The CJCS
- The appropriate operational commanders
- The Service Chief of the forces concerned

The authority to assume OPCON of forces in the event of a major emergency will not be delegated. Unusual circumstances in wartime, emergencies, or crises other than war (such as a terrorist incident) may require a GCC to directly exercise COCOM through a shortened chain of command to forces assigned for the purpose of resolving the crisis. Additionally, the CCDR can assume COCOM, in the event of war or an emergency that prevents control through normal channels, of security assistance organizations within the commander's general geographic AOR, or as directed by the SecDef. All commanders bypassed in such exceptional command arrangements will be advised of all directives issued to and reports sent from elements under such exceptional command arrangements. Such arrangements will be terminated as soon as practicable, consistent with mission accomplishment.

GCC Authority for Force Protection Outside the U.S.

GCCs shall exercise authority for force protection over all DOD personnel (including their dependents) assigned, attached, transiting through, or training in the GCC's AOR, except for those for whom the chief of mission retains security responsibility. Transient forces do not come under the authority of the GCC solely by their movement across operational area boundaries, except when the GCC is exercising TACON authority for force protection purposes or in the event of a major emergency. This force protection authority enables GCCs to change, modify, prescribe, and enforce force protection measures for covered forces.

GCC Authority for Exercise Purposes

Unless otherwise specified by SecDef, and with the exception of the USNORTHCOM AOR, a GCC has TACON for exercise purposes whenever forces not assigned to that CCDR undertake exercises in that GCC's AOR. TACON begins when the forces enter the AOR. In this context, TACON provides directive authority over exercising forces for purposes relating to force protection and to that exercise only; it does not authorize operational employment of those forces.

Assumption of Interim Command

In the temporary absence of a CCDR from the command, interim command will pass to the deputy commander. If a deputy commander has not been designated, interim command will pass to the next senior officer present for duty who is eligible to exercise command, regardless of Service affiliation.

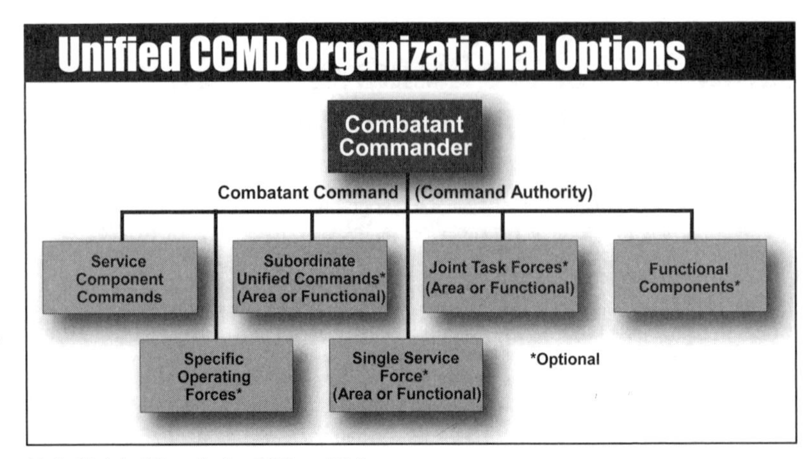

Ref: JP 1 (w/Chg 1), fig. IV-2, p. IV-6.

Command Structure

The commander of a unified command normally will adapt the command structure to exercise command authority through the commander of a subunified command, JTF, Service component, or functional component. Alternatively, the commander of a unified command may choose to exercise command authority directly through the commander of a single-Service force (e.g., task force, task group, MAGTF for a NEO) or a specific operational force (e.g., SOF for a direct action) who, because of the mission assigned and the urgency of the situation, must remain immediately responsive to the CCDR. The commander of a unified command normally assigns missions requiring a single-Service force to a Service component commander. These six options do not in any way limit the commander's authority to organize subordinate commands and exercise command authority over assigned forces as they see fit.

The commander of a unified command should not act concurrently as the commander of a subordinate command. For example, the commander of a unified command also should not act as a functional component commander without prior approval of the SecDef.

Primary Responsibilities

CCDRs are responsible for the development and production of joint OPLANs. During peacetime, they act to deter war through military engagement and security cooperation activities and prepare to execute other missions that may be required throughout the range of military operations. During war, they plan and conduct campaigns and major operations to accomplish assigned missions. Unified command responsibilities include the following:

- Planning and conducting military operations in response to crises, to include the security of the command and protection of the United States and its possessions and bases against attack or hostile incursion. The JSCP tasks the CCDRs to prepare joint OPLANs that may be one of four increasing levels of detail: commander's estimate, basic plan, concept plan, or OPLAN.
- Maintaining the preparedness of the command to carry out missions assigned to the command.
- Carrying out assigned missions, tasks, and responsibilities.
- Assigning tasks to, and directing coordination among, the subordinate commands to ensure unity of effort in the accomplishment of the assigned missions.

- Communicating directly with the following.
  - The Service Chiefs on single-Service matters as appropriate.
  - The CJCS on other matters, including the preparation of strategic, joint operation, and logistic plans; strategic and operational direction of assigned forces; conduct of combat operations; and any other necessary function of command required to accomplish the mission.
  - The SecDef, in accordance with applicable directives.
  - Subordinate elements, including the development organizations of the DOD agency or the Military Department directly supporting the development and acquisition of the CCDR's C2 system in coordination with the director of the DOD agency or secretary of the Military Department concerned.
- Keeping the CJCS promptly advised of significant events and incidents that occur in the functional area or area of operations responsibility, particularly those incidents that could create national or international repercussions.

II. Specified Combatant Command

There are currently no specified CCMDs designated. Because the option for the President through SecDef to create a specified CCMD still exists in Title 10, USC, Section 161, the following information is provided. A specified CCMD is a command that has broad continuing missions and is established by the President, through SecDef, with the advice and assistance of the CJCS (see Figure IV-3).

Although a specified CCMD normally is composed of forces from one Military Department, it may include units and staff representation from other Military Departments.

Transfer of Forces from Other Military Departments. When units of other Military Departments are transferred (assigned or attached) to the commander of a specified CCMD, the purpose and duration of the transfer normally will be indicated. Such transfer does not constitute the specified CCMD as a unified CCMD or a JTF. If the transfer is major and of long duration, a unified CCMD normally would be established in lieu of a specified CCMD.

The commander of a specified CCMD has the same authority and responsibilities as the commander of a unified CCMD, except that no authority exists to establish subordinate unified commands.

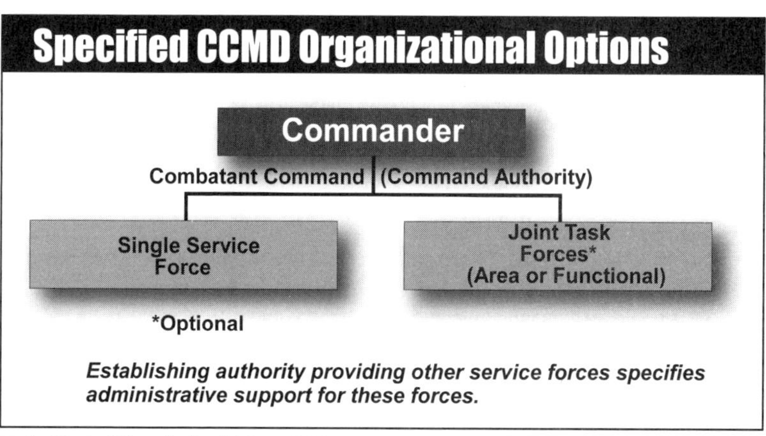

Ref: JP 1(w/Chg 1), fig. IV-3, p. IV-9.

III. Subordinate Unified Command

When authorized by SecDef through the CJCS, commanders of unified CCMDs may establish subordinate unified commands (also called subunified commands) to conduct operations on a continuing basis in accordance with the criteria set forth for unified CCMDs. A subordinate unified command (e.g., United States Forces Korea) may be established on a geographical area or functional basis. Commanders of subordinate unified commands have functions and responsibilities similar to those of the commanders of unified CCMDs and exercise OPCON of assigned commands and forces and normally over attached forces within the assigned joint operations area or functional area. The commanders of components or Service forces of subordinate unified commands have responsibilities and missions similar to those for component commanders within a unified CCMD. The Service component or Service force commanders of a subordinate unified command normally will communicate directly with the commanders of the Service component command of the unified CCMD on Service-specific matters and inform the commander of the subordinate unified command as that commander directs.

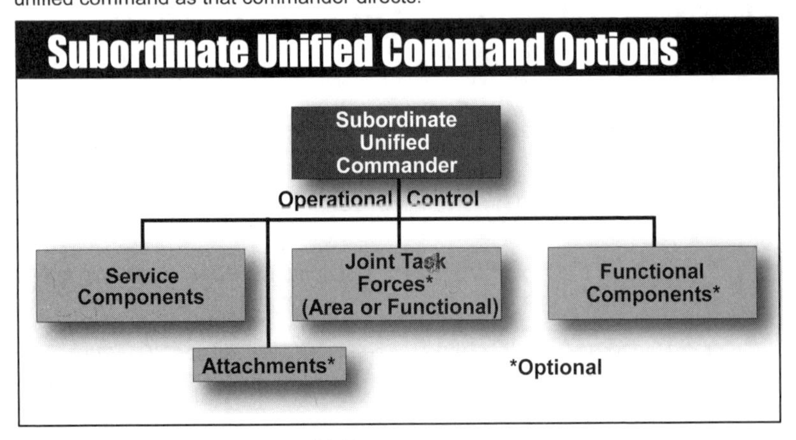

Ref: JP 1(w/Chg 1), fig. IV-4, p. IV-10.

IV. Joint Task Force (JTF)

A JTF is a joint force that is constituted and so designated by the SecDef, a CCDR, a subordinate unified commander, or an existing JTF commander.

A JTF may be established on a geographical area or functional basis when the mission has a specific limited objective and does not require overall centralized control of logistics. The mission assigned to a JTF should require execution of responsibilities involving a joint force on a significant scale and close integration of effort, or should require coordination within a subordinate area or coordination of local defense of a subordinate area. The proper authority dissolves a JTF when the purpose for which it was created has been achieved or when it is no longer required.

The authority establishing a JTF designates the commander, assigns the mission, designates forces, and delegates command authorities. Based on the decision of the establishing JFC, the commander of a JTF exercises OPCON over assigned (and normally over attached) forces, or may exercise TACON over attached forces. The JTF commander establishes command relationships between subordinate commanders and is responsible to the establishing commander for the proper employment of assigned and attached forces and for accomplishing such operational missions as may be assigned. JTF commanders also are responsible to the establishing commander for the conduct of joint training of assigned forces.

See chap. 5 for further discussion of joint task forces.

V. Joint Command and Control

Ref: JP 1 (w/Chg 1), Doctrine for the Armed Forces of the U.S. (Jul '17), chap. V.

I. Command Relationships

Command

Command is central to all military action, and unity of command is central to unity of effort. Inherent in command is the authority that a military commander lawfully exercises over subordinates including authority to assign missions and accountability for their successful completion. Although CDRs may delegate authority to accomplish missions, they may not absolve themselves of the responsibility for the attainment of these missions. Authority is never absolute; the extent of authority is specified by the establishing authority, directives, and law.

Unity of Command and Unity of Effort

Unity of command means all forces operate under a single CDR with the requisite authority to direct all forces employed in pursuit of a common purpose. Unity of effort, however, requires coordination and cooperation among all forces toward a commonly recognized objective, although they are not necessarily part of the same command structure. During multinational operations and interagency coordination, unity of command may not be possible, but the requirement for unity of effort becomes paramount. Unity of effort — coordination through cooperation and common interests — is an essential complement to unity of command. Unity of command requires that two CDRs may not exercise the same command relationship over the same force at any one time.

Command and Staff

JFCs are provided staffs to assist them in the decision-making and execution process. The staff is an extension of the JFC; its function is command support and its authority is delegated by the JFC. A properly trained and directed staff will free the JFC to devote more attention to directing subordinate commanders and maintaining a picture of the overall situation.

- Chain of command is the succession of commanding officers from a superior to a subordinate through which command is exercised.
- Staffing is the term used to describe the coordination between staffs at higher, adjacent, and subordinate headquarters. Higher headquarters staff officers exercise no independent authority over subordinate headquarters staffs, although staff officers normally respond to requests for information.

Levels of Authority

The specific command relationship (COCOM, OPCON, TACON, and support) will define the authority a commander has over assigned or attached forces.

See following pages (pp. 1-32 to 1-33) for further discussion.

Command Relationships Overview & Assignment and Transfer of Forces

Ref: JP 1 (w/Chg 1), Doctrine for the Armed Forces of the U.S. (Jul '17), pp. V-2 to V-4.

Levels of Authority

The specific command relationship (COCOM, OPCON, TACON, and support) will define the authority a commander has over assigned or attached forces. An overview of command relationships is shown in Figure V-1, below.

Command Relationships Synopsis

Combatant Command (Command Authority)

(Unique to Combatant Commander)

- Planning, programming, budgeting, and execution process input
- Assignment of subordinate commanders
- Relationships with Department of Defense agencies
- Directive authority for logistics

Operational control when delegated

- Authoritative direction for all military operations and joint training
- Organize and employ commands and forces
- Assign command functions to subordinates
- Establish plans and requirements for intelligence, surveillance, and reconnaissance activities
- Suspend subordinate commanders from duty

Tactical control when delegated

Local direction and control of movements or maneuvers to accomplish mission

Support relationship when assigned

Aid, assist, protect, or sustain another organization

Ref: JP 1 (w/Chg 1), fig. V-1, p. V-2.

Assignment and Transfer of Forces

All forces under the jurisdiction of the Secretaries of the Military Departments (except those forces necessary to carry out the functions of the Military Departments as noted in Title 10, USC, Section 162) are assigned to combatant commands or CDR, US Element NORAD (USELEMNORAD) by the SecDef in the "Forces for Unified Commands" memorandum. A force assigned or attached to a combatant command may be transferred from that command to another CCDR only when directed by the SecDef and under procedures prescribed by the SecDef and approved by the President. The command relationship the gaining CDR will exercise (and the losing CDR will relinquish) will be specified by the SecDef. Establishing authorities for subordinate unified commands and JTFs may direct the assignment or attachment of their forces to those subordinate commands and delegate the command relationship as appropriate.

Command Relationships Overview

- Forces, not command relationships, are transferred between commands. When forces are transferred, the command relationship the gaining commander will exercise (and the losing commander will relinquish) over those forces must be specified.
- When transfer of forces to a joint force will be permanent (or for an unknown but long period of time) the forces should be reassigned. Combatant commanders will exercise combatant command (command authority) and subordinate joint force commanders (JFCs), will exercise operational control (OPCON) over reassigned forces.
- When transfer of forces to a joint force will be temporary, the forces will be attached to the gaining command and JFCs, normally through the Service component commander, will exercise OPCON over the attached forces.
- Establishing authorities for subordinate unified commands and joint task forces direct the assignment or attachment of their forces to those subordinate commands as appropriate.

A. Combatant Command (COCOM) - Command Authority

COCOM provides full authority for a CCDR to perform those functions of command over assigned forces involving organizing and employing commands and forces, assigning tasks, designating objectives, and giving authoritative direction over all aspects of military operations, joint training (or in the case of USSOCOM, training of assigned forces), and logistics necessary to accomplish the missions assigned to the command. COCOM should be exercised through the commanders of subordinate organizations, normally JFCs, Service and/or functional component commanders.

See pp. 1-34 to 1-35 for further discussion.

B. Operational Control (OPCON)

OPCON is the command authority that may be exercised by commanders at any echelon at or below the level of CCMD and may be delegated within the command. It is the authority to perform those functions of command over subordinate forces involving organizing and employing commands and forces, assigning tasks, designating objectives, and giving authoritative direction over all aspects of military operations and joint training necessary to accomplish the mission.

See p. 1-36 for further discussion.

C. Tactical Control (TACON)

TACON is an authority over assigned or attached forces or commands, or military capability or forces made available for tasking, that is limited to the detailed direction and control of movements and maneuvers within the operational area necessary to accomplish assigned missions or tasks assigned by the commander exercising OPCON or TACON of the attached force.

See p. 1-37 for further discussion.

Support Relationships

Support is a command authority. A support relationship is established by a common superior commander between subordinate commanders when one organization should aid, protect, complement, or sustain another force. The support command relationship is used by SecDef to establish and prioritize support between and among CCDRs, and it is used by JFCs to establish support relationships between and among subordinate commanders. There are four defined categories of support: general support, mutual support, direct support, and close support.

See p. 1-38 for further discussion.

A. Combatant Command (COCOM) - Command Authority

Ref: JP 1 (w/Chg 1), Doctrine for the Armed Forces of the U.S. (Jul '17), pp. V-2 to V-6.

COCOM is the command authority over assigned forces vested only in the commanders of combatant commands by Title 10, USC, Section 164 (or as directed by the President in the UCP) and cannot be delegated or transferred.

Basic Authority

COCOM provides full authority for a CCDR to perform those functions of command over assigned forces involving organizing and employing commands and forces, assigning tasks, designating objectives, and giving authoritative direction over all aspects of military operations, joint training (or in the case of USSOCOM, training of assigned forces), and logistics necessary to accomplish the missions assigned to the command. COCOM should be exercised through the commanders of subordinate organizations, normally JFCs, Service and/or functional component commanders.

Combatant Commander (CCDR) Control

Unless otherwise directed by the President or the SecDef, the authority, direction, and control of the CCDR with respect to the command of forces assigned to that command includes the following.

- Exercise or delegate OPCON, TACON, or other specific elements of authority, establish support relationships among subordinate commanders over assigned or attached forces, and designate coordinating authorities, as described below.
- Exercise directive authority for logistic matters (or delegate directive authority for a common support capability to a subordinate commander via an establishing directive).
- Prescribe the chain of command to the commands and forces within the command.
- Organize subordinate commands and forces within the command as necessary to carry out missions assigned to the command.
- Employ forces within that command as necessary to carry out missions assigned to the command.
- Assign command functions to subordinate commanders.
- Coordinate and approve those aspects of administration, support, and discipline necessary to accomplish assigned missions.
- Plan, deploy, direct, control, and coordinate the actions of subordinate forces.
- Establish plans, policies, priorities, and overall requirements for the ISR activities of the command.
- Conduct joint exercises and training to achieve effective employment of the forces in accordance with joint and established training policies for joint operations. This authority also applies to forces attached for purposes of joint exercises and training.
- Assign responsibilities to subordinate commanders for certain routine operational matters that require coordination of effort of two or more commanders.
- Establish a system of control for local defense and delineate such operational areas for subordinate commanders.
- Delineate functional responsibilities and geographic operational areas of subordinate commanders.
- Give authoritative direction to subordinate commands and forces necessary to carry out missions assigned to the command, including military operations, joint training, and logistics.

- Coordinate with other CCDRs, USG departments and agencies, and organizations of other countries regarding matters that cross the boundaries of geographic areas specified in the UCP and inform USG departments and agencies or organizations of other countries in the AOR, as necessary, to prevent both duplication of effort and lack of adequate control of operations in the delineated areas.
- Unless otherwise directed by SecDef, function as the US military single point of contact and exercise directive authority over all elements of the command in relationships with other CCMDs, DOD elements, US diplomatic missions, other USG departments and agencies, and organizations of other countries in the AOR. Whenever a CCDR conducts exercises, operations, or other activities with the military forces of nations in another CCDR's AOR, those exercises, operations, and activities and their attendant command relationships will be mutually agreed to between the CCDRs.
- Determine those matters relating to the exercise of COCOM in which subordinates must communicate with agencies external to the CCMD through the CCDR.
- Establish personnel policies to ensure proper and uniform standards of military conduct.
- Submit recommendations through the CJCS to SecDef concerning the content of guidance affecting the strategy and/or fielding of joint forces.
- Participate in the planning, programming, budgeting, and execution process as specified in appropriate DOD issuances.
- Participate in the JSPS and the APEX system. CCDRs' comments are critical to ensuring that warfighting and peacetime operational concerns are emphasized in all planning documents.
- Concur in the assignment (or recommendation for assignment) of officers as commanders directly subordinate to the CCDR and to positions on the CCMD staff. Suspend from duty and recommend reassignment, when appropriate, any subordinate officer assigned to the CCMD.
- Convene general courts-martial in accordance with the UCMJ.
- In accordance with laws and national and DOD policies, establish plans, policies, programs, priorities, and overall requirements for the C2, communications system, and ISR activities of the command.

When directed in the UCP or otherwise authorized by SecDef, the commander of US elements of a multinational command may exercise COCOM of those US forces assigned to that command (e.g., United States Element, North American Aerospace Defense Command.

Directive Authority for Logistics (DAFL) *See also p. 4-11.*

CCDRs exercise directive authority for logistics and may delegate directive authority for a common support capability. The CCDR may delegate directive authority for as many common support capabilities to a subordinate JFC as required to accomplish the subordinate JFC's assigned mission. For some commodities or support services common to two or more Services, one Service may be given responsibility for management based on DOD EA designations or inter-Service support agreements. However, the CCDR must formally delineate this delegated directive authority by function and scope to the subordinate JFC or Service component commander. The exercise of directive authority for logistics by a CCDR includes the authority to issue directives to subordinate CDRs, including peacetime measures necessary to ensure the following: effective execution of approved OPLANs; effectiveness and economy of operation; and prevention or elimination of unnecessary duplication of facilities and overlapping of functions among the Service component commands. CCDRs will coordinate with appropriate Services before exercising directive authority for logistics or delegate authority for subordinate CDRs to exercise common support capabilities to one of their components.

B. Operational Control (OPCON)

Ref: JP 1 (w/Chg 1), Doctrine for the Armed Forces of the U.S. (Jul '17), pp. V-6 to V-7.

OPCON is the command authority that may be exercised by commanders at any echelon at or below the level of CCMD and may be delegated within the command.

Basic Authority

OPCON is able to be delegated from a lesser authority than COCOM. It is the authority to perform those functions of command over subordinate forces involving organizing and employing commands and forces, assigning tasks, designating objectives, and giving authoritative direction over all aspects of military operations and joint training necessary to accomplish the mission. It should be delegated to and exercised by the commanders of subordinate organizations; normally, this authority is exercised through subordinate JFCs, Service, and/or functional component commanders. OPCON provides authority to organize and employ commands and forces as the commander considers necessary to accomplish assigned missions. It does not include authoritative direction for logistics or matters of administration, discipline, internal organization, or unit training. These elements of COCOM must be specifically delegated by the CCDR. OPCON does include the authority to delineate functional responsibilities and operational areas of subordinate JFCs. Commanders of subordinate commands, including JTFs, will be given OPCON of assigned forces and OPCON or TACON of attached forces by the superior commander. OPCON includes the authority for the following:

- Exercise or delegate OPCON and TACON or other specific elements of authority and establish support relationships among subordinates, and designate coordinating authorities.
- Give direction to subordinate commands and forces necessary to carry out missions assigned to the command, including authoritative direction over all aspects of military operations and joint training.
- Prescribe the chain of command to the commands and forces within the command.
- With due consideration for unique Service organizational structures and their specific support requirements, organize subordinate commands and forces within the command as necessary to carry out missions assigned to the command.
- Employ forces within the command, as necessary, to carry out missions assigned to the command.
- Assign command functions to subordinate commanders.
- Plan for, deploy, direct, control, and coordinate the actions of subordinate forces.
- Establish plans, policies, priorities, and overall requirements for the ISR activities of the command.
- Conduct joint training exercises required to achieve effective employment of the forces of the command, in accordance with joint doctrine established by the CJCS, and establish training policies for joint operations required to accomplish the mission. This authority also applies to forces attached for purposes of joint exercises and training.
- Suspend from duty and recommend reassignment of any officer assigned to the command.
- Assign responsibilities to subordinate commanders for certain routine operational matters that require coordination of effort of two or more commanders.
- Establish an adequate system of control for local defense and delineate such operational areas for subordinate commanders as deemed desirable.
- Delineate functional responsibilities and geographic operational areas of subordinate commanders.

SecDef may specify adjustments to accommodate authorities beyond OPCON in an establishing directive when forces are transferred between CCDRs or when members and/or organizations are transferred from the Military Departments to a CCMD.

C. Tactical Control (TACON)

TACON is an authority over assigned or attached forces or commands, or military capability or forces made available for tasking, that is limited to the detailed direction and control of movements and maneuvers within the operational area necessary to accomplish assigned missions or tasks assigned by the commander exercising OPCON or TACON of the attached force.

Basic Authority

TACON is able to be delegated from a lesser authority than OPCON and may be delegated to and exercised by commanders at any echelon at or below the level of CCMD.

TACON provides the authority to:

- Give direction for military operations; and
- Control designated forces (e.g., ground forces, aircraft sorties, or missile launches).

TACON does not provide the authority to give or change the function of the subordinate commander. TACON provides sufficient authority for controlling and directing the application of force or tactical use of combat support assets within the assigned mission or task. TACON does not provide organizational authority or authoritative direction for administrative and logistic support. Functional component commanders typically exercise TACON over military capability or forces made available for tasking.

II. Support

Support is a command authority. A support relationship is established by a common superior commander between subordinate commanders when one organization should aid, protect, complement, or sustain another force. The support command relationship is used by SecDef to establish and prioritize support between and among CCDRs, and it is used by JFCs to establish support relationships between and among subordinate commanders.

Basic Authority

Support may be exercised by commanders at any echelon at or below the CCMD level. The designation of supporting relationships is important as it conveys priorities to commanders and staffs that are planning or executing joint operations. The support command relationship is, by design, a somewhat vague but very flexible arrangement. The establishing authority (the common JFC) is responsible for ensuring that both the supported commander and supporting commanders understand the degree of authority that the supported commander is granted.

The supported commander should ensure that the supporting commanders understand the assistance required. The supporting commanders will then provide the assistance needed, subject to a supporting commander's existing capabilities and other assigned tasks. When a supporting commander cannot fulfill the needs of the supported commander, the establishing authority will be notified by either the supported commander or a supporting commander. The establishing authority is responsible for determining a solution.

An establishing directive normally is issued to specify the purpose of the support relationship, the effect desired, and the scope of the action to be taken.

- The forces and resources allocated to the supporting effort;
- The time, place, level, and duration of the supporting effort;
- The relative priority of the supporting effort;
- The authority, if any, of the supporting commander to modify the supporting effort in the event of exceptional opportunity or an emergency; and
- The degree of authority granted to the supported commander over the supporting effort.

Categories of Support

Ref: JP 1 (w/Chg 1), Doctrine for the Armed Forces of the U.S. (Jul '17), p. V-9 to V-10.

There are four defined categories of support that a CCDR may direct over assigned or attached forces to ensure the appropriate level of support is provided to accomplish mission objectives.

General Support

That support which is given to the supported force as a whole rather than to a particular subdivision thereof.

Mutual Support

That support which units render each other against an enemy because of their assigned tasks, their position relative to each other and to the enemy, and their inherent capabilities.

Direct Support

A mission requiring a force to support another specific force and authorizing it to answer directly to the supported force's request for assistance.

Close Support

That action of the supporting force against targets or objectives that are sufficiently near the supported force as to require detailed integration or coordination of the supporting action with the fire, movement, or other actions of the supported force.

Support Relationships Between Combatant Commanders

SecDef establishes support relationships between the CCDRs for the planning and execution of joint operations. This ensures that the supported CCDR receives the necessary support. A supported CCDR requests capabilities, tasks supporting DOD components, coordinates with the appropriate USG departments and agencies (where agreements have been established), and develops a plan to achieve the common goal. As part of the team effort, supporting CCDRs provide the requested capabilities, as available, to assist the supported CCDR to accomplish missions requiring additional resources. The CJCS organizes the JPEC for joint operation planning to carry out support relationships between the CCMDs. The supported CCDR has primary responsibility for all aspects of an assigned task. Supporting CCDRs provide forces, assistance, or other resources to a supported CCDR. Supporting CCDRs prepare supporting plans as required. Under some circumstances, a CCDR may be a supporting CCDR for one operation while being a supported CCDR for another.

Support Relationships Between Component Commanders

The JFC may establish support relationships between component commanders to facilitate operations. Support relationships afford an effective means to prioritize and ensure unity of effort for various operations. Component commanders should establish liaison with other component commanders to facilitate the support relationship and to coordinate the planning and execution of pertinent operations. Support relationships may change across phases of an operation as directed by the establishing authority. When the commander of a Service component is designated as a functional component commander, the associated Service component responsibilities for assigned or attached forces are retained, but are not applicable to forces made available by other Service components. The operational requirements of the functional component commander's subordinate forces are prioritized and presented to the JFC by the functional component commander, relieving the affected Service component commanders of this responsibility, but the affected Service component commanders are not relieved of their administrative and support responsibilities.

III. Other Authorities

Other authorities outside the command relationships delineated above include:

A. Administrative Control (ADCON)

ADCON is the direction or exercise of authority over subordinate or other organizations with respect to administration and support, including organization of Service forces, control of resources and equipment, personnel management, logistics, individual and unit training, readiness, mobilization, demobilization, discipline, and other matters not included in the operational missions of the subordinate or other organizations. ADCON is synonymous with administration and support responsibilities identified in Title 10, USC. This is the authority necessary to fulfill Military Department statutory responsibilities for administration and support. ADCON may be delegated to and exercised by commanders of Service forces assigned to a CCDR at any echelon at or below the level of Service component command. ADCON is subject to the command authority of CCDRs. ADCON may be delegated to and exercised by commanders of Service commands assigned within Service authorities. Service commanders exercising ADCON will not usurp the authorities assigned by a CCDR having COCOM over commanders of assigned Service forces.

B. Coordinating Authority

Commanders or individuals may exercise coordinating authority at any echelon at or below the level of CCMD. Coordinating authority is the authority delegated to a commander or individual for coordinating specific functions and activities involving forces of two or more Military Departments, two or more joint force components, or two or more forces of the same Service (e.g., joint security coordinator exercises coordinating authority for joint security area operations among the component commanders). Coordinating authority may be granted and modified through an MOA to provide unity of effort for operations involving RC and AC forces engaged in interagency activities. The commander or individual has the authority to require consultation between the agencies involved but does not have the authority to compel agreement. The common task to be coordinated will be specified in the establishing directive without disturbing the normal organizational relationships in other matters. Coordinating authority is a consultation relationship between commanders, not an authority by which command may be exercised. It is more applicable to planning and similar activities than to operations. Coordinating authority is not in any way tied to force assignment. Assignment of coordinating authority is based on the missions and capabilities of the commands or organizations involved.

C. Direct Liaison Authorized (DIRLAUTH)

DIRLAUTH is that authority granted by a commander (any level) to a subordinate to directly consult or coordinate an action with a command or agency within or outside of the granting command. DIRLAUTH is more applicable to planning than operations and always carries with it the requirement of keeping the commander granting DIRLAUTH informed. DIRLAUTH is a coordination relationship, not an authority through which command may be exercised.

Command of National Guard and Reserve Forces

Ref: JP 1 (w/Chg 1), Doctrine for the Armed Forces of the U.S. (Jul '17), p. V-13 to V-14.

Mobilized Forces

When mobilized under Title 10, USC, authority, command of National Guard and Reserve forces (except those forces specifically exempted) is assigned by SecDef to the CCMDs. Those forces are available for operational missions when mobilized for specific periods or when ordered to active duty after being validated for employment by their parent Service. Normally, National Guard forces are under the commands of their respective governors in Title 32, USC, or state active duty status.

Training and Readiness

The authority CCDRs may exercise over assigned RC forces when not on active duty or when on active duty for training is training and readiness oversight (TRO). CCDRs normally will exercise TRO over assigned forces through the Service component commanders. TRO includes the authority to:

- Provide guidance to Service component commanders on operational requirements and priorities to be addressed in Military Department training and readiness programs;
- Comment on Service component program recommendations and budget requests;
- Coordinate and approve participation by assigned RC forces in joint exercises and other joint training when on active duty for training or performing inactive duty for training;
- Obtain and review readiness and inspection reports on assigned RC forces; and
- Coordinate and review mobilization plans (including post-mobilization training activities and deployability validation procedures) developed for assigned RC forces.

Additional Guidance

Unless otherwise directed by the SecDef, the following applies:

- Assigned RC forces on active duty (other than for training) may not be deployed until validated by the parent Service for deployment.
- CCDRs may employ RC forces assigned to their subordinate component CDRs in contingency operations only when the forces have been mobilized for specific periods in accordance with the law, or when ordered to active duty and after being validated for employment by their parent Service.
- RC forces on active duty for training or performing inactive-duty training may be employed in connection with contingency operations only as provided by law, and when the primary purpose is for training consistent with their mission or specialty.

CCDRs will communicate with assigned RC forces through the Military Departments when the RC forces are not on active duty or when on active duty for training.

CCDRs may inspect assigned RC forces in accordance with DODD 5106.04, Combatant Command Inspectors General, when such forces are mobilized or ordered to active duty (other than for training).

CDRUSSOCOM will exercise additional authority for certain functions for assigned RC forces and for all SOF assigned to other CCMDs in accordance with the current MOAs between CDRUSSOCOM and the Secretaries of the Military Departments.

Refer to DODI 1215.06, Uniform Reserve, Training, and Retirement Categories.

Chap 2

I. Fundamentals of Joint Operations

Ref: JP 3-0 (w/Chg 1), Joint Operations (Oct '18), chap. I and Executive Summary.

Joint Publication (JP) 3-0 is the keystone document in the joint operations series and is a companion to joint doctrine's capstone JP 1, Doctrine for the Armed Forces of the United States. It provides guidance to joint force commanders (JFCs) and their subordinates to plan, execute, and assess joint military operations. It also informs interagency and multinational partners, international organizations, nongovernmental organizations (NGOs), and other civilian decision makers of fundamental principles, precepts, and philosophies that guide the employment of the Armed Forces of the United States.

Joint Operations / Joint Force

The primary way the Department of Defense (DOD) employs two or more Services (from at least two Military Departments) in a single operation is through joint operations. **Joint operations** are military actions conducted by joint forces and those Service forces employed in specified command relationships with each other, which of themselves do not establish joint forces. **A joint force** is one composed of significant elements, assigned or attached, of two or more Military Departments operating under a single joint force commander (JFC).

I. Strategic Environment and National Security Challenges

The strategic environment consists of a variety of national, international, and global factors that affect the decisions of senior civilian and military leaders with respect to the employment of US instruments of national power in peace and periods of conflict. The strategic environment is uncertain, complex, and can change rapidly, requiring military leaders to maintain persistent military engagement with multinational partners. Although the basic character of war has not changed, the character of conflict has evolved.

Transregional, Multi-domain, and Multi-functional (TMM)

The military environment and the threats it presents are increasingly transregional, multi-domain, and multi-functional (TMM) in nature. By TMM, we mean the crises and ccontingencies joint forces face today cut across multiple combatant commands (CCMDs); the physical domains of land, maritime, air, and space; and the information environment (which includes cyberspace), as well as the electromagnetic spectrum (EMS), and involve conventional, special operations, ballistic missile, electronic warfare (EW), information, strike, cyberspace, and space capabilities. The strategic environment is fluid, with continually changing alliances, partnerships, and national and transnational threats that rapidly emerge, disaggregate, and reemerge. While it is impossible to predict precisely how challenges will emerge and what form they might take, we can expect that uncertainty, ambiguity, and surprise will persist. The commander's OE is influenced by the strategic environment.

By acquiring advanced technologies, adversaries are changing the conditions of warfare the US has become accustomed to in the past half century. Today's threats can increasingly synchronize, integrate, and direct operations and other elements of power to create lethal and nonlethal effects with greater sophistication and are less constrained by geographic, functional, legal, or phasing boundaries. Conflict is now, and will remain, inherently transregional as enemies' interests, influence,

II. Principles of Joint Operations

Ref: JP 3-0 (w/Chg 1), Joint Operations (Oct '18), pp. I-2 to I-3 and app. A.

Joint Warfare is Team Warfare. The Armed Forces of the United States—every military organization at all levels—are a team. The capacity of our Armed Forces to operate as a cohesive joint team is a key advantage in any operational environment (OE). Success depends on well-integrated command headquarters (HQ), supporting organizations, and forces that operate as a team. Integrating Service components' capabilities under a single JFC maximizes the effectiveness and efficiency of the force. However, a joint operation does not require that all forces participate merely because they are available; the JFC has the authority and responsibility to tailor forces to the mission.

Principles of War

Joint doctrine recognizes the nine principles of war. Experience gained in a variety of IW situations has reinforced the value of three additional principles—restraint, perseverance, and legitimacy. Together, they comprise the 12 principles of joint operations

Objective

The purpose of specifying the objective is to direct every military operation toward a clearly defined, decisive, and achievable goal. The purpose of military operations is to achieve specific objectives that support achievement of the overall strategic objectives identified to resolve the conflict.

Offensive

The purpose of an offensive action is to seize, retain, and exploit the initiative. Offensive operations are the means by which a military force seizes and holds the initiative while maintaining freedom of action and achieving decisive results.

Mass

The purpose of mass is to concentrate the effects of combat power at the most advantageous place and time to produce decisive results. To achieve mass, appropriate joint force capabilities are integrated and synchronized where they will have a decisive effect in a short period of time.

Maneuver

The purpose of maneuver is to place the enemy in a position of disadvantage through the flexible application of combat power. Maneuver is the movement of forces in relation to the enemy to secure or retain positional advantage, usually to deliver—or threaten delivery of—the direct and indirect fires of the maneuvering force.

Economy of Force

The purpose of economy of force is to expend minimum essential combat power on secondary efforts to allocate the maximum possible combat power on primary efforts.

Unity of Command

The purpose of unity of command is to ensure unity of effort under one responsible commander for every objective. Unity of command means all forces operate under a single commander with the requisite authority to direct all forces employed in pursuit of a common purpose..

Security

The purpose of security is to prevent the enemy from acquiring unexpected advantage.

Surprise

The purpose of surprise is to strike at a time or place or in a manner for which the enemy is unprepared.

Simplicity

The purpose of simplicity is to increase the probability that plans and operations will be executed as intended by preparing clear, uncomplicated plans and concise orders.

Additional Principles of Joint Operations

Restraint

The purpose of restraint is to prevent the unnecessary use of force.A single act could cause significant military and political consequences; therefore, judicious use of force is necessary. Restraint requires the careful and disciplined balancing of the need for security, the conduct of military operations, and national objectives. Excessive force antagonizes those parties involved, thereby damaging the legitimacy of the organization that uses it while potentially enhancing the legitimacy of the opposing party. Sufficiently detailed ROE the commander tailors to the specific circumstances of the operation can facilitate appropriate restraint.

Perseverance

The purpose of perseverance is to ensure the commitment necessary to attain the national strategic end state. Perseverance involves preparation for measured, protracted military operations in pursuit of the national strategic end state. Some joint operations may require years to reach the termination criteria. The underlying causes of the crisis may be elusive, making it difficult to achieve decisive resolution. The patient, resolute, and persistent pursuit of national goals and objectives often is essential to success. This will frequently involve diplomatic, economic, and informational measures to supplement military efforts.

Legitimacy

The purpose of legitimacy is to maintain legal and moral authority in the conduct of operations. Legitimacy, which can be a decisive factor in operations, is based on the actual and perceived legality, morality, and rightness of the actions from the various perspectives of interested audiences. These audiences will include our national leadership and domestic population, governments, and civilian populations in the operational area, and nations and organizations around the world. Committed forces must sustain the legitimacy of the operation and of the host government, where applicable. Security actions must be balanced with legitimacy concerns. All actions must be considered in the light of potentially competing strategic and tactical-level requirements and must exhibit fairness in dealing with competing factions where appropriate.

Common Operating Precepts

In addition to the principles of joint operations, 10 common operating precepts underlie successful joint operations. These precepts flow from broad challenges in the strategic environment to specific conditions, circumstances, and influences in a JFC's OE. The precepts can apply in all joint operations, although some may not be relevant activities, such as military engagement.

- Achieve and maintain unity of effort within the joint force and between the joint force and US Government, international, and other partners
- Leverage the benefits of operating indirectly through partners when strategic and operational circumstances dictate or permit
- Integrate joint capabilities to be complementary rather than merely additive
- Focus on objectives whose achievement suggests the broadest and most enduring results
- Ensure freedom of action
- Avoid combining capabilities where doing so adds complexity without compensating advantage
- Inform domestic audiences and shape the perceptions and attitudes of key foreign audiences as an explicit and continuous operational requirement
- Maintain operational and organizational flexibility
- Drive synergy to the lowest echelon at which it can be managed effectively
- Plan for and manage operational transitions over time and space

III. Levels of War

Ref: JP 3-0 (w/Chg 1), Joint Operations (Oct '18), pp. I-12 to I-14.

Three levels of warfare—strategic, operational, and tactical—model the relationship between national objectives and tactical actions. There are no fixed limits or boundaries between these levels, but they help commanders visualize a logical arrangement of operations, allocate resources, and assign tasks to appropriate commands. Echelon of command, size of units, types of equipment, and types and location of forces or components may often be associated with a particular level, but the strategic, operational, or tactical purpose of their employment depends on the nature of their task, mission, or objective. For example, intelligence and communications satellites, previously considered principally strategic assets, are also significant resources for tactical operations. Likewise, tactical actions can cause both intended and unintended strategic consequences, particularly in today's environment of pervasive and immediate global communications and networked threats.

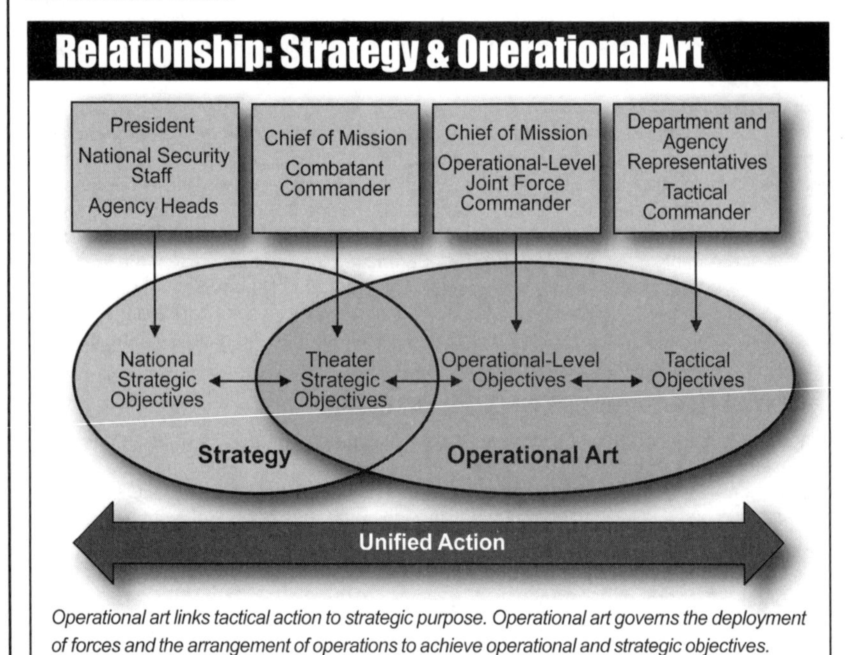

Operational art links tactical action to strategic purpose. Operational art governs the deployment of forces and the arrangement of operations to achieve operational and strategic objectives.

Ref: JP 3-0 (w/Chg 1), fig. I-4. Relationship Between Strategy and Operational Art.

A. Strategic Level

In the context of national interests, strategy develops an idea or set of ideas of the ways to employ the instruments of national power in a synchronized and integrated fashion to achieve national, multinational, and theater objectives. Through development of strategy (e.g., the NSS, DSR, and NMS), a nation's leader, often with other nations' leaders, determines national or multinational strategic objectives with specific guidance to shape and allocate national resources to achieve these objectives. The President, aided by the NSC, establishes policy and national strategic objectives. SecDef translates these objectives into strategic military objectives that facilitate theater strategic planning. CCDRs usually participate in strategic discussions with the President and SecDef through the CJCS. CCDRs also participate in strategic discussions with allies and multinational members. Thus, the CCDR's strategy relates to both US national strategy and operational-level activities within the theater. Military strategy, derived from national policy and strategy and informed by doctrine, provides a framework for conducting operations.

B. Operational Level

The operational level of warfare links the tactical employment of forces to national strategic objectives. The focus at this level is on the planning and execution of operations using operational art: the cognitive approach by commanders and staffs— supported by their skill, knowledge, and experience—to plan and execute (when required) strategies, campaigns, and operations to organize and employ military capabilities by integrating ends, ways, and available means. JFCs and component commanders use operational art to determine how, when, where, and for what purpose military forces will be employed, to influence the adversary's disposition before combat, to deter adversaries from supporting enemy activities, and to assure our multinational partners to achieve operational and strategic objectives.

Many factors affect relationships among leaders at these levels. Service and functional component commanders of a joint force do not plan the actions of their forces in a vacuum; they and their staffs collaborate with the operational-level JFC to plan the joint operation. This collaboration facilitates the components' planning and execution. Likewise, the operational-level JFC and staff typically collaborate with the CCDR to frame theater strategic objectives, as well as tasks the CCDR will eventually assign to the subordinate joint force.

C. Tactical Level

Tactics is the employment, ordered arrangement, and directed actions of forces in relation to each other. Joint doctrine focuses this term on planning and executing battles, engagements, and activities at the tactical level to achieve military objectives assigned to tactical units or task forces (TFs). An engagement can include a wide variety of noncombat tasks and activities and combat between opposing forces normally of short duration. A battle consists of a set of related engagements. Battles typically last longer than engagements, involve larger forces, and have greater potential to affect the course of a campaign.

Characterizing Military Operations and Activities

The US employs military capabilities in support of national security objectives in a variety of military operations and activities. The purpose of military action may be specified in a mission statement or implied from an order. Operations and activities are characterized as "joint" when they are conducted by a force composed of significant elements, assigned or attached, of two or more Military Departments operating under a single JFC.

Distinct military operations and activities may occur simultaneously with or independently of others even within the same OA. For example, a noncombatant evacuation operation (NEO) may be in the same OA where US forces are conducting COIN operations. Additionally, each may have different root causes and objectives.

See pp. 2-43 to 2-88, "Joint Operations Across the Conflict Continuum," for discussion of the simultaneous nature of theater operations and activities, and description of various joint operations and considerations in the context of the three broad areas of the range of military operations.

IV. Strategic Direction

Ref: JP 3-0 (w/Chg 1), Joint Operations (Oct '18), pp. I-5 to I-7.

National Strategic Direction

National strategic direction is governed by federal law, USG policy, internationally recognized law, and the national interest as represented by national security policy. This direction provides strategic context for the employment of the instruments of national power and defines the strategic purpose that guides employment of the military as part of a global strategy. Strategic direction is typically published in key documents, generally referred to as strategic guidance, but it may be communicated through any means available. Strategic direction may change rapidly in response to changes in the global environment, whereas strategic guidance documents are typically updated cyclically and may not reflect the most current strategic direction.

DOD derives its strategic-level documents from guidance in the NSS. The documents outline how DOD will support NSS objectives and provide a framework for other DOD policy and planning guidance, such as the National Defense Strategy; Guidance for Employment of the Force (GEF); Defense Planning Guidance; Global Force Management Implementation Guidance; and Chairman of the Joint Chiefs of Staff Instruction (CJCSI) 3110.01, (U) Joint Strategic Campaign Plan (JSCP) (simply known as the JSCP).

See p. 1-7 for further discussion.

National, Functional, and Theater-Strategic and Supporting Objectives

From this broad strategic guidance, more specific national, functional, and theater-strategic and supporting objectives help focus and refine the context and guide the military's joint planning and execution related to these objectives or a specific crisis. Integrated planning, coordination, and guidance among the Joint Staff, CCMD staffs, Service chiefs, and USG departments and agencies translate strategic priorities into clear planning guidance, tailored force packages, operational-level objectives, joint operation plans (OPLANs), and logistical support for the joint force to accomplish its mission.

See pp. 3-4 to 3-10 for further discussion.

The CCDR's Strategic Role

Based on guidance from the President and SecDef, GCCs and functional combatant commanders (FCCs) translate national security policy, strategy, and available military forces into theater and functional strategies to achieve national and theater strategic objectives. CCMD strategies are broad statements of the GCC's long-term vision for the AOR and the FCC's long-term vision for the global employment of functional capabilities guided by and prepared in the context of SecDef priorities outlined in the GEF and the CJCS's objectives articulated in the National Military Strategy (NMS). A prerequisite to preparing the theater strategy is development of a strategic estimate. It contains factors and trends that influence the CCMD's strategic environment and inform the ends, ways, means, and risk involved in pursuit of GEF-directed objectives.

Using their strategic estimates and theater or functional strategies, GCCs and FCCs develop CCPs consistent with guidance in the UCP, GEF, and JSCP, as well as in accordance with (IAW) planning architecture described in the Adaptive Planning and Execution (APEX) enterprise. In some cases, a CCDR may be required to develop a global campaign plan. FCCs develop operational support plans based on guidance in the UCP and their priorities and objectives in the GEF. FCCs may be responsible for developing functional-related global or subordinate campaign plans or both. As required, both GCCs and FCCs develop contingency plans, which are branch plans to the overarching CCP.

See pp. 3-8 to 3-10 and pp. 3-31 to 3-42 for further discussion.

capabilities, and reach extend beyond single areas of operation. Significant and emerging challenges include, but are not limited to, traditional armed conflict, attacks in cyberspace and the EMS, terrorism involving weapons of mass destruction (WMD), adversary information activities, and proliferation of adversary antiaccess (A2) and area denial (AD) capabilities. A2 capabilities, usually long range, prevent or inhibit an advancing force from entering an operational area (OA). If a force is able to overcome an enemy's A2 capabilities, additional AD capabilities can limit a force's freedom of action within an OA.

Enemies who attack the US homeland and US interests are likely to use asymmetric tactics and techniques. They will avoid hard (well-secured and heavily defended) targets and attack vulnerable ones. Vulnerable targets may include US and partner nations' (PNs') lines of communications (LOCs), ports, airports, staging areas, civilian populations, critical infrastructure, information centers, economic centers, and military and police personnel and facilities. Advances in information technology increase the tempo, lethality, and depth of warfare. Developments in cyberspace can provide the US military, its allies, and PNs leverage to improve economic and physical security. However, this also provides adversaries increased access to open-source information and intelligence, the Department of Defense information network (DODIN), critical infrastructure and key resources, and a limitless propaganda platform with global reach.

V. Instruments of National Power (DIME)

US instruments of national power are the national-level means our national leaders can apply in various ways to achieve strategic objectives (ends). Institutions that represent these instruments of national power are active continuously as the President directs along a conflict continuum that ranges from peace to war.

The ultimate purpose of the US Armed Forces is to fight and win the nation's wars. Although much of DOD's focus is on war and war preparation, opportunities also exist to prevent or mitigate the severity of conflict, legitimize US positions, reward PNs, provide expertise to multinational operations, and enhance the positive perception of the US. US national leaders can use military capabilities in a wide variety of activities, tasks, missions, and operations that vary in purpose, scale, risk, and combat intensity along the conflict continuum. The military's role increases relative to the other instruments as the need to compel an adversary through force increases. The potential range of military activities and operations extends from military engagement, security cooperation, and deterrence in times of relative peace up through major operations and campaigns that typically involve large-scale combat.

Acting alone in the strategic environment, the USG cannot resolve all crises or achieve all national objectives with just US resources. Under an umbrella of security cooperation, DOD supports USG strategic objectives by developing security relationships, building partner capacity and capability, and assuring access with selected PNs that enable them to act alongside, in support of, or in lieu of US forces around the globe. These strategic initiatives help advance national security objectives, promote stability, prevent conflicts, and reduce the risk of employing US military forces in a conflict. Security cooperation activities comprise an essential element of a geographic combatant commander's (GCC's) theater campaign plan (TCP).

The instruments of national power are: Diplomacy, Information, Military, and Economic (DIME). See pp. 1-4 to 1-5 for further discussion.

VI. Unified Action

Whereas the term joint operation focuses on the integrated actions of the Armed Forces of the United States, the term unified action has a broader connotation. Unified action refers to the synchronization, coordination, and integration of the activities of governmental and nongovernmental entities to achieve unity of effort. Failure to achieve unity of effort can cost lives, create conditions that enhance instability, and jeopardize mission accomplishment.

Unified action is based on national strategic direction, which is governed by the Constitution, federal law, and USG policy. Unified action is a comprehensive approach that focuses on coordination and cooperation of the US military and other interorganizational participants toward common objectives, even if the participants are not necessarily part of the same command or organization.

Enabled by the principle of unity of command, military leaders understand the effective mechanisms to achieve military unity of effort. The goal of unified action is to achieve a similar unity of effort between participants. This publication uses the term interorganizational participants to refer collectively to USG departments and agencies (i.e., interagency partners); state, territorial, local, and tribal agencies; foreign military forces and government agencies (i.e., multinational partners); NGOs; and the private sector.

The US Department of State (DOS) has a complementary approach, which defines unity of effort as a cooperative concept that refers to coordination and communication among USG organizations toward the same common goals for success.

See p. 1-8 for further discussion of unified action from JP 1-0.

The JFC's Role

JFCs are challenged to achieve and maintain operational coherence given the requirement to operate in conjunction with interorganizational partners. CCDRs play a pivotal role in unifying joint force actions, since all of the elements and actions that compose unified action normally are present at their level. However, subordinate JFCs also integrate and synchronize their operations directly with the operations of other military forces and the activities of nonmilitary organizations in the operational area to promote unified action.

Multinational Participation in Unified Action

Joint forces must be prepared to plan and execute operations with forces from PNs within the framework of an alliance or coalition under US or other-than-US leadership. US military leaders often are expected to play a central leadership role regardless of the US Armed Forces' predominance, capability, or capacity. Commanders should expect the military leaders of contributing member nations to emphasize common objectives, as well as to expect mutual support and respect. Although individual nations may place greater emphasis on some objectives than on others, the key is to find commonality within the objectives to promote synchronized progress to achieving the objectives. Cultivation and maintenance of personal relationships among counterparts enable success. Language and communication differences, cultural diversity, historical animosities, and the varying capabilities of allies and multinational partners are factors that complicate the integration and synchronization of activities during multinational operations. Likewise, differing national obligations derived from international treaties, agreements, and national legislation complicate multinational operations.

See chap. 7 for further discussion of multinational operations from JP 3-16.

Interorganizational Coordination in Unified Action

CCDRs and subordinate JFCs often interact with a variety of interorganizational participants. This interaction varies according to the nature of participant (capability, capacity, objectives, etc.) and type of operation. JFCs and planners consider the potential contributions of other agencies and determine which can best contribute to achieving specific objectives. Often, other interagency partners, primarily DOS, can facilitate a JFC's coordination with multinational and HN agencies, NGOs, and the private sector. DOD may support other agencies during operations; however, under US law, US military forces will remain under the DOD command structure. Federal lead agency responsibility may be prescribed by law or regulation, Presidential directive, policy, or agreement among or between agencies. Even then, because of its resources and well-established planning methods, the joint force will likely provide significant support to the lead agency.

See chap. 8 for further discussion of interorganizational cooperation from JP 3-08.

II. The Art of Joint Command

Ref: JP 3-0 (w/Chg 1), Joint Operations (Oct '18), chap. II.

> *"When all is said and done, it is really the commander's coup d'oeil, his ability to see things simply, to identify the whole business of war completely with himself, that is the essence of good generalship."*
>
> ***Carl von Clausewitz***
> ***On War***

I. Art of Command

Command is the authority that a commander in the armed forces lawfully exercises over subordinates by virtue of rank or assignment. Accompanying this authority is the responsibility to effectively organize, direct, coordinate, and control military forces to accomplish assigned missions. Command includes responsibility for health, welfare, morale, and discipline of assigned personnel.

While command authority stems from orders and other directives, the art of command resides in the commander's ability to use leadership to maximize performance. The combination of courage, ethical leadership, judgment, intuition, situational awareness, and the capacity to consider contrary views helps commanders make insightful decisions in complex situations. These attributes can be gained over time through training, education, and experience. Joint training and joint doctrine are designed to enable the conscious and skillful exercise of command authority through visualization, decision making, and leadership. Effective commanders combine judgment and visualization with information to determine whether a decision is required, when to decide, and what to decide with sufficient speed to maintain the initiative. Information management (IM), situational awareness, and a sound battle rhythm facilitate decision making.

II. Commander-Centric Leadership

A commander's perspective of challenges in the OE is broad and comprehensive due to the interaction with USG civilian leaders; senior, peer, subordinate, and supporting commanders; and interorganizational partners. Clear commander's guidance and intent, enriched by the commander's experience and intuition, enable joint forces to achieve objectives. Employing the "art of war," which has been the commander's central historical command role, remains critical regardless of technological and informational improvements in control—the "science of war."

The C2 function is commander-centric and network-enabled to facilitate initiative and decision making at the lowest appropriate level. Although joint forces have grown accustomed to communicating freely without fear of jamming or interception, US enemies and adversaries are likely to use technological advances in cyberspace and vulnerabilities in the EMS to conduct cyberspace or EMS attacks. Commanders should be prepared to operate in an environment degraded by electromagnetic interference. This is especially true at the lower echelons. If a commander loses reliable communications, mission command enables military operations through decentralized execution based on mission-type orders. Mission command is built on subordinate leaders at all echelons who exercise disciplined initiative and act aggressively and independently to accomplish the mission. Mission-type orders focus on the purpose of the operation rather than details of how to perform assigned tasks. Commanders delegate decisions to subordinates wherever possible, which minimizes detailed control and empowers subordinates' initiative to make decisions based on the commander's guidance rather than constant communications. Subordinates' understanding of the commander's intent at every level of command is essential to mission command.

Commanders should interact with other leaders to build personal relationships and develop trust and confidence. Developing these associations is a conscious, collaborative act. Commanders build trust through words and actions and continue to reinforce it not only during operations but also during training, education, and practice. Trust and confidence are essential to synergy and harmony, both within the joint force and with our interagency and multinational partners and other interorganizational stakeholders. Commanders may also interact with other political, societal, and economic leaders and other influential people who may influence joint operations. This interaction supports mission accomplishment. The JFC emphasizes the importance of key leader engagement (KLE) to subordinate commanders and encourages them to extend the process to lower levels, based on mission requirements.

Commanders should provide subordinate commands sufficient time to plan, particularly in a time-sensitive crisis situation. They do so by issuing a warning order to subordinates at the earliest opportunity and by collaborating with other commanders, agency leaders, and multinational partners to develop a clear understanding of the commander's mission, intent, guidance, and priorities. Commanders resolve issues that are beyond the staff's authority. Examples include highly classified, limited-access planning for sensitive operations and allowing multinational partners' planners restricted access to US classified information systems.

Commanders collaborate with their seniors and peers to resolve differences of interpretation of higher-level objectives and the ways and means to accomplish these objectives. Commanders generally expect their higher HQ has accurately described the OE, framed the problem, and devised a sound approach to achieve the best solution. Strategic guidance, however, can be vague, and the commander must interpret and clarify it for the staff. While national leaders and CCDRs may have a broader perspective of the problem, subordinate JFCs and their component commanders often have a better perspective of the situation at the operational level. Both perspectives are essential to a sound solution. During a commander's decision cycle, subordinate commanders should aggressively share their perspective with senior leaders to resolve issues at the earliest opportunity.

An essential skill of a JFC is the ability to assign missions and tasks that integrate the components' capabilities consistent with the JFC's envisioned CONOPS. Each component's mission should complement the others'. This enables each component to enhance the capabilities and limit the vulnerabilities of the others. Achieving this synergy requires more than just understanding the capabilities and limitations of each component. The JFC should also visualize operations holistically, identify the preconditions that enable each component to optimize its own contribution, and then determine how the other components might help to produce them. The JFC should compare alternative component missions and mixes solely from the perspective of combined effectiveness, unhampered by Service parochialism. This approach also requires mutual trust among commanders that the missions assigned to components will be consistent with their capabilities and limitations, those capabilities will not be risked for insufficient overall return, and components will execute their assignments.

Successful leaders encourage the exchange of information and ideas throughout their staffs to ensure decisions are based on the best understanding of the situation and available options. Such exchanges promote critical reviews of assumptions; facilitate consideration of all aspects of the situation, including cultural issues; stimulate broad consideration of military and nonmilitary alternatives; and emphasize efforts to minimize organizational and human sources of error and bias.

The JFC leads using operational art and operational design, joint planning, rigorous assessment of progress, and timely decision making.

III. Operational Art

Operational art is the cognitive approach by commanders and staffs— supported by their skill, knowledge, experience, creativity, and judgment—to develop strategies, campaigns, and operations to organize and employ military forces by integrating ends, ways, and means. It is a thought process to mitigate the ambiguity and uncertainty of a complex OE and develop insight into the problems at hand. Operational art also promotes unified action by enabling JFCs and staffs to consider the capabilities, actions, goals, priorities, and operating processes of interagency partners and other interorganizational participants, when they determine objectives, establish priorities, and assign tasks to subordinate forces. It facilitates the coordination, synchronization, and, where appropriate, the integration of military operations with activities of other participants, thereby promoting unity of effort.

The foundation of operational art encompasses broad vision; the ability to anticipate; and the skill to plan, prepare, execute, and assess. It helps commanders and their staffs organize their thoughts and envision the conditions necessary to accomplish the mission and reach the desired military end state in support of national objectives. Without operational art, campaigns and operations could be sets of disconnected events. Operational art informs the deployment of forces and the arrangement of operations to achieve military operational and strategic objectives.

Commander's Role

The commander is the central figure in operational art, not only due to education and experience but also because the commander's judgment and decisions guide the staff throughout joint planning and execution. Commanders leverage their knowledge, experience, judgment, and intuition to focus effort and achieve success. Operational art helps broaden perspectives to deepen understanding and enable visualization. Commanders compare similarities of the existing situation with their own experiences or history to distinguish unique features and then tailor innovative and adaptive solutions to each situation.

The commander's ability to think creatively enhances the ability to employ operational art in order to answer the following questions (Ends, Ways, Means):

Ends, Ways and Means

Operational art applies to all aspects of joint operations and integrates ends, ways, and means, while accounting for risk, across the levels of war. Among the many considerations, operational art requires commanders to answer the following questions.

- What are the **objectives and desired military end state?** (Ends)
- What **sequence of actions** is most likely to achieve those objectives and military end state? (Ways)
- What **resources** are required to accomplish that sequence of actions? (Means)
- What is the likely **chance of failure or unacceptable results** in performing that sequence of actions? (Risk)

Operational art encompasses operational design—the conception and construction of the framework that underpins a joint operation or campaign plan and its subsequent execution. Together, operational art and operational design strengthen the relationship between strategic objectives and the tactics employed to achieve them.

See pp. 3-43 to 3-68 for further discussion of operational art and operational design from JP 5-0.

IV. Operational Design

Operational design is the conception and construction of the framework that underpins a campaign or major operation plan and its subsequent execution. It extends operational art's vision with a creative process to help commanders and planners answer the ends-ways-means-risk questions. Commanders and staffs can use operational design when planning any joint operation.

Operational design supports operational art with a methodology designed to enhance understanding the situation and the problem. The methodology helps the JFC and staff identify broad solutions for mission accomplishment. Elements of operational design—such as objective, center of gravity (COG), line of operation (LOO), line of effort (LOE), and termination—are tools that help the JFC and the staff visualize and describe the broad operational approach to achieve objectives and accomplish the mission. These operational design elements are useful throughout the joint planning process (JPP).

Operational design works best when commanders encourage discourse and leverage dialogue and collaboration to identify and solve complex, ill-defined problems. To that end, the commander should empower organizational learning and develop methods to determine whether the operational approach should be modified during the course of an operation. This requires continuous assessment and reflection that challenge understanding of the existing problem and the relevance of actions addressing that problem.

Commanders and their staffs blend operational art, operational design, and JPP to produce plans and orders that drive joint operations. Effective operational design results in more efficient detailed planning and increases the chances of mission accomplishment.

See pp. 3-43 to 3-68 for further discussion of operational art and operational design from JP 5-0.

V. Joint Planning

Planning translates guidance into plans or orders to achieve a desired objective or attain an end state. The joint planning and execution community begins planning when a potential or actual event is recognized that may require a military response. Objectives provide a unifying purpose around which to focus actions and resources.

Joint Planning Process (JPP)

JPP aligns military activities and resources to achieve national objectives and enables leaders to examine cost-benefit relationships, risks, and trade-offs to determine a preferred course of action (COA) to achieve that objective or attain an end state. Joint planning occurs within the Joint Operation Planning and Execution System and the APEX enterprise, which encompasses department-level joint planning policies, processes, procedures, and reporting structures.

Joint planning consists of planning activities that help CCDRs and their subordinate commanders transform national objectives into actions that mobilize, deploy, employ, sustain, redeploy, and demobilize joint forces. It ties the employment of the Armed Forces to the achievement of national objectives during peacetime and war.

See pp. 3-69 to 3-118 for further discussion of the joint planning process from JP 5-0.

VI. Assessment

Assessment is a continuous process that measures the overall effectiveness of employing joint force capabilities during military operations. It involves monitoring and evaluating the current situation and progress toward mission completion. Assessments can help determine whether a particular activity contributes to progress with respect to a set of standards or desired objective or end state.

DOD and its components use a wide range of assessment tools and methods.

See pp. 3-125 to 3-136 for related discussion of operation assessment from JP 5-0.

Chap 2

III. Joint Functions

Joint Operations

Ref: JP 3-0 (w/Chg 1), Joint Operations (Oct '18), chap. III.

Joint functions are related capabilities and activities grouped together to help JFCs integrate, synchronize, and direct joint operations. Functions common to joint operations at all levels of warfare fall into seven basic groups—C2, information, intelligence, fires, movement and maneuver, protection, and sustainment. Some functions, such as C2, information, and intelligence, apply to all operations. Others, such as fires, apply as the JFC's mission requires. A number of subordinate tasks, missions, and related capabilities help define each function, and some could apply to more than one joint function.

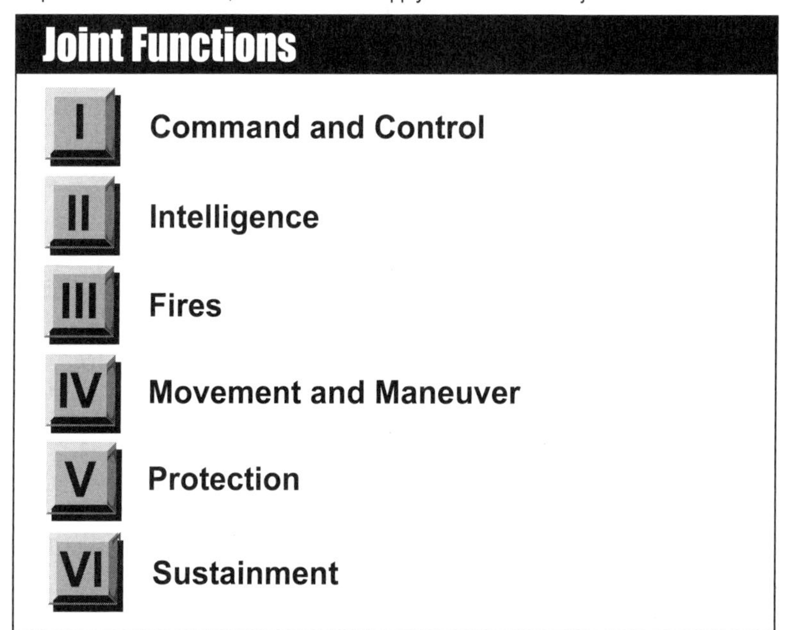

The joint functions reinforce and complement one another, and integration across the functions is essential to mission accomplishment. In any joint operation, the JFC can choose from a wide variety of joint and Service capabilities and combine them in various ways to perform joint functions and accomplish the mission. Plans describe how the JFC uses military capabilities (i.e., organizations, people, and systems) to perform tasks associated with each joint function. However, forces and other assets are not characterized by the functions for which the JFC is employing them. Individual Service capabilities can often support multiple functions simultaneously or sequentially while the joint force is executing a single task.

JFCs and staffs integrate, synchronize, employ, and assess a wide variety of information-related capabilities (IRCs) within and across joint functions, in concert with other actions to influence a target audience's decision making while protecting our own. IRCs constitute tools, techniques, or activities employed through the information environment that can be used to create effects, accomplish tasks, or achieve specific objectives at a specific time and place.

I. Command and Control (C2)

C2 encompasses the exercise of authority and direction by a commander over assigned and attached forces to accomplish the mission. The JFC provides operational vision, guidance, and direction to the joint force. The C2 function encompasses a number of tasks, including:

- Establish, organize, and operate a joint force HQ
- Command subordinate forces
- Prepare, modify, and publish plans, orders, and guidance
- Establish command authorities among subordinate commanders
- Assign tasks, prescribe task performance standards, and designate OAs
- Prioritize and allocate resources
- Manage risk
- Communicate and ensure the flow of information across the staff and joint force
- Assess progress toward accomplishing tasks, creating conditions, and achieving objectives
- Coordinate and control the employment of joint capabilities to create lethal and nonlethal effects
- Coordinate, synchronize, and, when appropriate, integrate joint operations with the operations and activities of other participants
- Ensure the flow of information and reports to and from higher authority

Command

Command includes both the authority and responsibility to use resources to accomplish assigned missions. Command at all levels is the art of motivating and directing people and organizations to accomplish missions. The C2 function supports an efficient decision-making process. Timely and relevant intelligence enables commanders to make decisions and execute those decisions more rapidly and effectively than the enemy. This decreases risk and allows the commander more control over the timing and tempo of operations.

Control is Inherent in Command

To control is to manage and direct forces and functions consistent with a commander's command authority. Control of forces and functions helps commanders and staffs compute requirements, allocate means, and integrate efforts. Control is necessary to determine the status of organizational effectiveness, identify variance from set standards, and correct deviations from these standards. Control permits commanders to acquire and apply means to support the mission and develop specific instructions from general guidance. Control provides the means for commanders to maintain freedom of action, delegate authority, direct operations from any location, and integrate and synchronize actions throughout the OA. Ultimately, it provides commanders a means to measure, report, and correct performance.

A. Command Authorities (and Support Relationships)

JFCs exercise various command authorities (i.e., combatant command [command authority] {COCOM}, OPCON, tactical control [TACON], and support) delegated to them by law or senior leaders and commanders over assigned and attached forces.

See pp. 1-31 to 1-40 for a listing and discussion of command and support relationships.

B. Building Shared Understanding

Ref: JP 3-0 (w/Chg 1), Joint Operations (Oct '18), pp. III-14 to III-15.

Unified and synchronized actions, narratives, and messaging are the most important products of the C2 function because they guide the force toward objectives and mission accomplishment. Commanders and staff require not only information to make these decisions but also the knowledge and shared understanding that aid in the wisdom essential to sound decision making (Figure III-2). Building shared understanding results from the effective exercise of leadership and the ability to influence and inspire others. To build shared understanding, commanders provide vision, guidance, and direction to the joint force. These collaborative processes and products vary across joint commands based on the commander's needs and preferences.

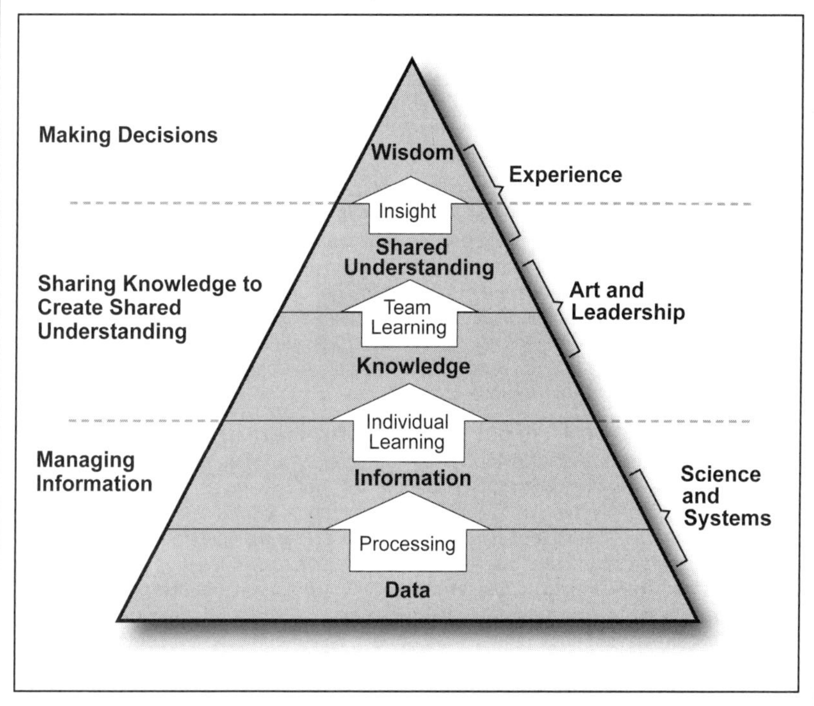

Ref: JP 3-0 (w/Chg 1), fig. III-2. Building Shared Understanding.

Facilitating Shared Understanding

Related to the concept of building shared understanding is the concept of facilitating shared understanding. The distinction between these concepts centers on the purpose for building shared understanding versus the processes that facilitate shared understanding. The purpose for building shared understanding is an element of C2, whereas the process (i.e., the methods) for facilitating shared understanding is an element of the information joint function.

Collaboration, knowledge sharing (KS), information management (IM), and information and intelligence sharing activities are essential to assist the commander in providing vision, guidance, and direction to the joint force

C. Command and Control System

JFCs exercise authority and direction through a C2 system, which consists of the facilities; equipment; communications; staff functions and procedures; and personnel essential for planning, preparing for, monitoring, and assessing operations. The C2 system must enable the JFC to maintain communication with higher, supporting, and subordinate commands in order to control all aspects of current operations while planning for future operations.

JFCs establish various maneuver and movement control, airspace coordination, fire support coordination, and communication measures to facilitate effective joint operations. These measures include boundaries, phase lines, objectives, coordinating altitudes to deconflict air operations, air defense areas, OAs, submarine operating patrol areas, no-fire areas, public affairs (PA) and other communication-related guidance, and others as required.

For additional guidance on C2 of air operations, refer to JP 3-30, Command and Control of Joint Air Operations. For additional guidance on control and coordination measures, refer to JP 3-09, Joint Fire Support, and JP 3-52, Joint Airspace Control. See Military Standard-2525, Department of Defense Interface Standard Joint Military Symbology, for additional guidance on the use and discussion of graphic control measures and symbols for the joint force.

The purpose of the joint communications system is to assist the JFC in C2 of military operations. Effective communication system planning is essential for effective C2 and integration and employment of the joint force's capabilities. The mission and structure of the joint force determine specific information flow and processing requirements. These requirements dictate the general architecture and specific configuration of the communications system. Therefore, communications system planning must be integrated and synchronized with joint planning. Through effective communications system planning, the JFC is able to apply capabilities at the critical time and place for mission success.

Commander's Critical Information Requirements (CCIR)

CCIRs are elements of information the commander identifies as critical to timely and effective decision making. CCIRs focus IM and help the commander assess the OE and identify decision points during operations. CCIRs belong exclusively to the commander. The CCIR list is normally short so the staff can focus its efforts and allocate scarce resources. But, the CCIR list is not static; JFCs, through operation assessment, add, delete, adjust, and update CCIRs throughout planning and execution based on the information they need to make decisions. At a minimum, CCIRs should be reviewed and updated throughout mission analysis, plan development, refinement, and adaptation and during each phase of order execution.

See p. 3-82 for further discussion of CCIR from JP 5-0.

Battle Rhythm

The HQ battle rhythm is its daily operations cycle of briefings, meetings, and report requirements. A stable battle rhythm facilitates effective decision making, efficient staff actions, and management of information within the HQ and with higher, supporting, and subordinate HQ. The commander and staff should develop a battle rhythm that minimizes meeting requirements while providing venues for command and staff interaction internal to the joint force HQ and with subordinate commands. Joint and component HQ's battle rhythms should be synchronized to accommodate operations in multiple time zones and the battle rhythm of higher, subordinate, and adjacent commands. Other factors such as planning, decision making, and operating cycles (i.e., intelligence collection, targeting, and joint air tasking cycle) influence the battle rhythm. Further, meetings of the staff organizations must be synchronized. The chief of staff normally manages the joint force HQ's battle rhythm.

Seep. 5-20 for related discussion and sample of a JTF battle rhythm.

D. Risk Management

Ref: JP 3-0 (w/Chg 1), Joint Operations (Oct '18), pp. III-19 to III-20. *See also p. 3-41.*

Risk management is the process to identifying and assessing hazards arising from operational factors and making decisions that balance risk cost with mission benefits. It assists organizations and individuals in making informed decisions to reduce or offset risk, thereby increasing operational effectiveness and the probability of mission success. The commander determines the level of risk that is acceptable, with respect to aspects of operations, and should state this determination in commander's intent. Risk is one of the key outputs of mission analysis and should be reviewed at every successive step in the JPP. The assessment of risk to mission includes an overall risk to mission analysis (e.g., low, moderate, significant, or high) along multiple criteria (e.g., authorities and permissions; policy; forces, basing, and agreements; resources; capabilities; PN contributions; and other USG support). To assist in risk management, commanders and their staffs may develop or institute a risk management process tailored to their mission or OA.

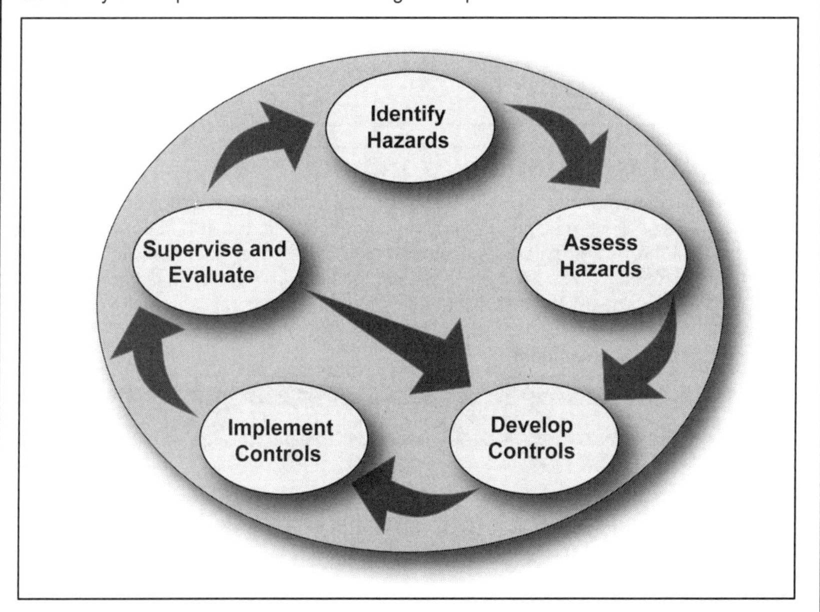

Ref: JP 3-0 (w/Chg 1), fig. III-3. Risk Management Process.

Risk management is a function of command and a key planning consideration that focuses on designing, implementing, and monitoring risk decisions. Risk management helps commanders preserve lives and resources; accept, avoid, or mitigate (reduce or transfer) unnecessary risk; identify feasible and effective control measures where specific standards do not exist; and develop valid COAs. Prevention of friendly fire incidents is a key consideration in risk management. However, risk management does not inhibit a commander's flexibility and initiative, remove risk altogether (or support a zero-defects mindset), dictate a go/no-go decision to take a specific action, sanction or justify violating the law, or remove the necessity for standard operating procedures (SOPs).

Safety preserves military power. High-tempo operations may increase the risk of injury and death due to mishaps. Command interest, discipline, risk mitigation measures, education, and training lessen those risks. The JFC reduces the chance of mishap by conducting risk assessments, assigning a safety officer and staff, implementing a safety program, and seeking advice from local personnel.

II. Information

All military activities produce **information**. Informational aspects are the features and details of military activities observers interpret and use to assign meaning and gain understanding. Those aspects affect the perceptions and attitudes that drive behavior and decision making. The JFC leverages informational aspects of military activities to gain an advantage; failing to leverage those aspects may cede this advantage to others. Leveraging the informational aspects of military activities ultimately affects strategic outcomes.

The **information function** encompasses the management and application of information and its deliberate integration with other joint functions to change or maintain perceptions, attitudes, and other elements that drive desired behaviors and to support human and automated decision making. The information function helps commanders and staffs understand and leverage the pervasive nature of information, its military uses, and its application during all military operations. This function provides JFCs the ability to integrate the generation and preservation of friendly information while leveraging the inherent informational aspects of military activities to achieve the commander's objectives and attain the end state.

A. The Information Environment

The information environment is the aggregate of individuals, organizations, and systems that collect, process, disseminate, or act on information. This environment consists of three interrelated dimensions which continuously interact with individuals, organizations, and systems. These dimensions are the physical, informational, and cognitive. The JFC's operational environment is the composite of the conditions, circumstances, and influences that affect employment of capabilities and bear on the decisions of the commander.

The Information Environment

The Physical Dimension

The Informational Dimension

The Cognitive Dimension

B. Information Function Activities

The information function includes activities that facilitate the JFC's understanding of the role of information in the OE, facilitate the JFC's ability to leverage information to affect behavior, and support human and automated decision making.

See pp. 2-22 to 2-23 for an overview and further discussion of these information function activities.

C. Information Operations (IO)

Ref: JP 3-0 (w/Chg 1), Joint Operations (Oct '18), pp. III-17 to III-22.

All military activities produce **information**. Informational aspects are the features and details of military activities observers interpret and use to assign meaning and gain understanding. Those aspects affect the perceptions and attitudes that drive behavior and decision making. The JFC leverages informational aspects of military activities to gain an advantage; failing to leverage those aspects may cede this advantage to others. Leveraging the informational aspects of military activities ultimately affects strategic outcomes.

The **information function** encompasses the management and application of information and its deliberate integration with other joint functions to change or maintain perceptions, attitudes, and other elements that drive desired behaviors and to support human and automated decision making.

The **instruments of national power** (diplomatic, informational, military, and economic) provide leaders in the US with the means and ways of dealing with crises around the world. Employing these means in the information environment requires the ability to securely transmit, receive, store, and process information in near real time. The nation's state and non-state adversaries are equally aware of the significance of this new technology, and will use information-related capabilities (IRCs) to gain advantages in the information environment, just as they would use more traditional military technologies to gain advantages in other operational environments. As the strategic environment continues to change, so does information operations (IO).

Regardless of its mission, the joint force considers the likely impact of all operations on **relevant actor** perceptions, attitudes, and other drivers of behavior. The JFC then plans and conducts every operation in ways that **create desired effects** that include maintaining or inducing relevant actor behaviors. These ways may include the timing, duration, scope, scale, and even visibility of an operation; the deliberately planned presence, posture, or profile of assigned or attached forces in an area; the use of signature management in deception operations; the conduct of activities and operations to similarly impact behavioral drivers; and the **employment of specialized capabilities** -- e.g., key-leader engagements (KLE), cyberspace operations (CO), military information support operations (MISO), electronic warfare (EW), and civil affairs (CA) -- to reinforce the JFC's efforts.

Inform activities involve the release of accurate information to domestic and international audiences to put joint operations in context; facilitate informed perceptions about military operations; and counter adversarial misinformation, disinformation, and propaganda. Inform activities help to assure the trust and confidence of the US population, allies, and partners and to deter and dissuade adversaries and enemies.

The joint force **attacks and exploits information, information networks, and systems** to affect the ability of relevant actors to leverage information in support of their own objectives. This includes the manipulation, modification, or destruction of information or disruption of the flow of information for the purpose of gaining a position of military advantage. This also includes targeting the credibility of information.

See following pages (pp. 2-20 to 2-21) for an overview of the integrating, coordinating, and planning functions of information operations (IO) from JP3-13.1.

Refer to Joint/Interagency SMARTbook 3: Information Operations (Multi-Domain Guide to IO & Information-Related Capabilities), when published. All military activities produce information. Informational aspects affect the perceptions and attitudes that drive behavior and decision making. The JFC leverages informational aspects of military activities to gain an advantage; failing to leverage those aspects may cede this advantage to others. Leveraging the informational aspects of military activities ultimately affects strategic outcomes.

Integrating, Coordinating, and Planning Functions of Information Operations (IO)

Ref: JP 3-13 (w/Chg 1), Information Operations (Nov '14).

Joint force commanders (JFCs) may establish an IO staff to provide command-level oversight and collaborate with all staff directorates and supporting organizations on all aspects of IO. Most combatant commands (CCMDs) include an IO staff to serve as the focal point for IO. Faced with an ongoing or emerging crisis within a geographic combatant commander's (GCC's) area of responsibility, a JFC can establish an IO cell to provide additional expertise and coordination across the staff and interagency.

Notional Information Operations Cell

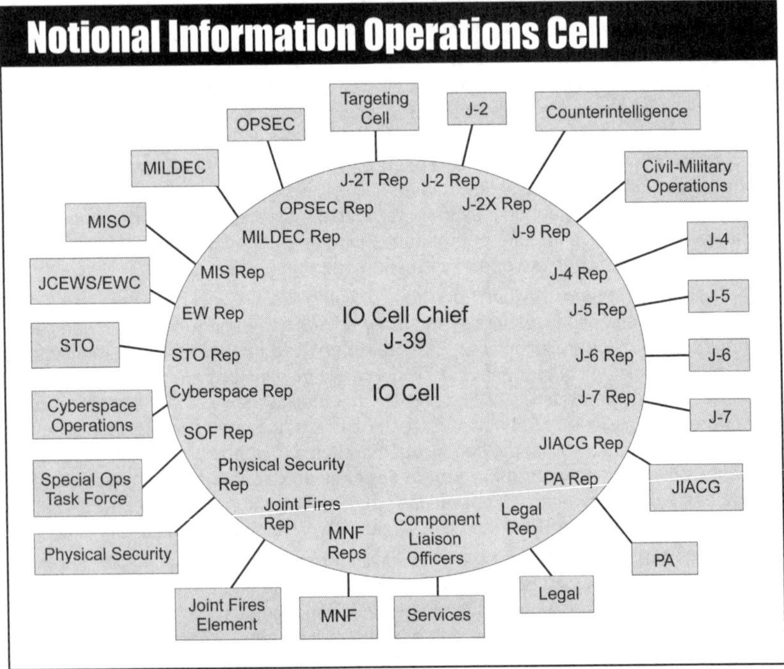

Ref: JP 3-13 w/change 1, fig. II-3, p. II-6.

In order to provide planning support, the IO staff includes IO planners and a complement of information-related capabilities (IRCs) specialists to facilitate seamless integration of IRCs to support the JFC's concept of operations (CONOPS). IRC specialists can include, but are not limited to, personnel from the EW, cyberspace operations (CO), military information support operations (MISO), civil-military operations (CMO), military deception (MILDEC), intelligence, and public affairs (PA) communities.

Refer to Joint/Interagency SMARTbook 3: Information Operations (Multi-Domain Guide to IO & Information-Related Capabilities), when published. All military activities produce information. Informational aspects affect the perceptions and attitudes that drive behavior and decision making. The JFC leverages informational aspects of military activities to gain an advantage; failing to leverage those aspects may cede this advantage to others. Leveraging the informational aspects of military activities ultimately affects strategic outcomes.

Information Operations Planning (Within the Seven Steps of the JPP)

Ref: JP 3-13 (w/Chg 1), Information Operations (Nov '14), fig. IV-1, p. IV-3.

Throughout the joint planning process, IRCs are integrated with the JFC's overall CONOPS.

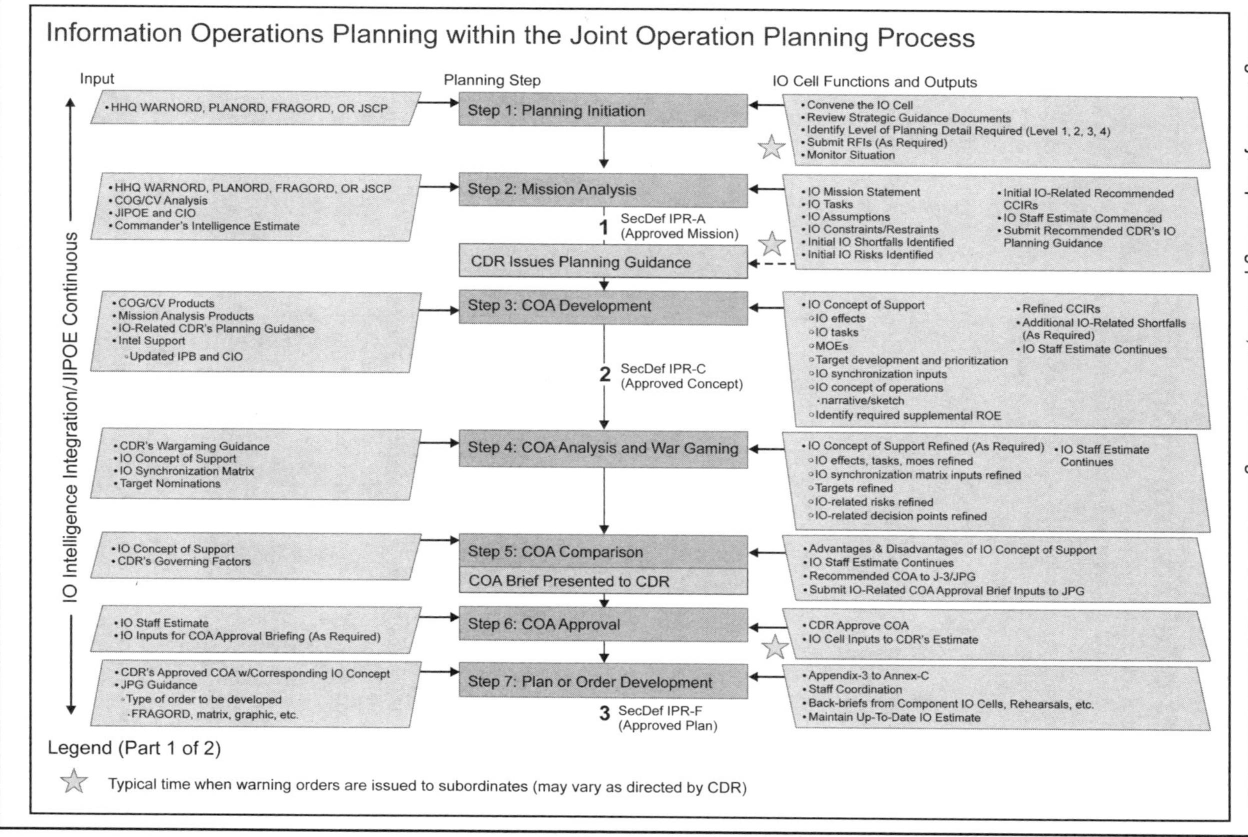

Information Function Activities

Ref: JP 3-0 (w/Chg 1), Joint Operations (Oct '18), pp. III-17 to III-22.

The information function includes activities that facilitate the JFC's understanding of the role of information in the OE, facilitate the JFC's ability to leverage information to affect behavior, and support human and automated decision making.

1. Understand Information in the Operational Environment (OE)

In conjunction with activities under the intelligence joint function, this activity facilitates the JFC's understanding of the pervasive nature of information in the OE, its impact on relevant actors, and its effect on military operations. It includes determining relevant actor perceptions, attitudes, and decision-making processes and requires an appreciation of their culture, history, and narratives, as well as knowledge of the means, context, and established patterns of their communication.

Information affects the perceptions and attitudes that drive the behavior and decision making of humans and automated systems. In order to affect behavior, the JFC must understand the perceptions, attitudes, and decision-making processes of humans and automated systems. These processes reflect the aggregate of social, cultural, and technical attributes that act upon and impact knowledge, understanding, beliefs, world views, and actions.

The human and automated systems whose behavior the JFC wants to affect are referred to as relevant actors. Relevant actors may include any individuals, groups, and populations, or any automated systems, the behavior of which has the potential to substantially help or hinder the success of a particular campaign, operation, or tactical action. For the purpose of military activities intended to inform audiences, relevant actors may include US audiences; however, US audiences are not considered targets for influence.

See pp. 2-38 to 2-39 and p. 3-50 for related discussion of the operational environment (OE).

Language, Regional, and Cultural Expertise

Language skills, regional knowledge, and cultural awareness enable effective joint operations. Deployed joint forces should understand and effectively communicate with HN populations; local and national government officials; multinational partners; national, regional, and international media; and other key stakeholders, including NGOs. This capability includes knowledge about the human aspects of the OE and the skills associated with communicating with foreign audiences. Knowledge about the human aspects of the OE is derived from the analysis of national, regional, and local culture, economy, politics, religion, and customs. Consequently, commanders should integrate training and capabilities for foreign language and regional expertise in contingency, campaign, and supporting plans and provide for them in support of daily operations and activities. Commanders should place particular emphasis on foreign language proficiency in technical areas identified as key to mission accomplishment.

For specific planning guidance and procedures regarding language and regional expertise, refer to CJCSI 3126.01, Language, Regional Expertise, and Culture (LREC) Capability Identification, Planning, and Sourcing.

2. Leverage Information to Affect Behavior

Tasks aligned under this activity apply the JFC's understanding of the impact information has on perceptions, attitudes, and decision-making processes to affect the behaviors of relevant actors in ways favorable to joint force objectives.

Influence Relevant Actors

Regardless of its mission, the joint force considers the likely impact of all operations on relevant actor perceptions, attitudes, and other drivers of behavior. The JFC then plans and conducts every operation in ways that create desired effects that include maintaining or inducing relevant actor behaviors. These ways may include the timing, duration, scope, scale, and even visibility of an operation; the deliberately planned presence, posture, or profile of assigned or attached forces in an area; the use of signature management in deception operations; the conduct of activities and operations to similarly impact behavioral drivers; and the employment of specialized capabilities (e.g., KLE, CO, military information support operations [MISO], EW, CA) to reinforce the JFC's efforts. Since some relevant actors will be located outside of the JFC's OA, coordination, planning, and synchronization of activities with other commands or mission partners is vital.

Inform Domestic, International, and Internal Audiences

Inform activities involve the release of accurate information to domestic and international audiences to put joint operations in context; facilitate informed perceptions about military operations; and counter adversarial misinformation, disinformation, and propaganda. Inform activities help to assure the trust and confidence of the US population, allies, and partners and to deter and dissuade adversaries and enemies.

Attack and Exploit Information, Information Networks, and Systems

The joint force attacks and exploits information, information networks, and systems to affect the ability of relevant actors to leverage information in support of their own objectives. This includes the manipulation, modification, or destruction of information or disruption of the flow of information for the purpose of gaining a position of military advantage. This also includes targeting the credibility of information.

3. Support Human and Automated Decision Making

The management aspect of the information joint function includes activities that facilitate shared understanding across the joint force and that protect friendly information, information networks, and systems to ensure the availability of timely, accurate, and relevant information necessary for JFC decision making.

Facilitating Shared Understanding

Facilitating shared understanding is related to building shared understanding in the C2 joint function. Where building shared understanding is an element of C2 and focuses on purpose (i.e., the commander's objective), facilitating shared understanding is concerned with process (i.e., the methods). Key components of facilitating understanding are collaboration, KS, and IM.

Protecting Friendly Information

Information Networks, and Systems. The information function reinforces the protection function and focuses on protecting friendly information, information networks, and systems. This aspect of the information function includes the preservation of friendly information across the staff and the joint force and any information shared with allies and partners. These activities reinforce the requirement to assure the flow of information important to the joint force, both by protecting the information and by assessing and mitigating risks to that information. The preservation of information includes both passive and active measures to prevent and mitigate adversary collection, manipulation, and destruction of friendly information, to include attempts to undermine the credibility of friendly information.

Joint Operations

D. Joint Force Capabilities, Operations, and Activities for Leveraging Information

Ref: JP 3-0 (w/Chg 1), Joint Operations (Oct '18), pp.) III-17 to III-26.

In addition to planning all operations to benefit from the inherent informational aspects of physical power and influence relevant actors, the JFC also has additional means with which to leverage information in support of enduring outcomes. The following capabilities, operations, and activities may reinforce the actions of assigned or attached forces, support LOOs or LOEs, or constitute the primary activity in a LOE.

See pp. 2-20 to 2-21) for an overview of the integrating, coordinating, and planning functions of information operations (IO) from JP3-13.1.

Key Leader Engagement (KLE)

Most operations require commanders and other leaders to engage key local and regional leaders to affect their attitudes and gain their support. Building relationships to the point of effective engagement and influence usually takes time. Language, regional expertise, and culture knowledge and skills are keys to successfully communicate with and, therefore, manage KLE. Commanders can be challenged to identify key leaders, develop messages, establish dialogue, and determine other ways and means of delivery, especially in societies where interpersonal relationships are paramount.

Public Affairs (PA)

PA contributes to the achievement of military objectives by countering incorrect information and propaganda through the dissemination of accurate information. PA personnel advise the JFCs on the possible direct and indirect effects of joint force actions on public perceptions, attitudes, and beliefs, and work to formulate and deliver timely and culturally attuned messages.

Civil-Military Operations (CMO)

CMO are activities that establish, maintain, influence, or exploit relationships between military forces and indigenous populations and institutions with the objective to reestablish or maintain stability in a region or HN. During all military operations, CMO can coordinate the integration of military and nonmilitary instruments of national power. CA support CMO by conducting military engagement and humanitarian and civic assistance to influence the populations of the HN and other PNs in the OA.

Military Deception (MILDEC)

Commanders conduct MILDEC to mislead enemy decision makers and commanders and cause them to take or not take specific actions. The intent is to cause enemy commanders to form inaccurate impressions about friendly force dispositions, capabilities, vulnerabilities, and intentions; misuse their intelligence collection assets; and fail to employ their combat or support units to best advantage. As executed by JFCs, MILDEC targets enemy leaders and decision makers through the manipulation of their intelligence collection, analysis, and dissemination systems. MILDEC depends on intelligence to identify deception targets, assist in developing credible stories, identify and orient on appropriate receivers (the readers of the story), and assess the effectiveness of the deception effort. Deception requires a thorough knowledge of the enemy and their decision-making processes.

Military Information Support Operations (MISO)

Psychological operations forces conduct MISO to convey selected information and indicators to foreign audiences to influence their emotions, motives, and objective reasoning and ultimately induce or reinforce foreign attitudes and behavior favorable to the originator's objectives. Psychological operations forces devise and execute psychological actions and craft persuasive messages using a variety of audio, visual, and audiovisual products, which can then be delivered to both targets and audiences.

Operations Security (OPSEC)

OPSEC uses a process to preserve friendly essential secrecy by identifying, controlling, and protecting critical information and indicators that would allow adversaries to identify and exploit friendly vulnerabilities. The purpose of OPSEC is to reduce vulnerabilities of US and multinational forces to adversary exploitation, and it applies to all activities that prepare, sustain, or employ forces.

Combat Camera (COMCAM)

Imagery is one of the most powerful tools available for informing internal and domestic audiences and for influencing foreign audiences. COMCAM forces provide imagery support in the form of a directed imagery capability to the JFC across the range of military operations. COMCAM imagery supports capabilities that use imagery for their products and efforts, including MISO, MILDEC, PA, and CMO, and provides critical operational documentation for sensitive site exploitation, legal and evidentiary requirements, and imagery for battle damage assessment/MOE analysis, as well as operational documentation and imagery for narrative development during foreign humanitarian assistance (FHA) operations and NEOs.

Space Operations

Space operations support joint operations throughout the OE by providing information in the form of ISR; missile warning; environment monitoring; satellite communications; and space-based positioning, navigation, and timing (PNT). Space operations also integrate offensive and defensive activities to achieve and maintain space superiority.

Special Technical Operations (STO)

STO should be deconflicted and synchronized with other activities. Detailed information related to STO and its contribution to joint force operations can be obtained from the STO planners at CCMD or Service component HQ.

*Cyberspace Operations (CO)

CO include the missions of OCO, DCO, and DODIN operations. These missions include the use of technical capabilities in cyberspace and cyberspace as a medium to leverage information in and through cyberspace.

See following pages (pp. 2-26 to 2-27) for an overview and further discussion of CO.

*Electronic Warfare (EW)

EW is the military action ultimately responsible for securing and maintaining freedom of action in the EMS for friendly forces while exploiting or denying it to adversaries. EW is an enabler for other activities that communicate or maneuver through the EMS, such as MISO, PA, or CO.

See following pages (pp. pp. 2-26 to 2-27) for an overview and further discussion.

Refer to Joint/Interagency SMARTbook 3: Information Operations (Multi-Domain Guide to IO & Information-Related Capabilities), when published. All military activities produce information. Informational aspects affect the perceptions and attitudes that drive behavior and decision making. The JFC leverages informational aspects of military activities to gain an advantage; failing to leverage those aspects may cede this advantage to others. Leveraging the informational aspects of military activities ultimately affects strategic outcomes.

E. Cyberspace Operations (CO) and Electronic Warfare (EW) Operations

Ref: JP 3-12, Cyberspace Operations (Jun '18).

United States armed forces operate in an increasingly network-based world. The proliferation of information technologies is changing the way humans interact with each other and their environment, including interactions during military operations. This broad and rapidly changing operational environment requires that today's armed forces must operate in cyberspace and leverage an electromagnetic spectrum that is increasingly competitive, congested, and contested.

Cyberspace

Cyberspace reaches across geographic and geopolitical boundaries and is integrated with the operation of critical infrastructures, as well as the conduct of commerce, governance, and national defense activities. Access to the Internet and other areas of cyberspace provides users operational reach and the opportunity to compromise the integrity of critical infrastructures in direct and indirect ways without a physical presence. The prosperity and security of our nation are significantly enhanced by our use of cyberspace, yet these same developments have led to increased exposure of vulnerabilities and a critical dependence on cyberspace, for the US in general and the joint force in particular.

Cyberspace Operations (CO)

Cyberspace Operations (CO) are the employment of cyberspace capabilities where the primary purpose is to achieve objectives in or through cyberspace. CO comprise the military, national intelligence, and ordinary business operations of DOD in and through cyberspace. Although commanders need awareness of the potential impact of the other types of DOD CO on their operations, the military component of CO is the only one guided by joint doctrine and is the focus of JP 3-12. CCDRs and Services use CO to create effects in and through cyberspace in support of military objectives. Military operations in cyberspace are organized into missions executed through a combination of specific actions that contribute to achieving a commander's objective. Various DOD agencies and components conduct national intelligence, ordinary business, and other activities in cyberspace.

The Relationship of Cyberspace Operations to Operations in the Information Environment

Cyberspace is wholly contained within the information environment. CO and other information activities and capabilities create effects in the information environment in support of joint operations. Their relationship is both an interdependency and a hierarchy; cyberspace is a medium through which other information activities and capabilities may operate. These activities and capabilities include, but are not limited to, understanding information, leveraging information to affect friendly action, supporting human and automated decision making, and leveraging information (e.g., military information support operations [MISO] or military deception [MILDEC]) to change enemy behavior. CO can be conducted independently or synchronized, integrated, and deconflicted with other activities and operations.

While commanders may conduct CO specifically to support information-specific operations, some CO support other types of military objectives and are integrated through appropriate cells and working groups. The lack of synchronized CO with other military operations planning and execution can result in friendly force interference and may counter the simplicity, agility, and economy of force principles of joint operations.

Cyberspace Missions

All actions in cyberspace that are not cyberspace-enabled activities are taken as part of one of three cyberspace missions: OCO, DCO, or DODIN operations. These three mission types comprehensively cover the activities of the cyberspace forces. The successful execution of CO requires integration and synchronization of these missions. Military cyberspace missions and their included actions are normally authorized by a military order (e.g., execute order [EXORD], operation order [OPORD], tasking order, verbal order), referred to hereafter as mission order, and by authority derived from DOD policy memorandum, directive, or instruction. Cyberspace missions are categorized as OCO, DCO, or DODIN operations based only on the intent or objective of the issuing authority, not based on the cyberspace actions executed, the type of military authority used, the forces assigned to the mission, or the cyberspace capabilities used.

Cyberspace Missions

Most DOD cyberspace actions use cyberspace to enable other types of activities, which employ cyberspace capabilities to complete tasks but are not undertaken as part of one of the three CO missions: OCO, DCO, or DODIN operations.

Cyberspace Actions

- A. Cyberspace Security
- B. Cyberspace Defense
- C. Cyberspace Exploitation
- D. Cyberspace Attack

Refer to CYBER1: The Cyberspace Operations & Electronic Warfare SMARTbook (Multi-Domain Guide to Offensive/Defensive CEMA and CO). Topics and chapters include cyber intro (global threat, contemporary operating environment, information as a joint function), joint cyberspace operations (CO), cyberspace operations (OCO/DCO/DODIN), electronic warfare (EW) operations, cyber & EW (CEMA) planning, spectrum management operations (SMO/JEMSO), DoD information network (DODIN) operations, acronyms/abbreviations, and a cross-referenced glossary of cyber terms.

III. Intelligence

Understanding the OE is Fundamental to Joint Operations

The intelligence function supports this understanding with analysis of the operational environment (OE) to inform JFCs about adversary capabilities, COGs, vulnerabilities, and future COAs and to help commanders and staffs understand and map friendly, neutral, and threat networks. Using the continuous JIPOE analysis process, properly tailored JIPOE products can enhance OE understanding and enable the JFC to act inside the enemy's decision cycle. Intelligence activities and assessments also occur while defending the homeland within the guidelines of applicable regulations and laws.

Understanding Human, Physical, and Informational Aspects of Military Operations

Intelligence is critical to the JFC's ability to leverage information to affect behavior. People and organizations other than the enemy may positively or negatively affect the friendly mission. These actors may include the population, HN government, and potential opposition leaders. Other relevant actors may include international organizations, non-state actors, and NGOs.

Joint Intelligence Preparation of the Operational Environment (JIPOE)

Tailored, continuous JIPOE products support JPP steps 2-7 and the four planning functions starting with an OE baseline characterization to facilitate planning. Throughout execution, tailored, continuous JIPOE products capture the dynamic OE in support of the assessment process to facilitate risk management and operations adjustments and to identify new opportunities.

See pp. 3-72 to 3-73 for an overview and further discussion of JIPOE.

JFCs use assigned and attached intelligence forces and coordinate with supporting interagency intelligence capabilities to develop a current intelligence picture and analyze the OE. These supporting capabilities include combat support agencies (e.g., National Security Agency, Defense Intelligence Agency, and National Geospatial-Intelligence Agency [NGA]) and national intelligence agencies (e.g., Central Intelligence Agency). National intelligence support may be provided to the J-2 as requested to integrate national intelligence capabilities into a comprehensive intelligence effort designed to support the joint force.

As crises emerge that potentially require military action, JFCs examine available intelligence estimates. As part of the JIPOE process, JFCs focus intelligence efforts to determine or confirm enemy COGs and refine estimates of enemy capabilities, dispositions, intentions, and probable COAs within the context of the current situation. They look for specific warning intelligence of imminent enemy activity that may require an immediate response or an acceleration of friendly decision cycles.

Key Considerations

The intelligence function encompasses the following intelligence process components:

- Planning and direction of intelligence activities.
- Collection of data
- Processing and exploitation of collected data to produce relevant information
- Analysis of information and production of intelligence
- Dissemination and integration of intelligence with operations
- Evaluation and feedback regarding intelligence effectiveness and quality

JFCs and their component commanders are the key players in planning and conducting intelligence tasks. Commanders are more than just consumers of intelligence. They are responsible for fully integrating intelligence into their plans and operations.

They are also responsible for distributing intelligence and information to subordinate commands, and when appropriate, to relevant participants through established protocols and systems. Commanders establish operational and intelligence requirements and continuous feedback to ensure optimum intelligence support to planning and operations. This interface supports the commander and operational planning and execution. It also mitigates surprise, assists friendly deception efforts, and enables joint operation assessment.

- **Surveillance and reconnaissance** support information collection across the OA. These activities focus on planned collection requirements, but are also sufficiently flexible to respond to time-sensitive and emerging requirements. Commanders will also require persistent surveillance of specific targets that are mission essential and support guidance and intent.
- **Counterintelligence (CI) and Human Intelligence (HUMINT).** CI and HUMINT both use human sources to collect information, and while their activities may at times be overlapping, each has its own distinct purpose and function.
- **Identity intelligence (I2)** is gathered from identity attributes of individuals, groups, networks, or populations of interest. Regional and global trends have placed greater requirements on the JFC to be able to recognize and differentiate one person from another to support protection and intelligence functions.

IV. Fires

To employ fires is to use available weapons and other systems to create a specific effect on a target. Joint fires are those delivered during the employment of forces from two or more components in coordinated action to produce desired results in support of a common objective. Fires typically produce destructive effects, but various other tools and methods can be employed with little or no associated physical destruction. This function encompasses the fires associated with a number of tasks, missions, and processes, including:

- **Conduct joint targeting**. This is the process of selecting and prioritizing targets and matching the appropriate response to them, taking account of command objectives, operational requirements, and capabilities. *See pp. 2-30 to 2-31.*
- **Provide joint fire support**. This task includes joint fires that assist joint forces to move, maneuver, and control territory, populations, space, cyberspace, airspace, and key waters.
- **Countering Air and Missile Threats.** This task integrates offensive and defensive operations and capabilities to achieve and maintain a desired degree of air superiority and force protection. These operations are planned to destroy or negate enemy manned and unmanned aircraft and missiles, both before and after launch.
- **Interdict Enemy Capabilities**. Interdiction diverts, disrupts, delays, or destroys the enemy's military surface capabilities before they can be used effectively against friendly forces or to otherwise achieve their objectives.
- **Conduct Strategic Attack**. This task includes offensive action against targets—whether military, political, economic, or other—which are selected specifically in order to achieve national or military strategic objectives.
- **Employ IRCs**. IRCs are tools, techniques, or activities employed within the information environment to create effects and operationally desirable conditions. In the context of the fires function, this task focuses on the integrated employment of IRCs in concert with other LOOs and LOEs, to influence, disrupt, corrupt, or usurp an enemy's decision making.
- **Assess the Results of Employing Fires**. This task includes assessing the effectiveness and performance of fires as well as their contribution to the larger operation or objective.

For additional guidance on joint fire support, refer to JP 3-09, Joint Fire Support.

Joint Targeting

Ref: JP 3-60, Joint Targeting (Jan '13).

Targeting is the process of selecting and prioritizing targets and matching the appropriate response to them, considering operational requirements and capabilities. Targeting is both a joint- and component-level function to create specific desired effects that achieve the JFC's objectives. Targeting selects targets that, when attacked, can create those effects, and selects and tasks the means to engage those targets. Targeting is complicated by the requirement to deconflict unnecessary duplication of target nominations by different forces or different echelons within the same force and to integrate the attack of those targets with other components of the joint force. An effective and efficient target development process coupled with the joint air tasking cycle is essential for the JFACC to plan and execute joint air operations. The joint targeting process should integrate the intelligence databases, analytical capabilities, and data collection efforts of national agencies, combatant commands, subordinate joint forces, and component commands.

Joint Targeting Cycle

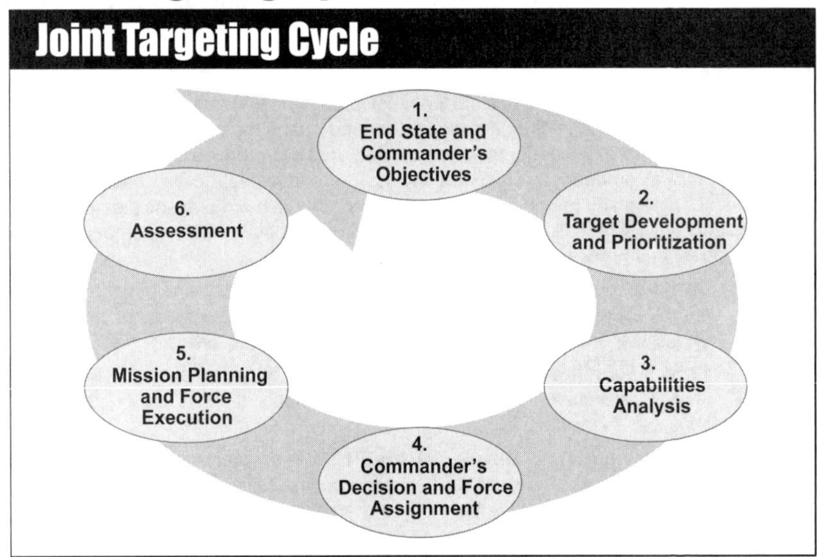

Ref: JP 3-60, Joint Targeting (Jan '13), fig. II-2. Joint Targeting Cycle.

The joint targeting cycle is an iterative process that is not time-constrained, and steps may occur concurrently, but it provides a helpful framework to describe the steps that must be satisfied to successfully conduct joint targeting. The deliberate and dynamic nature of the joint targeting process is adaptable through all phases of the air tasking cycle.

Refer to AFOPS2: The Air Force Operations & Planning SMARTbook, 2nd Ed. (Guide to Curtis E. LeMay Center & Joint Air Operations Doctrine). Topics and references of the 376-pg AFOPS2 include airpower fundamentals and principles (Vol 1), command and organizing (Vol 3); command and control (Annex 3-30/3-52), airpower (doctrine annexes), operations and planning (Annex 3-0), planning for joint air operations (JP 3-30/3-60), targeting (Annex 3-60), and combat support (Annex 4-0, 4-02, 3-10, and 3-34).

Target Development and Prioritization

Ref: JP 3-60, Joint Targeting (Jan '13), pp. II-5 to II-6.

Target development is the analysis, assessment, and documentation processes to identify and characterize potential targets that, when successfully engaged, support the achievement of the commander's objectives. A fully developed target must comply with national and command guidance, law of war, and the applicable ROE to be engaged. Phase 2 is comprised of three steps:

(a) Target system analysis;

(b) Entity-level target development; and

(c) Target list management (TLM).

Target developers systematically examine the enemy to the entities to the elements utilizing the targeting taxonomy, which hierarchically orders the adversary, its capabilities, and the targets which enable the capabilities into a clarifying framework.

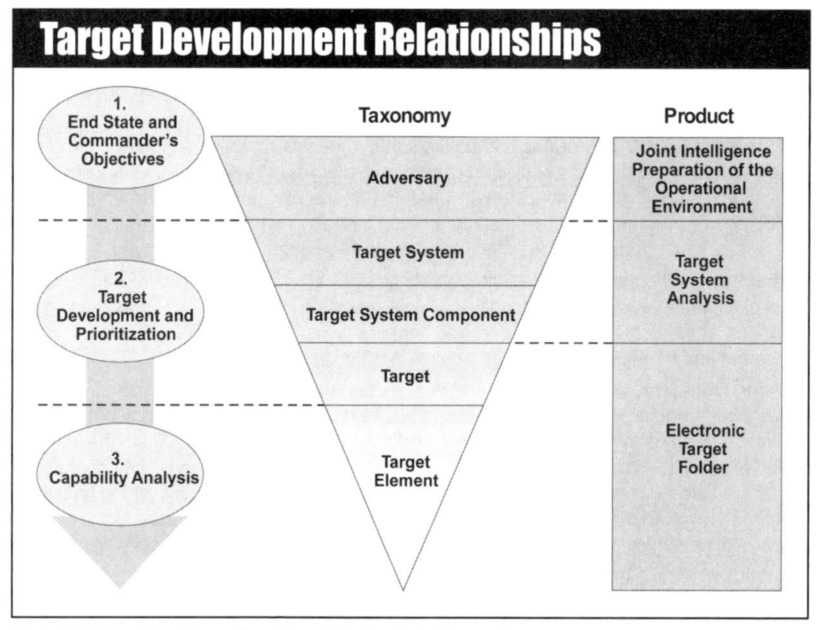

Ref: JP 3-60, Joint Targeting (Jan '13), fig. II-3. Target Development Relationships.

Target systems are typically a broad set of interrelated functionally associated components that generally produce a common output or have a shared mission. Target development always approaches adversary capabilities from a target systems perspective. This includes physical, logical, and complex social systems, and the interaction among them. While a single target may be significant because of its own characteristics, the target's real importance lies in its relationship to other targets within an operational system. A target system is most often considered as a collection of assets directed to perform a specific function or series of functions. While target systems are intra-dependent to perform a specific function, they are also interdependent in support of adversary capabilities. System-level target development links these multiple target systems and their components to reflect both their intra- and interdependency that, in aggregate, contribute to the adversary capabilities. JIPOE helps target developers prioritize an adversary's target systems based on how much each contributes to the adversary's ability to wage war.

Fires - Key Considerations

Ref: JP 3-0 (w/Chg 1), Joint Operations (Oct '18), pp. III-27 to III-33.

Targeting. Targeting supports the process of linking the desired effects of fires to actions and tasks at the component level. Commanders and their staffs must consider strategic and operational-level objectives, the potential for friendly fire incidents and other undesired fires effects, and operational limitations when making targeting decisions. Impact on all systems in the OE should be considered during this process.

Joint Fire Support. Joint fire support includes joint fires that assist air, land, maritime, cyberspace, and special operations forces to move, maneuver, and control territory, populations, airspace, cyberspace, EMS, and key waters.

Countering Air and Missile Threats. The JFC counters air and missile threats to ensure friendly freedom of action, provide protection, and deny enemy freedom of action. Counterair integrates offensive and defensive operations to attain and maintain a desired degree of air superiority and protection by neutralizing or destroying enemy aircraft and missiles, both before and after launch.

Interdiction. Interdiction is a powerful tool for JFCs. Interdiction operations are actions to divert, disrupt, delay, or destroy the enemy's military surface capability before it can be used effectively against friendly forces to achieve enemy objectives.

Strategic Attack (SA). A strategic attack is a JFC-directed offensive action against a target—whether military or other—that is selected to achieve national or military strategic objectives. These attacks seek to weaken the enemy's ability or will to engage in conflict or continue an action and as such, could be part of a campaign or major operation, or conducted independently as directed by the President or SecDef.

Global Strike. Global strike is the capability to rapidly plan and deliver extended-range attacks, limited in duration and scope, to create precision effects against enemy assets in support of national and theater commander objectives. Global strike missions employ lethal and non-lethal capabilities against a wide variety of targets.

Limiting Collateral Damage. Collateral damage is unintentional or incidental injury or damage to persons or objects that would not be lawful military targets based on the circumstances existing at the time. Causing collateral damage does not violate the law of war so long as the collateral damage caused is not excessive in relation to the concrete and direct military advantage anticipated from the attack. Under the law of war, the balancing of military necessity in relation to collateral damage is known as the principle of proportionality. Moreover, limiting collateral damage is often an operational or strategic imperative.

Capabilities That Can Create Nonlethal Effects. Some capabilities can generate nonlethal effects that limit collateral damage, reduce risk to civilians, and may reduce opportunities for enemy or adversary propaganda. They may also reduce the number of casualties associated with excessive use of force, limit reconstruction costs, and maintain the good will of the local populace. Some capabilities are nonlethal by design and include, but are not limited to, blunt impact and warning munitions, acoustic and optical warning devices, and vehicle and vessel stopping systems.

Cyberspace Attack. Cyberspace attack actions create various direct denial effects in cyberspace (i.e., degradation, disruption, or destruction) and manipulation that leads to denial that is hidden or that manifests in the physical domains.

Electronic Attack (EA). EA involves the use of electromagnetic energy, directed energy, or anti-radiation weapons to attack personnel, facilities, or equipment to degrade, neutralize, or destroy enemy combat capability.

Military Information Support Operations (MISO). MISO convey selected information and indicators to foreign audiences to influence their emotions, motives, and objective reasoning, and ultimately induce or reinforce foreign attitudes and behavior favorable to the originator's objectives.

V. Movement and Maneuver

This function encompasses the disposition of joint forces to conduct operations by securing positional advantages before or during combat operations and by exploiting tactical success to achieve operational and strategic objectives. This function includes moving or deploying forces into an OA and maneuvering them to operational depths for offensive and defensive purposes. It also includes assuring the mobility of friendly forces. The movement and maneuver function encompasses a number of tasks including:

- Deploy, shift, regroup, or move joint and/or component force formations within the operational area by any means or mode (i.e., air, land, or sea)
- Maneuver joint forces to achieve a position of advantage over an enemy
- Provide mobility for joint forces to facilitate their movement and maneuver without delays caused by terrain or obstacles
- Delay, channel, or stop movement and maneuver by enemy formations. This includes operations that employ obstacles (i.e., counter-mobility), enforce sanctions and embargoes, and conduct blockades
- Control significant areas in the operational area whose possession or control provides either side an operational advantage

Movement to Extend and Maintain Operational Reach

Forces, sometimes limited to those that are forward-deployed or even multinational forces formed specifically for the task at hand, can be positioned within operational reach of enemy COGs or decisive points to achieve decisive force at the appropriate time and place. Operational reach is the distance and duration across which a joint force can successfully employ its military capabilities. At other times, mobilization and deployment processes can be called up to begin the movement of reinforcing forces from the continental United States (CONUS) or other theaters to redress any unfavorable balance of forces and to achieve decisive force at the appropriate time and place.

Maneuver

Maneuver is the employment of forces in the OA through movement in combination with fires and information to gain a position of advantage in respect to the enemy. Maneuver of forces relative to enemy COGs can be key to the JFC's mission accomplishment. Through maneuver, the JFC can concentrate forces at decisive points to achieve surprise, psychological effects, and physical momentum. Maneuver may also enable or exploit the effects of massed or precision fires.

The principal purpose of maneuver is to place the enemy at a disadvantage through the flexible application of movement and fires. The goal of maneuver is to render opponents incapable of resisting by shattering their morale and physical cohesion (i.e., their ability to fight as an effective, coordinated whole) by moving to a point of advantage to deliver a decisive blow. This may be achieved by attacking enemy forces and controlling territory, airspace, EMS, populations, key waters, and LOCs through air, land, and maritime maneuvers.

There are multiple ways to gain positional advantage. An amphibious force with aircraft, cruise missiles, and amphibious assault capability, within operational reach of an enemy's COG, has positional advantage. In like manner, land and air expeditionary forces that are within operational reach of an enemy's COG and have the means and opportunity to strike and maneuver on such a COG also have positional advantage. Maintaining full-spectrum superiority contributes to positional advantage by facilitating freedom of action.

At all levels of warfare, successful maneuver requires not only fire and movement but also agility and versatility of thought, plans, operations, and organizations. It requires designating and then, if necessary, shifting the main effort and applying the principles of mass and economy of force.

Force Posture

Force posture (forces, footprints, and agreements) affects operational reach and is an essential maneuver-related consideration during theater strategy development and adaptive planning. Force posture is the starting position from which planners determine additional contingency basing requirements to support specific contingency plans and crisis responses.

Positional Advantage

JFCs should consider various ways and means to help maneuver forces attain positional advantage. Specifically, combat engineers provide mobility of the force by breaching obstacles, while simultaneously countering the mobility of enemy forces by emplacing obstacles, and minimizing the effects of enemy actions on friendly forces.

VI. Protection

The protection function encompasses force protection, force health protection (FHP), and other protection activities. The function focuses on force protection, which preserves the joint force's fighting potential in four primary ways. One way uses active defensive measures that protect the joint force, its information, its bases, necessary infrastructure, and LOCs from an enemy attack. Another way uses passive defensive measures that make friendly forces, systems, and facilities difficult to locate, strike, and destroy by reducing the probability of, and minimizing the effects of, damage caused by hostile action without the intention of taking the initiative. The application of technology and procedures to reduce the risk of friendly fire incidents is equally important. Finally, emergency management and response reduce the loss of personnel and capabilities due to isolating events, accidents, health threats, and natural disasters.

Force protection does not include actions to defeat the enemy or protect against accidents, weather, or disease. FHP complements force protection efforts by promoting, improving, preserving, or restoring the mental or physical well-being of Service members.

As the JFC's mission requires, the protection function also extends beyond force protection to encompass protection of US noncombatants.

The protection function encompasses a number of tasks, including:

- Provide air, space, and missile defense.
- Protect US civilians and contractors authorized to accompany the force.
- Conduct defensive countermeasure operations, including MILDEC in support of OPSEC, counterdeception, and counterpropaganda operations.
- Conduct OPSEC, cyberspace defense, cyberspace security, defensive EA, and electronic protection activities.
- Conduct PR operations.
- Establish antiterrorism programs.
- Establish capabilities and measures to prevent friendly fire incidents.
- Secure and protect combat and logistics forces, bases, JSAs, and LOCs.
- Provide physical protection and security for forces, to include conducting operations to mitigate the effects of explosive hazards.
- Provide chemical, biological, radiological, and nuclear (CBRN) defense.
- Minimize the effects of CBRN incidents through thorough planning, preparation, response, and recovery.
- Provide emergency management and response capabilities and services.
- Protect the DODIN using cyberspace security and cyberspace defense measures.
- Identify and neutralize insider threats.
- Conduct identity collection activities. These include security screening and vetting in support of identity intelligence (I2).

Protection - Key Considerations

Ref: JP 3-0 (w/Chg 1), Joint Operations (Oct '18), pp. III-37 to III-42.

Security of Forces and Means. Security of forces and means enhances force protection by identifying and reducing friendly vulnerability to hostile acts, influence, or surprise. Security operations protect combat and logistics forces, bases, JSAs, and LOCs. Physical security includes physical measures designed to safeguard personnel; to prevent unauthorized access to equipment, installations, material, and documents; and to safeguard them against espionage, sabotage, damage, and theft.

Defensive Counter Air (DCA). DCA supports protection using both active and passive air and missile defense measures.

Global Ballistic Missile Defense. Overarching characterization of cumulative planning and coordination for those defensive capabilities designed to neutralize, destroy, or reduce effectiveness of enemy ballistic missile attacks that cross AOR boundaries.

Information-Related Capabilities (IRCs). Defensive use of IRCs ensures timely, accurate, and relevant information access while denying enemies and adversaries opportunities to exploit friendly information and information systems for their own purposes. **OPSEC**, as an IRC, denies the adversary the information needed to correctly assess friendly capabilities and intentions. **Defensive Cyberspace Operations (DCO)** include passive and active CO to preserve friendly cyberspace capabilities and protect data, networks, and net-centric capabilities by monitoring, analyzing, detecting, and responding to unauthorized activity within DOD information systems and computer networks. **Cybersecurity** encompasses measures that protect computers, electronic communications systems and services, wired communications, and electronic communications.

Personnel Recovery (PR). PR missions use military, diplomatic, and civil efforts to recover and reintegrate isolated personnel. There are five PR task: report, locate, support, recover, and reintegrate.

CBRN Defense. Preparation for potential enemy use of CBRN weapons is integral to joint planning.

Antiterrorism. Antiterrorism programs support force protection by establishing defensive measures that reduce the vulnerability of individuals and property to terrorist acts, to include limited response and containment by local military and civilian forces.

Combat Identification (CID). CID is to accurately distinguish enemy objects and forces in the OE from others to support engagement decisions. CID supports force protection and enhances operations by helping minimize friendly fire incidents and collateral damage.

Critical Infrastructure Protection (CIP). Critical infrastructure protection programs support the identification and mitigation of vulnerabilities to defense critical infrastructure, which includes DOD and non-DOD domestic and foreign infrastructures essential to plan, mobilize, deploy, execute, and sustain US military operations on a global basis.

Counter-Improvised Explosive Device Operations (C-IED). C-IED operations are the organization, integration, and synchronization of capabilities and activities to reduce casualties and mitigate damage caused by IEDs.

Identify and Neutralize Insider Threats. Insider threats (sometimes referred to as "green-on-blue" or "inside-the-wire" threats) may include active shooters, bombers, spies, and other threats embedded within or working with US forces.

Force Health Protection (FHP). FHP complements force protection efforts, and includes all measures taken by the JFC and the Military Health System to promote, improve, and conserve the mental and physical well-being of Service members.

Protection of Civilians. Persons who are neither part of nor associated with an armed force or group, nor otherwise engaged in hostilities are classified as civilians and have protected status under the law of war.

Protection considerations affect planning in every joint operation. Campaigns and major operations involve large-scale combat against a capable enemy. These operations typically will require the full range of protection tasks, thereby complicating both planning and execution. Although the OA and joint force may be smaller for a crisis response or limited contingency operation, the mission can still be complex and dangerous, thus requiring a variety of protection considerations. Permissive environments associated with military engagement, security cooperation, and deterrence still require that commanders and their staffs consider protection measures commensurate with potential risks. These risks may include a wide range of threats such as terrorism, criminal enterprises, environmental threats and hazards, and cyberspace threats. Continuous research and access to accurate, detailed information about the OE, along with realistic training, can enhance protection activities.

Force Protection

Force protection includes preventive measures taken to prevent or mitigate enemy and insider threat actions against DOD personnel (to include family members and certain contractors), resources, facilities, and critical information. These actions preserve the force's fighting potential so it can be applied at the decisive time and place and incorporate integrated and synchronized offensive and defensive measures that enable the effective employment of the joint force while degrading opportunities for the enemy. Force protection is achieved through the tailored selection and application of multilayered active and passive measures commensurate with the level of risk.

VII. Sustainment

Sustainment is the provision of logistics and personnel services to maintain operations through mission accomplishment and redeployment of the force. Sustainment provides the JFC the means to enable freedom of action and endurance and to extend operational reach. Sustainment determines the depth to which the joint force can conduct decisive operations, allowing the JFC to seize, retain, and exploit the initiative. The sustainment function includes tasks to:

- Coordinate the supply of food, operational energy (fuel and other energy requirements), arms, munitions, and equipment.
- Provide for maintenance of equipment.
- Coordinate and provide support for forces, including field services; personnel services support; health services; mortuary affairs; religious support; postal support; morale, welfare, and recreational support; financial support; and legal services.
- Build and maintain contingency bases.
- Assess, repair, and maintain infrastructure.
- Acquire, manage, and distribute funds.
- Provide common-user logistics support to other government agencies, international organizations, NGOs, and other nations.
- Establish and coordinate movement services.
- Establish large-scale detention compounds and sustain enduring detainee operations

JFCs should identify sustainment capabilities early in planning. Sustainment should be a priority consideration when the timed-phased force and deployment data is built. Sustainment provides JFCs with flexibility to develop branches and sequels and to refocus joint force efforts. Given mission objectives and adversary threats, the ultimate goal is for planners to develop a feasible, supportable, and efficient CONOPS that takes into account the threat and defense of logistical forces.

See chap. 4 for discussion of joint logistics.

IV. Organizing for Joint Operations

Ref: JP 3-0 (w/Chg 1), Joint Operations (Oct '18), chap. IV.

Organizing for joint operations involves many considerations. Most can be associated in three primary groups related to organizing the joint force, organizing the joint force headquarters, and organizing operational areas (OAs) to help control operations.

I. Organizing the Joint Force

How JFCs organize their assigned or attached forces affects the responsiveness and versatility of joint operations. The JFC's mission and operational approach, as well as the principle of unity of command and a mission command philosophy, are guiding principles to organize the joint force for operations. Joint forces can be established on a geographic or functional basis. JFCs may centralize selected functions within the joint force, but should not reduce the versatility, responsiveness, and initiative of subordinate forces. JFCs should allow Service and special operations tactical and operational forces, organizations, and capabilities to function generally as they were designed. All Service components contribute distinct capabilities to joint operations that enable joint effectiveness. Joint interdependence is the purposeful reliance by one Service on another Service's capabilities to maximize the complementary and reinforcing effects of both. The degree of interdependence varies with specific circumstances.

Joint Force Options

- **Combatant Commands (CCMD).** A CCMD is a unified or specified command with a broad continuing mission under a single commander established and so designated by the President, through SecDef, and with the advice and assistance of the CJCS. *See pp. 1-17 to 1-22.*
- **Subordinate Unified Commands.** When authorized by SecDef through the CJCS, commanders of unified (not specified) commands may establish subordinate unified commands to conduct operations on a continuing basis IAW the criteria set forth for unified commands. A subordinate unified command may be established on a geographic area or functional basis. *See p. 1-30.*
- **Joint Task Forces (JTFs).** A JTF is a joint force constituted and designated by SecDef, a CCDR, a subordinate unified command commander, or an existing commander, joint task force (CJTF) to accomplish missions with specific, limited objectives, and which do not require centralized control of logistics. *See chap. 5 for further discussion.*

Component Options

CCDRs and subordinate unified commanders conduct either single-Service or joint operations to accomplish a mission.

- **Service Components**. Regardless of the organization and command arrangements within joint commands, Service component commanders retain responsibility for certain Service-specific functions and other matters affecting their forces, including internal administration, personnel support, training, sustainment, and Service intelligence operations. *See p. 5-14.*
- **Functional Components.** The JFC can establish functional component commands to conduct operations when forces from two or more Services must operate in the same physical domain or accomplish a distinct aspect of the assigned mission. *See pp. 5-14 to 5-16.*

II. Understanding an Operational Environment (OE)

Ref: JP 3-0 (w/Chg 1), Joint Operations (Oct '18), pp. IV-1 to IV-4.

Factors that affect joint operations extend far beyond the boundaries of the JFC's assigned JOA. The JFC's OE is the composite of the conditions, circumstances, and influences that affect employment of capabilities and bear on the decisions of the commander. It encompasses physical areas of the air, land, maritime, and space domains; the information environment (which includes cyberspace); the EMS; and other factors.

Included within these are enemy, friendly, and neutral systems that are relevant to a specific joint operation. The nature and interaction of these systems will affect how the commander plans, organizes for, and conducts joint operations.

A Systems Perspective (PMESII)

A system is a functionally, physically, or behaviorally related group of regularly interacting or interdependent elements forming a unified whole. One way to think of the OE is as a set of complex and constantly interacting political, military, economic, social, information, and infrastructure (PMESII) systems as depicted in Figure IV-1. The interaction of these systems can then be viewed as a network or networks based on the participants. The nature and interaction of these systems affect how the commander plans, organizes, and conducts joint operations. The JFC's intergovernmental partners and other civilian participants routinely focus on systems other than military, so the JFC and staff should understand these systems and how military operations affect them. Equally important is understanding how elements in other PMESII systems can help or hinder the JFC's mission. A commonly shared understanding among stakeholders in the operation can influence actions beyond the JFC's directive authority and promote a unified approach to achieve objectives.

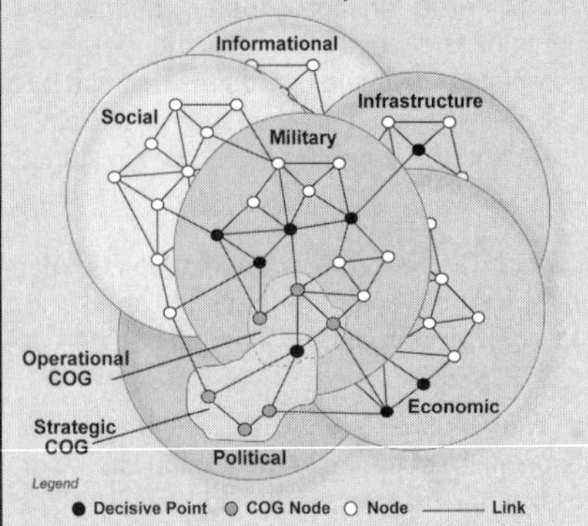

Ref: JP 3-0 (w/Chg 1), fig. IV-1. A Systems Perspective of the Operational Environment.

See pp. 3-52 to 3-53 for discussion of using PMESII as a tool in analyzing the current and future operational environment.

Physical Areas and Factors

Physical Areas

The fundamental physical area in the OE is the JFC's assigned Operational Area (OA). This term encompasses more descriptive terms for geographic areas in which joint forces conduct military operations. *See following pages (pp. 2-40 to 2-41) for further discussion.*

Physical Factors

The JFC and staff must consider many factors associated with operations in the air, land, maritime, and space domains and the information environment (which includes cyberspace). These factors include terrain (including urban settings), population, weather, topography, hydrology, EMS, and other environmental conditions in the OA; distances associated with the deployment to the OA and employment of joint capabilities; the location of bases, ports, and other supporting infrastructure; the physical results of combat operations; and both friendly and enemy forces and other capabilities. Combinations of these factors affect operations and sustainment.

Information Environment

The information environment comprises and aggregates numerous social, cultural, cognitive, technical, and physical attributes that act upon and impact knowledge, understanding, beliefs, world views, and, ultimately, actions of an individual, group, system, community, or organization. The information environment also includes technical systems and their use of data. The information environment directly affects all OEs.

See p. 2-18 to 2-19 for related discussion of the information environment.

Information is Pervasive throughout the OE

To operate effectively requires understanding the interrelationship of the informational, physical, and human aspects that are shared by the OE and the information environment. Informational aspects reflect the way individuals, information systems, and groups communicate and exchange information. Physical aspects are the material characteristics of the environment that create constraints on and freedoms for the people and information systems that operate in it. Finally, human aspects frame why relevant actors perceive a situation in a particular way. Understanding the interplay between the informational, physical, and human aspects provides a unified view of the OE.

Cyberspace

Cyberspace is a global domain within the information environment. It consists of the interdependent network of information technology infrastructures and resident data, including the Internet, telecommunications networks, computer systems, and embedded processors and controllers. Most aspects of joint operations rely in part on cyberspace, which reaches across geographic and geopolitical boundaries—much of it residing outside of US control—and is integrated with the operation of critical infrastructures, as well as the conduct of commerce, governance, and national security. Commanders must consider their critical dependencies on information and cyberspace, as well as factors such as degradations to confidentiality, availability, and integrity of information and information systems, when they plan and organize for operations.

Cyberspace Operations (CO)

Commanders conduct CO to retain freedom of maneuver in cyberspace, accomplish the JFC's objectives, deny freedom of action to enemies and adversaries, and enable other operational activities. CO include DODIN operations to secure and operate DOD cyberspace. CO rely on links and nodes that reside in the physical domains and perform functions in cyberspace and the physical domains. Similarly, activities in the physical domains can create effects in and through cyberspace by affecting the EMS or the physical infrastructure.

Electromagnetic Spectrum (EMS)

The EMS is the range of all frequencies of electromagnetic radiation. Electromagnetic radiation consists of oscillating electric and magnetic fields characterized by frequency and wavelength. The EMS is usually subdivided into frequency bands based on certain physical characteristics and includes radio waves, microwaves, millimeter waves, infrared radiation, visible light, ultraviolet radiation, x-rays, and gamma rays.

See pp. 2-26 to 2-27 for related discussion of cyberspace operations and electronic warfare.

III. Organizing Operational Areas (OAs)

Ref: JP 3-0 (w/Chg 1), Joint Operations (Oct '18), pp. IV-9 to IV-13.

Except for AORs, which are assigned in the UCP, GCCs and other JFCs designate smaller operational areas (e.g., JOA and AO) on a temporary basis. OAs have physical dimensions comprised of some combination of air, land, maritime, and space domains. While domains are useful constructs for visualizing and characterizing the physical environment in which operations are conducted (the OA), the use of the term "domain" is not meant to imply or mandate exclusivity, primacy, or C2 of any domain. Specific authorities and responsibilities within an operational area are as specified by the appropriate JFC. JFCs define these areas with geographical boundaries, which help commanders and staffs coordinate, integrate, and deconflict joint operations among joint force components and supporting commands.

Operational Areas Within a Theater

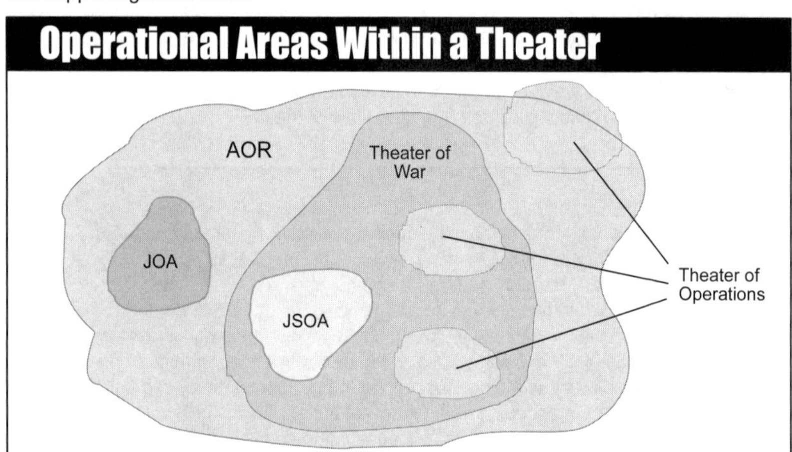

This example depicts a combatant commander's AOR, also known as a theater. Within the AOR, the combatant commander has designated a theater of war. Within the theater of war are two theaters of operations and a JSOA. To handle a situation outside the theater of war, the combatant commander has established a theater of operations and a JOA, within which a joint task force will operate. JOAs could also be established within the theater of war or theaters of operations.

Ref: JP 3-0 (w/Chg 1), fig. IV-3. Operational Areas within a Theater.

A. Combatant Command-Level Areas

GCCs conduct operations in their assigned AORs. When warranted, the President, SecDef, or GCCs may designate a theater of war and/or theater of operations for each operation. GCCs can elect to control operations directly in these OAs, or may establish subordinate joint forces for that purpose, while remaining focused on the broader AOR.

Area of Responsibility (AOR)

An AOR is an area established by the UCP that defines geographic responsibilities for a GCC. A GCC has authority to plan for operations within the AOR and conduct those operations approved by the President or SecDef. CCDRs may operate forces wherever required to accomplish approved missions. All cross-AOR operations must be coordinated among the affected GCCs.

Theater of War

A theater of war is a geographical area established by the President, SecDef, or GCC for the conduct of major operations and campaigns involving combat. A theater of war is established primarily when there is a formal declaration of war or it is necessary to encompass more than one theater of operations (or a JOA and a separate theater of operations) within a single boundary for the purposes of C2, sustainment, protection, or mutual support. A theater of war does not normally encompass a GCC's entire AOR, but may cross the boundaries of two or more AORs.

Theater of Operations

A theater of operations is an OA defined by the GCC for the conduct or support of specific military operations. A theater of operations is established primarily when the scope of the operation in time, space, purpose, and/or employed forces exceeds what a JOA can normally accommodate. More than one joint force HQ can exist in a theater of operations. A GCC may establish one or more theaters of operations. Different theaters will normally be focused on different missions. A theater of operations typically is smaller than a theater of war, but is large enough to allow for operations in depth and over extended periods of time. Theaters of operations are normally associated with major operations and campaigns and may cross the boundary of two AORs.

Continued on next page

B. Operational- and Tactical-Level Areas

For operations somewhat limited in scope and duration, or for specialized activities, the commander can establish the following OAs.

Joint Operations Area (JOA)

A JOA is an area of land, sea, and airspace, defined by a GCC or subordinate unified commander, in which a JFC (normally a CJTF) conducts military operations to accomplish a specific mission. JOAs are particularly useful when operations are limited in scope and geographic area or when operations are to be conducted on the boundaries between theaters.

Joint Special Operations Area (JSOA)

A JSOA is an area of land, sea, and airspace assigned by a JFC to the commander of SOF to conduct special operations activities. It may be limited in size to accommodate a discreet direct action mission or may be extensive enough to allow a continuing broad range of unconventional warfare (UW) operations. A JSOA is defined by a JFC who has geographic responsibilities. JFCs may use a JSOA to delineate and facilitate simultaneous conventional and special operations. The JFSOCC is the supported commander within the JSOA.

Joint Security Area (JSA)

A JSA is a specific surface area, designated by the JFC as critical, that facilitates protection of joint bases and supports various aspects of joint operations such as LOCs, force projection, movement control, sustainment, C2, airbases/airfields, seaports, and other activities. JSAs are not necessarily contiguous with areas actively engaged in combat. JSAs may include intermediate support bases and other support facilities intermixed with combat elements. JSAs may be used in both linear and nonlinear situations.

Amphibious Objective Area (AOA)

The AOA is a geographic area within which is located the objective(s) to be secured by the amphibious force. This area must be of sufficient size to ensure accomplishment of the amphibious force's mission and must provide sufficient area for conducting necessary sea, air, and land operations.

Area of Operations (AO)

JFCs may define AOs for land and maritime forces. AOs do not typically encompass the entire OA of the JFC, but should be large enough for component commanders to accomplish their missions (to include a designated amount of airspace) and protect their forces. Component commanders with AOs typically designate subordinate AOs within which their subordinate forces operate. These commanders employ the full range of joint and Service control measures and graphics as coordinated with other component commanders and their representatives to delineate responsibilities, deconflict operations, and achieve unity of effort.

Continued on next page

C. Contiguous and Noncontiguous Operational Areas

Ref: JP 3-0 (w/Chg 1), Joint Operations (Oct '18), pp. IV-12 to IV-13.

Operational areas (OAs) may be contiguous or noncontiguous. When they are contiguous, a boundary separates them. When OAs are noncontiguous, subordinate commands do not share a boundary. The higher HQ retains responsibility for the unassigned portion of its OA.

In some operations, a Service or functional component (typically the ground component) could have such a large OA that the component's subordinate units operate in a noncontiguous manner, widely distributed and beyond mutually supporting range of each other. In these cases, the JFC should consider options whereby joint capabilities can be pushed to lower levels and placed under control of units that can use them effectively.

Contiguous and Noncontiguous Op Areas

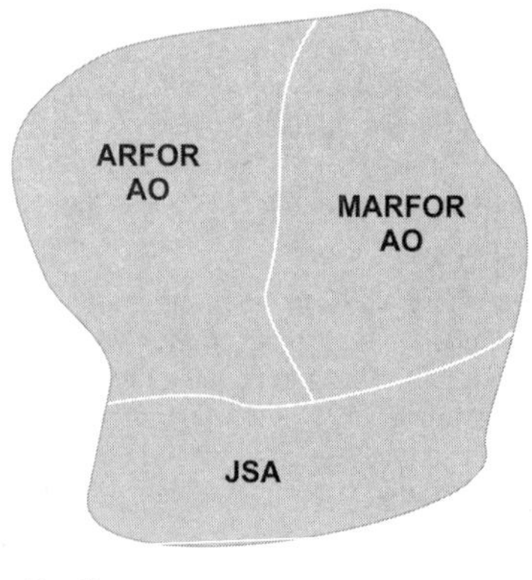

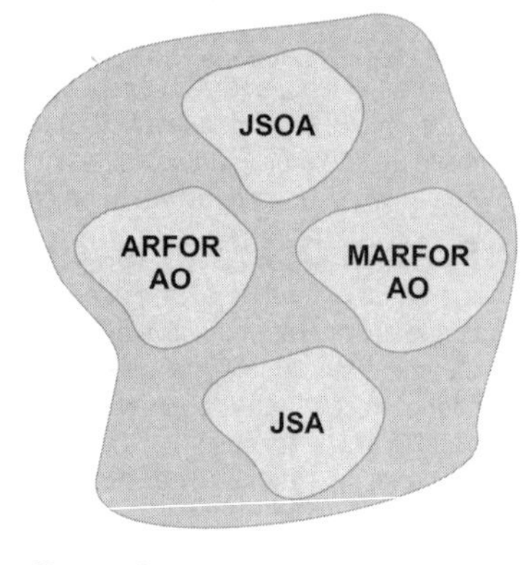

Contiguous

Adjacent, subordinate command's operational areas share boundaries. In this case, the higher headquarters has assigned all of its operational area to subordinate commands.

Noncontiguous

Subordinate commands are assigned operational areas that do not share boundaries. The higher headquarters retains responsibility for the portion of its operational area not assigned to subordinate commands.

Ref: JP 3-0 (w/Chg 1), fig. IV-4. Contiguous and Noncontiguous Operational Areas.

Considerations When Assuming Responsibility of an Operational Area

The establishing commander should activate an assigned OA at a specified date and time based on mission and situation considerations addressed during COA analysis and wargaming. Among others, common considerations include C2, the information environment, intelligence requirements, communications support, protection, security, LOCs, terrain management, movement control, airspace control, surveillance, reconnaissance, air and missile defense, PR, targeting and fires, interorganizational coordination, and environmental issues.

Refer to JP 3-33, Joint Task Force Headquarters, for specific guidance on assuming responsibility for an OA.

Continued from previous page

Continued from previous page

V. Joint Ops Across the Conflict Continuum

Ref: JP 3-0 (w/Chg 1), Joint Operations (Oct '18), chap. V.

Threats to US and allied interests throughout the world can sometimes only be countered by US forces able to respond to a wide variety of challenges along a conflict continuum that spans from peace to war. Our national interests and the nature of crises that can occur along this continuum require our nation's Armed Forces to be proficient in a wide variety of activities, tasks, missions, and operations that vary in purpose, scale, risk, and combat intensity.

I. The Range of Military Operations (ROMO)

The range of military operations is a fundamental construct that helps relate military activities and operations in scope and purpose. The potential range of military activities and operations extends from military engagement, security cooperation, and deterrence in times of relative peace up through large-scale combat operations.

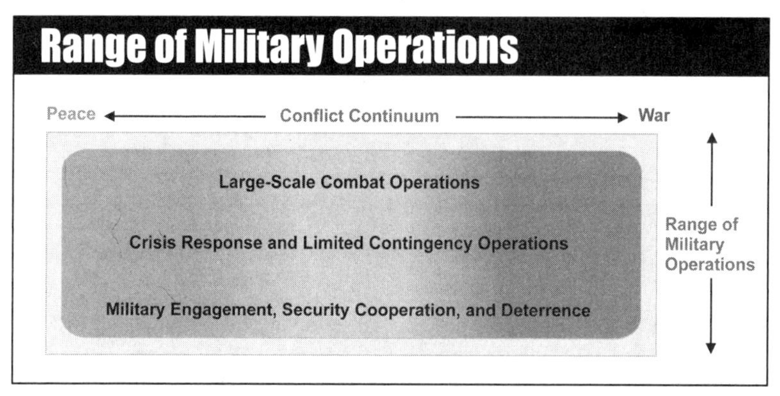

Ref: JP 3-0 (w/Chg 1), fig. V-2. Notional Operations Across the Conflict Continuum. See p. 2-59 for related discussion of the conflict continuum.

Military engagement, security cooperation, and deterrence activities develop local and regional situational awareness, build networks and relationships with partners, shape the OE, keep day-to-day tensions between nations or groups below the threshold of armed conflict, and maintain US global influence. Many missions associated with crisis response and limited contingencies, such as DSCA and FHA, may not require combat. But others, such as Operation RESTORE HOPE in Somalia, can be dangerous and may require combat operations to protect US forces. Large-scale combat often occurs in the form of major operations and campaigns that achieve national objectives or contribute to a larger, long-term effort (e.g., OEF).

The complex nature of the strategic environment may require US forces to conduct different types of joint operations and activities simultaneously across the conflict continuum. Although this publication discusses specific types of operations and activities under the various categories in the range of military operations, each type is not doctrinally fixed and could shift within that range.

- **Military Engagement, Security Cooperation, and Deterrence**. *See pp. 2-57 to 2-62.*
- **Crisis Response and Limited Contingency Operations**. *See pp. 2-63 to 2-68.*
- **Large-Scale Combat Operations.** *See pp. 2-69 to 2-88.*

II. Military Operations and Related Missions, Tasks, and Actions

In general, a military operation is a set of actions intended to accomplish a task or mission. Although the US military is organized, trained, and equipped for sustained, large-scale combat anywhere in the world, the capabilities to conduct these operations also enable a wide variety of other operations and activities. In particular, opportunities exist prior to large-scale combat to shape the OE to prevent, or at least mitigate, the effects of war. Characterizing the employment of military capabilities (people, organizations, and equipment) as one or another type of military operation has several benefits. For example, publications can be developed that describe the nature, tasks, and tactics associated with specific types of diverse operations, such as NEO and COIN. These publications provide the basis for related joint training and joint professional military education that help joint forces conduct military operations as effectively and efficiently as possible even in difficult and dangerous circumstances. Characterizations also help military and civilian leaders explain US military involvement in various situations to the US and international public and news media. Likewise, such characterizations, supplemented by operational experience, can clarify the need for specific capabilities that enhance certain operations. For example, facial recognition software associated with biometric capabilities helps military and law enforcement personnel identify terrorists and piece together their human networks as part of combating terrorism.

Military operations are often categorized by their focus, as shown in Figure V-1. In some cases, the title covers a variety of missions, tasks, and activities. Many activities accomplished by single Services, such as tasks associated with security cooperation, do not constitute a joint operation. Nonetheless, most of these occur under a joint "umbrella," because they contribute to achievement of CCDRs' CCP objectives.

Examples of Military Operations and Activities

- Stability activities
- Defense support of civil authorities
- Foreign humanitarian assistance
- Recovery
- Noncombatant evacuation
- Peace operations
- Countering weapons of mass destruction
- Chemical, biological, radiological, and nuclear response
- Foreign internal defense
- Counterdrug operations
- Combating terrorism
- Counterinsurgency
- Homeland defense
- Mass atrocity response
- Security cooperation
- Military engagement

Ref: JP 3-0 (w/Chg 1), fig. V-1. Examples of Military Operations and Activities.

For further discussion and an overview of these types of military operations and activities, see:

- **Military Engagement, Security Cooperation, and Deterrence**. *See pp. 2-57 to 2-62.*
- **Crisis Response and Limited Contingency Operations**. *See pp. 2-63 to 2-68.*
- **Large-Scale Combat Operations.** *See pp. 2-69 to 2-88.*

III. The Combatant Command Campaign

Military operations, actions, and activities in a GCC's AOR, from security cooperation through large-scale combat, are conducted in the context of the GCC's ongoing theater campaign.

Combatant Command Campaign Plan (CCP)

The CCDR's campaign is the overarching framework that synchronizes all activities and operations within the theater to achieve theater and national strategic objectives. A CCP operationalizes the GCC's strategy and approach to achieve these objectives within two to five years by organizing and aligning available resources. CCPs also support the campaign objectives of other CCDRs responsible for synchronizing collaborative DOD planning. As Figure V-3 shows, CCPs encompass all ongoing and planned operations across the range of military operations, continuously adjusted in response to changes in the OE. The CCP's long-term and persistent and preventative activities are intended to identify and deter, counter, or otherwise mitigate an adversary's actions before escalation to combat. Many of these activities are conducted with DOD in support of the diplomatic, economic, and informational efforts of USG partners and PNs. The CCDR adjusts these activities as required for the occasional execution of a contingency plan or response to a crisis.

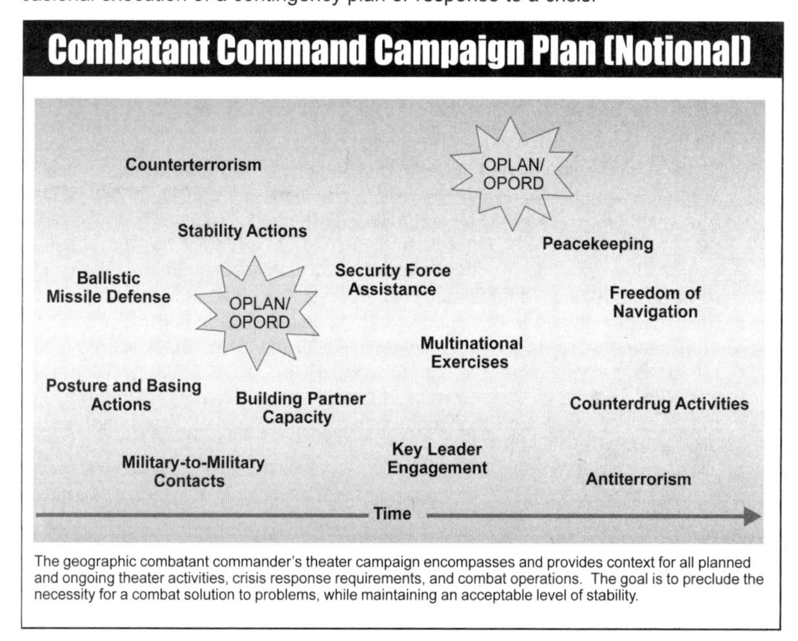

The geographic combatant commander's theater campaign encompasses and provides context for all planned and ongoing theater activities, crisis response requirements, and combat operations. The goal is to preclude the necessity for a combat solution to problems, while maintaining an acceptable level of stability.

Ref: JP 3-0 (w/Chg 1), fig. V-3. Notional Joint Operations in a Combatant Command Campaign Context.

The CCP also provides context for ongoing crisis response and contingency operations to facilitate execution of contingency plans as branch plans to the CCP. These are plans to respond to potential crises such as natural or man-made disasters and military aggression. Also linked to the GCC's CCP and subordinate campaign plans are designated DOD global campaign plans that address integrated execution of global security priorities.

GCCs prepare CCPs to achieve military objectives and GEF-directed prioritized campaign and DOD objectives for their AORs. GCCs integrate planning in their CCPs to support designated missions assigned to the CCDRs responsible for collaborative DOD planning. Similar to GCCs, FCCs prepare global campaigns to achieve global security objectives as required in the GEF. FCCs also prepare CCPs to achieve military objectives and GEF-directed objectives for their missions. FCCs synchronize planning for designated missions across CCMDs, Services, and DOD agencies.

A GCC can simultaneously conduct multiple joint operations with different objectives within their AOR. The GCC might initiate one or more OPLANs while security cooperation activities are ongoing in the same or another part of the theater. Further, a crisis response or contingency operation could occur separately or as part of a campaign or major operation (e.g., the NEO in Somalia during Operation DESERT SHIELD in 1991). In the extreme, a major operation or a subordinate campaign may occur within a theater concurrently with a separate or related campaign. CCDRs should synchronize and integrate the activities of assigned, attached, and allocated forces with subordinate and supporting JFCs so they complement rather than compete in achieving national or theater-strategic objectives. Due to the transregional nature of various enemies or adversaries (e.g., insurgents, terrorists, drug cartels, pirates), coordination and synchronization requirements may also extend to adjacent GCCs. CCDRs and subordinate JFCs must work with DOS regional and functional bureaus, individual country chiefs of mission, and other USG departments and agencies to better integrate military operations in unified action with the diplomatic, economic, and informational instruments of national power.

Some military operations may be conducted for one purpose. For example, FHA is focused on a humanitarian purpose (e.g., Operation TOMODACHI, an assistance operation to support Japan in disaster relief following the 2011 Tohoku earthquake and tsunami). A strike may be conducted for the specific purpose of compelling or deterring an action (e.g., Operation EL DORADO CANYON, the 1986 operation to coerce Libya to conform with international laws against terrorism). Often, however, military operations will have multiple purposes (based on strategic and operational-level objectives) and will be influenced by a fluid and changing situation. Branch and sequel events may produce additional tasks for the force, challenging the command with multiple missions (e.g., Operations PROVIDE COMFORT in Iraq and RESTORE HOPE in Somalia were PEO that evolved from FHA efforts). Joint forces must strive to meet such challenges with clearly defined objectives addressing diverse purposes.

IV. A Joint Operation Model

Most individual joint operations share certain activities or actions in common. These include forming a joint HQ, deploying and redeploying capabilities (forces, materiel, etc.), and interacting with other interorganizational participants. Some activities can also characterize specific types of operations such as large-scale combat, FHA, and COIN. For example, Figure V-4 shows six general groups of military activities that may typically occur in preparation for and during a single large-scale joint combat operation.

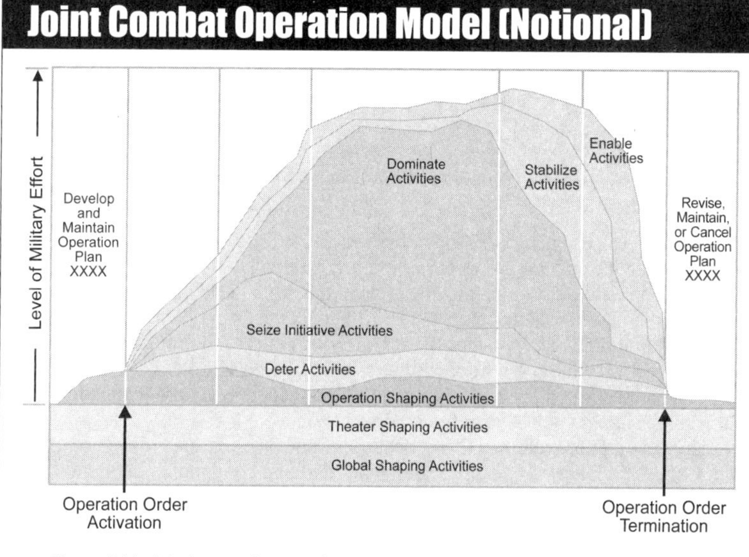

- The model depicts six general groups of military activities that typically comprise a single joint combat operation. The model applies to a large-scale combat operation as well as to a combat operation relatively limited in scope and duration. It shows that emphasis on activity types shifts as an operation progresses.
- Operation shaping activities may begin during plan development to help set conditions for successful execution. They may continue after the operation ends if the command continues to maintain an operation plan.
- Theater and global shaping activities occur continuously to support theater and global requirements. Specific theater and global shaping activities may support a specific joint operation plan during its execution.

Ref: JP 3-0 (w/Chg 1), fig. V-4. A Notional Joint Combat Operation Model.

The nature of operations and activities during a typical joint combat operation will change from its beginning (when the CJCS issues the execute order) to the operation's end (when the joint force disbands and components return to a pre-operation status). Shaping activities are usually ongoing and may continue during and after the operation. The purpose of shaping activities is to help set the conditions for successful execution of the operation. Figure V-4 shows, that from deter through enable civil authority, the operations and activities in these groups vary in magnitude—time, intensity, forces, etc.—as the operation progresses (the relative magnitudes in the figure are notional). At various points in time, each specific group might characterize the main effort of the joint force. For example, dominate activities would characterize the main effort after the joint force seizes the initiative until the enemy no longer is able to effectively resist. Even so, activities in the other groups would usually occur concurrently at some level of effort.

See following pages for further discussion and an overview of these activities.

Joint Operation Model (Example Activities)

Ref: JP 3-0 (w/Chg 1), Joint Operations (Oct '18), pp. V-17 to V-20.

Example Joint Operation Activities

A. Shape
B. Deter
C. Seize Initiative
D. Dominate
E. Stabilize
F. Enable Civil Authority

A. Shape

In general, shaping activities help set conditions for successful theater operations. Shaping activities include long-term persistent and preventive military engagement, security cooperation, and deterrence actions to assure friends, build partner capacity and capability, and promote regional stability. They help identify, deter, counter, and/or mitigate competitor and adversary actions that challenge country and regional stability. A GCC's CCP provides these and other activities tasked by SecDef/CJCS strategic guidance in pursuit of national objectives. Likewise, CCDRs may direct more focused geographic and functional shaping activities at the potential execution of specific contingency plans for various types of operations. In the best case, shaping activities may avert or diminish conflict. At the least, shaping provides a deeper, and common, understanding of the OE. Preparatory intelligence activities inform operation assessment, planning, and execution to improve the JFC's understanding of the OE.

Shaping activities are largely conducted through other interorganizational participants (e.g., USG departments and agencies, PNs), with DOD in a supporting role. Where US and PN interests converge, cooperation is possible. Some partners are quite capable already; others may benefit from US assistance. When a nation shares our interests and has the capacity to absorb US training, regional security can be increased. Military engagement and security cooperation activities are executed continuously to enhance international legitimacy and gain multinational cooperation. These activities should improve perceptions and influence adversaries' and allies' behavior, develop allied and friendly military capabilities for self-defense and multinational operations, improve information exchange and intelligence sharing, provide US forces with peacetime and contingency access, and positively affect conditions that could lead to a crisis. These activities prepare the OE in advance to facilitate access, should contingency operations be required. The joint community, in concert with multinational and interagency partners, must maintain and exercise strong regional partnerships as essential shaping activities in peacetime to ensure operational access during plan execution. For example, obtaining and maintaining rights of navigation and overflight help ensure global reach and rapid projection of military power.

B. Deter

Successful deterrence prevents an adversary's undesirable actions, because the adversary perceives an unacceptable risk or cost of acting. Deterrent actions are generally weighted toward protection and security activities that are characterized by preparatory actions to protect friendly forces, assets, and partners, and indicate the intent to execute subsequent phases of the planned operation. A number of FDOs, FROs, and force enhancements could be implemented to support deterrence. The nature of these options varies according to the nature of the adversary (e.g., traditional or irregular, state or non-state), the adversary's actions, US national objectives, and other factors. Once a crisis is defined, these actions may include mobilization, tailoring of forces, and other predeployment activities; initial deployment into a theater; employment of intelligence collection assets; and development of mission-tailored C2, intelligence, force protection, and logistic requirements to support the JFC's CONOPS. CCDRs continue to conduct military engagement with multinational partners to maintain access to areas, thereby providing the basis for further crisis response. Many deterrent actions build on security cooperation activities. They can also be part of stand-alone operations.

C. Seize Initiative

JFCs seek to seize the initiative in all situations through decisive use of joint force capabilities. In combat, this involves both defensive and offensive operations at the earliest possible time, forcing the enemy to culminate offensively and setting the conditions for decisive operations. Rapid application of joint combat power may be required to delay, impede, or halt the enemy's initial aggression and to deny the enemy its initial objectives. Operations to gain access to theater infrastructure and expand friendly freedom of action continue during this phase, while the JFC seeks to degrade enemy capabilities with the intent of resolving the crisis at the earliest opportunity.

D. Dominate

These actions focus on breaking the enemy's will to resist or, in noncombat situations, to control the OE. Successful domination depends on overmatching enemy capabilities at critical times and places. Joint force options include attacking weaknesses at the leading edge of the enemy's defensive perimeter to roll enemy forces back, and striking in depth to threaten the integrity of the enemy's A2/AD, offensive weapons and force projection capabilities, and defensive systems. Operations can range from large-scale combat to various stability actions depending on the nature of the enemy. Dominating activities may establish the conditions to achieve strategic objectives early or may set the conditions for transition to a subsequent phase of the operation.

E. Stabilize

These actions and activities are typically characterized by a shift in focus from sustained combat operations to stability activities. These operations help reestablish a safe and secure environment and provide essential government services, emergency infrastructure reconstruction, and humanitarian relief. The intent is to help restore local political, economic, and infrastructure stability. Civilian officials may lead operations during part or all of this period, but the JFC will typically provide significant supporting capabilities and activities. The joint force may be required to perform limited local governance (i.e., military government) and integrate the efforts of other supporting interagency and multinational partners until legitimate local entities are functioning. The JFC continuously assesses the impact of operations on the ability to transfer authority for remaining requirements to a legitimate civil entity.

F. Enable Civil Authority

Joint force support to legitimate civil governance typically characterizes these actions and activities. The commander provides this support by agreement with the appropriate civil authority. In some cases, especially for operations within the US, the commander provides this support under direction of the civil authority. The purpose is to help the civil authority regain its ability to govern and administer the services and other needs of the population. The military end state is typically reached during this phase, signaling the end of the joint operation. CCMD involvement with other nations and other government agencies beyond the termination of the joint operation, such as lower-level stability activities and FHA, may be required to achieve national objectives.

For additional discussion and alternative activity examples, see p. 2-56.

Joint Operations

V. Phasing a Joint Operation

The six general groups of activity in Figure V-4 provide a convenient basis for thinking about a joint operation in notional phases, as Figure V-7 depicts. A phase is a definitive stage or period during a joint operation in which a large portion of the forces and capabilities are involved in similar or mutually supporting activities for a common purpose that often is represented by intermediate objectives. Phasing, which can be used in any operation regardless of size, helps the JFC organize large operations by integrating and synchronizing subordinate operations. Phasing helps JFCs and staffs visualize, plan, and execute the entire operation and define requirements in terms of forces, resources, time, space, and purpose. It helps them systematically achieve military objectives that cannot be achieved all at once by arranging smaller, focused, related operations in a logical sequence. Phasing also helps commanders mitigate risk in the more dangerous or difficult portions of an operation.

Figure V-7 shows one phasing alternative. Actual phases of an operation will vary (e.g., compressed, expanded, or omitted entirely) according to the nature of the operation and the JFC's decisions. For example, UW operations normally use a seven-phase model. During planning, the JFC establishes conditions, objectives, and events for transitioning from one phase to another and plans sequels and branches for potential contingencies. Phases may be conducted sequentially, but some activities from a phase may begin in a previous phase and continue into subsequent phases. The JFC adjusts the phases to exploit opportunities presented by the enemy and operational situation or to react to unforeseen conditions.

Phasing an Operation

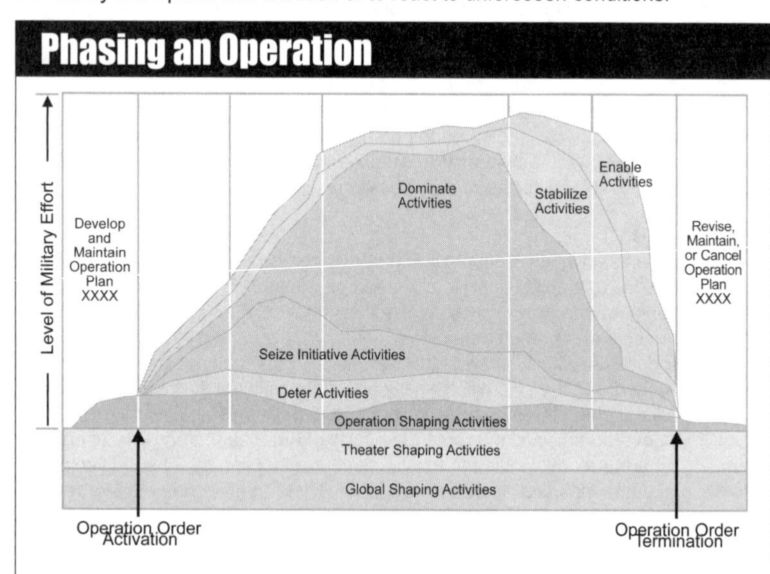

- The six general groups of activities provide a basis for thinking about a joint operation in notional phases.
- Phasing can be used in any joint operation regardless of size.
- Phasing helps joint force commanders and staffs visualize, plan, and execute the entire operation and define requirements in terms of forces, resources, time, space, and purpose to achieve objectives.

Ref: JP 3-0 (w/Chg 1), fig. V-7. Phasing an Operation Based on Predominant Military Activities.

The use of groups of activities for the purpose of phasing applies only to planning and executing individual joint operations, not to a GCC's theater campaign or strategy development.

A GCC's campaign encompasses all operations and activities for which the GCC is responsible, from relatively benign security cooperation activities through ongoing large-scale combat operations. All six groups of joint operation activities may be present in the GCC's AOR. However, use of the groups of activities for the purpose of phasing applies only to planning and executing individual operations, whether small-scale contingencies or large-scale campaigns that support the GCC's campaign. The groups of military activities associated with phases in Figure V-7 can serve as a frame of reference that facilitates common understanding among interagency and multinational partners and supporting commanders of how a JFC intends to execute a specific joint operation, as well as progress during execution.

Transitions

During execution, a transition marks a change between phases or between the ongoing operations and execution of a branch or sequel. This shift in focus by the joint force often is accompanied by changes in command or support relationships and priorities of effort. Transitions require planning and preparation well before their execution. The activities that predominate during a given phase rarely align with neatly definable breakpoints. The need to move into another phase normally is identified by assessing that a set of objectives has been achieved or that the enemy has acted in a manner that requires a major change for the joint force. Thus, the transition to a new phase is usually driven by events rather than time. An example is the shift from sustained combat operations to stability activities to enable civil authority. Through continuous assessment, the staff measures progress toward planned transitions so that the force prepares for and executes them.

Sometimes, however, the situation facing the JFC will change unexpectedly and without apparent correlation to a planned transition. The JFC may choose to shift operations to address unanticipated critical changes. The JFC must recognize fundamental changes in the situation and respond quickly and smoothly. Failure to do so can cause the joint force to lose momentum, miss important opportunities, experience setbacks, or even fail to accomplish the mission. Conversely, successful transitions enable the joint force to seize the initiative and quickly and efficiently garner favorable results. The JFC should anticipate transformations, as well as plan shifts, during operations.

COMMON OPERATING PRECEPT

Plan for and manage operational transitions over time and space.

VI. The Balance of Offense, Defense, and Stability Activities

Ref: JP 3-0 (w/Chg 1), Joint Operations (Oct '18), pp. V-15 to V-17.

Combat missions and tasks can vary widely depending on context of the operation and the objective. Most combat operations will require the commander to balance offensive, defensive, and stability activities. This is particularly evident in a campaign or major operation, where combat can occur during several phases and stability activities may occur throughout. Figure V-8 depicts notional proportions of offensive, defensive, and stability activities through the phases of a joint operation.

Notional Balance of Activities

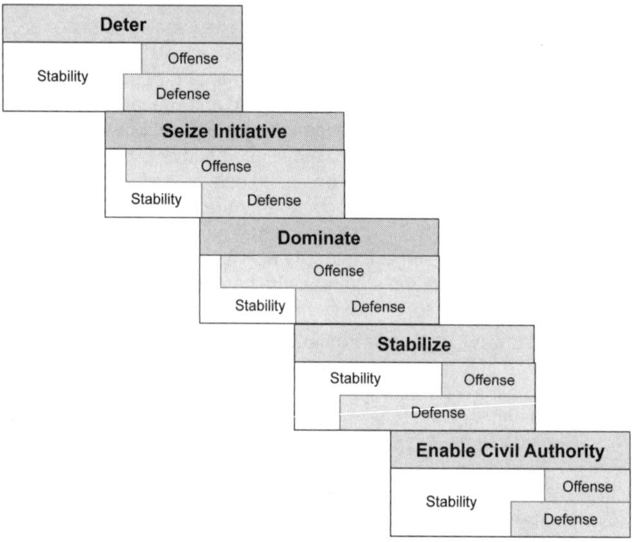

Deter
Offense
Stability
Defense

Seize Initiative
Offense
Stability
Defense

Dominate
Offense
Stability
Defense

Stabilize
Stability
Offense
Defense

Enable Civil Authority
Offense
Stability
Defense

NOTES:
The figure reflects a single operation.

Stability activities are conducted outside the United States. Department of Defense provides similar support to US civil authorities for homeland defense and other operations in the US through **defense support of civil authorities** operations.

Ref: JP 3-0 (w/Chg 1), fig. V-8. Notional Balance of Offense, Defense, and Stability Activities.

Offensive and Defensive Operations

Major operations and campaigns, whether they involve large-scale combat, normally include both offensive and defensive components (e.g., interdiction, maneuver, forcible entry, fire support, countering air and missile threats, DCO, base defense). Although defense may be the stronger form, offense is normally decisive in combat. To achieve military objectives quickly and efficiently, JFCs normally seek the earliest opportunity to conduct decisive offensive operations. Nevertheless, during a sustained offensive, selected elements of the joint force may need to pause, defend, resupply, or reconstitute, while other forces continue the attack. Accordingly, certain defensive measures and protection activities (e.g., OPSEC) are required throughout each joint operation phase. Joint forces at all levels should be capable of rapid transition between offense and defense and vice versa. The relationship between offense and defense, then, is a complementary one. Defensive operations enable JFCs to conduct or prepare for decisive offensive operations.

Stability Activities

Commanders conduct stability activities to maintain or reestablish a safe and secure environment and provide essential governmental services, emergency infrastructure reconstruction, and humanitarian relief. To achieve objectives and reach the desired military end state, JFCs integrate and synchronize stability activities with offense and defense, as necessary, during the phases of an operation. Stability activities support USG stabilization efforts and contribute to USG initiatives to build partnerships. These initiatives set the conditions to interact with partner, competitor, or adversary leaders, military forces, or relevant populations by developing and presenting information and conducting activities to affect their perceptions, will, behavior, and capabilities. The JFC will likely conduct stability activities in coordination with interorganizational participants and the private sector in support of HN authorities. Stability activities are conducted outside the US. DOD can provide similar support to US civil authorities through DSCA.

Balance and Simultaneity

Commanders strive to apply the many dimensions of military power simultaneously across the depth, breadth, and height of the OA. The challenge of balance and simultaneity affects all operations involving combat, particularly campaigns, due to their scope. Consequently, JFCs often concentrate in some areas or on specific functions and require economy of force in others. However, plans for major operations and campaigns will normally exhibit a balance between offense and defense and stability activities in various phases. Therefore, planning for stability activities should begin when joint operation planning begins.

Figure V-8 relates to Figure V-7 and the phasing explanation in paragraph 6, "Phasing a Joint Operation." Figure V-8 illustrates the notional balance between offensive and defensive actions and stability activities as an operation progresses. Since the focus of the CCMD's ongoing campaign is on prevention and preparation, any stability activities in the JFC's proposed OA might continue, and combat (offense and defense) may be limited or absent. Defensive measures might be limited to providing an increased level of security. A similar balance applies to deterrence activities, whether conducted as part of the CCP or on initiation of an OPLAN, since the intent is to limit escalation in the OA. A JFC might begin to limit stability activities if an adversary's potential combat actions are imminent. In combat operations, seize the initiative and dominate phases focus on offense and defense. Stability activities are likely restricted to parts of the OA away from immediate combat or might not occur at all. As the joint force achieves objectives and combat abates, the focus shifts to actions to stabilize and enable civil authority. Stability activities resume and will usually increase in proportion to the decrease in combat.

Planning for the transition from sustained combat operations to assumption of responsibility by civil authority, should begin during plan development and continue during all phases of a joint operation. Planning for redeployment should be considered early and continue throughout the operation and is best accomplished in the same time-phased process in which deployment was accomplished. An unnecessarily narrow focus on planning offensive and defensive operations in the dominate phase may threaten full development of the stabilize and enable civil authority phases and negatively affect joint operation momentum.

VII. Linear and Nonlinear Operations

Ref: JP 3-0 (w/Chg 1), Joint Operations (Oct '18), pp. V-17 to V-20.

A. Linear Operations

In linear operations, each commander directs and sustains combat power toward enemy forces in concert with adjacent units. Linearity refers primarily to the conduct of operations with identified forward lines of own troops (FLOTs). In linear operations, emphasis is placed on maintaining the position of friendly forces in relation to other friendly forces. From this relative positioning of forces, security is enhanced and massing of forces can be facilitated. Also inherent in linear operations is the security of rear areas, especially LOCs between sustaining bases and fighting forces. Protected LOCs, in turn, increase the endurance of joint forces and ensure freedom of action for extended periods. A linear OA organization may be best for some operations or certain phases of an operation. Conditions that favor linear operations include those where US forces lack the information needed to conduct nonlinear operations or are severely outnumbered. Linear operations also are appropriate against a deeply arrayed, echeloned enemy force or when the threat to LOCs reduces friendly force freedom of action. In these circumstances, linear operations allow commanders to concentrate and synchronize combat power more easily.

B. Nonlinear Operations

In nonlinear operations, forces orient on objectives without geographic reference to adjacent forces. Nonlinear operations typically focus on creating specific effects on multiple decisive points. Nonlinear operations emphasize simultaneous operations along multiple LOOs from selected bases (ashore or afloat). Simultaneity overwhelms opposing C2 and allows the JFC to retain the initiative. In nonlinear operations, sustaining functions may depend on sustainment assets moving with forces or aerial delivery. Noncombatants and the fluidity of nonlinear operations require careful judgment in clearing fires, both direct and indirect. Situational awareness, coupled with precision fires, frees commanders to act against multiple objectives. Swift maneuver against several decisive points supported by precise, concentrated fire can induce paralysis and shock among enemy troops and commanders.

During nonlinear offensive operations, attacking forces must focus offensive actions against decisive points, while allocating the minimum essential combat power to defensive operations. Reserves must have a high degree of mobility to respond where needed. JFCs may be required to dedicate combat forces to provide for LOC and base defense. Vulnerability increases as operations extend and attacking forces are exposed over a larger OA. Linkup operations, particularly those involving vertical envelopments, require extensive planning and preparation. The potential for friendly fire incidents increases due to the fluid nature of the nonlinear OA and the changing disposition of attacking and defending forces. The presence of civilians in the OA further complicates operations.

During nonlinear defensive operations, defenders focus on destroying enemy forces, even if it means losing physical contact with other friendly units. Successful nonlinear defenses require all friendly commanders to understand the JFCs intent and maintain a common operational picture (COP). Noncontiguous defenses are generally mobile defenses; however, some subordinate units may conduct area defenses to hold key terrain or canalize attackers into engagement areas. Nonlinear defenses place a premium on reconnaissance and surveillance to maintain contact with the enemy, produce relevant information, and develop and maintain a COP. The defending force focuses almost exclusively on defeating the enemy force rather than retaining large areas. Although less challenging than in offensive operations, LOC and sustainment security will still be a test and may require allocation of combat forces to protect LOCs and other high-risk functions or bases. The JFC must establish clear command relationships to properly account for the added challenges to base, base cluster, and LOC security

Combinations of Areas of Operations and Linear/Nonlinear Operations

JFCs consider incorporating combinations of contiguous and noncontiguous AOs with linear and nonlinear operations as they conduct operational design. They choose the combination that fits the OE and the purpose of the operation. Association of contiguous and noncontiguous AOs with linear and nonlinear operations creates these combinations:

Linear Operations in Contiguous AOs

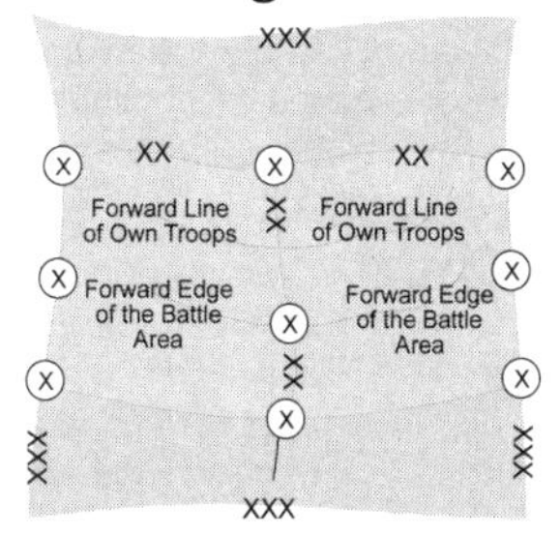

- Operation OVERLORD
- DESERT STORM

Linear operations in contiguous AOs (upper left-hand pane in Figure V-9) typify sustained offensive and defensive operations against powerful, echeloned, and symmetrically organized forces. The contiguous areas and continuous FLOT focus combat power and protect sustainment functions.

Linear Operations in Noncontiguous AOs

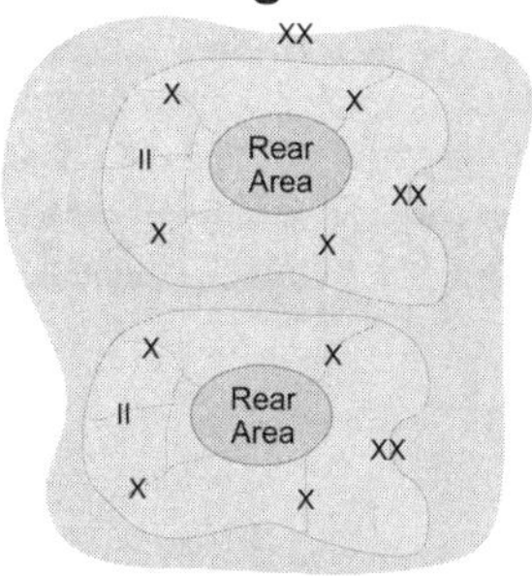

- Hue City
- URGENT FURY

This figure depicts a JFC's OA with subordinate component commanders conducting linear operations in noncontiguous AOs. In this case, the JFC retains responsibility for that portion of the OA outside the subordinate commanders' AOs.

Nonlinear Operations in Contiguous AOs

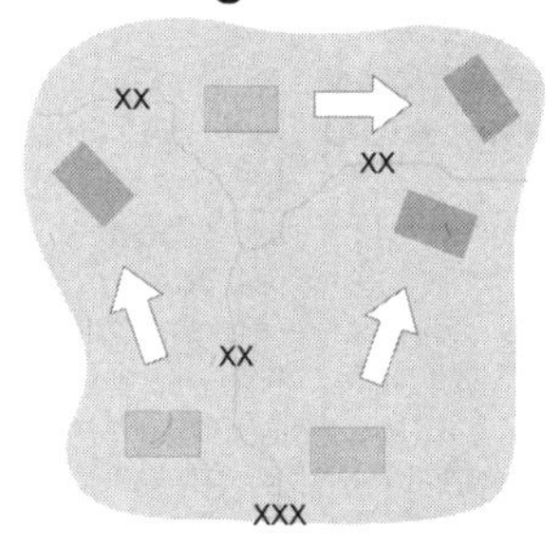

- Joint Task Force Andrew
- Vietnam 1965-73

This figure illustrates the JFC's entire assigned OA divided into subordinate AOs. Subordinate component commanders are conducting nonlinear operations within their AOs. This combination typically is applied in stability activities and DSCA actions.

Nonlinear Operations in Noncontiguous AOs

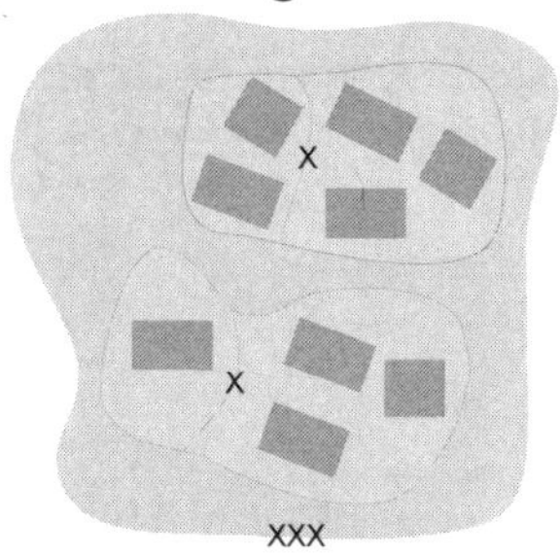

- JUST CAUSE
- RESTORE HOPE

This figure depicts a JFC's OA with subordinate component commanders conducting nonlinear operations in noncontiguous AOs. In this case, the JFC retains responsibility for that portion of the operational area outside the subordinate commanders' AOs.

Joint Operations

Additional Activity Examples

Ref: JP 3-0 (w/Chg 1), Joint Operations (Oct '18)w, pp. V-11 to V-12.

Some joint operations below the level of large-scale combat will have distinguishable groups of activity. However, activities may be compressed or absent entirely according to the nature of the operation. For example, deployment of forces associated with seize the initiative activities may have a deterrent effect sufficient to dissuade an enemy from conducting further operations, returning the OE to a more stable state. Likewise, although FID and NEO may occur as supporting operations to larger combat operations in the OA, they will have no evident dominating activities.

Figure V-5 shows a notional successful joint strike, which did not require follow-on operations. Figure V-6 shows a notional FHA operation that required predominantly stabilize and enable civil authority activities.

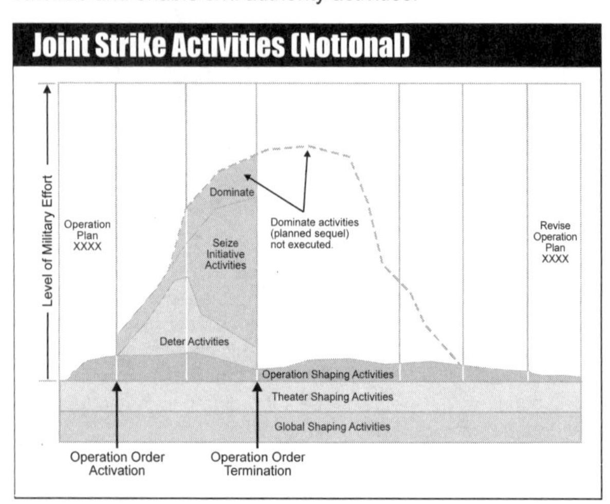

Ref: JP 3-0 (w/Chg 1), fig. V-5. Notional Balance of Activities for a Joint Strike.

This is a notional example of the balance of military activities in a successful operation to coerce the enemy to stop unacceptable behavior (e.g., state-sponsored terrorism, pursuit of nuclear weapons).

Examples include Operations EL DORADO CANYON (Libya, Apr 1986) and DESERT THUNDER (Iraq, Dec 1998).

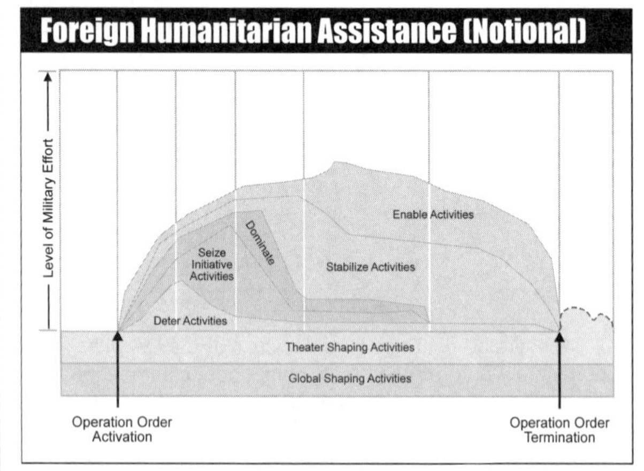

Ref: JP 3-0 (w/Chg 1), fig. V-6. Notional Balance of Activities for a Foreign Humanitarian Assistance Operation.

This is a notional example of the balance of military activities by a joint task force (JTF) responding to one type of crisis (foreign humanitarian assistance). There is no pre-existing operation plan in this example.

An example is Operation UNIFIED RESPONSE (Haiti, Jan-Mar 2010).

VI. Engagement, Security Cooperation & Deterrence

Ref: JP 3-0 (w/Chg 1), Joint Operations (Oct '18), chap. VI.

Military engagement, security cooperation, and deterrence missions, tasks, and actions encompass a wide range of actions where the military instrument of national power is tasked to support other instruments of national power as represented by interagency partners, as well as cooperate with international organizations (e.g., UN, NATO) and other countries to protect and enhance national security interests, deter conflict, and set conditions for future contingency operations. This may also involve domestic operations that include supporting civil authorities. These activities generally occur continuously in all GCCs' AORs regardless of other ongoing joint operations. Military engagement, security cooperation, and deterrence activities usually involve a combination of military forces and capabilities separate from but integrated with the efforts of interorganizational participants. These activities are conducted as part of a CCDR's routine CCP and country plan objectives and may support deterrence. Because DOS is frequently the major player in these activities, JFCs should maintain a working relationship with the DOS regional bureaus in coordination with the chiefs of the US diplomatic missions and country teams in their area. Commanders and their staffs should establish and maintain dialogue with HN government, multinational partners, and leaders of other organizations pertinent to their operation.

Projecting US military force invariably requires extensive use of international waters, international airspace, space, and cyberspace. Military engagement, security cooperation, and deterrence help assure operational access for crisis response and contingency operations despite changing US overseas defense posture and the growth of A2/AD capabilities around the globe. The more a GCC can promote favorable access conditions in advance across the AOR and in potential OAs, the better. Relevant activities include KLEs; security cooperation activities, such as bilateral and multinational exercises to improve multinational operations; missions to train, advise, and equip foreign forces to improve their national ability to contribute to access; negotiations to secure basing and transit rights, establish relationships, and formalize support agreements; the use of grants and contracts to improve relationships with and strengthen PNs; and planning conferences to develop multinational plans.

I. Military Engagement

Military engagement is the routine contact and interaction between individuals or elements of the Armed Forces of the United States and those of another nation's armed forces, or foreign and domestic civilian authorities or agencies, to build trust and confidence, share information, coordinate mutual activities, and maintain influence. Military engagement occurs as part of security cooperation but also extends to interaction with domestic civilian authorities. GCCs seek out partners and communicate with adversaries to discover areas of common interest and tension. This military

Refer to TAA2: Military Engagement, Security Cooperation & Stability SMARTbook (Foreign Train, Advise, & Assist) for further discussion. Topics include the Range of Military Operations (JP 3-0), Security Cooperation & Security Assistance (Train, Advise, & Assist), Stability Operations (ADRP 3-07), Peace Operations (JP 3-07.3), Counterinsurgency Operations (JP & FM 3-24), Civil-Military Operations (JP 3-57), Multinational Operations (JP 3-16), Interorganizational Cooperation (JP 3-08), and more.

engagement increases the knowledge base for subsequent decisions and resource allocation. Such military engagements can reduce tensions and may preclude conflict or, if conflict is unavoidable, allow a more informed USG to enter into it with stronger alliances or coalitions.

II. Security Cooperation

Security cooperation involves all DOD interactions with foreign defense establishments to build defense relationships that promote specific US security interests, develop allied and friendly military capabilities for self-defense and multinational operations, and provide US forces with peacetime and contingency access to the HN. The policy on which security cooperation is based resides in Presidential Policy Directive-23, Security Sector Assistance. This directive refers to the policies, programs, and activities the US uses to work with foreign partners and help shape their policies and actions in the security sector; help foreign partners build and sustain the capacity and effectiveness of legitimate institutions to provide security, safety, and justice for their people; and enable foreign partners to contribute to efforts that address common security challenges.

Security cooperation is a key element of global and theater shaping activities and critical aspect of communication synchronization. GCCs shape their AORs through security cooperation and stability activities by continually employing military forces to complement and support other instruments of national power that typically provide development assistance or humanitarian assistance to PNs. The GCC's CCP provides a framework within which CCMDs conduct cooperative security cooperation activities and development with PNs. Ideally, security cooperation activities mitigate the causes of a potential crisis before a situation deteriorates and requires US military intervention. Security assistance and security force assistance (SFA) normally provide some of the means for security cooperation activities.

III. Deterrence

Deterrence prevents adversary action through the presentation of a credible threat of unacceptable counteraction and belief that the cost of the action outweighs the perceived benefits. The nature of deterrent options varies according to the nature of the adversary (e.g., traditional or irregular, state or non-state), the adversary's actions, US national objectives, and other factors. Deterrence stems from an adversary's belief that the opponent's actions have created or can create an unacceptable risk to the adversary's achievement of objectives (i.e., the contemplated action cannot succeed or the costs are too high). Thus, a potential aggressor chooses not to act for fear of failure, risk, or consequences. Ideally, deterrent forces should be able to conduct decisive operations immediately. However, if available forces lack the combat power to conduct decisive operations, they conduct defensive operations while additional forces deploy. Effective deterrence requires a CCP and a coordinated communication effort that emphasize security cooperation activities with PNs that support US interests, DOD force posture planning, and contingency plans that prove the willingness of the US to employ forces in defense of its interests. Various joint operations (e.g., show of force and enforcement of sanctions) support deterrence by demonstrating national resolve and willingness to use force when necessary. Other CCP actions that help maintain or set the CCDR's desired conditions support deterrence by enhancing a climate of peaceful cooperation and FHA, thus promoting stability. Joint actions such as antiterrorism, DOD support to CD operations, show of force operations, and arms control are applied to meet military engagement, security cooperation, and deterrence objectives.

Sustained presence contributes to deterrence and promotes a secure environment in which diplomatic, economic, and informational programs designed to reduce the causes of instability can perform as designed. Presence can take the form of forward basing, forward deploying, or pre-positioning assets.

The Conflict Continuum

Ref: JP 3-0 (w/Chg 1), Joint Operations (Oct '18),pp. VI-1 to VI-3.

Military engagement, security cooperation, and deterrence activities provide the foundation of the CCDR's theater campaign. The goal is to prevent and deter conflict by keeping adversary activities within a desired state of cooperation and competition. The joint operation model described in section V, "Joint Operations Across the Conflict Continuum," has limited application with respect to phasing these activities for normal cooperative and competitive environments. Figure VI-1 shows a notional depiction of activities in an environment of cooperation and competition. DOD forces, as part of larger whole-of-government efforts, conduct operations with partners to prevent, deter, or turn back escalatory activity by adversaries.

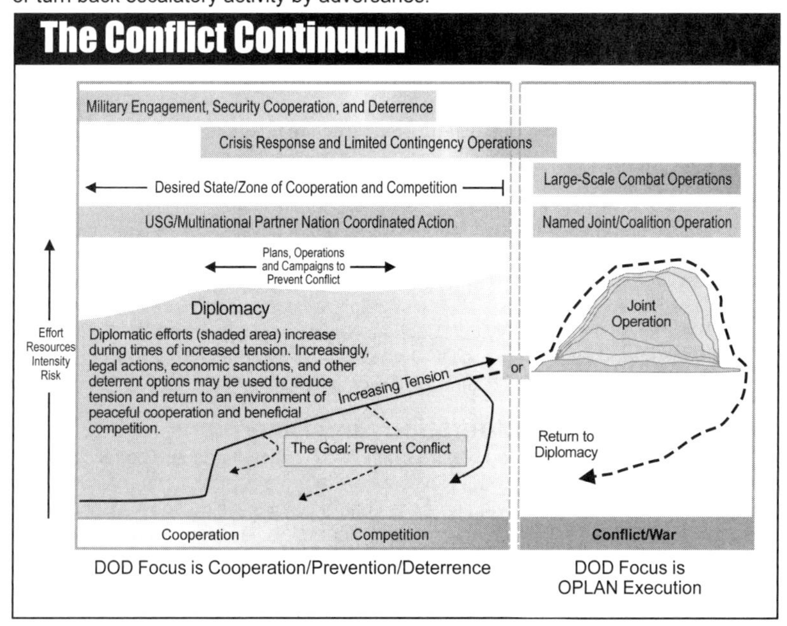

Ref: JP 3-0 (w/Chg 1), fig. VI-1. The Conflict Continuum.

Global and theater shaping increases DOD's depth of understanding of an environment, a partner's viewpoint of that environment, and where the US and PN have common interests. This understanding allows the US, through the relationships that have been developed, to shape the OE. These initiatives help advance national security objectives, promote stability, prevent conflicts (or limit their severity), and reduce the risk of employing US military forces in a conflict.

In an environment that is more competitive, tensions increase. A partner's resources can enhance USG understanding of an adversary's capabilities and intent and expand options against the adversary. In the best case, conflict can be averted or diminished by coordinated USG/PN action.

Despite the efforts to prevent or mitigate conflict, an armed conflict may occur. As conditions and objectives become more defined, GCCs may transition to the notional phasing construct for execution of a specific contingency operation as Figure VI-1 depicts. However, time spent "to the left" allows DOD to develop a deeper understanding of the environment to see and act ahead of conflict flashpoints, develop options, and maximize the efficiency of resources.

IV. Typical Operations & Activities

Ref: JP 3-0 (w/Chg 1), Joint Operations (Oct '18), pp. VI-5 to VI-13.

Typical operations include:

Military Engagement Activities

Numerous routine missions (e.g., security cooperation) and continuing operations or tasks (e.g., freedom of navigation) occur globally on a continuing basis under the general heading of military engagement. These activities build strong relationships with partners, increase regional awareness and knowledge of a PN's capabilities and capacity, and can be used to influence events in a desirable direction. Military engagement activities can also increase understanding of an adversary's capabilities, capacity, and intentions and can provide forewarning of undesirable events. In some cases, what begins as a military engagement activity (e.g., limited support to COIN through a security assistance program) can expand to a limited contingency operation or even a major operation when the President commits US forces. Military engagement activities are generally governed by various directives and agreements and do not require a joint OPLAN or OPORD for execution.

Emergency Preparedness

Emergency preparedness consists of measures taken in advance of an emergency to reduce the loss of life and property and to protect a nation's institutions from all types of hazards through a comprehensive emergency management program of preparedness, mitigation, response, and recovery. At the strategic level, emergency preparedness encompasses those planning activities, such as continuity of operations and continuity of government, undertaken to ensure DOD processes, procedures, and resources are in place to support the President and SecDef in a designated national security emergency.

Arms Control, Nonproliferation, and Disarmament

Arms control, nonproliferation, and disarmament are not synonymous. The following are examples of US military personnel involvement in arms control, nonproliferation, and disarmament activities: verifying an arms control treaty; seizing and securing WMD; escorting authorized deliveries of weapons and other materials (e.g., enriched uranium) to preclude loss or unauthorized use of these assets; conducting and hosting site inspections; participating in military data exchanges; implementing armament reductions; or dismantling, destroying, or disposing of weapons and hazardous material.

Combating Terrorism

Combating terrorism involves actions to oppose terrorism from all threats. It encompasses antiterrorism—defensive measures taken to reduce vulnerability to terrorist acts—and CT—offensive measures to prevent, deter, preempt, and respond to terrorism. Antiterrorism involves defensive measures used to reduce the vulnerability of individuals and property to terrorist acts, to include limited response and containment by local military forces and civilians. Counterterrorism (CT) is primarily a special operations core activity and consists of activities and operations taken to neutralize terrorists and their organizations and networks to render them incapable of using violence to instill fear and coerce governments or societies to achieve their goals.

Support to Counterdrug (CD) Operations

DOD supports federal, state, and local LEAs in their effort to disrupt the transport and/or transfer of illegal drugs into the US. Specific DOD authorities that pertain to a CD are contained in Title 10, USC, Sections 124 and 371 to 382.

Sanction Enforcement

Sanction enforcement is any operation that employs coercive measures to control the movement of designated items into or out of a nation or specified area. Maritime interception operations are efforts to monitor, query, and board merchant vessels in international waters to enforce sanctions against other nations such as those in support of UN Security Council resolutions and/or prevent the transport of restricted goods. These operations serve both strategic and military purposes. The strategic objective is to compel a country or group to conform to the objectives of the initiating body, while the military objective focuses on establishing a selective barrier that allows only authorized goods to enter or exit. Depending on the geography, sanction enforcement normally involves some combination of air and surface forces. Assigned forces should be capable of complementary mutual support and full communications interoperability.

Enforcement of Exclusion Zones

A sanctioning body establishes an exclusion zone to prohibit specified activities in a specific geographic area. Exclusion zones usually are imposed due to breaches of international standards of human rights or flagrant violations of international law regarding the conduct of states. Situations that may warrant such action include persecution of civil populations by a government and efforts by a hostile nation to acquire territory by force. Exclusion zones can be established in the air (no-fly zones), sea (maritime), or on land (no-drive zones). An exclusion zone's purpose may be to persuade nations or groups to modify their behavior to meet the desires of the sanctioning body or face continued imposition of sanctions or threat or use of force. Such measures are usually imposed by the UN or another international body of which the US is a member, although they may be imposed unilaterally by the US.

Freedom of Navigation and Overflight

Freedom of navigation operations are conducted to protect US navigation, overflight, and related interests on, under, and over the seas, against excessive maritime claims. Freedom of navigation is a sovereign right accorded by international law.) International law has long recognized that a coastal state may exercise jurisdiction and control within its territorial sea in the same manner it can exercise sovereignty over its own land territory. International law accords the right of "innocent" passage to ships of other nations through a state's territorial waters. Freedom of navigation by aircraft through international airspace is a well-established principle of international law.

Protection of Shipping

When necessary, US forces provide protection of US flag vessels, US citizens (whether embarked in US or foreign vessels), and US property against unlawful violence in and over international waters (such as Operation EARNEST WILL, in which Kuwaiti ships were reflagged under the US flag in 1987). This protection may be extended to foreign flag vessels under international law and with the consent of the flag state. Actions to protect shipping include coastal sea control, harbor defense, port security, countermine operations, and environmental defense, in addition to operations on the high seas. Protection of shipping, which is a critical element in the fight against piracy, requires the coordinated employment of surface, air, space, and subsurface units, sensors, and weapons, as well as a command structure both ashore and afloat and a logistic base. Protection of shipping may require a combination of operations to be successful. These actions can include area operations, escort duties, mine countermeasures, and environmental defense missions.

Show of Force Operations

Show of force operations are designed to demonstrate US resolve. They involve the appearance of a credible military force in an attempt to defuse a situation that, if allowed to continue, may be detrimental to US interests. These operations also underscore US commitment to our multinational partners.

Continued on next page

Continued on next page

Typical Operations & Activities (Cont.)

Continued from previous page

Foreign Assistance

Foreign assistance is civil or military assistance rendered to a nation by the USG within that nation's territory based on agreements mutually concluded between the US and that nation (e.g., Operation PROMOTE LIBERTY, in 1990, following Operation JUST CAUSE in Panama). Foreign assistance supports the HN by promoting sustainable development and growth of responsive institutions. The goal is to promote long-term regional stability.

Security Assistance

Security assistance is a group of programs by which the US provides defense articles, military training, and other defense-related services to foreign nations by grant, loan, credit, or cash sales in furtherance of national policies and objectives. These programs are funded and authorized by DOS to be administered by DOD and the Defense Security Cooperation Agency. They are an element of security cooperation.

Security Force Assistance (SFA)

SFA is DOD's contribution to unified action by the USG to support the development of the capacity and capability of foreign security forces (FSF) and their supporting institutions, to achieve objectives shared by the USG. SFA is conducted with and through FSF.

Foreign Internal Defense (FID)

FID encompasses participation by civilian and military agencies of a government in the action programs taken by another government or other designated organization to free and protect its society from subversion, lawlessness, insurgency, terrorism, and other threats to its security. USG support to FID can include diplomatic, economic, informational, and military development assistance to HN security sector and collaborative planning with multinational and HN authorities to anticipate, preclude, and counter those threats.

Humanitarian Assistance

Humanitarian assistance programs are governed by Title 10, USC, Section 401. This assistance may be provided in conjunction with military operations and exercises, but must fulfill unit training requirements that incidentally create humanitarian benefit to the local populace.

Support to Insurgency

The US may support insurgencies that oppose oppressive regimes. The US coordinates this support with its friends and allies. US military support is typically through UW, which includes activities to enable a resistance movement or insurgency to coerce, disrupt, or overthrow a government or occupying power by operating with an underground, auxiliary, and guerrilla force in a denied area.

Continued from previous page

Counterinsurgency Operations (COIN)

COIN operations include civilian and military efforts designed to support a government in the military, paramilitary, political, economic, psychological, and civic actions it undertakes to defeat insurgency and address its root causes. COIN operations often include security assistance programs such as foreign military sales, foreign military financing, and international military education and training. Such support may also include FID and SFA.

Refer to TAA2: Military Engagement, Security Cooperation & Stability SMARTbook (Foreign Train, Advise, & Assist) for further discussion. Topics include the Range of Military Operations (JP 3-0), Security Cooperation & Security Assistance (Train, Advise, & Assist), Stability Operations (ADRP 3-07), Peace Operations (JP 3-07.3), Counterinsurgency Operations (JP & FM 3-24), Civil-Military Operations (JP 3-57), Multinational Operations (JP 3-16), Interorganizational Cooperation (JP 3-08), and more.

VII. Crisis Response & Limited Contingency Ops

Ref: JP 3-0 (w/Chg 1), Joint Operations (Oct '18), chap. VII.

Crisis response and limited contingency operations typically are focused in scope and scale and conducted to achieve a very specific strategic or operational-level objective in an OA. They may be conducted as a stand-alone response to a crisis (e.g., NEO) or executed as an element of a larger, more complex operation. Joint forces conduct crisis response and limited contingency operations to achieve operational and, sometimes, strategic objectives.

CCDRs plan for various situations that require military operations in response to natural disasters, terrorists, subversives, or other contingencies and crises as directed by appropriate authority. The level of complexity, duration, and resources depends on the circumstances. Limited contingency operations ensure the safety of US citizens and US interests while maintaining and improving the ability to operate with multinational partners to deter hostile ambitions of potential aggressors. Many of these operations involve a combination of military forces and capabilities operating in close cooperation with interorganizational participants. APEX integrates planning into one unified construct to facilitate unity of effort and transition from planning to execution. Planning functions can be performed in series over a period of time or they can be compressed, performed in parallel, or truncated as appropriate.

Initial Response

When crises develop and the President directs, CCDRs respond. If the crisis revolves around external threats to a regional partner, CCDRs employ joint forces to deter aggression and signal US commitment (e.g., deploying joint forces to train in Kuwait). If the crisis is caused by an internal conflict that threatens regional stability, US forces may intervene to restore or guarantee stability (e.g., Operation RESTORE DEMOCRACY, the 1994 intervention in Haiti). If the crisis is within US territory (e.g., natural or man-made disaster, deliberate attack), US joint forces will conduct DSCA and HD operations as directed by the President and SecDef. Prompt deployment of sufficient forces in the initial phase of a crisis can preclude the need to deploy larger forces later. Effective early intervention can also deny an adversary time to set conditions in their favor, achieve destabilizing objectives, or mitigate the effects of a natural or man-made disaster.

Strategic Aspects

Two important aspects about crisis response and foreign limited contingency operations stand out. First, understanding the strategic objective helps avoid actions that may have adverse diplomatic or political effects. It is not uncommon in some operations, such as peacekeeping, for junior leaders to make decisions that have significant strategic implications. Second, commanders should remain aware of changes not only in the operational situation, but also in strategic objectives that may warrant a change in military operations.

Economy of Force

The strategic environment requires the US to maintain and prepare joint forces for crisis response and limited contingency operations simultaneously with other operations, preferably in concert with allies and/or PNs when appropriate. This approach recognizes these operations will vary in duration, frequency, intensity, and the number of personnel required. The burden of many crisis response and limited contingency operations may lend themselves to using small elements like SOF in coordination with allied nations or PNs.

I. Typical Crisis Response Operations

Ref: JP 3-0 (w/Chg 1), Joint Operations (Oct '18), pp. VII-2 to VII-5.

A. Noncombatant Evacuation Operations (NEO)

NEOs are operations directed by DOS or other appropriate authority, in conjunction with DOD, whereby noncombatants are evacuated from locations within foreign countries to safe havens designated by DOS when their lives are endangered by war, civil unrest, or natural disaster. Although principally conducted to evacuate US citizens, NEOs may also include citizens from the HN, as well as citizens from other countries. Pursuant to Executive Order 12656, Assignment of Emergency Preparedness Responsibilities, DOS is responsible for the protection and evacuation of US citizens abroad and for safeguarding their property. This order also directs DOD to advise and assist DOS to prepare and implement plans for the evacuation of US citizens. The US ambassador, or chief of the diplomatic mission, prepares the emergency action plans that address the military evacuation of US citizens and designated foreign nationals from a foreign country. The GCC conducts military operations to assist in the implementation of emergency action plans as directed by SecDef.

B. Peace Operations (PO)

PO are multiagency and multinational operations involving all instruments of national power—including international humanitarian and reconstruction efforts and military missions—to contain conflict, restore the peace, and shape the environment to support reconciliation and rebuilding and facilitate the transition to legitimate governance.

For the Armed Forces of the United States, PO encompass PKO, predominantly military PEO, predominantly diplomatic PB actions, PM processes, and conflict prevention. PO are conducted in conjunction with the various diplomatic activities and humanitarian efforts necessary to secure a negotiated truce and resolve the conflict. PO are tailored to each situation and may be conducted in support of diplomatic activities before, during, or after conflict. PO support national/multinational strategic objectives. Military support improves the chances for success in the peace process by lending credibility to diplomatic actions and demonstrating resolve to achieve viable political settlements.

- **Peacekeeping Operations (PKO).** PKO are military operations undertaken with the consent of all major parties to a dispute, designed to monitor and facilitate implementation of an agreement (cease fire, truce, or other such agreements) and support diplomatic efforts to reach a long-term political settlement. Such actions are often taken under the authority of Chapter VI, Pacific Settlement of Disputes, of the UN Charter.
- **Peace Enforcement Operations (PEO).** PEO are the application of military force or threat of its use, normally pursuant to international authorization, to compel compliance with resolutions or sanctions designed to maintain or restore peace and order. PEO may include the enforcement of sanctions and exclusion zones, protection of FHA, restoration of order, and forcible separation of belligerent parties or parties to a dispute. Unlike PKO, such operations do not require the consent of the states involved or of other parties to the conflict.
- **Peace Building (PB).** PB consists of stability actions (predominantly diplomatic, economic, and security related) that strengthen and rebuild governmental infrastructure and institutions, build confidence, and support economic reconstruction to prevent a return to conflict. Military support to PB may include rebuilding roads, reestablishing or creating government entities, or training defense forces.
- **Peacemaking (PM).** This is the process of diplomacy, mediation, negotiation, or other forms of peaceful settlement that arranges an end to a dispute or resolves issues that led to conflict. It can be an ongoing process, supported by military, economic, diplomatic, and informational instruments of national power. The purpose is to instill in the parties an understanding that reconciliation is a better alternative

to fighting. The military can assist in establishing incentives, disincentives, and mechanisms that promote reconciliation. Military activities that support PM include military-to-military exchanges and security assistance.

- **Conflict Prevention.** Conflict prevention consists of diplomatic and other actions taken in advance of a predictable crisis to prevent or limit violence, deter parties, and reach an agreement before armed hostilities. These actions are normally conducted under Chapter VI, Pacific Settlement of Disputes, of the UN Charter. However, military deployments designed to deter and coerce parties will need to be credible, and this may require a combat posture and an enforcement mandate under the principles of Chapter VII, Action with Respect to Threats to the Peace, Breaches of the Peace, and Acts of Aggression, of the UN Charter.

C. Foreign Humanitarian Assistance (FHA)

FHA operations relieve or reduce human suffering, disease, hunger, or privation in countries outside the US. These operations are different from foreign assistance primarily because they occur on short notice as a contingency operation to provide aid in specific crises or similar events rather than as more deliberate foreign assistance programs to promote long-term stability. DOS or the chief of mission in country is responsible for confirming the HN's declaration of a foreign disaster or situation that requires FHA. FHA provided by US forces is generally limited in scope and duration; it is intended to supplement or complement efforts of HN civil authorities or agencies with the primary responsibility for providing assistance. DOD provides assistance when the need for relief is gravely urgent and when the humanitarian emergency dwarfs the ability of normal relief agencies to effectively respond.

D. Recovery Operations

Recovery operations may be conducted to search for, locate, identify, recover, and return isolated personnel, sensitive equipment, items critical to national security, or human remains (e.g., JTF Full Accounting, which had the mission to achieve the fullest possible accounting of Americans listed as missing or prisoners of war from all past wars and conflicts). Regardless of the recovery purpose, each type of recovery operation is generally a sophisticated activity requiring detailed planning in order to execute. Recovery operations may be clandestine, covert, or overt depending on whether the OE is hostile, uncertain, or permissive.

E. Strikes and Raids

- **Strikes** are attacks conducted to damage or destroy an objective or a capability. Strikes may be used to punish offending nations or groups, uphold international law, or prevent those nations or groups from launching their own attacks (e.g., Operation EL DORADO CANYON conducted against Libya in 1986, in response to the terrorist bombing of US Service members in Berlin). Although often tactical in nature with respect to the ways and means used and duration of the operation, strikes can achieve strategic objectives as did the strike against Libya.
- **Raids** are operations to temporarily seize an area, usually through forcible entry, in order to secure information, confuse an enemy, capture personnel or equipment, or destroy an objective or capability. Raids end with a planned withdrawal upon completion of the assigned mission.

Refer to TAA2: Military Engagement, Security Cooperation & Stability SMARTbook (Foreign Train, Advise, & Assist) for further discussion. Topics include the Range of Military Operations (JP 3-0), Security Cooperation & Security Assistance (Train, Advise, & Assist), Stability Operations (ADRP 3-07), Peace Operations (JP 3-07.3), Counterinsurgency Operations (JP & FM 3-24), Civil-Military Operations (JP 3-57), Multinational Operations (JP 3-16), Interorganizational Cooperation (JP 3-08), and more.

II. Unique Considerations (Crisis Response and Limited Contingency Operations)

Ref: JP 3-0 (w/Chg 1), Joint Operations (Oct '18), pp. VII-6 to VII-8.

Duration and End State

Crisis response and limited contingency operations may be relatively short in duration (e.g., NEO, strike, raid) or last for an extended period to achieve the national objective (such as US participation with ten other nations in the independent [non-UN] peace-keeping operation, Multinational Force and Observers, in the Sinai Peninsula since 1982). Short duration operations are not always possible, particularly in situations where destabilizing conditions have existed for years or where conditions are such that a long-term commitment is required to achieve national strategic objectives. Nevertheless, it is imperative to have clear national objectives for all types of contingencies.

Intelligence

As soon as practical, JFCs and their staffs determine intelligence requirements to support the anticipated operation. Intelligence planners also consider the capability for a unit to receive external intelligence support, the capability to store intelligence data, the timeliness of intelligence products, the availability of intelligence publications, and the possibility of using other agencies and organizations as intelligence sources. In some contingencies (e.g., PKO), the term information collection is used rather than the term intelligence because of the operation's sensitivity.

- **Human Intelligence (HUMINT).** HUMINT may often provide the most useful source of information and is essential to understanding an enemy or adversary. If a HUMINT infrastructure is not in place when US forces arrive, it needs to be established as quickly as possible. HUMINT also complements other intelligence sources with information not available through technical means. For example, while overhead imagery may graphically depict the number of people gathered in a town square, it cannot gauge the motivations or enthusiasm of the crowd. Additionally, in underdeveloped areas, belligerent forces may not rely heavily on radio communications, thereby denying US forces intelligence derived through signal intercept.
- **Open-Source Intelligence (OSINT).** Where there is little USG or US military presence, open-source intelligence (OSINT) may be the best immediately available information to prepare US forces to operate in a foreign country. OSINT from broadcasts, print media, and social networks may be the best immediately available information to provide tip-offs for HUMINT and other intelligence and information collection methods.
- **Open-Source Intelligence (GEOINT).** GEOINT consists of the exploitation and analysis of imagery and geospatial information to describe, assess, and visually depict physical features and geographically referenced activities. GEOINT consists of imagery, imagery intelligence, and geospatial information.

Tailored products based on continuous JIPOE can promote timely and comprehensive understanding of all aspects of the OE needed for crisis response and limited contingency operations.

Intelligence organizations (principally at the JTF HQ level) should include foreign area officers. Due to extensive training and experience working in foreign countries as defense attachés and in defense support to US embassy operations, foreign area officers add valuable cultural awareness and insights to intelligence products.

Operational Limitations

A JFC tasked with conducting or supporting a crisis response or limited contingency operation may face numerous constraints, restraints, and ROE based on the specific circumstances. For example, international acceptance of each operation may be extremely important, not only because military forces may be used to support international sanctions but also because of the probability of involvement by international organizations. As a consequence, legal and fiscal constraints unique to the operation should be addressed in detail by the CCDR's staff. Also, operational limitations imposed on any agency or organization involved in the operation should be clarified for other agencies and organizations to facilitate coordination.

Force Protection

Limited contingency operations may involve a requirement to protect nonmilitary personnel. In the absence of the rule of law, the JFC must address when, how, and to what extent he will extend force protection to civilians and what that protection means.

Training

Participation in certain types of smaller-scale contingencies may preclude normal mission-related training. For example, infantry units or fighter squadrons conducting certain protracted PO may not have the time, facilities, or environment in which to maintain individual or unit proficiency for traditional missions. In these situations, commanders should develop programs that enable their forces to maintain proficiency in their core competencies/mission essential tasks to the greatest extent possible.

Interorganizational Response Considerations

JP 3-08 describes the joint force commander's (JFC's) coordination with various external organizations that may be involved with, or operate simultaneously with, joint operations. This coordination includes the Armed Forces of the United States; United States Government (USG) departments and agencies; state, territorial, local, and tribal government agencies; foreign military forces and government agencies; international organizations; nongovernmental organizations (NGOs); and the private sector. Interagency coordination describes the interaction between USG departments and agencies and is a subset of interorganizational cooperation.

Although planning is conducted in anticipation of future events, there may be crisis situations that call for an immediate US military response (e.g., noncombatant evacuation operation or FHA). CCDRs frequently develop courses of action (COAs) based on recommendations and considerations originating in DOS joint/regional bureaus or in one or more US embassies. The country team provides resident agency experience and links through the CCMD and by extension to agency HQ in Washington, DC. Emergency action plans at every US embassy cover a wide range of contingencies and crises and can assist the commanders in identifying COAs, options, and constraints to military actions and support activities. The GCC's staff also consults with JS and other organizations to coordinate military operations and synchronize actions at the national strategic and theater strategic levels.

See pp. 8-7 to 8-14 for a discussion of joint planning and interorganizational cooperation considerations and pp. 8-19 to 8-28 for a discussion of foreign considerations to include USG structure in foreign countries, foreign operations, stakeholders, crisis action organization, joint task force considerations and civil-military teaming.

III. Homeland Defense and Defense Support of Civil Authorities (DSCA)

Ref: JP 3-0 (w/Chg 1), Joint Operations (Oct '18), pp. VII-5 to VII-6.

Security and defense of the US homeland is the USG's top responsibility and is conducted as a continuous, cooperative effort among all federal agencies, as well as state, tribal, and local government. Military operations inside the US and its territories, though limited in many respects, are conducted to accomplish two missions—HD and DSCA.

Commander, US Northern Command, and Commander, US Pacific Command, have specific responsibilities for HD and DSCA. These responsibilities include conducting operations to deter, prevent, and defeat threats and aggression aimed at the US, its territories, and interests within their assigned AORs, as directed by the President or SecDef. However, DOD support to HD is global in nature and is often conducted by all CCDRs beginning at the source of the threat. In the forward regions outside US territories, the objective is to detect and deter threats to the homeland before they arise and to defeat these threats as early as possible when so directed.

Homeland Defense (HD)

HD is the protection of US sovereignty, territory, domestic population, and critical defense infrastructure against external threats and aggression or other threats as directed by the President. DOD is the federal agency with lead responsibility, supported by other agencies, to defend against external threats and aggression.

However, against internal threats DOD may be in support of another USG department or agency. When ordered to conduct HD operations within US territory, DOD will coordinate closely with other government agencies. Consistent with laws and policy, the Services will provide capabilities to support CCDR requirements against a variety of threats to national security. These include invasion, cyberspace attack, and air and missile attacks.

Defense Support of Civil Authorities (DSCA)

DSCA is support provided by US federal military forces; DOD civilians, DOD contract personnel, DOD component assets, DOD agencies, and National Guard forces (when SecDef, in coordination with the governors of the affected states, elects and requests to use those forces in Title 32, USC status) in response to requests for assistance from civil authorities for domestic emergencies, law enforcement support, and other domestic activities, or from qualifying entities for special events. For DSCA operations, DOD supports and does not supplant civil authorities. The majority of DSCA operations are conducted IAW the NRF, which establishes a comprehensive, national, all-hazards approach to domestic incident response. Within a state, that state's governor is the key decision maker and commands the state's National Guard forces when they are not in federal Title 10, USC, status. When the governor mobilizes the National Guard, it will most often be under state active duty when supporting civil authorities.

Other DSCA operations can include CD activities, support to national special security events, or other support to civilian law enforcement IAW specific DOD policies and US law. Commanders and staffs must carefully consider the legal and policy limits imposed on intelligence activities in support of LEAs, and on intelligence activities involving US citizens and entities by intelligence oversight regulations, policies, and executive orders.

Refer to The Homeland Defense & DSCA SMARTbook (Protecting the Homeland / Defense Support to Civil Authority) for complete discussion. Topics and references include homeland defense (JP 3-28), defense support of civil authorities (JP 3-28), Army support of civil authorities (ADRP 3-28), multi-service DSCA TTPs (ATP 3-28.1/MCWP 3-36.2), DSCA liaison officer toolkit (GTA 90-01-020), key legal and policy documents, and specific hazard and planning guidance.

VIII. Large-Scale Combat Operations

Ref: JP 3-0 (w/Chg 1), Joint Operations (Oct '18), chap. VIII.

Campaign

Traditionally, campaigns are the most extensive joint operations, in terms of the amount of forces and other capabilities committed and duration of operations. In the context of large-scale combat, a campaign is a series of related major operations aimed at achieving strategic and operational objectives within a given time and space.

Major Operations

A major operation is a series of tactical actions, such as battles, engagements, and strikes, and is the primary building block of a campaign. Major operations and campaigns typically include multiple phases (e.g., the 1990-1991 Operations DESERT SHIELD and DESERT STORM and 2003 OIF). Campaign planning is appropriate when the contemplated military operations exceed the scope of a single major operation.

Campaigns can occur across the continuum of conflict. In campaigns characterized by combat, the general goal is to prevail against the enemy as quickly as possible; conclude hostilities; and establish conditions favorable to the HN, the US, and its multinational partners. Establishing these conditions may require joint forces to conduct stability activities to restore security, provide essential services and humanitarian relief, and conduct emergency reconstruction. Some crisis response or contingency operations may not involve large-scale combat but could meet the definition of a major operation or campaign based on their scale and duration (e.g., the Tsunami relief efforts in Indonesia or Hurricane Katrina relief efforts in the US, both in 2005).

Campaigns are joint in nature—functional and Service components of the joint force conduct supporting operations, not independent campaigns. Within a campaign, forces of a single or several Services, coordinated in time and space, conduct operations to achieve strategic or operational objectives in one or more OAs. Forces operate simultaneously or sequentially IAW a common plan, and are controlled by a single Service commander or the JFC.

Combatant Command (CCMD) Planning

The CCMD strategy links national strategic guidance to development of CCMD campaign and contingency plans. A CCMD strategy is a broad statement of the GCC's long-term vision for the AOR and the FCC's long-term vision for the global employment of functional capabilities. CCDRs prepare these strategies in the context of SecDef's priorities outlined in the GEF and the CJCS's objectives articulated in the NMS. However, the size, complexity, and anticipated duration of operations typically magnify the planning challenges. There are three categories of campaigns, which differ generally in scope and focus.

CCDRs document the full scope of their campaigns in the set of plans that includes the CCP or FCP and all of its GEF- and JSCP-directed plans, subordinate and supporting plans, posture or master plans, country plans (for the geographic CCMDs), OPLANs of operations currently in execution, and contingency plans.

See chap. 3 for further discussion.

Setting Conditions for Theater Operations

Ref: JP 3-0 (w/Chg 1), Joint Operations (Oct '18), pp. VIII-2 to VIII-6.

CCDRs and JFCs execute their campaigns and operations in pursuit of US national objectives and to shape the OE. In pursuit of national objectives, these campaigns and operations also seek to prevent, prepare for, or mitigate the impact of a crisis or contingency. In many cases, these actions enhance bonds between potential multinational partners, increase understanding of the region, help ensure access when required, and strengthen the capability for future multinational operations, all of which help prevent crises from developing.

Organizing and Training Forces

Organizing and, where possible, training forces to conduct operations throughout the OA can be a deterrent. JTFs and components that are likely to be employed in theater operations should be exercised regularly during peacetime.

Staffs should be identified and trained for planning and controlling joint and multinational operations. The composition of joint force staffs should reflect the composition of the joint force to ensure those employing joint forces have thorough knowledge of their capabilities and limitations. When possible, JFCs and their staffs should invite non-DOD agencies to participate in training to facilitate a common understanding and to build a working relationship prior to actual execution. Commanders must continue to refine interactions with interagency partners they will work with most often and develop common procedures to improve interoperability. When it is not possible to train forces in the theater of employment, as with US-based forces with multiple tasks, commanders should make maximum use of regularly scheduled and ad hoc exercise opportunities. The training focus for all forces and the basis for exercise objectives should be the CCDR's joint mission-essential tasks.

Rehearsals

Rehearsal provides an opportunity to learn, understand, and practice a plan in the time available before actual execution. Rehearsing key combat and sustainment actions allows participants to become familiar with the operation, visualize the plan, and identify possible friction points. This process orients joint and multinational forces to surroundings and to other units during execution. Rehearsals also provide a forum for subordinate leaders to analyze the plan, but they must exercise caution in adjusting the plan. Changes must be coordinated throughout the chain of command to prevent errors in integration and synchronization. HQ at the tactical level often conduct rehearsals involving participation of maneuver forces positioned on terrain that mirrors the OE. HQ at the operational level rehearse key aspects of a plan using command post exercises, typically supported by computer-aided simulations. While the joint force may not be able to rehearse an entire operation, the JFC should identify essential elements for rehearsal.

Maintaining Operational Access

JFCs must overcome the enemy's A2/AD capabilities to establish and maintain access to OAs where they are likely to operate, ensuring forward presence, basing (to include availability of airfields and seaports and adequate sustainment), resiliency of combat power after enemy action, freedom of navigation, and cooperation with allied and/or coalition nations to enhance operational reach. In part, this effort is national or multinational, involving maintenance of intertheater (between theaters) air, land, sea, space, EMS, and cyberspace LOCs. Supporting CCDRs can greatly enhance this effort.

Space Considerations

Space operations support all joint operations. Prior to and during conflict, commanders need to ensure US, allied, and/or multinational forces gain and maintain space superiority. Commanders must anticipate and mitigate hostile actions that may affect friendly space operations. Commanders should also anticipate the proliferation and increasing sophistication of commercial space capabilities and products available that the commander can leverage but which may also be available to enemies and adversaries. USSTRATCOM plans and conducts space operations. The GCC, in coordination with other USG departments and agencies, conducts certain aspects of theater space operations, to include planning for, supporting, and conducting the recovery of astronauts, space vehicles, space payloads, and objects as directed. They may also request the CDRUSSTRATCOM's assistance in integrating space forces, capabilities, and considerations into each phase of campaign and major OPLANs. Global and theater space operations require robust planning and skilled employment to synchronize and integrate space operations with the joint operation. It is therefore incumbent upon the GCCs to coordinate as required to minimize conflicts. Space capabilities help shape the OE in a variety of ways, including providing intelligence and communications necessary to keep commanders and leaders informed worldwide. JFCs and their components should request space support early in the planning process to ensure effective and efficient use of space assets.

Electromagnetic Spectrum (EMS) Considerations

The joint force is critically dependent on the EMS for operations across all joint functions and throughout the OE. For example, modern C2 requires operation of EMS-dependent sensing and communication systems, while advanced weapons rely on PNT information transmitted through the EMS. Therefore, the joint force should strive for local EMS superiority prior to executing joint operations. EMS superiority is that degree of dominance in the EMS that permits the conduct of operations at a given time and place without prohibitive interference, while affecting an adversary's ability to do the same. Achieving EMS superiority is complicated by increasing joint EMS-use requirements, EME congestion, and proliferation of EMS threats. Joint forces execute JEMSO, facilitated by electromagnetic battle management (EMBM), to achieve the necessary unity of effort for EMS superiority.

Stability Activities

Activities in the shape phase may focus on continued planning and preparation for anticipated stability activities in the subsequent phases. These activities should include conducting collaborative interagency planning to synchronize the civil-military effort, confirming the feasibility of pertinent military objectives and the military end state, and providing for adequate intelligence, an appropriate force mix, and other capabilities. US military support to stabilization efforts in this phase may be required as part of the USG's security sector assistance, purposed to quickly restore security and infrastructure or provide humanitarian relief in select portions of the OA to dissuade further adversary actions or to help gain and maintain access and future success.

Physical Environment

Weather, terrain, and sea conditions can significantly affect operations and sustainment support of the joint force and should be carefully assessed before and during sustained combat operations. Urban areas possess all of the characteristics of the natural landscape, coupled with man-made construction and the associated infrastructure, resulting in a complicated and dynamic environment that influences the conduct of military operations in many ways. Control of the littoral area often is essential to maritime superiority. The electromagnetic spectrum (EMS), which has become increasingly complex, contested, and congested as technology has advanced, can significantly affect joint force operations.

I. Considerations for Deterrence

Deterrence is characterized by preparatory actions that indicate resolve to commit resources and respond to the situation. These actions begin when a CCDR or JFC identifies that routine operations may not achieve desired objectives due to an adversary's actions. This requires the commander to have identified CCIRs and assessed whether additional resources, outside those currently allocated and assigned for ongoing operations, are required to defuse the crisis, reassure partners, demonstrate the intent to deny the adversary's goals and execute subsequent phases of the operation. Deterrence should be based on capability (having the means to influence behavior), credibility (maintaining a level of believability that the proposed actions may actually be employed), and communication (transmitting the intended message to the desired audience) to ensure greater effectiveness (effectiveness of deterrence must be viewed from the perspective of the agent/actor that is to be deterred). Before hostilities begin, the JFC and staff analyze and assess the adversary's goals and decision-making process to determine how, where, and when these can be affected and what friendly actions (military and others) can influence events and act as a deterrent.

Isolating the Enemy

With Presidential and SecDef approval, guidance, and national support, JFCs strive to isolate enemies by denying them allies and sanctuary. The intent is to strip away as much enemy support or freedom of action as possible while limiting the enemy's potential for horizontal or vertical escalation. JFCs may also be tasked by the President and SecDef to support diplomatic, economic, and informational actions.

The JFC also seeks to isolate the main enemy force from both its strategic leadership and its supporting infrastructure. Such isolation can be achieved through the use of information-related activities and the physical interdiction of LOCs or resources affecting the enemy's ability to conduct or sustain military operations.

Flexible Deterrent Options/ Flexible Response Options (FDOs/FROs)

FDOs and FROs are executed on order and provide scalable options to respond to a crisis. Both provide the ability to scale up (escalate) or de-escalate based on continuous assessment of an adversary's actions and reaction. While FDOs are primarily intended to prevent the crisis from worsening and allow for de-escalation, FROs are generally punitive in nature.

See following pages (pp. 2-74 to 2-75) for further discussion.

Protection

JFCs must protect their forces and their freedom of action to accomplish their mission. This dictates that JFCs not only provide force protection but be aware of and participate as appropriate in the protection of interagency and regional multinational capabilities and activities. JFCs may spend as much time on protection to assure partners to preserve coalition resolve and maintain access as on direct preparation of their forces for combat.

Space Operations

JFCs depend upon and exploit the advantages of space capabilities. During the deter phase, space forces are limited to already fielded and immediately deployable assets and established priorities for service.

GEOINT Support to Operations

Geospatial products or services—including maps, charts, imagery products, web services, and support data—must be fully coordinated with JFC components, as well as with the Joint Staff, Office of the Secretary of Defense, and the NGA through the JFC's GEOINT cell.

Preparing the Operational Area (Deterrence)

Ref: JP 3-0 (w/Chg 1), Joint Operations (Oct '18), pp. VIII-7 to VIII-8.

Special Operations (SOF)

SOF play a major role in preparing and shaping the operational area and environment by setting conditions which mitigate risk and facilitate successful follow-on operations. The regional focus, cross-cultural/ethnic insights, language capabilities, and relationships of SOF provide access to and influence in nations where the presence of conventional US forces is unacceptable or inappropriate. SOF contributions can provide operational leverage by gathering critical information, undermining an adversary's will or capacity to wage war, and enhancing the capabilities of conventional US, multinational, or indigenous/surrogate forces.

Stability Activities

Joint force planning and operations conducted prior to commencement of hostilities should establish a sound foundation for operations in the stabilize and enable civil authority phases. JFCs should anticipate and address how to fill the power vacuum created when sustained combat operations wind down. Considerations include:

- Limit the damage to key infrastructure (water, energy, medical) and services.
- Assist with the restoration and development of power generation facilities.
- Establish the intended disposition of captured leadership and demobilized military and paramilitary forces.
- Provide for the availability of cash or other means of financial exchange.
- Determine the proper force mix (e.g., combat, military police, CA, engineer, medical, multinational).
- Assess availability of HN law enforcement and health and medical resources.
- Secure key infrastructure nodes and facilitate HN law enforcement and first responder services.
- Develop and disseminate information necessary to suppress potential new enemies and promote new governmental authority.

Civil Affairs (CA)

CA forces have a variety of specialty skills that may support the joint operation being planned. CA forces conduct military engagement, humanitarian and civic assistance, and nation assistance to influence HN and foreign nation populations. CA forces assess impacts of the population and culture on military operations, assess impact of military operations on the population and culture, and facilitate interorganizational coordination. Establishing and maintaining civil-military relations may include interaction among US, allied, multinational, and HN forces, as well as other government agencies, international organizations, and NGOs.

Sustainment

Thorough planning for logistic and personnel support is critical. Planning must include active participation by all deploying and in-theater US and multinational forces, as well as interagency personnel. This planning is done through theater distribution plans (TDPs) in support of the GCCs' TCPs. Setting the conditions enables the JFCs to address global, end-to-end distribution requirements and identify critical capabilities, infrastructure, and relationships required to be resourced and emplaced in a timely manner to sustain and enable global distribution operations.

See chap. 4 for further discussion.

Flexible Deterrent Options (FDOs)

Ref: JP 3-0 (w/Chg 1), Joint Operations (Oct '18), pp. VIII-8 to VIII-9. (FDO examples from JP 5-0, Joint Planning (Jun '17), app. F.)

Flexible Deterrent Options (FDOs) are preplanned, deterrence-oriented actions carefully tailored to bring an issue to early resolution without armed conflict. Both military and nonmilitary FDOs can be used to dissuade actions before a crisis arises or to deter further aggression during a crisis. FDOs are developed for each instrument of national power, but they are most effective when used in combination.

- **Military FDOs** can be initiated before or after unambiguous warning of enemy action. Deployment timelines, combined with the requirement for a rapid, early response, generally require economy of force; however, military FDOs should not increase risk to the force that exceeds the potential benefit of the desired effect. Military FDOs must be carefully tailored regarding timing, efficiency, and effectiveness. They can rapidly improve the military balance of power in the OA, especially in terms of early warning, intelligence gathering, logistic infrastructure, air and maritime forces, MISO, and protection without precipitating armed response from the adversary. Care should be taken to avoid undesired effects such as eliciting an armed response should adversary leadership perceive that friendly military FDOs are being used as preparation for a preemptive attack.
- **Nonmilitary FDOs** are preplanned, preemptive actions taken by other government agencies to dissuade an adversary from initiating hostilities. Nonmilitary FDOs need to be coordinated, integrated, and synchronized with military FDOs to focus all instruments of national power.

Example Diplomatic FDOs

- Alert and introduce special teams (e.g., public diplomacy)
- Reduce international diplomatic ties
- Initiate noncombatant evacuation opns
- Restrict activities of diplomatic missions
- Prepare to withdraw or withdraw US embassy personnel
- Take actions to gain international support
- Restrict travel of US citizens
- Gain support through the UN

Example Military FDOs

- Upgrade alert status
- Increase ISR activities
- Initiate show-of-force actions
- Increase training & exercise activities
- Take steps to increase U.S. public support
- Increase defense support to public diplomacy
- Deploy forces into or near the potential operational area

Example Informational FDOs

- Increase public awareness of the problem and potential for conflict
- Interrupt satellite downlink transmissions
- Publicize violations of international law
- Publicize increased force presence, joint exercises, military capability
- Increase informational efforts
- Implement meaconing, interference, jamming, and intrusion of enemy informational assets
- Maintain dialogue with the news media

Example Economic FDOs

- Freeze or seize real property in the United States where possible
- Freeze monetary assets in the U.S.
- Freeze international assets where possible
- Encourage US and international corporations to restrict transactions
- Embargo goods and services
- Enact trade sanctions
- Enact restrictions on technology transfer

Flexible Response Options (FROs)

Ref: JP 3-0 (w/Chg 1), Joint Operations (Oct '18), pp. VIII-8 to VIII-9 and JP 5-0, Joint Planning (Jun '17), pp. F-5 to F-8.

Flexible Response Options (FROs), usually used in response to terrorism, can also be employed in response to aggression by a competitor or adversary. Like FDOs, the discussion should include indicators of their effectiveness and probability of consequences, desired and undesired. The basic purpose of FROs is to preempt and/or respond to attacks against the US and/or US interests. FROs are intended to facilitate early decision making by developing a wide range of prospective actions carefully tailored to produce desired effects, congruent with national security policy objectives. An FRO is the venue in which various military capabilities are made available to the President and SecDef, with actions appropriate and adaptable to existing circumstances, in reaction to any threat or attack.

Flexible Response Option Scalability

Rapid Response Demonstrate Resolve	Limited Response Target Those Directly Responsible	Decisive Response Defeat Violent Extremist Organization
Priority of Effort: • Speed	**Priority of Effort:** • Legitimacy via attribution	**Priority of Effort:** • Direct attack on enemy center of gravity
Advantages: • Demonstrate resolve • Least impact of current operations	**Advantages:** • Response aimed directly at those responsible • Demonstrates restraint • International cooperation more likely	**Advantages:** • Proactive vice reactive • Targets critical enemy vulnerabilities • Greater impact on enemy
Disadvantages: • Limited strategic effect • More likely lethal in nature • Probable negative international reaction • More likely unilateral action	**Disadvantages:** • Uncertain timeline • Persistent operation may require reallocation of resources • US remains vulnerable to other extremist organization elements	**Disadvantages:** • Potential to destabilize region of focus • Perception of US overreaction • Higher risk • Unintended consequences

Ref: *JP 5-0, Joint Planning (Jun '17), fig. F-6. Flexible Response Option Scalability.*

Disrupt

Disrupt is used to address both specific, transregional threats and nonspecific, heightened threats. Disrupt options are developed to preempt enemy attacks.

Respond

Respond contingencies are triggered as a result of a successful or unsuccessful attack against the US or its interests. If efforts fail to preempt, disrupt, or defeat a major attack, respond options rapidly provide flexible and scalable options to respond with global operations against the entire scope of the enemy (see Figure F-6). The following are examples of FRO scalability. Operations in each category can be executed individually, concurrently, or sequentially.

II. Considerations for Seizing the Initiative

As operations commence, the JFC needs to exploit friendly advantages and capabilities to shock, demoralize, and disrupt the enemy immediately. The JFC seeks decisive advantage through the use of all available elements of combat power to seize and maintain the initiative, deny the enemy the opportunity to achieve its objectives, and generate in the enemy a sense of inevitable failure and defeat. Additionally, the JFC coordinates with other USG departments and agencies to facilitate coherent use of all instruments of national power in achieving national strategic objectives.

Force Projection (and Forcible Entry)

Projecting US military force invariably requires extensive use of the international waters, international airspace, space, cyberspace, and the EMS to gain operational access. Our ability to freely maneuver to position and sustain our forces is vital to our national interests and those of our PNs. US forces may gain operational access to areas through invitation by an HN to establish an operating base in or near the conflict or by the use of forcible entry operations. Treaties, agreements, and activities that occur during the shape and deter phases may aid in the invitation to establish a base or support facility. However, gaining and maintaining operational access requires the ability to defeat the enemy's A2/AD actions and capabilities.

See following pages (pp. 2-78 to 2-79) for an overview and further discussion.

Unit Integrity During Deployment

US military forces normally train as units and are best able to accomplish a mission when deployed intact. By deploying as an existing unit, forces are able to continue to operate under established procedures, adapting them to the mission and situation, as required. When personnel and elements are drawn from various commands, effectiveness may be decreased. By deploying without established operating procedures, an ad hoc force takes more time to form and adjust to requirements of the mission. This not only complicates mission accomplishment but may also have an impact on force protection.

Even if diplomatic/political restraints on an operation dictate that a large force cannot be deployed intact, commanders should select elements for deployment that have established internal procedures and structures, have trained and operated together, and possess appropriate joint force combat capabilities. To provide a JFC with needed versatility, it may not be possible to preserve complete unit integrity. In such cases, units must be prepared to send elements that are able to operate independently of parent units. Attachment to a related unit is the usual mode. In this instance, units not accustomed to having attachments may be required to provide administrative and logistic support to normally unrelated units.

The CCDR, in coordination with Commander, USTRANSCOM; subordinate JFCs; and the Service component commanders, needs to carefully balance the desire to retain unit integrity through the deployment process with the effective use of strategic lift platforms. While maximizing unit integrity may reduce JRSOI requirements and allow combat units to be employed more quickly, doing so will often have a direct negative impact on the efficient use of the limited strategic lift. In some cases, this negative impact on strategic lift may have a negative effect on DOD deployment and sustainment requirements beyond the GCC's AOR. A general rule of thumb is that unit integrity is much more important for early deploying units than for follow-on forces.

While access operations focus on enabling access to the OA, entry operations focus on actions within the OA. Joint forces conduct entry operations for various purposes, including to defeat threats to the access and use of portions of the OE; to control, defeat, disable, and/or dispose of specific WMD threats; to assist populations and groups; to establish a lodgment; and to conduct other limited duration missions.

Full-Spectrum Superiority (Seizing the Initiative)

Ref: JP 3-0 (w/Chg 1), Joint Operations (Oct '18), pp. VIII-16 to VIII-17.

Mission success in large-scale combat requires full-spectrum superiority; the cumulative effect of achieving superiority in the air, land, maritime, and space domains; the information environment; and the EMS. Such superiority permits the conduct of joint operations without effective opposition or prohibitive interference. JFCs seek superiority throughout the OE to accomplish the mission as rapidly as possible. The JFC may have to initially focus all available joint forces on seizing the initiative. A delay at the outset of combat may damage US credibility, lessen coalition support, and provide incentives for other adversaries to begin conflicts elsewhere.

- JFCs normally strive to achieve air and maritime superiority early. Air and maritime superiority allows joint forces to conduct operations without prohibitive interference from opposing air and maritime forces. Control of the air is a critical enabler because it allows joint forces both freedom from attack and freedom to attack. Using both defensive and offensive operations, JFCs employ complementary weapon systems and sensors to achieve air and maritime superiority.
- Land forces can be moved quickly into an area to deter the enemy from inserting forces, thereby precluding the enemy from gaining an operational advantage. The rapid deployment and employment of land forces (with support of other components) enable sustained operations, more quickly contribute to the enemy's defeat, and help restore stability in the OA.
- Space superiority must be achieved early to support freedom of action. Space superiority allows the JFC access to communications, environmental monitoring, PNT warning, and intelligence collection assets without prohibitive interference by the opposing force. Space control operations are conducted by joint and allied and/or coalition forces to gain and maintain space superiority.
- Early superiority in the information environment (which includes cyberspace) is vital in joint operations. It degrades the enemy's C2 while allowing the JFC to maximize friendly C2 capabilities. Information superiority also allows the JFC to better understand the enemy's intentions, capabilities, and actions, as well as influence foreign attitudes and perceptions of the operation.
- Control of the EME must be achieved early to support freedom of action. This control is important for superiority across the physical domains and information environment.

Attack of Enemy COGs

As part of creating decisive advantages early, joint force operations may be directed immediately against enemy COGs using conventional forces and SOF if COGs are vulnerable and sufficient friendly force capabilities are available. These attacks may be decisive or may begin offensive operations throughout the enemy's depth that can create dilemmas causing paralysis and destroying cohesion.

Force Projection and Forcible Entry (Seizing the Initiative)

Ref: JP 3-0 (w/Chg 1), Joint Operations (Oct '18), pp. VIII-12 to VIII-16.

Force Projection

Projecting US military force invariably requires extensive use of the international waters, international airspace, space, cyberspace, and the EMS to gain operational access. Our ability to freely maneuver to position and sustain our forces is vital to our national interests and those of our PNs. US forces may gain operational access to areas through invitation by an HN to establish an operating base in or near the conflict or by the use of forcible entry operations. Treaties, agreements, and activities that occur during the shape and deter phases may aid in the invitation to establish a base or support facility. However, gaining and maintaining operational access requires the ability to defeat the enemy's A2/AD actions and capabilities.

The President and SecDef may direct a CCDR to resolve a crisis quickly, employing immediately available forces and appropriate FDOs as discussed above to preclude escalation. When these forces and actions are not sufficient, follow-on strikes and/or the deployment of forces from CONUS or another theater and/or the use of multinational forces may be necessary. Consequently, the CCDR must sequence, enable, and protect the deployment of forces to create early decisive advantage. The CCDR should not overlook enemy A2/AD capabilities that may affect the deployment of combat and logistic forces from bases to ports of embarkation. The CCDR may have to adjust the time-phased force and deployment data to meet a changing OE. The deployment of forces may be either opposed or unopposed by an enemy.

Opposed

Initial operations may be designed to suppress enemy A2/AD capabilities. For example, the ability to generate sufficient combat power through long-range air operations or from the sea can provide for effective force projection in the absence of timely or unencumbered access. Other opposed situations may require a forcible entry capability. In other cases, force projection can be accomplished rapidly by forcible entry operations coordinated with strategic air mobility, sealift, and pre-positioned forces. For example, the seizure and defense of lodgment areas by amphibious forces would then serve as initial entry points for the continuous and uninterrupted flow of forces and materiel into the theater. Both efforts demand a versatile mix of forces that are organized, trained, equipped, and poised to respond quickly.

Unopposed

Unopposed deployment operations provide the JFC and subordinate components a more flexible OE to efficiently and effectively build combat power, train, rehearse, acclimate, and otherwise establish the conditions for successful combat operations. In unopposed entry, JFCs arrange the flow of forces, to include significant theater opening logistics forces, that best facilitates the CONOPS. In these situations, logistics forces may be a higher priority for early deployment than combat forces, as determined by the in-theater protection requirements.

Commanders should brief deploying forces on the threat and force protection requirements prior to deployment and upon arrival in the OA. Also, JFCs and their subordinate commanders evaluate the timing, location, and other factors of force deployment in each COA for the impact of sabotage, criminal activity, and terrorist acts and their impact on joint reception, staging, onward movement, and integration (JRSOI) and the follow-on CONOPS. The threat could involve those not directly supporting or sympathetic to the enemy, but those seeking to take advantage of the situation.

Forcible Entry

Entry operations may be unopposed or opposed. Unopposed entry operations often, but not always, follow unopposed access. These circumstances generally allow orderly deployment into the OA in preparation for follow-on operations. Forcible entry is a joint military operation conducted either as a major operation or a part of a larger campaign to seize and hold a military lodgment in the face of armed opposition for the continuous landing of forces. Forcible entry operations can strike directly at the enemy COGs and can open new avenues for other military operations.

Forcible entry operations may include amphibious, airborne, and air assault operations, or any combination thereof. Forcible entry operations can create multiple dilemmas by creating threats that exceed the enemy's capability to respond. Commanders will employ distributed, yet coherent, operations to attack the objective area or areas. The net result will be a coordinated attack that overwhelms the enemy before they have time to react. A well-positioned and networked force enables the defeat of any enemy reaction and facilitates follow-on operations, if required.

Forcible entry is normally complex and risky and should, therefore, be kept as simple as possible in concept. These operations require extensive intelligence, detailed coordination, innovation, and flexibility. Schemes of maneuver and coordination between forces need to be clearly understood by all participants. Forces are tailored for the mission and echeloned to permit simultaneous deployment and employment. When airborne, amphibious, and air assault operations are combined, unity of command is vital. Rehearsals are a critical part of preparation for forcible entry. Participating forces need to be prepared to fight immediately upon arrival and require robust communications and intelligence capabilities to move with forward elements.

The forcible entry force must be prepared to immediately transition to follow-on operations and should plan accordingly. Joint forcible entry actions occur in both singular and multiple operations. These actions include establishing forward presence, preparing the OA, opening entry points, establishing and sustaining access, receiving follow-on forces, conducting follow-on operations, sustaining the operations, and conducting decisive operations.

Successful OPSEC and MILDEC may confuse the enemy and ease forcible entry operations. OPSEC helps foster a credible MILDEC. Additionally, the actions, themes, and messages portrayed by all friendly forces must be consistent if MILDEC is to be believable.

SOF may precede forcible entry forces to identify, clarify, establish, or modify conditions in the lodgment. SOF may conduct the assaults to seize small, initial lodgments such as airfields or seaports. They may provide or assist in employing fire support and conduct other operations in support of the forcible entry, such as seizing airfields or conducting reconnaissance of landing zones or amphibious landing sites. They may conduct special reconnaissance and direct action well beyond the lodgment to identify, interdict, and destroy forces that threaten the conventional entry force.

The sustainment requirements and challenges for forcible entry operations can be formidable, but must not be allowed to become such an overriding concern that the forcible entry operation itself is jeopardized. JFCs must carefully balance the introduction of sustainment forces needed to support initial combat with combat forces required to establish, maintain, and protect the lodgment as well as forces required to transition to follow-on operations.

For additional and detailed guidance on forcible entry operations, refer to JP 3-18, Joint Forcible Entry Operations.

C2 in Littoral Areas

Controlled littoral areas often offer the best positions from which to begin, sustain, and support joint operations, especially in OAs with limited or poor infrastructure for supporting US joint operations ashore. JFCs can gain and maintain the initiative through the ability to project fires and employ forces from sea-based assets in combination with C2, intelligence collection, and IRCs. Maritime forces operating in littoral areas can dominate coastal areas and rapidly generate high intensity offensive power at times and in locations required by JFCs. Maritime forces' relative freedom of action enables JFCs to position these capabilities where they can readily strike opponents. Maritime forces' very presence, if made known, can pose a threat that the enemy cannot ignore.

JFCs can operate from a HQ platform at sea. Depending on the nature of the joint operation, a maritime commander can serve as the JFC or function as a JFACC while the operation is primarily maritime and shift that command ashore if the operation shifts landward IAW the JFC's CONOPS. A sea base provides JFCs with the ability to command and control forces and conduct select functions and tasks at sea without dependence on infrastructure ashore. In other cases, a maritime HQ may serve as the base of the joint force HQ, or subordinate JFCs or other component commanders may use the C2 and intelligence facilities aboard ship.

Transferring C2 from sea to shore requires detailed planning, active liaison, and coordination throughout the joint force. Such a transition may involve a simple movement of flags and supporting personnel, or it may require a complete change of joint force HQ. The new joint force HQ may use personnel and equipment, especially communications equipment, from the old HQ, or it may require augmentation from different sources. One technique is to transfer C2 in several stages. Another technique is for the JFC to satellite off the capabilities of one of the components ashore until the new HQ is fully prepared.

SOF-Conventional Force Integration

The JFC, using SOF independently or integrated with conventional forces, gains an additional and specialized capability to achieve objectives that might not otherwise be attainable. Integration enables the JFC to take fullest advantage of conventional and SOF core competencies. SOF are most effective when special operations are fully integrated into the overall plan and the execution of special operations is through proper SOF C2 elements in a supporting or supported relationship with conventional forces. Joint SOF C2 elements are provided to conduct a specific special operation or prosecute special operations in support of a joint campaign or operation. Special operations commanders also provide liaison to component commands to integrate, coordinate, and deconflict SOF and conventional force operations. Exchange of SOF and conventional force LNOs is essential to enhance situational awareness and reduce risk of friendly fire incidents.

Stability Activities

Combat in this phase provides an opportunity to begin various stability activities that will help achieve military strategic and operational-level objectives and create the conditions for the later stability and enable civil authority phases. Operations to neutralize or eliminate potential stabilize phase enemies may be initiated. National and local HN authorities may be contacted and offered support. Key infrastructure may be seized or otherwise protected. Civil IM, which is broadly tasked to support the overall intelligence collection on the status of enemy infrastructure, government organizations, and humanitarian needs, should be increased. MISO used to influence the behavior of approved foreign target audiences in support of military strategic and operational objectives can ease the situation encountered when sustained combat is concluded. In coordination with interorganizational participants, the JFC must arrange for necessary financial support of these operations well in advance.

Protection

JFCs must strive to conserve the fighting potential of the joint/multinational force at the onset of combat operations. Further, HN infrastructure and logistic support key to force projection and sustainment of the force must be protected.

JFCs counter the enemy's fires and maneuver by making personnel, systems, and units difficult to locate, strike, and destroy. They protect their force from enemy maneuver and fires by using various physical and informational measures. OPSEC and MILDEC are key elements of this effort. Operations to gain air, space, maritime, and EMS superiority; defensive use of IO; PR; and protection of airports and seaports, LOCs, and friendly force lodgment also contribute significantly to force protection at the onset of combat operations.

Prevention of Friendly Fire Incidents

JFCs must make every effort to reduce the potential for the killing or wounding of friendly personnel by friendly fire. The destructive power and range of modern weapons, coupled with the high intensity and rapid tempo of modern combat, increase the potential for friendly fire incidents. Commanders must be aware of those situations that increase the risk of friendly fire incidents and institute appropriate preventive measures. The primary mechanisms for reducing friendly fire incidents are command emphasis, disciplined operations, close coordination among component commands and multinational partners, SOPs, training and exercises, technology solutions (e.g., identify friend or foe, blue force tracking), rehearsals, effective CID, and enhanced awareness of the OE. Commanders should seek to minimize friendly fire incidents while not limiting boldness and initiative. CCMDs should consult with USAID when it has a mission presence to determine locations of friendly international organizations, NGOs, and local partners operating in the targeted area to avoid friendly fire incidents.

III. Considerations for Dominance

JFCs conduct sustained combat operations when a swift victory is not possible. During sustained combat operations, JFCs simultaneously employ conventional forces and SOF throughout the OA. The JFC may designate one component or LOO to be the main effort, with other components providing support and other LOOs as supporting efforts. When conditions or plans change, the main effort might shift. Some missions and operations (i.e., strategic attack, interdiction, and IO) continue throughout to deny the enemy sanctuary, freedom of action, or informational advantage. These missions and operations, when executed concurrently with other operations, degrade enemy morale and physical cohesion and bring the enemy closer to culmination. When prevented from concentrating, opponents can be attacked, isolated at tactical and operational levels, and defeated in detail. At other times, JFCs may cause their opponents to concentrate their forces, facilitating their attack by friendly forces. In some circumstances (e.g., regime change, ensuring stability prior to transition to civil authority), the JFC may be required to maintain a temporary military occupation of enemy territory while continuing offensive actions. If the occupation is extended and a country's government is not functioning, the JFC may be required to establish a military government through the designation of a transitional military authority.

Operating in the Littoral Areas

Even when joint forces are firmly established ashore, littoral operations provide JFCs with excellent opportunities to gain leverage over the enemy by operational maneuver from the sea. Such operations can introduce significant size forces over relatively great distances in short periods of time into the rear or flanks of the enemy. The mobility and fire support capability of maritime forces at sea, coupled with the ability to rapidly land operationally significant forces, can be key to achieving military operational objectives. These capabilities are further enhanced by operational flexibility and the ability to identify and take advantage of fleeting opportunities.

Attack on Enemy COGs

Attacks on enemy COGs typically continue during sustained operations. JFCs should time their actions to coincide with actions of other operations of the joint force and vice versa to achieve military strategic and operational-level objectives. As with all joint force operations, direct and indirect attacks of enemy COGs should be planned to achieve the required military strategic and operational-level objectives per the CONOPS, while limiting potential undesired effects on operations in follow-on phases.

Synchronizing and/or Integrating Maneuver and Interdiction

Synchronizing and integrating air, land, maritime, and cyberspace interdiction and maneuver, enabled by JEMSO and space-based capabilities, provides one of the most dynamic concepts available to the joint force. Interdiction and maneuver usually are not considered separate operations against a common enemy, but rather are complementary operations planned to achieve the military strategic and operational-level objectives. Moreover, maneuver by air, land, or maritime forces can be conducted to interdict enemy military potential. Potential responses to integrated and synchronized maneuver and interdiction can create a dilemma for the enemy. If the enemy attempts to counter the maneuver, enemy forces may be exposed to unacceptable losses from interdiction. If the enemy employs measures to reduce such interdiction losses, enemy forces may not be able to counter the maneuver. The synergy achieved by integrating and synchronizing interdiction and maneuver throughout the OA assists commanders in optimizing leverage at the operational level.

As a guiding principle, JFCs should exploit the flexibility inherent in joint force command relationships, joint targeting procedures, and other techniques to resolve the issues that can arise from the relationship between interdiction and maneuver. When interdiction and maneuver are employed, JFCs need to carefully balance the needs of surface maneuver forces, area-wide requirements for interdiction, and the undesirability of fragmenting joint force capabilities. The JFC's objectives, intent, and priorities, reflected in mission assignments and coordinating arrangements, enable subordinates to fully exploit the military potential of their forces while minimizing the friction generated by competing requirements. Effective targeting procedures in the joint force also alleviate such friction. As an example, interdiction requirements will often exceed interdiction means, requiring JFCs to prioritize requirements. Land and maritime force commanders responsible for integrating and synchronizing maneuver and interdiction within their AOs should be knowledgeable of JFC priorities and the responsibilities and authority assigned and delegated to commanders designated by the JFC to execute theater- and/or JOA-wide functions. JFCs alleviate this friction through the CONOPS and clear statements of intent for interdiction conducted relatively independent of surface maneuver operations. In doing this, JFCs rely on their vision as to how the major elements of the joint force contribute to achieving theater-strategic objectives. JFCs then employ a flexible range of techniques to assist in identifying requirements and applying capabilities to meet them. JFCs must define appropriate command relationships, establish effective joint targeting procedures, and make apportionment decisions.

All commanders should consider how their operations can complement interdiction. These operations may include actions such as MILDEC, withdrawals, lateral repositioning, and flanking movements that are likely to cause the enemy to reposition surface forces, making them better targets for interdiction. Likewise, interdiction operations need to conform to and enhance the JFC's scheme of maneuver. This complementary use of maneuver and interdiction places the enemy in the operational dilemma of either defending from disadvantageous positions or exposing forces to interdiction strikes during attempted repositioning.

Weapons of Mass Destruction (WMD) During Operations

Ref: JP 3-0 (w/Chg 1), Joint Operations (Oct '18), pp. VIII-22 to VIII-24.

Locating WMD and WMD Materials. Since an enemy's use of WMD can quickly change the character of an operation or campaign, joint forces may be required to track, seize, and secure any WMD and materials used to develop WMD discovered or located in an OA. Once located, resources may be required to secure and inventory items for subsequent exploitation. If WMD sites are located, but joint forces are unable to seize and secure them, the JFC should plan to strike the sites if required to prevent WMD from being used or falling into enemy control. The desired effects of strikes are to minimize collateral effects and deny access to WMD. If sites are not under enemy control or in imminent jeopardy of falling to the enemy, monitor them persistently until sites can be seized and secured. During combat operations, exploitation, secure transport of WMD, and safe transport of technical personnel for disposition may depend upon a permissive OE.

Enemy Employment

The use or the threatened use of WMD can cause large-scale shifts in strategic and operational-level objectives, phases, and COAs. Multinational operations become more complicated with the threatened employment of these weapons. An enemy may use WMD against friendly force multinational partners, especially those with little or no defense against these weapons, to disintegrate the alliance or coalition. The enemy may also use chemical, biological or radiological weapons as part of an A2/AD plan.

Friendly Employment

When directed by the President and SecDef, CCDRs will plan for the employment of nuclear weapons by US forces in a manner consistent with national policy and strategic guidance. The employment of such weapons signifies an escalation of the war and is a presidential decision. USSTRATCOM's capabilities to lead in the collaborative planning of all nuclear missions are available to support nuclear weapon employment. If directed to plan for the use of nuclear weapons, JFCs typically have two escalating objectives.

Countering WMD (CWMD)

JFC's should be prepared to conduct activities to curtail the conceptualization, development, possession, proliferation, use, and effects of WMD. When planning or executing operations and activities to counter WMD, JFCs coordinate and cooperate with not only other USG departments and agencies, but also local, tribal, and state organizations, in addition to multinational partners. With numerous stakeholders in the CWMD mission area, it is critical JFCs understand and consider the capabilities and responsibilities of various interorganizational partners when defining command relationships and coordinating interorganizational activities. Operations to counter WMD may require formation of a functional JTF for that purpose.

Within the JOA, all joint force component operations must contribute to achievement of the JFC's objectives. To facilitate these operations, JFCs may establish AOs within their OA. Synchronization and/or integration of maneuver and interdiction within land or maritime AOs is of particular importance, particularly when JFCs task component commanders to execute theater- and/or JOA-wide functions.

JFCs need to pay particular attention and give priority to activities impinging on and supporting the maneuver and interdiction needs of all forces. In addition to normal target nomination procedures, JFCs establish procedures through which land or maritime force commanders can specifically identify those interdiction targets they are unable to engage with organic assets within their OAs that could affect planned or ongoing maneuver. These targets may be identified individually or by category, specified geographically, or tied to a desired effect or time period. Interdiction target priorities within the land or maritime OAs are considered along with theater and JOA-wide interdiction priorities by JFCs and reflected in the air apportionment decision. The JFACC uses these priorities to plan, coordinate, and execute the theater- and/or JOA-wide air interdiction effort. The purpose of these procedures is to afford added visibility to, and allow JFCs to give priority to, targets directly affecting planned maneuver by air, land, or maritime forces.

Stability Activities

Stability tasks and activities that began in previous phases may continue during this phase. These activities may focus on stability tasks that will help achieve strategic and operational-level objectives and create the conditions for the later stabilization and enable civil authority phases. Minimum essential stability activities should focus on protecting and facilitating the personal security and well-being of the civilian population. Stability activities provide minimum levels of security, food, water, shelter, and medical treatment. If no civilian or HN agency is present, capable, and willing, then JFCs and their staffs must resource these minimum essential stability tasks. When demand for resources exceeds the JFC's capability, higher level joint commanders should provide additional resources. These resources may be given to the requesting JFC or the mission may be given to follow-on forces to expeditiously conduct the tasks.

IV. Considerations for Stabilization

Operations in a stabilize phase typically begin with significant military involvement, to include some combat and the potential for longer-term occupation. Operations then move increasingly toward transitioning to an interim civilian authority and enabling civil authority as the threat wanes and civil infrastructures are reestablished. The JFC's mission accomplishment requires fully integrating US military operations with the efforts of interorganizational participants in a comprehensive approach to accomplish assigned and implied tasks. As progress is made, military forces will increase their focus on supporting the efforts of HN authorities, other USG departments and agencies, international organizations, and/or NGOs. National Security Presidential Directive-44, Management of Interagency Efforts Concerning Reconstruction and Stabilization, assigns DOS the responsibility to plan and coordinate USG efforts in stabilization and reconstruction. The Secretary of State coordinates with SecDef to ensure harmonization with planned and ongoing operations. Military support to stabilization efforts within the JOA are the responsibility of the JFC.

See following pages (pp. 2-86 to 2-87) for an overview and discussion of stability activities.

Lines of Operation (LOOs)

Several LOOs may be initiated immediately (e.g., providing FHA, establishing security). In some cases, the scope of the problem set may dictate using other nonmilitary entities which are uniquely suited to address the problems. The goal of these military and civil efforts should be to eliminate root causes or deficiencies that create the problems (e.g., strengthen legitimate civil authority, rebuild government institutions, foster a sense of confidence and well-being, and support the conditions for economic reconstruction). With this in mind, the JFC may need to address how to coordinate CMO with the efforts of participating other government agencies, international organizations, NGOs, and HN assets.

Forces and Capabilities Mix

The JFC may need to realign forces and capabilities or adjust force structure to begin stability activities in some portions of the OA even while sustained combat operations still are ongoing in other areas. For example, CA forces and HUMINT capabilities are critical to supporting stabilize phase operations and often involve a mix of forces and capabilities far different than those that supported the previous phases. Planning and continuous assessment will reveal the nature and scope of forces and capabilities required. These forces and capabilities may be available within the joint force or may be required from another theater or from the Reserve Component.

V. Considerations for Enabling Civil Authority

In this phase, the joint operation is assessed and enabling objectives are established for transitioning from large-scale combat operations to FID and security cooperation. The catalyst for transition is that a legitimate civil authority has been established to manage the situation without further outside military intervention. The new government obtains legitimacy, and authority is transitioned from an interim civilian authority or transitional military authority to the new indigenous government. This situation may require a change in the joint operation as a result of an extension of the required stability activities in support of US diplomatic, HN, international organization, and/or NGO stabilization efforts.

Peacebuilding (PB)

The transition from military operations to full civilian control may involve ongoing operations that have a significant combat component, including COIN operations, antiterrorism, and CT. Even while combat operations are ongoing, the operation will include a large stability component that is essentially a PB mission. PB, transitioning to a DOS-led effort, provides the reconstruction and societal rehabilitation that offers hope to the HN populace. Stability measures establish the conditions that enable PB to succeed. PB promotes reconciliation, strengthens and rebuilds civil infrastructures and institutions, builds confidence, and supports economic reconstruction to prevent a return to conflict. The ultimate measure of success in PB is political, not military. Therefore, JFCs seek a clear understanding of the national/PN objectives and how military operations support that end state.

Transfer to Civil Authority

In many cases, the US will transfer responsibility for the political and military affairs of the HN to another authority (e.g., UN observers, multinational peacekeeping force, or NATO) consistent with established termination criteria. This will probably occur after an extended period of conducting joint or multinational stability activities and PB missions as described above. Overall, transfer will likely occur in stages (e.g., HN sovereignty, PO under UN mandate, termination of all US military participation). Joint force support to this effort may include the following:

- **Support to Truce Negotiations.** This support may include providing intelligence, security, transportation and other logistic support, and linguists for all participants.
- **Transition to Civil Authority.** This transfer could be to a local or HN government or to an international authority facilitated by the UN. For example, an interim government (Northern Alliance) assumed governance during OEF and then transferred governance to a legitimate (newly elected) national government in Afghanistan.

Stability Activities

Ref: JP 3-0 (w/Chg 1), Joint Operations (Oct '18), pp. VIII-26 to VIII-28.

As sustained combat operations conclude, military forces will shift their focus to stability activities as the military instrument's contribution to the more comprehensive stabilization efforts by all instruments of national power. Force protection will continue to be important, and combat operations might continue, although with less frequency and intensity than in the dominate phase. Of particular importance will be CMO, initially conducted to secure and safeguard the populace, reestablish civil law and order, protect or rebuild key infrastructure, and restore public services. US military forces should be prepared to lead the activities necessary to accomplish these tasks, especially if conducting a military occupation, and restore rule of law when indigenous civil, USG, multinational or international capacity does not exist or is incapable of assuming responsibility. Once legitimate civil authority is prepared to conduct such tasks, US military forces may support such activities as required/necessary. SFA plays an important part during stability activities by supporting and augmenting the development of the capacity and capability of FSFs and their supporting institutions. Likewise, the JFC's communication synchronization will play an important role in providing public information to foreign populations during this period.

The military's predominant presence and its ability to command and control forces and logistics under extreme conditions may give it the de facto lead in stabilization efforts normally governed by other agencies that lack such capacities. However, most stability activities will likely be in support of, or transition to support of, US diplomatic, UN, or HN efforts. Integrated civilian and military efforts are key to success and military forces need to work competently in this environment while properly supporting the agency in charge. To be effective, planning and conducting stabilization efforts require a variety of perspectives and expertise and the cooperation and assistance of other USG departments and agencies, other Services, and alliance or multinational partners. Military forces should be prepared to work in integrated civilian-military teams that could include representatives from other US departments and agencies, foreign governments and security forces, international organizations, NGOs, and members of the private sector with relevant skills and expertise. Typical military support includes emergency infrastructure reconstruction, engineering, logistics, law enforcement, health services, and other activities to restore essential services.

- **Civil Affairs (CA)** forces are organized and trained to perform CA operations that support CMO conducted in conjunction with stability activities. MISO forces will develop, produce, and disseminate products to gain and reinforce popular support for the JFC's objectives. Complementing conventional forces' IW efforts, SOF will conduct FID to assess, train, advise, assist, and equip foreign military and paramilitary forces as they develop the capacity to secure their own lands and populations.
- **Counterintelligence (CI)** activities safeguard essential elements of friendly information. This is particularly pertinent in countering adversary HUMINT efforts. HN authorities, international organizations, and NGOs working closely with US forces may pass information (knowingly or unknowingly) to adversary elements that enables them to interfere with stability activities. Members of the local populace, who might actually be belligerents, often gain access to US military personnel and their bases by providing services such as laundry and cooking. They can then pass on information gleaned from that interaction to seek favor with a belligerent element or to avoid retaliation from belligerents. Identity activities, coupled with all-source intelligence analysis of the collected data, can support verification and deconfliction of HUMINT source identities and assist the JFC to take appropriate actions to counter potential compromise.
- **Public Affairs (PA)** activities support stability activities by providing public information about progress to internal and external audiences.

Joint Operations

Protection

In the stabilize phase, commanders must consider protection from virtually any person, element, or group hostile to US interests. Personnel should stay alert even in an operation with little or no perceived risk. JFCs must take measures to prevent complacency and be ready to counter activity that could bring harm to units or jeopardize the operation. However, security requirements should be balanced with the military operation's nature and objectives. During some stability activities, the use of certain security measures, such as carrying arms, wearing helmets and protective vests, or using secure communications may cause military forces to appear more threatening than intended, which may degrade the force's legitimacy and hurt relations with the local population.

Restraint

During the stabilize phase, military capability must be applied even more prudently since the support of the local population is essential for success. The actions of military personnel and units are framed by the disciplined application of force, including specific ROE. These ROE often will be more restrictive and detailed when compared to those for sustained combat operations due to national policy concerns. Moreover, these rules may change frequently during operations. The principle of restraint does not preclude the application of overwhelming force, when appropriate and authorized, to display US resolve and commitment. The reasons for the restraint often need to be understood by the individual Service member, because a single act could cause adverse diplomatic/political consequences.

Perseverance

Some operations may move quickly through the stabilize phase and transition smoothly to the enable civil authority phase. Other situations may require years of stabilization activities before this transition occurs. Therefore, the patient, resolute, and persistent pursuit for as long as necessary of the conditions desired to reach national objectives is often the requirement for success.

Legitimacy

Military activities must sustain the legitimacy of the operation and of the emerging or host government. During operations where a government does not exist, extreme caution should be used when dealing with individuals and organizations to avoid inadvertently legitimizing them. Implementation of strategic guidance through the CCS process can enhance perceptions of the legitimacy of stabilization efforts.

OPSEC

Although there may be no clearly defined threat, the essential elements of US military operations should be safeguarded. The uncertain nature of the situation, coupled with the potential for rapid change, requires that OPSEC be an integral part of all military operations. They can then pass on information gleaned from that interaction or provide other support to a belligerent element to seek favor or to avoid retaliation. The JFC must consider these and similar possibilities and take appropriate actions to counter potential compromise. OPSEC planners must consider the effect of media coverage and the possibility coverage may compromise essential security or disclose critical information.

Refer to TAA2: Military Engagement, Security Cooperation & Stability SMARTbook (Foreign Train, Advise, & Assist) for further discussion. Topics include the Range of Military Operations (JP 3-0), Security Cooperation & Security Assistance (Train, Advise, & Assist), Stability Operations (ADRP 3-07), Peace Operations (JP 3-07.3), Counterinsurgency Operations (JP & FM 3-24), Civil-Military Operations (JP 3-57), Multinational Operations (JP 3-16), Interorganizational Cooperation (JP 3-08), and more.

Redeployment

Ref: JP 3-0 (w/Chg 1), Joint Operations (Oct '18), pp. VIII-29 to VIII-30.

Redeployment is the transfer of forces and materiel to support another JFC's operational requirements, or to return personnel, equipment, and materiel to home/demobilization stations for reintegration and out-processing. Redeployment is normally conducted in stages—the entire joint force will likely not redeploy in one relatively short period. It may include waste disposal, port operations, closing of contracts and other financial obligations, disposition of contracting records and files, clearing and marking of minefields and other explosive ordnance disposal activities, and ensuring appropriate units remain in place until their missions are complete. Redeployment must be planned and executed in a manner that facilitates the use of redeploying forces and supplies to meet new missions or crises.

Redeployment planning is the responsibility of the losing supported commander, when personnel, equipment, and materiel are redeployed to home or demobilization stations. The gaining supported commander is responsible for this planning when the redeployment is to a new OA.

Upon redeployment, units or individuals may require refresher training prior to reassuming more traditional roles and missions. Service members and leaders may also require follow-on schooling to ensure normal career progression. Due to this, redeployment planning must be a collaborative and synchronized effort between supported and supporting commanders.

Redeployment to Other Contingencies

Due to competing demands for limited forces, the joint force coordinator (for conventional forces) or joint force provider (for SOF and mobility forces) may source recommendations for allocating a force from one CCDR to another higher priority mission if the risks warrant. If SecDef approves the sourcing recommendation, the allocation will be ordered in a deployment order. Commanders and their staffs should consider how they would extricate forces and ensure they are prepared for the new contingency. This might include such things as a prioritized redeployment schedule, identification of aerial ports for linking intra- and intertheater airlift, the most recent intelligence assessments and supporting GEOINT products for the new contingency, and some consideration to achieving the national objectives of the original contingency through other means.

Redeployment in Support of Rotational Requirements

Due to Service or other rotational requirements, forces may be relieved in place and redeployed to home station for reconstitution or regeneration. Commanders and their staffs must consider security and protective measures during the relief in place between incoming and outgoing forces.

Chap 3

I. Joint Planning

Ref: JP 5-0, Joint Planning (Jun '17), chap. I.

Joint Planning

Joint planning is the deliberate process of determining how (the ways) to use military capabilities (the means) in time and space to achieve objectives (the ends) while considering the associated risks. Ideally, planning begins with specified national strategic objectives and military end states to provide a unifying purpose around which actions and resources are focused. The joint planning and execution community (JPEC) conducts joint planning to understand the strategic and operational environment (OE) and determines the best method for employing the Department of Defense's (DOD's) existing capabilities to achieve national objectives. Joint planning identifies military options the President can integrate with other instruments of national power (diplomatic, economic, informational) to achieve those national objectives. In the process, joint planning identifies likely benefits, costs, and risks associated with proposed military options. In the absence of specified national objectives and military end states, combatant commanders (CCDRs) may propose objectives and military end states for the President's and/or the Secretary of Defense's (SecDef's) consideration before beginning detailed planning. The Chairman of the Joint Chiefs of Staff (CJCS), as the principal military advisor to the President and SecDef, may offer military advice on the proposed objectives and military end states as a part of this process.

The Transregional, Multi-Domain, and Multi-Functional (TMM) Environment

The strategic environment is uncertain, complex, and changes rapidly. While the nature of war has not changed, the character of warfare has evolved. Military operations will increasingly operate in a transregional, multi-domain, and multi-functional (TMM) environment. TMM operations will cut across multiple combatant commands (CCMDs) and across land, maritime, air, space, and cyberspace. Effective planning provides leadership with options that offer the highest probability for success at acceptable risk and enables the efficient use of limited resources, including time, to achieve objectives in this global environment. When specific objectives are not identified, planning identifies options with likely outcomes and risks to enable leaders at all levels to make informed decisions, without unnecessary expenditure of resources.

I. Joint Planning Purposes

At the CCMD level, joint planning serves two critical purposes. At the strategic level, joint planning provides the President and SecDef options, based on best military advice, on use of the military in addressing national interests and achieving the objectives in the National Security Strategy (NSS) and Defense Strategy Review (DSR).

At the operational level, once strategic guidance is given, planning translates this guidance into specific activities aimed at achieving strategic and operational-level objectives and attaining the military end state. This level of planning ties the training, mobilization, deployment, employment, sustainment, redeployment, and demobilization of joint forces to the achievement of military objectives that contribute to the achievement of national security objectives in the service of enduring national interests.

II. Joint Planning Principles

Ref: JP 5-0, Joint Planning (Jun '17), executive summary.

Focuses on the End State

Joint planning is end state oriented: plans and actions positively contribute to achieving national objectives. Planning begins by identifying the desired national and military end states. The commander and staff derive their understanding of those end states by evaluating the strategic guidance, their analysis of the OE, and coordination with senior leadership. Joint planners must ensure plans are consistent with national priorities and are directed toward achieving national objectives. Planning must also determine and articulate the correct problem set to which military effort might be applied. The CCDR and staff work with DOD leadership in this effort.

Globally Integrated and Coordinated

Planning considers that operations take place throughout the OE irrespective of geographic, political, or domain boundaries. Planning, therefore, must look across CCMD, Service, and even DOD or US boundaries to ensure effective support for national objectives.

Resource Informed

Joint planning is resource informed and time constrained. It provides a realistic assessment of the application of forces, given current readiness, availability, location, available transportation, and speed of movement. Planning assumes that an operation will employ forces and capabilities currently available—not future capabilities or capacities.

Risk Informed

Assessing and articulating risks and opportunities while identifying potential mitigation strategies are fundamental to joint planning. Planning provides decision makers an honest assessment of the costs and potential consequences of military actions. Planning identifies the impact of all assumptions whether proven valid or invalid, as well as the impact of constraints and restraints imposed on the operation.

Framed within the Operational Environment (OE)

Planning requires an understanding of the OE as it exists and changes. Unlike concepts and future development, adaptive planning is based on continuous monitoring and understanding of actual conditions affecting the OE such as current friendly and adversary force postures, readiness, geopolitical conditions, and adversary perceptions. Adaptive planning accommodates changes aimed at improving probability of success or mitigating risk (e.g., additional forces, partner nation contributions, agreements, or access, basing, and overflight permission needed; preparation activities, including prepositioning).

Informs Decision Making

Planning, even constrained by time, identifies issues and assumptions required for planning to continue, likely resource requirements, costs and cost-benefit trade-offs, and risks associated with different COAs. Discussions on these topics enable key leaders to make informed decisions that best serve the national interests.

Adaptive and Flexible

Planning is an adaptive process. It occurs in a networked, collaborative environment that requires dialogue among senior leaders; concurrent plan development; and collaboration across strategic, operational, and tactical planning levels.

II(a). Strategic Guidance & Coordination

Ref: JP 5-0, Joint Planning (Jun '17), chap. II, section a.

The President, SecDef, and CJCS provide their orders, intent, strategy, direction, and guidance via strategic direction to the military to pursue national interests within legal and constitutional limitations. They generally communicate strategic direction to the military through written documents, but it may be communicated by any means available. Strategic direction is contained in key documents, generally referred to as strategic guidance. Strategic direction may change rapidly in response to changing situations, whereas strategic guidance documents are typically updated cyclically and may not reflect the most current strategic direction.

I. National & Department of Defense Guidance

The National Security Council (NSC) develops and recommends national security policy options for Presidential approval. The NSC is the President's principal forum for considering national security and foreign policy matters with senior national security advisors and cabinet officials. NSC decisions may be directed to any of the member departments or agencies. The President chairs the NSC. Its regular attendees (both statutory and nonstatutory) are the Vice President, Secretary of State, Secretary of the Treasury, SecDef, Secretary of Homeland Security, and Assistant to the President for National Security Affairs. CJCS is the statutory military advisor to the NSC, and the Director of National Intelligence is the intelligence advisor. For DOD, the President's decisions drive SecDef's strategic guidance, which CJCS may refine. To carry out Title 10, United States Code (USC), statutory responsibilities, the CJCS utilizes the Joint Strategic Planning System (JSPS) to provide a formal structure in aligning ends, ways, and means, and to identify opportunities and mitigate risk for the military in shaping the best assessments, advice, and direction of the Armed Forces for the President and SecDef.

A. Strategic Guidance and Direction

The President provides strategic guidance through the NSS, Presidential policy directives (PPDs), executive orders, and other strategic documents in conjunction with additional guidance and refinement from the NSC. The President also signs the Unified Command Plan (UCP) and the contingency planning guidance in the SecDef-signed GEF, which are both developed by DOD.

SecDef has authority, direction, and control over DOD. SecDef oversees the development of broad defense policy goals and priorities for the deployment, employment, and sustainment of US military forces based on the NSS. For planning, SecDef provides guidance to ensure military action supports national objectives. SecDef approves assignment and allocation of forces.

USD(P) assists SecDef with preparing written policy guidance for the preparation of plans, reviewing plans, and assisting SecDef with other duties.

The CJCS provides independent assessments; serves as principal military advisor to the President, SecDef, and the NSC; and assists the President and SecDef with providing unified strategic direction to the Armed Forces. In this capacity, the CJCS develops the NMS and the JSCP, which provide military implementation strategies and planning direction.

See pp. 1-7 to 1-10 for discussion on national strategic direction from JP-1.

B. National Security Council (NSC) System

The NSC system is the principal forum for interagency deliberation of national security policy issues requiring Presidential decision. In addition to NSC meetings chaired by the President, the current NSC organization includes the Principals Committee, deputies committee, and interagency policy committees. Specific issue interagency working groups support these higher-level committees.

See p. 8-9 for further discussion of the National Security Council.

C. National Security Strategy (NSS)

The NSS is required annually by Title 50, USC, Section 3043. It is prepared by the Executive Branch of the USG for Congress and outlines the major national security concerns of the US and how the administration plans to address them using all instruments of national power.

D. Department of State (DOS) & the United States Agency for International Development (USAID)

The Department of State (DOS) is the lead US foreign affairs agency within the Executive Branch and the lead institution for the conduct of American diplomacy. The Secretary of State is the President's principal foreign policy advisor. The Secretary of State implements the President's foreign policies worldwide through DOS and its employees. USAID is an independent federal agency that receives overall foreign policy guidance from the Secretary of State.

See facing page for further discussion.

E. Department of Defense

Defense Strategy Review (DSR)

The DSR is legislatively mandated by Congress per Title 10, USC, Section 118, and required every four years. The DSR articulates a defense strategy consistent with the most recent NSS by defining force structure, modernization plans, and a budget plan allowing the military to successfully execute the full range of missions within that strategy for the next 20 years.

Unified Command Plan (UCP)

The UCP, signed by the President, establishes CCMD missions and CCDR responsibilities, addresses assignment of forces, delineates geographic AORs for GCCs, and specifies responsibilities for FCCs. The unified command structure identified in the UCP is flexible and changes as required to accommodate evolving US national security needs.

See pp. 1-19 to 1-24 for further discussion of the unified command plan.

Guidance for Employment of the Force (GEF)

The GEF, signed by SecDef, and its associated Contingency Planning Guidance, signed by the President, convey the President's and the SecDef's guidance for contingency force management, security cooperation, and posture planning. The GEF translates NSS objectives into prioritized and comprehensive planning guidance for the employment of DOD forces.

Global Force Management Implementation Guidance (GFMIG)

The GFMIG provides SecDef's direction for global force management (GFM) to manage forces from a global perspective. It provides the specific direction for force assignment, apportionment, and allocation processes enabling SecDef to make risk-informed decisions regarding the distribution of US Armed Forces among the CCDRs.

See following page (p. 3-6) for further discussion of GFMIG.

Department of State and the United States Agency for International Development

Ref: JP 5-0, Joint Planning (Jun '17), p. II-3.

The Department of State (DOS) is the lead US foreign affairs agency within the Executive Branch and the lead institution for the conduct of American diplomacy. The Secretary of State is the President's principal foreign policy advisor. The Secretary of State implements the President's foreign policies worldwide through DOS and its employees. USAID is an independent federal agency that receives overall foreign policy guidance from the Secretary of State.

Quadrennial Diplomacy and Development Review

The Quadrennial Diplomacy and Development Review provides a blueprint for advancing America's interests in global security, inclusive economic growth, climate change, accountable governance, and freedom for all. As a joint effort of DOS and USAID, the review identifies major global and operational trends that constitute threats or opportunities, delineates priorities, and reforms to ensure our civilian institutions are in the strongest position to shape and respond to a rapidly changing world.

DOS-USAID Joint Strategic Plan

This DOS-USAID Joint Strategic Plan is a blueprint for investing in America's future and achieving the goals the President laid out in the NSS and those in the Quadrennial Diplomacy and Development Review. It lays out strategic goals and objectives for four years and includes key performance goals for each objective.

Key DOS/USAID Planning Documents

The following are key DOS/USAID planning documents that commanders and planners must consult when developing theater plans.

Joint Regional Strategies

A joint regional strategy is a three-year regional strategy developed jointly by the regional bureaus of DOS and USAID. It identifies the priorities, goals, and areas of strategic focus within the region. Joint regional strategies provide a forward-looking and flexible framework within which bureaus and missions prioritize desired end states, supporting resources, and response to unanticipated events.

Integrated Country Strategies

An integrated country strategy is a three-year strategy developed by a DOS country team for a particular country. It articulates a common set of USG priorities and goals by setting the mission goals and objectives through a coordinated and collaborative planning effort. It provides the basis for the development of the annual mission resource requests. The chief of mission leads the development process and has final approval authority.

Country Development Cooperation Strategy

The country development cooperation strategy is a five-year country-level strategy that focuses on USAID-implemented assistance, including non emergency humanitarian and transition assistance and related USG non-assistance tools.

See chap. 8, Interorganizational Cooperation, for further discussion.

II. Joint Strategic Planning System (JSPS)

Ref: JP 5-0, Joint Planning (Jun '17), pp. II-6 to II-8.

The JSPS is the primary system by which the CJCS carries out USC-assigned statutory responsibilities. The JSPS enables the CJCS to conduct assessments; provide military advice to the President, SecDef, NSC, and Homeland Security Council (HSC); and assist the President and SecDef in providing strategic direction to the US Armed Forces. The NMS and JSCP are core strategic guidance documents that provide CJCS direction and policy essential to the achievement of NSS objectives by augmenting the strategic direction provided in the UCP, GEF, and other Presidential directives. Other elements of JSPS, such as the CJCS risk assessment, the joint strategy review, and the annual joint assessment (AJA), inform decision making and identify new contingencies that may warrant planning and the commitment of resources.

See fig. II-1 (facing page), Providing for the Direction of the Armed Forces, for an overview of these relationships. The JSPS is described in detail in CJCSI 3100.01B, Joint Strategic Planning System.

A. National Military Strategy (NMS)

The NMS, derived from the NSS and DSR, prioritizes and focuses the efforts of the Armed Forces of the United States while conveying the CJCS's direction with regard to the OE and the necessary military actions to protect national security interests. The NMS defines the national military objectives (ends), how to accomplish these objectives (ways), and addresses the military capabilities required to execute the strategy (means). The NMS provides focus for military activities by defining a set of interrelated military objectives and joint operating concepts from which the Service Chiefs and CCDRs identify desired capabilities and against which the CJCS assesses risk.

B. Joint Strategic Capabilities Plan (JSCP)

The JSCP is the primary document in which the CJCS carries out his statutory responsibility for providing unified strategic direction to the Armed Forces. The JSCP provides military strategic and operational guidance to CCDRs, Service Chiefs, CSAs, and applicable DOD agencies for preparation of plans based on current military capabilities. It implements the planning guidance provided in the GEF and the joint planning activities and products that accomplish that guidance. In addition to communicating to the CCMDs' specific planning guidance necessary for planning, the JSCP operationalizes the strategic vision described in the NMS and nests with the strategic direction delineated by the NSS, DSR, and the DOD's planning and resourcing guidance provided in the GEF. The JSCP also provides integrated planning guidance and direction for planners

The JSCP is described in detail in CJCSI 3110.01G, Joint Strategic Capabilities Plan.

C. Global Force Management Implementation Guidance (GFMIG)

The GFMIG documents force planning and execution guidance and show assignment of forces in support of the UCP. GFM aligns force assignment, apportionment, and allocation methodologies in support of the DSR and GEF, joint force availability requirements, and joint force assessments. It provides comprehensive insights into the global availability of US military resources and provides senior decision makers a process to quickly and accurately assess the impact and risk of proposed changes in force assignment, apportionment, and allocation. JS prepares the document for SecDef approval, with the Joint Staff J-8 [Director for Force Structure, Resource, and Assessment] overseeing the assignment and apportionment of forces and the Joint Staff J-3 [Operations Directorate] overseeing the allocation of forces.

Refer to "Joint/Interagency SMARTbook 1: Joint Strategic and Operational Planning" for additional information and discussion.

Providing for the Direction of the Armed Forces

Ref: JP 5-0, Joint Planning (Jun '17), p. II-7.

The President, SecDef, and CJCS use strategic direction to communicate their broad goals and issue-specific guidance to DOD. It provides the common thread that integrates and synchronizes the planning activities and operations of the JS, CCMDs, Services, joint forces, combat support agencies (CSAs), and other DOD agencies. It provides purpose and focus to the planning for employment of military force. Strategic direction identifies a desired military objective or end state; national-level planning assumptions; and national-level constraints, limitations, and restrictions. In addition to previously mentioned documents, additional strategic direction will emerge as orders or as part of the iterative plans dialogue reflected in APEX.

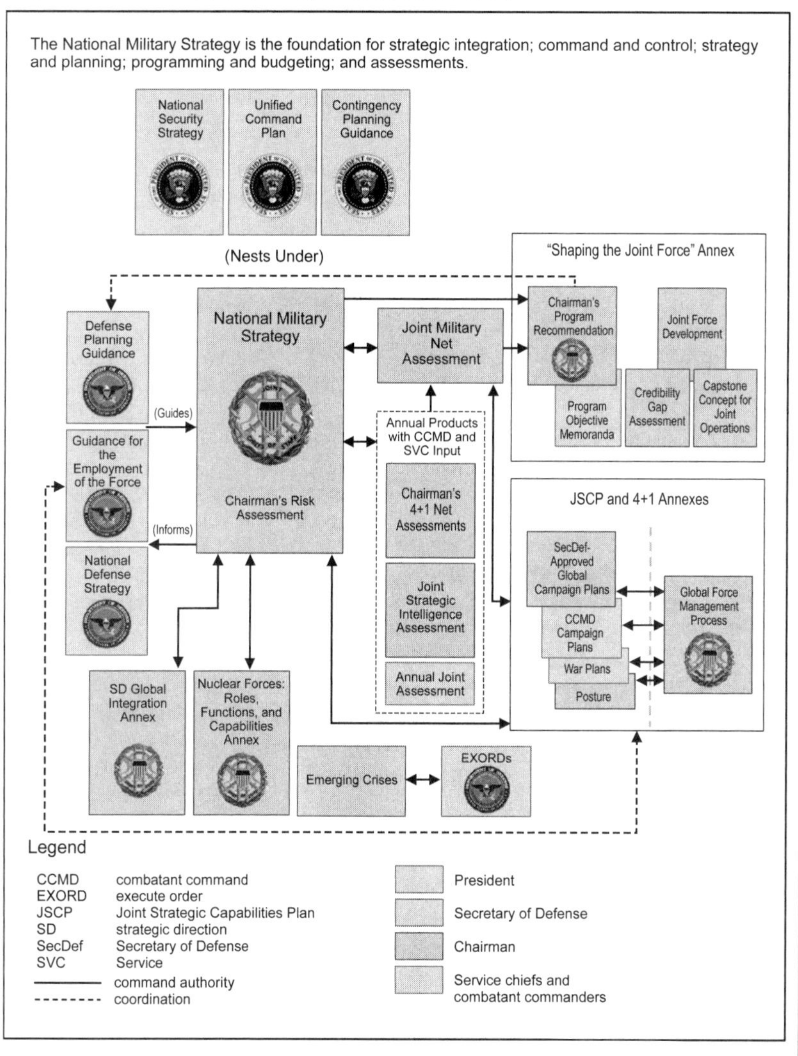

Ref: JP 5-0, Joint Planning, fig. II-1. Providing for the Direction of the Armed Forces.

III. Combatant Commanders (CCDRs)

See pp. 1-19 to 1-24 for an overview of the combatant commands.

At the CCMD level, a joint planning group (JPG), operational planning group, or operational planning team (OPT) is typically established to direct planning efforts across the command, including implementation of plans and orders.

A. Strategic Estimate

The CCDR and staff, with input from subordinate commands and supporting commands and agencies, prepare a strategic estimate by analyzing and describing the political, military, economic, social, information, and infrastructure (PMESII) factors and trends, and the threats and opportunities that facilitate or hinder achievement of the objectives over the timeframe of the strategy.

The strategic estimate is a tool available to commanders as they develop plans. CCDRs use strategic estimates developed in peacetime to facilitate the employment of military forces across the range of military operations. The strategic estimate is more comprehensive in scope than estimates of subordinate commanders, encompasses all aspects of the CCDR's OE, and is the basis for the development of the GCC's theater strategy.

The CCDR, the CCDR's staff, and supporting commands and agencies evaluate the broad strategic-level factors that influence the theater strategy.

The estimate should include an analysis of strategic direction received from the President, SecDef, or the authoritative body of a MNF; an analysis of all states, groups, or organizations in the OE that may threaten or challenge the CCMD's ability to advance and defend US interests in the region; visualization of the relevant geopolitical, geoeconomic, and cultural factors in the region; an evaluation of major strategic and operational challenges facing the CCMD; an analysis of known or anticipated opportunities the CCMD can leverage; and an assessment of risks inherent in the OE.

The result of the strategic estimate is a visualization and better understanding of the OE to include allies, partners, neutrals, enemy combatants, and adversaries. The strategic estimate process is continuous and provides input used to develop strategies and implement plans. The broad strategic estimate is also the starting point for conducting the commander's estimate of the situation for a specific operation.

Supported and supporting CCDRs and subordinate commanders all prepare strategic estimates based on assigned tasks. CCDRs who support multiple JFCs prepare estimates for each supporting operation.

See facing page for a notional strategic estimate format.

B. Combatant Command (CCMD) Strategies

A strategy is a broad statement of the commander's long-term vision. It is the bridge between national strategic guidance and the joint planning required to achieve national and command objectives and attain end states. Specifically, it links CCMD activities, operations, and resources to USG policy and strategic guidance. A strategy should describe the ends as directed in strategic guidance and the ways and means to attain them. A strategy should begin with the strategic estimate. Although there is no prescribed format for a strategy, it may include the commander's vision, mission, challenges, trends, assumptions, objectives, and resources. CCDRs employ strategies to align and focus efforts and resources to mitigate and prepare for conflict and contingencies, and support and advance US interests. To support this, strategies normally emphasize security cooperation activities, force posture, and preparation for contingencies. Strategies typically employ military engagement, close cooperation with DOS, embassies, and other USG departments and agencies. A strategy should be informed by the means or resources available to support the attainment of designated end states and may include military resources, programs, policies, and available funding. CCDRs

Notional Strategic Estimate Format

Ref: JP 5-0, Joint Planning (Jun '17), app. B.

1. Strategic Direction

This section analyzes broad policy, strategic guidance, and authoritative direction to the theater situation and identifies theater strategic requirements in global and regional dimensions.

a. **US Policy Goals.** *Identify the US national security or military objectives and strategic tasks assigned to or coordinated by the CCMD.*

b. **Non-US/Multinational Policy Goals.** *Multinational (alliance/coalition) security or military objectives and strategic tasks that may also be assigned to, or coordinated by the CCMD.*

c. **Opposition Policy Goals and Desired End State**.

d. **End State(s).** *Describe the strategic end state[s] and related military end state[s] to be maintained or accomplished.*

2. Strategic Environment

a. **AOR.** *Provide a visualization of the relevant geographic, political, economic, social, demographic, historic, and cultural factors in the AOR assigned to the CCDR.*

b. **Area of Interest.** *Describe the area of interest to the commander, including the area of influence and adjacent areas and extending into adversary territory. This area also includes areas occupied by enemy forces that could jeopardize the accomplishment of the mission.*

c. **Adversary Forces.** *Identify all states, groups, or organizations expected to be hostile to, or that may threaten, US and partner nation interests, and appraise their general objectives, motivations, and capabilities. Provide the information essential for a clear understanding of the magnitude of the potential threat.*

d. **Friendly Forces.** *Identify all relevant friendly states, forces, and organizations. These include assigned US forces, regional allies, and anticipated multinational partners. Describe the capabilities of the other instruments of power [diplomatic, economic, and informational], US military supporting commands, and other agencies that could have a direct and significant influence on the operations in this AOR.*

e. **Neutral Forces.** *Identify all other relevant states, groups, or organizations in the AOR and determine their general objectives, motivations, and capabilities. Provide the information essential for a clear understanding of their motivations and how they may impact US and friendly multinational operations.*

3. Assessment of Major Strategic and Operational Challenges

a. This is a continuous appreciation of the major challenges in the AOR with which the CCDR may be tasked to deal.

b. These may include a wide range of challenges, from direct military confrontation, peace operations, and security cooperation (including building partner capacity and capability), to providing response to atrocities, humanitarian assistance, disaster relief, and stability activities.

4. Potential Opportunities

a. This is an analysis of known or anticipated circumstances, as well as emerging situations, that the CCMD may use as positive leverage to improve the theater strategic situation and further US or partner nation interests.

b. Each potential opportunity must be carefully appraised with respect to existing strategic guidance and operational limitations.

5. Assessment of Risks

Risk is the probability and severity of loss linked to hazards.

a. This assessment matches a list of the potential challenges with anticipated capabilities in the operational environment

b. Risks associated with each major challenge should be analyzed separately and categorized according to significance or likelihood (most dangerous or most likely)

c. The CCMD staff should develop a list of possible mitigation measures to these risks.

publish strategies to provide guidance to subordinates and supporting commands/agencies and improve coordination with other USG departments and agencies and regional partners. A CCDR operationalizes a strategy through a campaign plan (see Figure II-2, below).

Additional Sources of Strategic Guidance

- National Security Strategy
- National Strategy for Combating Terrorism
- National Strategy for Public Diplomacy and Strategic Communication
- National Counterintelligence Strategy
- National Intelligence Strategy
- National Strategy to Combat Terrorist Travel
- National Strategy to Secure Cyberspace
- National Strategy for Homeland Security
- National Strategy for Maritime Security
- National Strategy for Information Sharing
- National Strategy for Pandemic Influenza
- National Strategy for Physical Protection of Critical Infrastructure
- National Strategy for Countering Biological Threats

List is not all inclusive.

Ref: JP 5-0, Fig. II-2. Additional Sources of Strategic Guidance.

C . Commander's Communication Synchronization

Commander's communication synchronization is the process to coordinate and synchronize narratives, themes, messages, images, operations, and actions to ensure their integrity and consistency to the lowest tactical level across all relevant communication activities.

Within the USG, DOS has primary responsibility for communication synchronization oversight. It is led by the Under Secretary for Public Diplomacy and Public Affairs and is the overall mechanism by which the USG coordinates public diplomacy across the interagency community. A key product of this committee is the US National Strategy for Public Diplomacy and Strategic Communication. This document provides USG-level guidance, intent, strategic imperatives, and core messages under which DOD can nest its themes, messages, images, and activities.

The US military plays an important supporting role in communication synchronization, primarily through commander's communication synchronization, public affairs, and defense support to public diplomacy. Communication synchronization considerations should be included in all joint planning for military operations from routine, recurring military activities in peacetime through major operations.

In addition to synchronizing the communication activities within the joint force, an effective communication synchronization effort is developed in concert with other USG departments and agencies, partner nations, and NGOs as appropriate. CCDRs should develop staff procedures for implementing communication synchronization guidance into all joint planning and targeting processes as well as collaborative processes for integrating communication synchronization activities with nonmilitary partners and subject matter experts.

See pp. 2-18 to 2-27 for related discussion of information as a joint function, the information environment, information function activities and information operations.

II(b). Application of Guidance

Ref: JP 5-0, Joint Planning (Jun '17), chap. II, section b.

I. Joint Planning and Execution Community (JPEC)

The headquarters, commands, and agencies involved in joint planning or committed to a joint operation are collectively termed the JPEC. Although not a standing or regularly meeting entity, the JPEC consists of the stakeholders shown below.

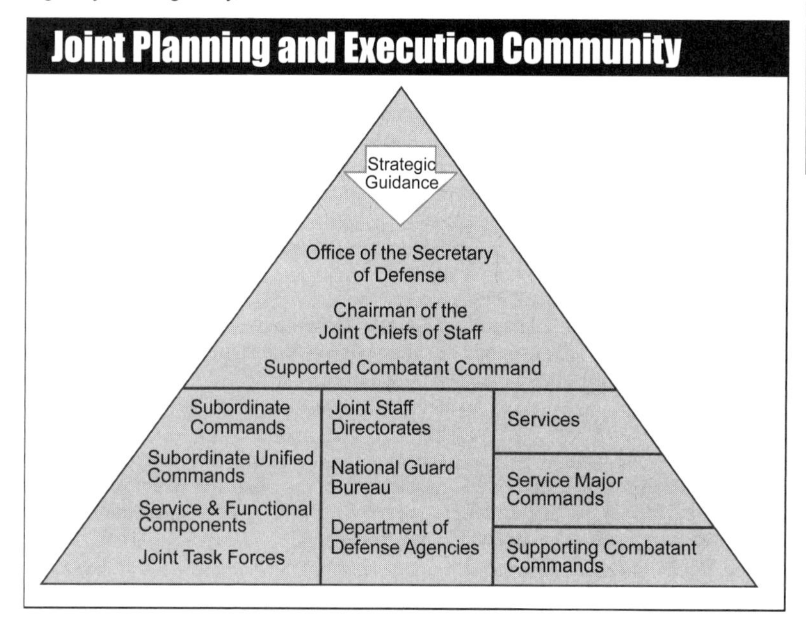

Ref: JP 5-0, fig. II-3. Joint Planning and Execution Community.

The President, with the advice and assistance of the NSC and CJCS, issues policy and strategic direction to guide the planning efforts of DOD and other USG departments and agencies that represent all of the instruments of national power. SecDef, with the advice and assistance of the CJCS, organizes the JPEC for joint planning by establishing appropriate command relationships among the CCDRs and by establishing appropriate support relationships between the CCDRs and the CSAs for that portion of their missions involving support for operating forces.

A supported commander is identified for specific planning tasks, and other JPEC stakeholders are designated as appropriate. This process provides for increased unity of command in the planning and execution of joint operations and facilitates unity of effort within the JPEC.

Supported CCDR

The supported CCDR has primary responsibility for all aspects of a task assigned by the GEF, the JSCP, or other joint planning directives. In the context of joint planning, the supported commander can initiate planning at any time based on command authority or in response to direction or orders from the President, SecDef, or CJCS. The designated supporting commanders provide planning assistance, forces, or other resources to a supported commander, as directed.

Supporting Commander

Supporting commanders provide forces, assistance, or other resources to a supported commander in accordance with the principles set forth in JP 1. Supporting commanders prepare supporting plans as required. A commander may be a supporting commander for one operation while being a supported commander for another.

II. Adaptive Planning and Execution Enterprise (APEX)

APEX integrates the planning activities of the JPEC and facilitates the transition from planning to execution. The APEX enterprise operates in a networked, collaborative environment, which facilitates dialogue among senior leaders, concurrent and parallel plan development, and collaboration across multiple planning levels. Strategic direction and continuous dialogue between senior leaders and planners facilitate an early understanding of the situation, problems, and objectives. The intent is to develop plans that contain military options for the President and SecDef as they seek to shape the environment and respond to contingencies. This facilitates responsive plan development that provides up-to-date planning and plans for civilian leaders. The APEX enterprise also promotes involvement with other USG departments and agencies and multinational partners.

While joint planning has the inherent flexibility to adjust to changing requirements, APEX incorporates policies and procedures to facilitate a more responsive planning process. APEX fosters a shared understanding of the current OE and planning through frequent dialogue between civilian and military leaders to provide viable military options to the President and SecDef. Continuous assessment and collaborative technology provide increased opportunities for consultation and updated guidance during the planning and execution processes.

APEX encompasses four operational activities, four planning functions, seven execution functions, and a number of related products. Each of these planning functions will include IPRs as necessary throughout planning and execution. IPR participants are based on the requirements of the plan. For example, plans directed by the GEF or JSCP generally require SecDef-level review, while plans directed by a CCDR may require only CCDR-level review.

IPRs are an iterative dialogue among civilian and military leaders at the strategic level to gain a shared understanding of the situation, inform leadership, and influence planning. Topics such as planning assumptions, interagency and multinational participation guidance, supporting and supported activity requirements, desired objectives, key capability shortfalls, acceptable levels of risk, and SecDef decisions are typically discussed. Further, IPRs expedite planning by ensuring the plan addresses the most current strategic assessments and objectives.

Refer to CJCS Guide 3130, Adaptive Planning and Execution Overview and Policy Framework, for a more complete discussion of the APEX enterprise. CJCSI 3141.01, Management and Review of Joint Strategic Capabilities Plan (JSCP)-Tasked Plans, discusses IPRs in more detail.

Joint Planning Activities, Functions and Products

Ref: JP 5-0, Joint Planning (Jun '17), fig. II-4, p. II-14.

APEX integrates the planning activities of the JPEC and facilitates the transition from planning to execution. The APEX enterprise operates in a networked, collaborative environment, which facilitates dialogue among senior leaders, concurrent and parallel plan development, and collaboration across multiple planning levels.

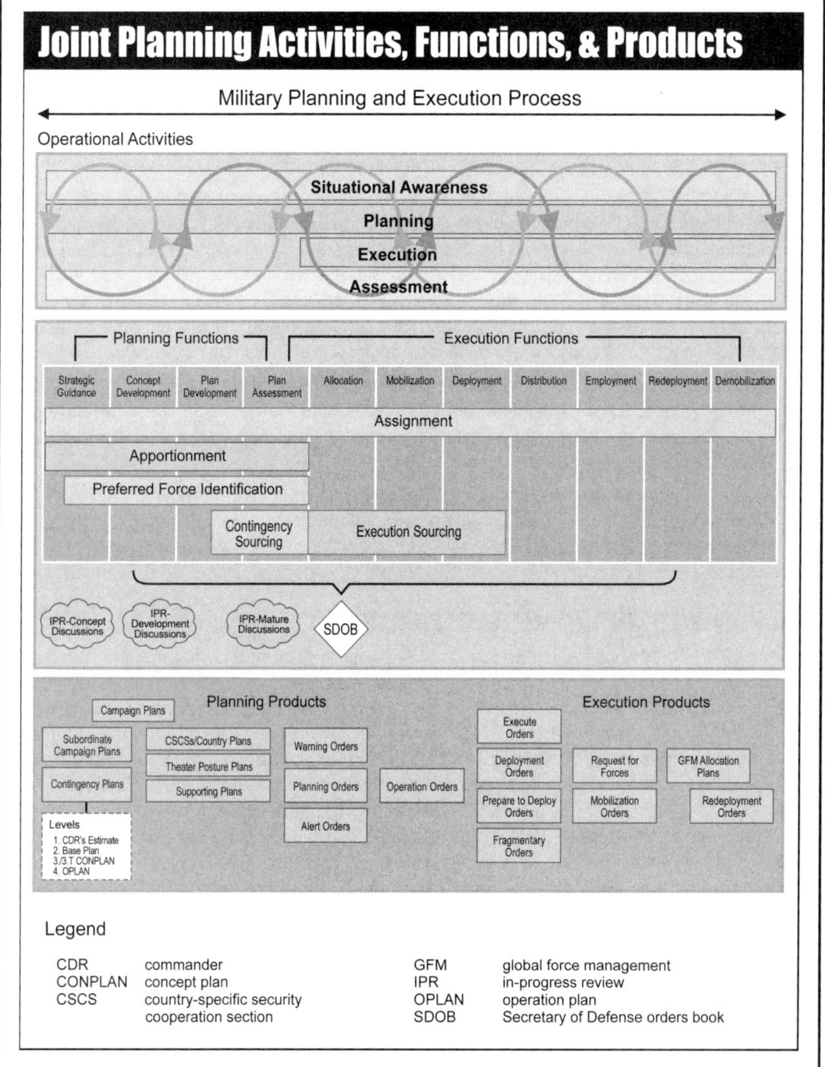

Legend

CDR	commander	GFM	global force management
CONPLAN	concept plan	IPR	in-progress review
CSCS	country-specific security cooperation section	OPLAN	operation plan
		SDOB	Secretary of Defense orders book

APEX encompasses four operational activities, four planning functions, seven execution functions, and a number of related products (see Figure II-4). Each of these planning functions will include IPRs as necessary throughout planning and execution.

Joint Planning

III. Planning Functions

Ref: JP 5-0, Joint Planning (Jun '17), pp. II-18 to II-30.

The four planning functions of strategic guidance, concept development, plan development, and plan assessment are generally sequential, although often run simultaneously in order to deepen the dialogue between civilian and military leaders and accelerate the overall planning process. SecDef, CJCS, or the CCDR may direct the planning staff to refine or adapt a plan by reentering the planning process at any of the earlier functions. The time spent accomplishing each activity and function depends on the circumstances. In time-sensitive cases, planning functions may be compressed and decisions reached in an open forum. Orders may be combined and initially communicated orally.

A. Strategic Guidance

Strategic guidance initiates planning, provides the basis for mission analysis, and enables the JPEC to develop a shared understanding of the issues, OE, objectives, and responsibilities.

- The CCDR provides input through sustained civilian-military dialogue that may include IPRs. The CCDR crafts objectives that support national strategic objectives with the guidance and consent of the SecDef; if required, the CJCS offers advice to SecDef. This process begins with an analysis of existing strategic guidance such as the GEF and JSCP or a CJCS WARNORD, PLANORD, or ALERTORD issued in a crisis. It includes mission analysis, threat assessment, and development of assumptions, which as a minimum, will be briefed to SecDef during the strategic guidance IPR.
- Some of the primary end products of the strategic guidance planning function are assumptions, ID of available/acceptable resources, conclusions about the strategic and OE (nature of the problem), strategic and military objectives, and the supported commander's mission.
- The CCDR will maintain dialogue with DOD leadership to ensure a common understanding of the above topics and alignment of planning to date. This step can be iterative, as the CCDR consults with the staff to identify concerns with or gaps in the guidance., referred to as IPR A, the CCDR should consider discussing USG SC guidance.

B. Concept Development

During planning, the commander develops several COAs, each containing an initial CONOPS that should identify major capabilities and authorities required and task organization, major operational tasks to be accomplished by components, a concept for employment and sustainment, and assessment of risk. Each COA may contain embedded multiple alternatives to accomplish designated objectives as conditions change (e.g., OE, problem, strategic direction). In time-sensitive situations, a WARNORD may not be issued, and a PLANORD, ALERTORD, or EXORD might be the first directive the supported commander receives with which to initiate planning. Using the guidance included in the directive and the CCDR's mission statement, planners solicit input from supporting and subordinate commands to develop COAs based upon the outputs of the strategic guidance planning function.

During concept development, if an IPR is required, the commander outlines the COA(s) and makes a recommendation to higher authority for approval and further development.

The commander recommends a COA that is most appropriate for the situation. Concept development should consider a range of COAs that integrate robust options to provide greater flexibility and to expedite transition during a crisis. CCDRs should be prepared to continue to develop multiple COAs to provide national-level leadership options should the crisis develop. For CCMD campaign plans, CCDRs should address resource requirements, expected changes in the environment, and how each COA supports achieving national objectives. The commander also requests SecDef's guidance on interorganizational planning and coordination and makes appropriate recommendations based on the interorganizational requirements identified during mission analysis and COA development.

One of the main products from the concept development planning function is approval for continued development of one or more COAs. Detailed planning begins upon COA approval in the concept development function.

C. Plan Development

This function is used to develop a feasible plan or order that is ready to transition into execution. This function fully integrates mobilization, deployment, employment, sustainment, conflict termination, redeployment, and demobilization activities through all phases of the plan. When the CCDR believes the plan is sufficiently developed to become a plan of record, the CCDR briefs the final plan to SecDef (or a designated representative) for approval.

Refer to CJCSI 3141.01, Management and Review of Joint Strategic Capabilities Plan (JSCP)-Tasked Plans, for discussion of topics to be discussed during reviews at each stage.

D. Plan Assessment (RATE)

Commanders continually review and evaluate the plan; determine one of four possible outcomes: refine, adapt, terminate, or execute; and then act accordingly. Commanders and the JPEC continue to evaluate the situation for any changes that would require changes in the plan. The CCDR will brief SecDef during routine plan update IPRs of modifications and updates to the plan based on the CCDR's assessment of the situation, changes in resources or guidance, and the plan's ability to achieve the objectives and attain the end states.

R - Refine

During all planning efforts, plan refinement typically is an orderly process that follows plan development and is part of the assessment function. Refinement is facilitated by continuous operation assessment to confirm changing OE conditions related to the plan or potential contingency. In a crisis, continuous operation assessment accommodates the fluidity of the crisis and facilitates continuous refinement throughout plans or OPORD development. Planners frequently adjust the plan or order based on evolving commander's guidance, results of force planning, support planning, deployment planning, shortfall ID, adversary or MNF actions, changes to the OE, or changes to strategic guidance. Based on continuous operation assessment, refinement continues throughout execution, with changes typically transmitted in the form of fragmentary orders (FRAGORDs) rather than revised copies of the plan or order.

A - Adapt

Planners adapt plans when major modifications are required, which may be driven by one or more changes in the following: strategic direction, OE, or the problem facing the JFC. Planners continually monitor the situation for changes that would necessitate adapting the plan, to include modifying the commander's operational approach and revising the CONOPS. When this occurs, commanders may need to recommence the IPR process.

T - Terminate

Commanders may recommend termination of a plan when it is no longer relevant or the threat no longer exists. For GEF- or JSCP-tasked plans, SecDef, with advice from the CJCS, is the approving authority to terminate a planning requirement.

E - Execute

"Execution."

Joint Planning

IV. Operational Activities

Operational activities are comprised of a sustained cycle of situational awareness, planning, execution, and assessment activities that occur continuously to support leader decision-making cycles at all levels of command.

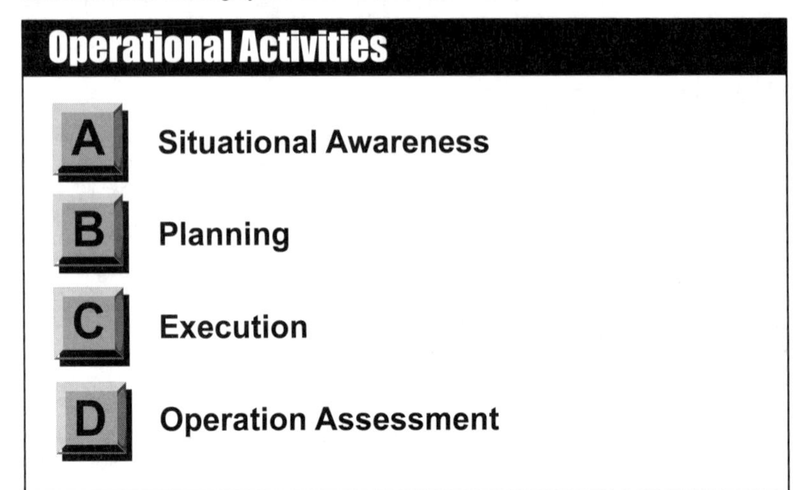

A. Situational Awareness

Situational awareness addresses procedures for describing the OE, including threats to national security. This occurs during continuous monitoring of the national and international political and military situations so CCDRs, JFCs, and their staffs can determine and analyze emerging crises, notify decision makers, and determine the specific nature of the threat. Persistent or recurring theater military engagement activities contribute to maintaining situational awareness.

Situational awareness encompasses activities such as monitoring the global situation, identifying that an event has occurred, recognizing the event is a problem or a potential problem, reporting the event, and reviewing enduring and emerging warning concerns and the CCMD's running intelligence estimate (based on continuous joint intelligence preparation of the operational environment [JIPOE]). An event is a national or international occurrence assessed as unusual and viewed as potentially having an adverse impact on US national interests and national security. The recognition of the event as a problem or potential problem follows from the observation.

B. Planning

Planning translates strategic guidance and direction into campaign plans, contingency plans, and OPORDs. Joint planning may be based on defined tasks identified in the GEF and the JSCP. Alternatively, joint planning may be based on the need for a military response to an unforeseen current event, emergency, or time-sensitive crisis.

Planning for contingencies is normally tasked in the JSCP based on the GEF or other directive. Planners derive assumptions needed to continue planning and reference the force apportionment tables to provide the number of forces expected.

Planning for crises is initiated to respond to an unforeseen current event, emergency, or time-sensitive crisis. It is based on planning guidance, typically communicated in orders (e.g., ALERTORD, WARNORD, PLANORD), and actual circumstances. Supported commanders evaluate the availability of assigned and currently allocated forces to respond to the event. They also determine what other force requirements are needed and begin putting together a rough order of magnitude force list.

C. Execution

Execution begins when the President or SecDef authorizes the initiation of a military operation or other activity. An execute order (EXORD), or other authorizing directive, is issued by the CJCS at the direction of the President or SecDef to initiate or conduct the military operations. Depending upon time constraints, an EXORD may be the only order a CCDR or subordinate commander receives. The EXORD defines the time to initiate operations and conveys guidance not provided earlier.

The CJCS monitors the deployment and employment of forces, makes recommendations to SecDef to resolve shortfalls, and tasks directed actions by SecDef and the President to support the successful execution of military operations. Execution continues until the operation is terminated or the mission is accomplished. In execution, based on continuous assessment activities, the planning process is repeated as circumstances and missions change.

The supported CCDR monitors the deployment, distribution, and employment of forces; measures task performance and progress toward mission accomplishment; and adapts and adjusts operations as required to achieve the objectives and attain the end state. This continual assessment and adjustment of operations creates an organizational environment of learning and adaptation. This adaptation can range from minor operational adjustments to a radical change of approach. When fundamental changes have occurred that challenge existing understanding or indicate a shift in the OE/problem, commanders and staffs may develop a new operational approach that recognizes that the initial problem has changed, thus requiring a different approach to solving the problem. The change to the OE could be so significant that it may require a review of the global strategic, theater strategic, and military end states and discussions with higher authority to determine whether the end states are still viable.

Changes to the original plan may be necessary because of tactical, intelligence, or environmental considerations; force and non-unit cargo availability; availability of strategic lift assets; and port capabilities. Therefore, ongoing refinement and adjustment of deployment requirements and schedules and close coordination and monitoring of deployment activities are required.

The CJCS-issued EXORD defines D-day [the unnamed day on which operations commence or are scheduled to commence] and H-hour [the specific time an operation begins] and directs execution of the OPORD. Date-time groups are expressed in universal time. While OPORD operations commence on the specified D-day and H-hour, deployments providing forces, equipment, and sustainment to support such are defined by C-day [an unnamed day on which a deployment operation begins], and L-hour [a specific hour on C-day at which a deployment operation commences or is to commence]. The CJCS's EXORD is a record communication that authorizes execution of the COA approved by the President or SecDef and detailed in the supported commander's OPORD. It may include further guidance, instructions, or amplifying orders. In a fast-developing crisis, the EXORD may be the first record communication generated by the CJCS. The record communication may be preceded by a voice authorization. The issuance of the EXORD is time sensitive. The format may differ depending on the amount of previous correspondence and the applicability of prior guidance. CJCSM 3122.01, Joint Operation Planning and Execution System (JOPES) Volume I (Planning Policies and Procedures), contains the format for the EXORD. Information already communicated in previous orders should not be repeated unless previous orders were not made available to all concerned. The EXORD need only contain the authority to execute the operation and any additional essential guidance, such as D-day and H-hour.

The supported commander issues an OPORD to subordinate and supporting commanders prior to or upon receipt of an EXORD issued by the CJCS at the direction of the President or SecDef. It may provide detailed planning guidance resulting from updated or amplifying orders, instructions, or guidance that the EXORD does not cover. Supporting commanders may develop OPORDs in support of the supported commander's OPORD. The supported commander also implements an operation assessment, which evaluates the progress toward or achievement of military objectives. This assessment informs the commanders' recommendation to the President and SecDef of when to terminate a military operation. If significant changes in the OE or the problem are identified which call into question viability of the current operational approach or objectives, the supported commander should consult with subordinate and supporting commanders and higher authority.

Following the GFM allocation process as detailed in CJCSM 3130.06, (U) Global Force Management Allocation Policies and Procedures, the supported CCDR's approved and validated force requests that have been allocated by the SecDef's decision are entered in the GFMAP annexes. The JFPs subsequently release GFMAP annex schedules reflecting specific deployment directions. The GFMAP Annexes (A and D) and Annex Schedules (B and C) serve as the deployment order (DEPORD) for specific FPs to allocate forces.

GCCs coordinate with USTRANSCOM, other supporting CCDRs, JS, and FPs to provide an integrated transportation plan from origin to destination. The transportation component commands of USTRANSCOM (Army Military Surface Deployment and Distribution Command, and Navy Military Sealift Command) coordinate common-user land and sea movements while the Air Force Air Mobility Command coordinates common user air movements for the supported GCC's time-phased force and deployment data (TPFDD). The GCCs control the flow of requirements into and out of their theater, using the appropriate TPFDD validation process, in which both supporting and supported CCMDs' staff and Service components validate unit line numbers throughout the flow.

D. Operation Assessment

Assessment determines the progress of the joint force toward mission accomplishment. Throughout the four planning functions, assessment involves comparing desired conditions of the OE with actual conditions to determine the overall effectiveness of the campaign or operation. More specifically, assessment helps the JFC measure task performance, determine progress toward or regression from accomplishing a task, creating an effect, achieving an objective, or attaining an end state; and issue the necessary guidance for change to guide forward momentum.

Assessment is a continuous operation activity in both planning and execution functions and informs the commander's decision making. It determines whether current actions and conditions are creating the desired effects and changes in the OE towards the desired objectives. Before changes in the OE can be observed, a baseline or initial assessment is required. As follow-on assessments occur, historical trends can aid the analysis and provide more definitive and reliable measures and indicators of change. Assessment helps commands analyze changes in the OE, strategic guidance, or the challenges facing the joint force in order to adapt and update plans and orders to effectively achieve desired objectives.

During planning, analysis associated with assessment helps facilitate greater understanding of the current conditions of the OE as well as identify how the command will determine the achievement of objectives if the plan is executed.

During execution, assessment helps the command evaluate the progress or regression toward mission accomplishment, and then adapt and adjust operations as required to reach the desired end state (or strategic objectives). This analysis and adjustment of operations creates an organizational environment of learning and adaptation.

Adaptation can range from minor operational adjustments to a radical change of approach, including termination of the operation. When fundamental changes have occurred that challenge existing understanding or indicate a shift in the OE, commanders and staffs may develop a new operational approach that recognizes that the initial problem has changed, thus requiring a different approach toward the solution. The change to the OE could be so significant that it may require a review of the national strategic, theater strategic, and military objectives and discussions with higher authority to determine whether the military objectives or national strategic end states are still viable.

For more information on operational assessment, see pp. 3-125 to 3-136.

V. CCMD Campaign and Contingency Planning

Contingency and CCMD campaign planning encompasses the preparation of plans that occur in non-crisis situations with a timeline generally not driven by external events. It is used to develop plans for a broad range of activities based on requirements identified in the GEF, JSCP, or other planning directives. CCMD campaign plans are the centerpiece of DOD's planning construct. They provide the means to translate strategic guidance into CCMD strategies and subsequently into executable activities. CCMD campaign plans provide the vehicle for linking current operations to contingency plans.

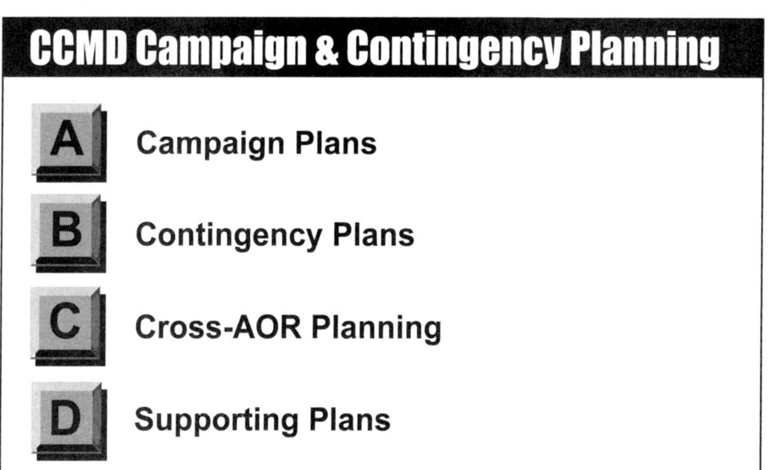

A. Campaign Plans

A campaign is a series of related military operations aimed at accomplishing strategic and operational objectives within a given time and space. Planning for a campaign is appropriate when the contemplated military operations exceed the scope of a single operation. Thus, campaigns are often the most extensive joint operations in terms of time and other resources. CCDRs document the full scope of their campaigns in the set of plans that includes the campaign plan and all of its subordinate and supporting plans.

CCDRs plan and conduct campaigns and operations, while Service and functional components conduct subordinate campaigns, operations, activities, battles, and engagements, not independent campaigns. GCCs or FCCs can plan and conduct subordinate campaigns or operations in support of another CCMD's campaign. While intended primarily to guide the use of military power, discussions and decisions at the national strategic level provide guidance for employing the different instruments of national power and should be included in the campaign plan; as should the efforts of various interorganizational partners, to achieve national strategic objectives.

Campaign plans implement a CCDR's strategy by comprehensively and coherently integrating all its activities (actual) and contingency (potential) operations. A CCDR's strategy and resultant campaign plan should be designed to achieve prioritized campaign objectives and serve as the integrating framework that informs and synchronizes all subordinate and supporting planning and operations. Campaign plans also help the CCDR in identifying resources required for achieving the objectives and tasks directed in the GEF and JSCP for input into budget and force allocation requests.

Daily operations and activities should be designed to achieve national strategic objectives; to deter and prepare for crises identified in the GEF, JSCP, and other guidance documents; and to mitigate the potential impacts of a contingency. The campaign plan is the primary vehicle for organizing, integrating, and executing security cooperation activities.

Under this construct, plans developed to respond to contingencies are best understood as branches to the overarching campaign plan (functional or theater). They address scenarios that put one or more US strategic end states in jeopardy and leave the US no recourse other than to address the problem through military actions (Figure II-5). Military actions can be in response to many scenarios, including armed aggression, regional instability, a humanitarian crisis, or a natural disaster. Contingency plans should provide a range of military options, to include flexible deterrent options (FDOs) or flexible response options (FROs), and should be coordinated with the total USG response.

B. Contingency Plans

Contingency plans are branches of campaign plans that are planned for potential threats, catastrophic events, and contingent missions without a crisis at-hand, pursuant to the strategic guidance in the UCP, GEF, and JSCP, and of the CCDR. A contingency is a situation that likely would involve military operations in response to natural and man-made disasters, terrorism, military operations by foreign powers, or other situations as directed by the President or SecDef.

Planners develop plans from the best available information, using available forces and capabilities per the GFMIG, quarterly GFM apportionment tables, existing contracts, and task orders. Planning for contingencies is based on hypothetical situations and therefore relies heavily on assumptions regarding the circumstances that will exist when a crisis arises. Planning for a contingency encompasses the activities associated with the development of plans for the deployment, employment, sustainment, and redeployment of forces and resources in response to potential crises identified in joint strategic planning documents. An existing plan with a similar scenario may be used to initiate planning in an emergent crisis situation. To accomplish this, planners develop a CONOPS that details the assumptions; adversary forces; operation phases; prioritized missions; and force requirements, deployment, and positioning. Detailed, wargamed planning supports force requirements and training in preparation for the most likely operational requirements. It also enables rapid comparison of the hypothetical conditions, operation phases, missions, and force requirements of existing contingency plans to the actual requirements of an emergent crisis. Contingency planning allows the JPEC to develop understanding, as well as the analytical and planning expertise that can be useful during a crisis.

Levels of Detail (Contingency Plans)

Ref: JP 5-0, Joint Planning (Jun '17), pp. II-23 to II-27.

There are four levels of planning detail for contingency plans, with an associated planning product for each level.

Level 1 Planning Detail—Commander's Estimate

This level of planning involves the least amount of detail and focuses on producing multiple COAs to address a contingency. The product for this level can be a COA briefing, command directive, commander's estimate, or a memorandum with a required force list. The commander's estimate provides SecDef with military COAs to meet a potential contingency. The estimate reflects the commander's analysis of the various COAs available to accomplish an assigned mission and contains a recommended COA.

Level 2 Planning Detail—Base Plan (BPLAN)

A BPLAN describes the CONOPS, major forces, concepts of support, and anticipated timelines for completing the mission. It normally does not include annexes. A BPLAN may contain alternatives, including FDOs, to provide flexibility in addressing a contingency as it develops or to aid in developing the situation. D.

Level 3 Planning Detail—Concept Plan (CONPLAN)

A CONPLAN is an OPLAN in an abbreviated format that may require considerable expansion or alteration to convert it into a complete and detailed Level 4 OPLAN or an OPORD. It includes a plan summary, a BPLAN, and usually includes the following annexes: A (Task Organization), B (Intelligence), C (Operations), D (Logistics), J (Command Relations), K (Communications), S (Special Technical Operations), V (Interagency Coordination), and Z (Distribution). If the development of a TPFDD is directed for the CONPLAN, the planning level is designated as 3T. A troop list and TPFDD would also require that an Annex E (Personnel) and Annex W (Operational Contract Support) be prepared.

For more information on OPLAN/CONPLAN format, see CJCSM 3130.03, Adaptive Planning and Execution (APEX) Planning Formats and Guidance, and Appendix A, "Joint Operation Plan Format."

Level 4 Planning Detail—OPLAN

An OPLAN is a complete and detailed plan containing a full description of the CONOPS, all applicable annexes to the plan including a time-phased force and deployment list (TPFDL), and a transportation-feasible notional TPFDD. The notional TPFDD phases unit requirements in the theater of operations at the times and places required to support the CONOPS. The OPLAN identifies the force requirements, functional support, and resources required to execute the plan and provide closure estimates for their flow into the theater. An OPLAN is normally prepared when:

- The contingency is critical to national security and requires detailed prior planning.
- The magnitude or timing of the contingency requires detailed planning.
- Detailed planning is required to support multinational planning.
- Detailed planning is necessary to determine force deployment, employment, sustainment, and redeployment requirements; determine available resources to fill identified requirements; and validate shortfalls.

When directed by the President or SecDef through the CJCS, CCDRs convert level 1, 2, and 3 plans into level 4 OPLANs or into fully developed OPORDs for execution.

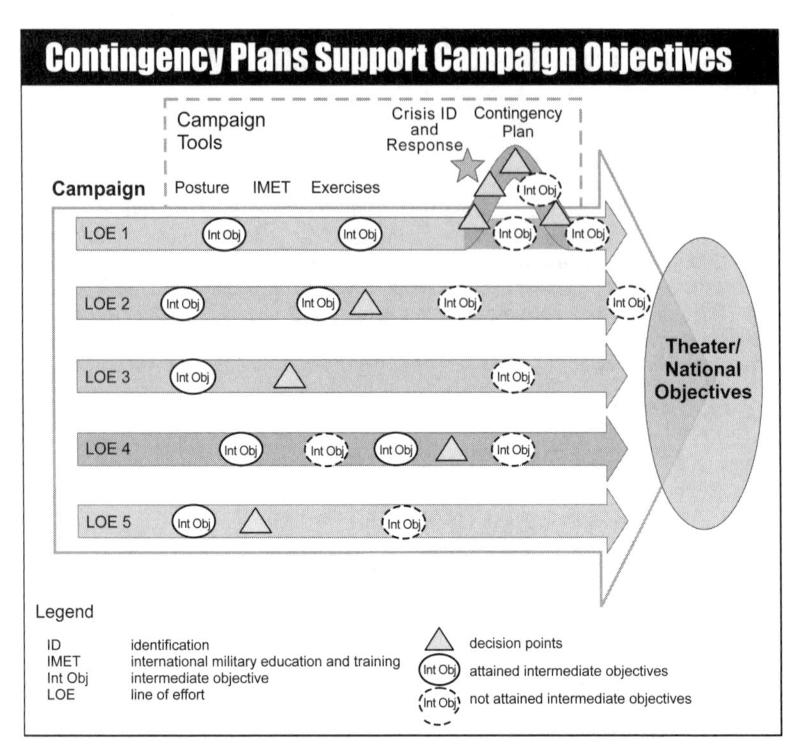

Ref: JP 5-0, fig. II-5. Contingency Plans Support Campaign Objectives.

If a situation develops outside of the strategic guidance development cycle of the GEF and JSCP that warrants a new plan that was not anticipated, the President or SecDef may issue direction through an SGS in response to the new situation. The CJCS implements the President's or SecDef's planning guidance into the appropriate orders or policy to direct the initiation of planning.

Contingency plans are produced, reviewed, and updated periodically to ensure relevancy. This planning most often addresses contingencies where military options focus on combat operations. However, these plans also account for other types of joint military operations. In addition to plans addressing all phases, including those where military action may support other agencies, planning addresses contingencies where the military is in support from the onset. These include defense support of civil authorities, support to stabilization efforts, and foreign humanitarian assistance.

Contingency plans are created as part of a collaborative process with SecDef, OSD, CJCS, JCS, CCDRs, Services, and staffs of the entire JPEC for all contingencies identified in the GEF, JSCP, and other planning directives. Planning includes JPEC concurrent and collaborative joint planning activities. The JPEC reviews those plans tasked in the JSCP for SecDef approval. The USD(P) also reviews those plans for policy considerations in parallel with their review by the CJCS. CCDRs may request a JPEC review for any tasked or untasked plans that pertain to their AOR. CCDRs may direct the development of additional plans by their commands to accomplish assigned or implied missions.

When directed by the President or SecDef through the CJCS, CCDRs convert level 1, 2, and 3 plans into level 4 OPLANs or into fully developed OPORDs for execution.

C. Cross-AOR Planning

When the scope of contemplated military operations exceeds the authority or capabilities of a single CCDR to plan and execute, the President, SecDef, or the CJCS, when designated by the President or SecDef, identify a CCDR to lead the planning for the designated strategic challenge or threat. The commander's assessment supporting this decision could be either the assessments of multiple CCDRs addressing a similar threat in their AORs or a single threat assessment from a CCDR addressing the threat from a global, cross-AOR, or functional perspective. Situations that may trigger this assessment range from combat operations that span UCP-designated boundaries to the threat of asymmetric attack that transits CCMD boundaries and functions and requires the strategic integration of the campaigns and operations of two or more CCDRs.

Per Title 10, USC, SecDef may exercise responsibilities for overseeing the activities of the CCMDs through the CJCS. Such assignment by SecDef does not confer any command authority on the CJCS and does not alter CCDRs' responsibilities prescribed in Title 10, USC, Section 164(b)(2).

When designated, the CJCS or delegated CCDR, with the authority of SecDef, issues a planning directive to the JPEC and may be tasked to lead the planning effort. The CJCS or delegated CCDR performs a mission analysis; issues initial global planning guidance based on national strategic objectives and priorities; and develops COAs in coordination with the affected CCMDs, Services, and CSAs. This COA mitigates operational gaps, seams, and vulnerabilities from a global perspective and develops an improved understanding of how actions in one AOR impact ongoing or potential plans and operations in other AORs. This will be achieved through a recommendation for the optimal allocation, prioritization, or reallocation of forces and capabilities required to develop a cohesive global CONOPS. These planning procedures will detail how CCDRs will employ forces and capabilities in support of another CCDR. The COA will be based largely on recommendations of the affected CCDRs. However, it should also assess the cumulative risk beyond a limited time horizon from a global perspective. These COAs may require refinement as initial planning apportionments are adjusted across the global CONOPS. Planners must be aware of competing requirements for potentially scarce strategic resources such as intelligence, surveillance, and reconnaissance (ISR) capabilities and transportation and ensure global planning is coordinated with GFM procedures.

All planning should be collaborative and integrated. Integrated planning addresses complex threats that span multiple AORs and functional responsibilities and provides the President and SecDef a clear understanding of how the entire military, not just a portion, will respond to those threats. The CJCS or delegated CCDR is required to mitigate operational gaps, seams, and vulnerabilities and resolve the conflict over forces, resources, capabilities, or priorities from a global perspective. Employment of space, cyberspace, and special capabilities must be informed by risks, benefits, and trade-off considerations. Early ID and submission of requests for forces and authorities and clear articulation of intent and risk can expedite decision making associated with employment of these capabilities.

When directing the execution of a contingency plan or OPORD, the President or SecDef will also select a CCDR as the supported commander for implementation of the plan. The designated supported commander has primary responsibility for all aspects of a mission. In the context of planning, the supported commander leads integrated planning with supporting CCDRs to prepare plans or orders in response to higher headquarters requirements.

D. Supporting Plans

Supporting CCDRs, subordinate JFCs, component commanders, and CSAs prepare supporting plans as tasked by the JSCP or other planning guidance. Commanders and staffs prepare supporting plans in CONPLAN/OPLAN format that follow the supported commander's concept and describe how the supporting commanders intend to achieve their assigned objectives and/or tasks. Supporting commanders and staffs develop these plans in collaboration with the supported commander's planners.

CJCSI 3141.01, Management and Review of Joint Strategic Capabilities Plan (JSCP)-Tasked Plans, governs the formal review and approval process for campaign plans and level 1–4 plans.

VI. Planning in Crises

A crisis is an incident or situation that typically develops rapidly and creates a condition of such diplomatic, economic, or military importance that the President or SecDef considers a commitment of US military forces and resources to achieve or defend national objectives. It may occur with little or no warning. It is fast-breaking and requires accelerated decision making. Sometimes a single crisis may spawn another crisis elsewhere, or there may be multiple crises occurring that concurrently impact two or more CCDRs. Furthermore, there may be a single threat with cross-AOR implications that simultaneously threaten two or more CCDRs. In this situation, supported and supporting command relationships may be fluid. Forces and capabilities committed to mitigate the emergent threat will require dynamic reallocation or reprioritization. These situations, which are increasingly the norm, further highlight the key role of integrated planning. While the planning and thought process are the same, planning in response to a crisis generally results in the publication of an order and the execution of an operation.

Planning initiated in response to an emergent event or crisis uses the same construct as all other planning. However, steps may be compressed to enable the time-sensitive development of OPLANs or OPORDs for the deployment, employment, and sustainment of forces and capabilities in response to a situation that may result in actual military operations. While planning for contingencies is based on hypothetical situations and normally is conducted in anticipation of future events, planning in a crisis is based on circumstances that exist at the time planning occurs. When possible, planners should use previously prepared plans when the emergent crisis is similar. If unanticipated circumstances occur, and no previously developed plan proves adequate for the operational circumstances, then planning would begin from scratch. Regardless of whether a plan already exists, a similar plan will be modified, or planning for the emergent crisis will begin from scratch, for those crisis situations where the problem or threat affects more than one CCDR, the basic tenets of integrated planning would still apply. There are always situations arising in the present that might require a US military response. Such situations may approximate those previously planned for, although it is unlikely they would be identical, and sometimes they will be completely unanticipated. The time available to plan responses to such real-time events can be short. In as little as a few days, commanders and staffs may need to develop and approve a feasible COA with a notional TPFDD; publish the plan or order; prepare forces; make certain that scarce assets such as communications systems, lift, precision munitions, and ISR are sufficient; develop and execute an integrated intelligence plan [Annex B (Intelligence)]; and arrange sustainment for the employment of US military forces.

Figure II-6 (facing page) provides a comparison of planning for future contingencies and planning in a crisis.

Contingency and Crisis Comparison

Ref: JP 5-0, Joint Planning (Jun '17), fig. II-6, p. II-28.

Planning initiated in response to an emergent event or crisis uses the same construct as all other planning. However, steps may be compressed to enable the time-sensitive development of OPLANs or OPORDs for the deployment, employment, and sustainment of forces and capabilities in response to a situation that may result in actual military operations. While planning for contingencies is based on hypothetical situations and normally is conducted in anticipation of future events, planning in a crisis is based on circumstances that exist at the time planning occurs.

	Planning for a CONTINGENCY	Planning in a CRISIS
Time available	As defined in authoritative directives (normally 6+ months)	Situation dependent (hours, days, up to 12 months)
Environment	Distributed, collaborative planning	Distributed, collaborative planning and execution
Facts and assumptions	Significant use of assumptions	Rely on facts and minimal use of assumption
JPEC involvement	Full JPEC participation (Note: JPEC participation may be limited for security reasons.)	Full JPEC participation (Note: JPEC participation may be limited for security reasons.)
APEX operational activities	Situational awareness Planning Assessment	Situational awareness Planning Execution Assessment
APEX functions	Strategic guidance Concept development Plan development Plan assessment	Strategic guidance Concept development Plan development Plan assessment
Document assigning planning task	CJCS issues: 1. JSCP 2. Planning directive 3. WARNORD (for short suspense planning)	CJCS issues: 1. WARNORD 2. PLANORD 3. SecDef-approved ALERTORD
Forces for planning	Apportioned in JSCP	Allocated in WARNORD, PLANORD, or ALERTORD.
Planning guidance	CJCS issues JSCP or WARNORD. CCDR issues PLANDIR and TPFDD LOI.	CJCS issues WARNORD, PLANORD, or ALERTORD. CCDR issues WARNORD, PLANORD, or ALERTORD and TPFDD LOI to subordinates, supporting commands, and supporting agencies.
COA selection	CCDR prepares COAs and submits to CJCS and SecDef for review. Specific COA may or may not be selected.	CCDR develops commander's estimate with recommended COA.
CONOPS approval	SecDef approves planning or directs additional planning or changes.	President/SecDef approve COA, disapproves or approves further planning.
Final planning product	Campaign plan. Level 1–4 contingency plan.	OPORD
Final planning product approval	CCDR submits final plan to CJCS for review and SecDef for approval.	CCDR submits final plan to President/SecDef for approval.
Execution document	Not applicable.	CJCS issues SecDef-approved EXORD. CCDR issues EXORD.
Output	Plan	Execution

Legend

ALERTORD alert order
APEX Adaptive Planning and Execution
CCDR combatant commander
CJCS Chairman of the Joint Chiefs of Staff
COA course of action
CONOPS concept of operations
EXORD execute order
JPEC joint planning and execution community
JSCP Joint Strategic Campaign Plan
LOI letter of instruction
OPORD operations order
PLANDIR planning directive
PLANORD planning order
SecDef Secretary of Defense
TPFDD time-phased force and deployment data
WARNORD warning order community

A. APEX Planning Functions

Ref: JP 5-0, Joint Planning (Jun '17), pp. II-27 to II-30.

APEX planning functions, whether performed deliberately or in response to a crisis, use the same construct to facilitate unity of effort and the transition from planning to execution. These planning functions can be compressed or truncated in time sensitive conditions to enable the rapid exchange of information and analysis, the timely preparation of military COAs for consideration by the President or SecDef, and the prompt transmission of their decisions to the JPEC. Planning activities may be performed sequentially or concurrently, with supporting and subordinate plans or OPORDs being developed concurrently. The exact flow of activities is largely determined by the time available to complete the planning and by the significance of the crisis.

The following paragraphs summarize the activities and interaction that occur in a compressed planning process such as a crisis. Refer to the CJCSM 3130 and 3122 series of publications, which address planning policies and procedures, for detailed procedures.

1. When the President, SecDef, or CJCS decides to develop military options, the CJCS issues a planning directive to the JPEC initiating the development of COAs and requesting that the supported commander submit a commander's estimate of the situation with a recommended COA to resolve the situation. Normally, the directive will be a WARNORD, but a PLANORD or ALERTORD may be used if the nature and timing of the crisis warrant accelerated planning. In a quickly evolving crisis, the initial WARNORD may be communicated verbally with a follow-on record copy to ensure the JPEC is kept informed. If the directive contains a force deployment preparation order or DEPORD, SecDef approval is required.

2. The amount of detail included in the WARNORD depends on the known facts and time available when issued. The WARNORD should describe the situation, establish command relationships, and identify the mission and any planning constraints. It may identify forces and strategic mobility resources, or it may request that the supported commander develop these factors. It may establish tentative dates and times to commence mobilization, deployment, or employment, or it may solicit the recommendations of the supported commander regarding these dates and times. The WARNORD should also identify any planning assumptions, restraints, or constraints the President or SecDef have identified to shape the response. If the President, SecDef, or CJCS directs development of a specific option or especially a COA, the WARNORD will describe the COA and request the supported commander's assessment. A WARNORD sample is in the CJCSM 3130.03, Adaptive Planning and Execution (APEX) Planning Formats and Guidance.

3. In response to the WARNORD, the supported commander, in collaboration with subordinate and supporting commanders and the rest of the JPEC, reviews existing joint contingency plans for applicability and develops, analyzes, and compares COAs and prepares a commander's estimate that provides recommendations and advice to the President, SecDef, or higher headquarters for COA selection. Based on the supported commander's guidance, supporting commanders begin their planning activities.

4. Although an existing plan almost never completely aligns with an emerging crisis, it can be used to facilitate rapid COA development and be modified to fit the specific situation. TPFDDs developed for specific plans are stored in the Joint Operation Planning and Execution System (JOPES) database and are made available to the JPEC for review.

5. The CJCS, in consultation with other members of the JCS and JPEC, reviews and evaluates the supported CCDR's estimate and provides recommendations and advice to the President and SecDef for COA selection. The supported CCDR's COAs

may be accepted, refined, or revised, or a new COA(s) may have to be developed. The President or SecDef selects a COA and directs that detailed planning be initiated.

6. Upon receiving directions from the President or SecDef, the CJCS issues a SecDef-approved ALERTORD to the JPEC. The order is a record communication stating the President or SecDef has approved the detailed development of a military plan to help resolve the crisis. The contents of an ALERTORD may vary depending upon the crisis and amount of prior planning accomplished, but it should always describe the selected COA in sufficient detail to allow the supported commander, in collaboration with other members of the JPEC, to conduct the detailed planning required to deploy, employ, and sustain forces. However, the ALERTORD does not authorize execution of the approved COA.
7. The supported commander then develops an OPORD using the approved COA. The speed with which the OPORD is developed depends upon the amount of prior planning and the planning time available. The supported commander and subordinate commanders identify force requirements, contracted support requirements and management, existing contracts and task orders, and mobility resources, and describe the CONOPS in OPORD format. The supported commander reviews available assigned and allocated forces that can be used to respond to the situation and if a gap exists, submits a request for forces (RFF) to the JS for forces to be allocated. For a detailed description of the GFM allocation process refer to CJCSM 3130.06, (U) Global Force Management Allocation Policies and Procedures.
8. The supported CCDR submits the completed OPORD for approval to SecDef or the President via the CJCS. The President or SecDef may decide to begin deployment in anticipation of executing the operation or as a show of resolve, execute the operation, place planning on hold, or cancel planning pending resolution by some other means. Detailed planning may transition to execution as directed or become realigned with continuous situational awareness, which may prompt planning product adjustments and/or updates.
9. Plan development continues after the President or SecDef's execution decision. When the crisis does not lead to execution, the CJCS provides guidance regarding continued planning.

Abbreviated APEX Planning Procedures

The preceding discussion describes the activities sequentially. During a crisis, they may be conducted concurrently or compressed, depending on prevailing conditions. It is also possible that the President or SecDef may decide to commit forces shortly after an event occurs, thereby significantly compressing planning activities. Although the allocation process has standard timelines, they may be accelerated. No specific length of time can be associated with any particular planning activity. Severe time constraints may require crisis participants to pass information verbally, including the decision to commit forces. Verbal orders are followed up, as soon as practical, with written orders.

B. Joint Orders

Ref: JP 5-0, Joint Planning (Jun '17), pp. II-30 to II-32.

Upon approval, CCDRs and Services issue orders directing action (see Figure II-7). Formats for orders can be found in CJCSM 3130.03, Adaptive Planning and Execution (APEX) Planning Formats and Guidance. By the CJCS's direction, the JS J-3 develops, coordinates, and prepares APEX orders. Subsequently, the JS J-3 is responsible for preparing and coordinating the Secretary of Defense Orders Book to present recommendations to SecDef for decision.

Joint Orders

	Order Type	Intended Action	Secretary of Defense Approval Required
Warning order	WARNORD	Initiates development and evaluation of COAs by supported commander. Requests commander's estimate be submitted.	No. Required when WARNORD includes deployment or deployment preparation actions.
Planning order	PLANORD	Begins planning for anticipated President or SecDef-selected COA. Directs preparation of OPORDs or contingency plan.	No. Conveys anticipated COA selection by the President or SecDef.
Alert order	ALERTORD	Begins execution planning on President or SecDef-selected COA. Directs preparation of OPORD or contingency plan.	Yes. Conveys COA selection by the President or SecDef.
Operation order	OPORD	Effect coordinated execution of an operation.	Specific to the OPORD.
Prepare to deploy order	PTDO	Increase/decrease deployability posture of units.	Yes (if allocates force). Refers to five levels of deployability posture.
Deployment/ redeployment order	DEPORD	Deploy/redeploy forces. Establish C-day/L-hour. Increase deployability. Establish joint task force.	Yes (if allocates force). Required for movement of unit personnel and equipment into combatant commander's AOR.
Execute order	EXORD	Implement President or SecDef decision directing execution of a COA or OPORD.	Yes.
Fragmentary order	FRAGORD	Issued as needed after an OPORD to change or modify the OPORD execution.	No.

Legend

AOR area of responsibility
C-day unnamed day on which a deployment operation begins
COA course of actions
L-hour specific hour on C-day at which deployment operation commences or is to commence
SecDef Secretary of Defense

Ref: JP 5-0, fig. II-7. Joint Orders.

See pp. 3-119 to 3-124 for a notional operation plan format from JP 5-0. The CJCSM 3130 series volumes describe joint planning interaction among the President, SecDef, CJCS, the supported joint commander, and other JPEC members, and provides models of planning messages and estimates. CJCSM 3130.03, Adaptive Planning and Execution (APEX) Planning Formats and Guidance, provides the formats for joint plans.

Warning Order (WARNORD)

A WARNORD, issued by the CJCS and/or commander, is a planning directive that initiates the development and evaluation of military COAs by a supported commander and requests that the supported commander submit a commander's estimate. If the order contains the deployment of forces, SecDef's authorization is required.

Planning Order (PLANORD)

A PLANORD is a planning directive that provides essential planning guidance and directs the initiation of plan development before the directing authority approves a military COA.

Alert Order (ALERTORD)

An ALERTORD is a planning directive that provides essential planning guidance and directs the initiation of plan development after the directing authority approves a military COA. An ALERTORD does not authorize execution of the approved COA.

Prepare to Deploy Order (PTDO)

PTDOs are approved by SecDef for allocated forces and contained in the GFMAP. The supported CCDR may order their assigned forces to deploy or order them to be prepared to deploy via a DEPORD. A PTDO is an order to prepare a unit to increase the deployability posture of units on a specified timeline.

Deployment Order (DEPORD)

A planning directive from SecDef, issued by the CJCS, authorizes the transfer and allocation of all forces among CCMDs, Services, and DOD agencies and specifies the authorities the gaining CCDR will exercise over specified forces to be transferred. The GFMAP is a global DEPORD for all allocated forces. FPs deploy or prepare forces to deploy on a specified timeframe as directed in the GFMAP. CJCSM 3130.06, (U) Global Force Management Allocation Policies and Procedures, and GFMIG discuss the DEPORD in more detail.

Execute Order (EXORD)

An EXORD is a directive to implement an approved military CONOPS. Only the President and SecDef have the authority to approve and direct the initiation of military operations. The CJCS, by the authority of and at the direction of the President or SecDef, may subsequently issue an EXORD to initiate military operations. Supported and supporting commanders and subordinate JFCs use an EXORD to implement the approved CONOPS.

Operation Order (OPORD)

An OPORD is a directive issued by a commander to subordinate commanders for the purpose of effecting the coordinated execution of an operation. Joint OPORDs are prepared under joint procedures in prescribed formats during a crisis.

See pp. 3-119 to 3-124 for a notional operation plan format from JP 5-0.

Fragmentary Order (FRAGORD)

A FRAGORD is a modification to any previously issued order. It is issued as needed to change an existing order or to execute a branch or sequel of an existing OPORD. It provides brief and specific directions that address only those parts of the original order that have changed.

In a crisis, situational awareness is continuously fed by the latest all-source intelligence and operations reports as part of the continuous assessment of operational activities. An adequate and feasible military response in a crisis demands flexible procedures that consider time available, rapid and effective communications, and relevant previous planning products whenever possible.

In a crisis or time-sensitive situation, the CCDR reviews previously prepared plans for suitability. The CCDR may refine or adapt these plans into an executable OPORD or develop an OPORD from scratch when no useful contingency plan exists.

Chap 3

III. Strategy & Campaign Development

Ref: JP 5-0, Joint Planning (Jun '17), chap. III.

Joint Planning

DOD is tasked to conduct operations on a daily basis to aid in achieving national objectives. In turn, CCDRs are tasked to develop strategies and campaigns to shape the OE in a manner that supports those strategic objectives. They conduct their campaigns primarily through military engagement, operations, posture, and other activities that seek to achieve US national objectives and prevent the need to resort to armed conflict while setting conditions to transition to contingency operations when required. The CCMD strategies and campaign plans are nested within the framework of the NSS, DSR, and NMS and are conducted in conjunction with the other instruments of national power. Specific guidance to the commanders is found in the UCP, GEF, and JSCP. Strategy prioritizes resources and actions to achieve future desired conditions. It acknowledges the current conditions as its start point, but must look past the current conditions and envision a future, then plot the road to get there. Plans address detailed execution to implement the strategy. National strategy prioritizes the CCMD's efforts within and across theaters, functional, and global responsibilities; and considers all means and capabilities available in the CCMD's operations, activities, and investments to achieve the national objectives and complement related USG efforts over a specified timeframe (currently five years). In this construct, the CCDRs and their planners develop strategy and plan campaigns to integrate joint operations with national-level resource planning and policy formulation and in conjunction with other USG departments and agencies.

Vision

The CCDR develops a long-range vision that is consistent with the national strategy and US policy and policy objectives. The vision is usually not constrained by time or resources, but is bounded by the national policy.

Strategy

Strategy is a broad statement of the CCDR's long-term vision guided by and prepared in the context of SecDef's priorities and within projected resources. Strategy links national strategic guidance to joint planning.

The CCDR's strategy prioritizes the ends, ways, and means within the limitations established by the budget, GFM processes, and strategic guidance/direction. The strategy must address risk and highlight where and what level risk will be accepted and where it will not be accepted. The strategy's objectives are directly linked to the achievement of national objectives.

Strategy includes a description of the factors and trends in the OE key to achieving the CCMD's objectives, the CCDR's approach to applying military power in concert with the other instruments of national power in pursuit of the objectives and the risks inherent in implementation.

Strategy must be flexible to respond to changes in the OE, policy, and resources. Commanders and their staff assess the OE, as well as available ways, means, and risk then update the strategy as needed. It also recognizes when ends need updating either because the original ones have been attained or they are no longer applicable.

I. Purpose of the CCDRs' Campaign Plans

The CCDRs' campaigns operationalize the CCDRs' strategies by organizing and aligning operations, activities, and investments with resources to achieve the CCDRs' objectives and complement related USG efforts in the theaters or functional areas.

CCDRs translate the strategy into executable actions to accomplish identifiable and measurable progress toward achieving the CCDRs' objectives, and thus the national objectives. The achievement of these objectives is reportable to DOD leadership through IPRs and operation assessments (such as the CCDRs' annual input to the AJA).

CCMD campaign plans integrate posture, resources, requirements, subordinate campaigns, operations, activities, and investments that prepare for, deter, or mitigate identified contingencies into a unified plan of action.

The purpose of CCMD campaigns is to shape the OE, deter aggressors, mitigate the effects of a contingency, and/or execute combat operations in support of the overarching national strategy.

Shaping the OE is changing the current conditions within the OE to conditions more favorable to US interests. It can entail both combat and noncombat operations and activities to establish conditions that support future US activities or operations, or validate planning assumptions.

Deterrence activities, as part of a CCMD campaign, are those actions or operations executed specifically to alter adversaries' decision calculus. These actions or operations may demonstrate US commitment to a region, ally, partner, or principle. They may also demonstrate a US capability to deny an adversary the benefit of an undesired action. Theater posture and certain exercises are examples of possible deterrent elements of a campaign. These actions are the most closely tied elements of the campaign to contingency plans directed in the GEF and JSCP. Additional deterrence activities are associated with early phases of a contingency plan, usually directed and executed in response to changes in threat posture.

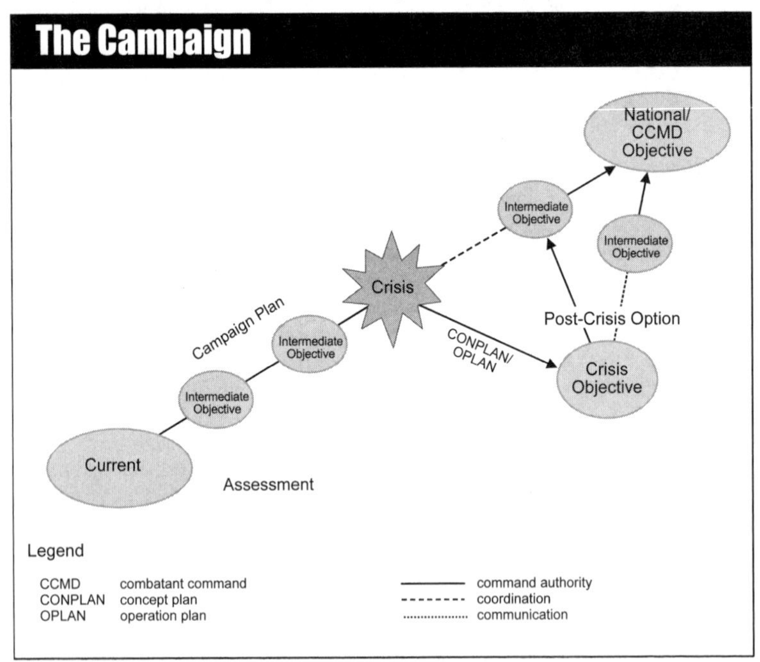

Ref: JP 5-0, fig. III-1. The Campaign.

Differences Between CCMD Campaign Plans and Contingency Plans

Ref: JP 5-0, Joint Planning (Jun '17), pp. III-3 to III-5.

CCMD Campaigns Plans

CCMD campaigns plans seek to shape the OE and achieve national objectives. They establish operations, activities, and investments the command undertakes to achieve specific objectives (set conditions) in support of national policy and objectives.

CCMD campaigns are proactive and rarely feature a single measure of military success implying victory in the traditional sense. The campaign may include operations across the spectrum of conflict, to include ongoing combat operations, such as counterterrorism operations. In the event a contingency operation is executed, that operation is subsumed into the campaign and becomes an element the CCDR considers when identifying the impact of US operations on the OE, the opportunities to favorably affect the OE to achieve national-level and theater-level objectives, and examining MOEs that may impact the campaign's intermediate objectives.

Campaign plans seek to capitalize on the cumulative effect of multiple coordinated and synchronized operations, activities, and investments that cannot be accomplished by a single major operation.

Contingency Plans

Contingency plans identify how the command might respond in the event of a crisis or the inability to achieve objectives. Contingency plans specifically seek to favorably resolve a crisis that either was not or could not be deterred or avoided by directing operations toward achieving specified objectives.

Contingency plans have specified end states that seek to re-establish conditions favorable to the US. They react to conditions beyond the scope of the CCMD campaign plan. Contingency plans have an identified military objective and set of termination criteria. Upon terminating a contingency plan, military operations return to campaign plan execution. However, the post-contingency OE may require different or additional military activities to sustain new security conditions.

Although campaign plan operations, activities, and investments can have deterrent effects, contingency plan deter activities specifically refer to actions for which separate and unique resourcing and planning are required. These actions are executed on order of the President or SecDef and generally entail specific orders for their execution and require additional resources allocated through GFM processes.

Campaign plans will have some similarities with contingency plans:

Measurable and Time-Bound

Campaign plans, like contingencies, must have measurable objectives and a process for associating CCMD actions to the changes in the OE. The commander must be able to identify within a directed time-span the ability to effect change and whether or not given actions successfully affected an associated change.

Changeable and Flexible

Campaigns must adapt to changes in the OE, other actors actions, and changes in resourcing and priorities based on national and defense priorities. However, a campaign should not necessarily change every time a commander or staff changes. Well-designed campaigns can withstand changes in foreseeable national leadership fluctuations in the US and by the countries addressed in the campaign.

A campaign can also set conditions that mitigate the impact of a possible contingency (see Figure III-1). Activities conducted as part of the campaign, such as posture and security cooperation (e.g., military engagement with allies and partners or building partner capacity and capability) can set the stage for more rapid, successful execution of a contingency plan if conflict erupts. Campaign activities can also validate planning assumptions used in the contingencies.

A campaign can support stabilization. Where US national security objectives depend upon maintaining or reestablishing stability, stabilization links the application of joint force combat power and security assistance capabilities with the achievement of strategic and policy objectives. Stabilization efforts focus on the root causes of instability and mitigating the drivers of conflict for an affected host nation (HN), thus helping the HN reach a sustainable political settlement that allows societal conflicts to be resolved peacefully.

II. Campaign Planning

Principles of Joint Operations

Campaigns and campaign planning follow the principles of joint operations while synchronizing efforts throughout the OE with all participants. Examples include:

- **Objective.** Clear campaign objectives must be articulated and understood across the joint force. Objectives are specified to direct every military operation toward a clearly defined, decisive, and achievable goal. Objectives may change as national and military leaders gain a better understanding of the situation, or they may occur because the situation itself changes. The JFC should remain sensitive to shifts in political goals necessitating changes in the military objectives toward attainment of the strategic end states.
- **Unity of Command.** Unity of command means all forces operate under a single commander with the requisite authority to direct all forces employed in pursuit of a common purpose. During multinational operations and interagency coordination, unity of command may not always be possible, but unity of effort, the coordination and cooperation toward common objectives, becomes paramount for successful unified action.
- **Economy of Force.** Economy of force is the judicious employment and distribution of forces to achieve campaign objectives.
- **Legitimacy.** Legitimacy maintains legal and moral authority in the conduct of operations. Legitimacy is based on the actual and perceived legality, morality, and rightness of actions from the perspectives of interested audiences.

See pp. 2-2 to 2-3 for discussion of the principles of joint operations.

Lines of Effort (LOEs)

In GEF- and JSCP-directed campaigns, it is often easier to organize the campaign along LOEs. A LOE links multiple tasks and missions using the logic of purpose—cause and effect—to focus efforts toward establishing operational and strategic conditions. LOEs link intermediate objectives on a path to the military and hence the campaign objective. LOEs are used to visualize the relationships between conditions, campaign objectives and, by inference, the theoretical end state. Because a campaign is conditions-based and must be adaptive to events, LOEs indicate a route rather than a precise timetable of events. They indicate how, and in what order (and with what dependencies), it is envisaged that the activities of the joint force will contribute to the achievement of desired objectives.

LOEs may intersect and interact. The campaign should identify how success or failure along a LOE will impact the lines of operation (LOOs) and other LOEs and, if necessary, how resources can be redirected to respond to unexpected effects (successes, failures, or unintended consequences) of operations on both its own and other LOEs.

Development of Campaign Plans

Ref: JP 5-0, Joint Planning (Jun '17), pp. III-5 to III-6.

Campaign plans are informed by **operation assessments** that continuously measure progress or regression regarding clearly defined, measurable, and attainable intermediate objectives nested under campaign objectives. During the planning functions, planners can use a combination of operational design and JPP that asks four questions:

- What are the current conditions of the OE (where are you)?
- What are the future conditions you want to establish (where do you want to go; what is the objective)?
- How will you get there (resources and authorities)?
- How will you know that you have been successful (assessment)? (Assessment is not just measuring achievement of an intermediate or campaign objective. It also requires measuring the performance and the effects of joint activities to determine whether they can or will generate the desired effects or establish the desired conditions.)

Campaigns are informed by **strategic guidance** and the requirement to be ready to execute contingency plans. Throughout the four planning functions, beginning with mission analysis within JPP, the CCDR and staff develop and update the commander's critical information requirements (CCIRs). This concurrently complements assessment activities by including information requirements critical to addressing key assessment indicators, required contingency preparations, deterrent opportunities, and the critical vulnerabilities of all actors within the OE. Through backward planning, CCMDs identify precursor actions, campaign activities, and necessary authorities that should be executed (or provided) as part of the campaign to deter, prepare for, or mitigate contingencies outside of crisis conditions. If successfully conducted, the campaign mitigates the risk for conflict in the context of the directed contingency plan, sets conditions for more rapid and successful transition of the contingency plan to execution if conflict proves unavoidable, and sets conditions to forestall future crises.

The same construct of APEX operational activities and planning functions, processes, procedures, and tools is used by planners to develop contingency plans and campaign plans. The applications of these can be tailored.

Because there is no military end state or termination criteria for a GEF- or JSCP-directed campaign, the objectives established in the plan are guideposts rather than goalposts and map a route in support of US objectives. The GEF- and JSCP-directed campaign plans do not seek to defeat an enemy in combat but to improve the OE in support of US national interests. As one objective is achieved, another should be designated.

The frame of reference for the campaign plan must be critically examined. When trying to map a complex system, planners tend to map it from their point of view. The relationships and logic chains developed during planning will reflect their perspective. Other participants in the system, to include allies, partners, and adversaries, often come from different backgrounds with different rules and relationships, so the effects of US actions may not result in the desired conditions. What may seem like cooperation from a US perspective may appear to be coercion from the partner's perspective.

Rather than having an enemy COG, the CCMD campaign plan may identify several COGs or areas the command may affect to achieve its objectives. Since the campaign addresses a large, complex problem, it may not be a single issue, but a confluence of several issues interacting that affect the OE.

See pp. 3-125 to 3-136 for a more complete discussion of operation assessment.

A. DOD-Wide Campaign Plans

When a campaign addresses a persistent threat that spans multiple commands, such as terrorism, threats to space and cyberspace assets or capabilities, or distribution operations, the President or SecDef may designate coordinating authority to one CCDR to lead the planning effort, with execution accomplished across multiple CCMDs. CCMDs may identify those activities that support the overall plan through the development of a separate subordinate campaign plan or through inclusion in their overall campaign plan.

The CCDR with coordinating authority coordinates planning efforts of CCDRs, Services, and applicable DOD agencies in support of the designated DOD global campaign plan. The phrase "coordinated planning" pertains specifically to planning efforts only and does not, by itself, convey authority to execute operations or direct execution of operations. Unless directed by SecDef, the CCDR responsible for leading the planning effort is responsible for aligning and harmonizing the CCMD campaign plans. Execution of the individual plans remains the responsibility of the GCC or FCC in whose UCP authority it falls.

CCDRs develop subordinate campaign plans to satisfy the planning requirements of DOD global campaign plans. While these plans are designated subordinate plans, this designation does not alter current command relationships. GCCs remain the supported commanders for the execution of their plans unless otherwise directed by SecDef.

See facing page for discussion of lead and supporting responsibilities for DOD-wide campaign plans.

B. Conditions, Objectives, Effects, and Tasks Linkage

For CCMD campaign plans, the CCDR develops military objectives to aid in focusing the strategy and campaign plan. CCDRs' strategies establish long-range objectives to provide context for intermediate objectives. Achieving intermediate objectives sets conditions to achieve the command's objectives. The CCDR and planners update the CCMD's strategy and TCP based on changes to national objectives, achievement of TCP objectives, and changes in the OE.

Conditions

Conditions describe the state of the OE. These are separate from the objective, as an objective may be achieved, but fail to set the desired conditions.

Objectives

Objectives are clearly defined, measurable, and attainable. Intermediate objectives serve as waypoints against which the CCMD can measure success in attaining GEF-directed and national objectives.

Tasks

Tasks direct friendly actions to create desired effect(s). These are the discrete activities directed in the campaign plan used to influence the OE. The execution of a task will result in an effect.

DOD-Wide Campaign Plans (Responsibilities)

Ref: JP 5-0, Joint Planning (Jun '17), pp. III-8 to III-9.

If directed to develop or synchronize a DOD-wide campaign plan, the lead CCMD:

- Provides a common plan structure and strategic framework to guide and inform development of CCDR subordinate campaign plans, Services, CSAs, the National Guard Bureau, and other DOD agencies supporting plans and mitigate seams and vulnerabilities from a global perspective.
- Establishes a common process for the development of subordinate and supporting plans.
- Organizes and executes coordination and collaboration conferences in support of the global campaign to enhance development of subordinate and supporting plans consistent with the established strategic framework and to coordinate and conduct synchronization activities.
- Disseminates lessons learned to CCDRs, Services, and applicable DOD agencies.
- Reviews and coordinates all subordinate and supporting plans to align them with the DOD global campaign plan.
- Assesses and provides recommendations to senior military and civilian leadership on the allocation of forces to coordinate the supported and supporting plans from a global perspective.
- Assesses supported and supporting plans and presents integrated force and capability shortfalls with potential sourcing options. These shortfalls and options inform SecDef of the challenges to executing the plan and the decisions that will likely be required should the plan transition to execution.
- Provides advice and recommendations to CCDRs, JS, and OSD to enhance integration and coordination of subordinate and supporting plans with the DOD global campaign plan.
- Accompanies supporting CCDRs as they brief their supporting plans through final approval, as required. To ensure coordination, all plans should be briefed at the same time.
- Develops assessment criteria and timelines. Collects and collates assessments, and provides feedback on plan success (e.g., accomplishment of intermediate objectives, milestones) through IPRs and the AJA process.
- In coordination with the JS, makes recommendations for communication annex.
- The JSCP may provide additional guidance on coordinating authority based on specific planning requirements.

Supporting CCDRs, Services, the National Guard Bureau, and applicable DOD agencies:

- Provide detailed planning support to the lead CCMD to assist in development of the DOD-wide campaign plan.
- Support plan conferences and planning efforts.
- Develop supporting plans consistent with the strategic framework, planning guidance, and process established by the lead planner.
- Provide subordinate or supporting plans to the lead planner prior to IPRs with enough time for the lead CCMD to review and propose modifications prior to the IPR.

The SecDef may also direct the CJCS to support global campaign planning. This designation will not change command relationships, but takes advantage of the CJCS's position to look across the CCMDs and provide a global perspective of opportunities and risk in developing globally integrated plans.

III. Elements of a CCMD Campaign Plan

Ref: JP 5-0, Joint Planning (Jun '17), pp. III-12 to III-14.

The CCMD campaign plan consists of all plans contained within the established theater or functional responsibilities to include contingency plans, subordinate and supporting plans, posture plans, country-specific security cooperation sections/country plans (for geographic commands), and operations in execution.

A. Campaign Plan

The campaign plan operationalizes the CCDR's strategy by organizing operations, activities, and investments within the assigned and allocated resources to achieve the GEF- and JSCP-directed objectives, as well as additional CCDR-determined objectives within the time frame established by the GEF or JSCP.

The campaign plan should show the linkages between operations, activities, investments, and expenditures and the campaign objective and associated end states that available resources will support. The campaign plan should identify the assessment process by which the command ascertains progress toward or regression from the national security objectives.

Refer to CJCSM 3130.01, Campaign Planning Procedures and Responsibilities, for additional information on how to develop campaign plans.

B. Posture Plan

The posture plan is the CCMD's proposal for forces, footprint, and agreements required and authorized to achieve the command's objectives and set conditions for accomplishing assigned missions.

C. Theater Logistics and Distribution Plans

See chap. 4, Joint Logistics, pp. 4-34 to 4-37 for further discussion of joint logistics planning process outputs.

Theater Distribution Plan (TDP)

The TDP provides detailed theater mobility and distribution analysis to ensure sufficient capacity or planned enhanced capability throughout the theater and synchronization of distribution planning throughout the global distribution network. The TDP includes a comprehensive list of references, country data, and information requirements necessary to plan, assess, and conduct theater distribution and joint reception, staging, onward movement, and integration (JRSOI) operations. The GCCs develop their TDPs using the format in USTRANSCOM's Campaign Plan for Global Distribution, JSCP, and JSCP Logistics Supplement. TDPs and TPPs complement each other by posturing forces, footprints, and agreements that will interface with the GCC's theater distribution network in order to provide a continuous flow of material and equipment into the AOR. This synchronization enables a GCC's theater distribution pipeline to have sufficient capacity and capability to support development of TCPs, OPLANs, and CONPLANs.

For more information, refer to JP 5-0, Appendix J, "Theater Distribution Plans."

Theater Logistics Overview (TLO)

The TLO codifies the GCC's theater logistics analysis (TLA) within the TPP. The TLO provides a narrative overview, with supporting matrices of key findings and capabilities from the TLA, which is included in the TPP as an appendix.

Theater Logistics Analysis (TLA)

The TLA provides detailed country by country analysis of key infrastructure by location or installation (main operating base [MOB]/forward operating site [FOS]/cooperative security location [CSL]), footprint projections, HN agreements, existing contracts, and task orders required to logistically support theater campaigns and their embedded contingency operations.

D. Regional and Country-Specific Security Cooperation Sections/Country Plans

As needed or directed, CCDRs prepare country-specific security cooperation sections/ country plans within their campaign plans for each country where the CCMD intends to apply significant time, money, and/or effort. CCDRs may also prepare separate regional plans. These are useful to identify and call out activities directed toward specific regional or country objectives and provide focus for the command.

Regional and country-specific security cooperation sections/country plans can also serve to better harmonize activities and investments with other agencies. By isolating the desired objectives, planners can more easily identify supporting efforts and specific assessment measures toward achieving US objectives.

Where the US has identified specific objectives with a country or region (through strategic guidance or policy), separate regional or country-specific security cooperation sections/ country plans help to identify resource requirements and risk associated with resource limitations that may be imposed.

E. Subordinate, Supporting, and Campaign Support Plans

Subordinate Campaign Plan

JFCs subordinate to a CCDR or other JFC may develop subordinate campaign plans in support of the higher plan to better synchronize operations in time and space. It may, depending upon the circumstances, transition to a supported or supporting plan in execution.

Supporting Plans

Supporting plans are prepared by a supporting commander, a subordinate commander, or the head of a department or agency to satisfy the requests or requirements of the supported commander's plan.

Campaign Support Plans

Campaign support plans are developed by the Services, National Guard Bureau, and DOD agencies that integrate the appropriate USG activities and programs, describe how they will support the CCMD campaigns, and articulate institutional or component-specific guidance.

F. Contingency Plans

Contingency plans are branch plans to the campaign plan that are based upon hypothetical situations for designated threats, catastrophic events, and contingent missions outside of crisis conditions. The campaign plan should address those known issues in the contingencies that can be addressed prior to execution to establish conditions, conduct deterrence, or address assumptions. As planners develop contingency plans, issues and concerns in the contingency should be included as an element of the campaign.

IV. Resource-Informed Planning (Capability Assignment, Apportionment, Allocation)

GEF- and JSCP-directed campaigns, unlike contingency plans, are plans in execution. They are constrained by the readiness and availability of resources and authorities and forecast future requirements based on projected results of current on-going operations and activities.

CCDRs are responsible for planning, assessing, and executing their GEF- and JSCP-directed campaign plans. The CCMDs, however, receive limited budgeting and rely on the Services and the CCMD component commands to budget for and execute campaign activities. As such, the components, Joint Force Coordinator and JFPs must be involved during the planning process to identify resources and tools that are likely to be made available to ensure the campaign plan is executable.

Campaign planning requires planning across four resource timeframes (see fig. III-2).

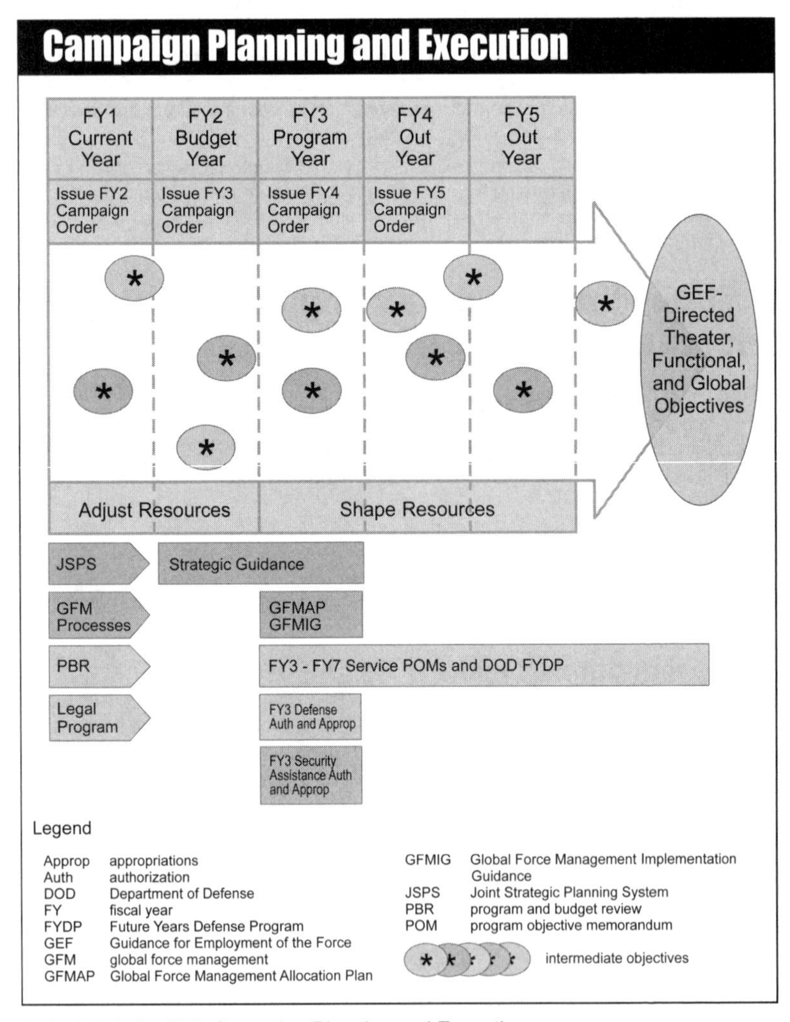

Ref: JP 5-0, fig. III-2. Campaign Planning and Execution.

V. Risk

Ref: JP 5-0, Joint Planning (Jun '17), pp. III-14 to III-15.

GCCs assess how strongly US interests are held within their respective areas, how those interests can be threatened, and their ability to execute assigned missions to protect them and achieve US national objectives. This is documented in the CCDR's strategic estimate and in the annual submission to the AJA.

CCDRs and DOD's senior leaders work together to reach a common understanding of integrated risk (the strategic risk assessed at the CCMD level combined with the military risk), decide what risk is acceptable, and minimize the effects of accepted risk by establishing appropriate risk controls.

Strategic Risk

For strategic risk, CCDRs identify the probability and consequence of near (0-2 years) and mid-term (3-7 years) strategic events or crises that could harm US national interests, and they identify the impacts of long-term (8-20 years) trends and future adversary capabilities.

Military Risk

For military risk, CCMDs evaluate the impact of the difference between required and available capability, capacity, readiness, plans, and authorities on their ability to execute assigned missions. Military risk is composed of the risk to mission assessed by the CCMD, risk to the force assessed by the Services, and risk to potential future operations. Assessments include, but are not limited to:

- FYDP budgetary priorities, trade-offs, or fiscal constraints.
- Deficiencies and strengths in force capabilities identified during preparation and review of campaign and contingency plans.
- Projected readiness of forces required to execute the campaign in future years.
- Assumptions or plans about contributions or support of:
- Other USG departments and agencies.
- Alliances, coalitions, and other friendly nations.
- Operational contract support (OCS).
- Changes in adversary capabilities identified during the preparation of the strategic estimate and other intelligence products.

Commanders must be willing to stop unproductive and minimally productive activities. Although there is currently no proven cost-benefit analysis for strategic assessment, the commanders should be willing to try new activities to see if there are better or less risky methods to achieve theater and national objectives.

For additional information on risk, see CJCSM 3105.01, Joint Risk Analysis.

See p. 2-17 for related discussion of risk management from JP 3-0.

VI. Assessing Theater and Functional Campaign Plans

Campaign plan assessments determine the progress toward accomplishing a task, creating a condition, or achieving an objective. Campaign assessments enable the CCDR and supporting organizations to refine or adapt the campaign plan and supporting plans to achieve the campaign objectives or, with SecDef approval, to adapt the GEF-directed objectives to changes in the strategic and OEs.

The campaign assessment is also DOD's bridging mechanism from the CCDR's strategy to the strategic, resource, and authorities planning processes, informing DOD's strategic direction; assignment of roles and missions; and force employment, force posture, force management, and force development decision making. Through the AJA, the campaign assessment also informs the CJCS's risk assessment and the SecDef's risk mitigation plan.

The campaign assessment provides the CCDR's input to DOD on the capabilities needed to accomplish the missions in the contingency plans of their commands over the planning horizon of the CCDR's strategy, taking into account expected changes in threats and the strategic and OEs.

Assessments enable the CCDR to make the case for additional resources or to recommend re-allocating available resources to the highest priorities. The assessment allows SecDef and senior leaders to do the same across all CCMDs and to make the case to Congress to add or re-allocate resources through the Future Years Defense Program (FYDP).

See pp. 3-125 to 3-136 for related discussion of operation assessment.

VII. Opportunity

CCDRs need to identify opportunities they can exploit to influence the situation in a positive direction. Limited windows of opportunity may open and the CCDR must be ready to exploit these to set the conditions that will lead to successful transformation of the conflict and thus to transition. This should be done in collaboration with interagency partners, international partners, and partner nations who may have assessment tools that look for opportunities to enhance resilience and mitigate conflict.

It is important to comprehend dynamics such as evolving strategic guidance and mandates, the type of conflict, the strategic logic of perpetrators, the impact of operations, and changing vulnerabilities and threats that relate to protection of civilians, resiliencies, and emerging opportunities, to enhance positive changes in the OE or among the actors.

Assessing the OE from the perspective of the root causes and immediate drivers of instability is essential to identify and create opportunities for longer-term processes to deal with the root causes.

Successful conflict transformation relies on the ability of the joint force along with the other intervening actors and local stakeholders to identify and resolve the primary sources of instability by focusing on the underlying sources of that instability, while also managing its visible symptoms. In countries seeking to transition from war to peace, a limited window of opportunity exists to mitigate sources of instability. This may include deterring adversaries and mitigating their effects on local populaces and institutions, as well as developing approaches that include marginalized groups, consensus-building mechanisms, checks and balances on power, and transparency measures.

IV. Operational Art & Operational Design

Ref: JP 5-0, Joint Planning (Jun '17), chap. IV.

The JFC and staff develop plans and orders through the application of operational art and operational design in conjunction with JPP. They combine art and science to develop products that describe how (ways) the joint force will employ its capabilities (means) to achieve military objectives (ends), given an understanding of unacceptable consequences of employing capabilities as intended (risk).

The purpose of operational design and operational art is to produce an operational approach, allowing the commander to continue JPP, translating broad strategic and operational concepts into specific missions and tasks and produce an executable plan.

Developing the Operational Approach

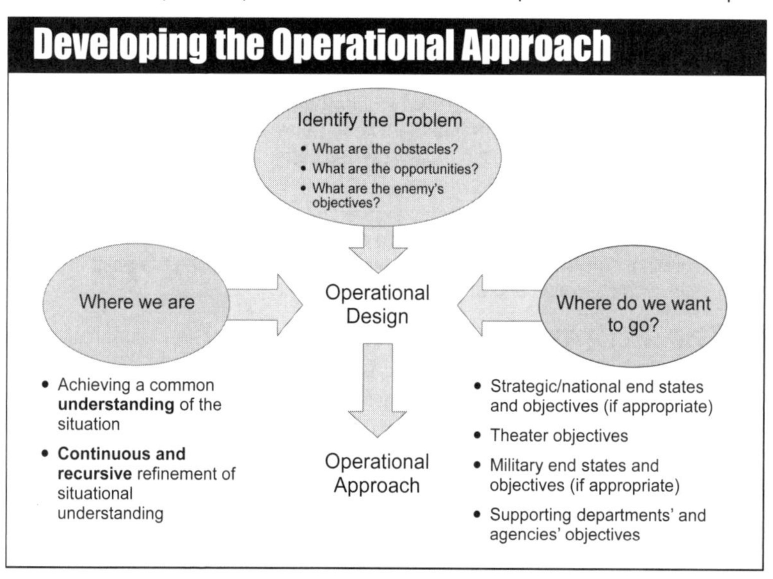

Ref: JP 5-0, fig. IV-1. Developing the Operational Approach.

Operational Art

Operational art is the cognitive approach by commanders and staffs— supported by their skill, knowledge, experience, creativity, and judgment—to develop strategies, campaigns, and operations to organize and employ military forces by integrating ends, ways, means, and risks. Operational art is inherent in all aspects of operational design.

Operational Design

Operational design is the conception and construction of the framework that underpins a campaign or operation and its subsequent execution. The framework is built upon an iterative process that creates a shared understanding of the OE; identifies and frames problems within that OE; and develops approaches, through the application of operational art, to resolving those problems, consistent with strategic guidance and/or policy. The operational approach, a primary product of operational design, allows the commander to continue JPP, translating broad strategic and operational concepts into specific missions and tasks (see Figure IV-1) to produce an executable plan.

Operational design is one of several tools available to help the JFC and staff understand the OE and develop broad solutions for mission accomplishment and understand the uncertainty in a complex OE. Additionally, it supports a recursive and ongoing dialogue concerning the nature of the problem and an operational approach to achieve the desired objectives.

Operational design and operational art enable understanding. Understanding is more than just knowledge of the capabilities and capacities of the relevant actors (individuals and organizations) or the nature of the OE, it provides context for decision making and how the many facets of the problem are likely to interact, allowing commanders and planners to identify consequences, opportunities, and recognize risk. The tools described in this chapter are meant to aid commanders in conducting robust analysis, particularly in handling unexpected events or those events outside of their previous experience or understanding. Robust analysis will aid in better understanding and ultimately better decision making.

Implementation is based on the commander's and planners' experience and time available. Different commanders and planners will need different tools to help them as each person has inherent strengths, weaknesses, and prejudices. Similarly, every problem is different and may require different tools to analyze and address it. The tools chosen by the planner should be appropriate for the problem and should complement the planners' strengths and weaknesses.

The amount of data readily available today can quickly overwhelm the planning process. Planners and commanders need to understand that a good timely decision with incomplete information may present a better solution than waiting until all information is available.

In the complex social systems that are an integral part of military operations, additional data can greatly increase the complexity of the problem without aiding understanding. Operational art aids the commander in identifying the point of diminishing returns in collection and analysis.

I. The Commander's Role

The commander is the central figure in operational design due to knowledge and experience, and because the commander's judgment and decisions are required to guide the staff through the process. Generally, the more complex a situation, the more critical the role of the commander early in planning. Commanders draw on operational design to mitigate the challenges of complexity and uncertainty, as well as leveraging their knowledge, experience, judgment, intuition, responsibility, and authority to generate a clearer understanding of the conditions needed to focus effort and achieve success.

Commanders distinguish the unique features of their current situations to enable development of innovative or adaptive solutions. They understand that each situation requires a solution tailored to the context of the problem. Through the use of operational design and the application of operational art, commanders develop innovative, adaptive alternatives to solve complex challenges. These broad alternatives are the operational approach *(fig. IV-1 on previous page)*.

Commanders use the knowledge and understanding gained from operational design, along with any additional guidance from higher headquarters, to provide commander's guidance that directs and guides the staff through JPP in preparing detailed plans and orders. Developing meaningful touch-points throughout the planning process with the supported and supporting commanders and other stakeholders enables a shared understanding of the operational environment (OE).

Operational design requires the commander to encourage discourse and leverage dialogue and collaboration to identify complex, ill-defined problems. To that end, the commander must empower organizational learning and develop methods to determine whether modifying the operational approach is necessary during the course of

Red Teaming

Ref: JP 5-0, Joint Planning (Jun '17), pp. IV-3 to IV-4.

Gathering and analyzing information—along with discerning the perceptions of adversaries, enemies, partners, and other relevant actors—is necessary to correctly frame the problem, which enables planning. A red team, an independent group that challenges an organization to improve its effectiveness, can aid a commander and the staff to think critically and creatively; see things from varying perspectives; challenge their thinking; avoid false mind-sets, biases, or group thinking; and avoid the use of inaccurate analogies to frame the problem.

Red teaming provides an independent capability to fully explore alternatives in plans and operations in the context of the OE and from the perspective of adversaries and other relevant actors.

Commanders use red teams to aid them and their staffs to provide insights and alternatives during planning, execution, and assessment to:

- Broaden the understanding of the OE.
- Assist the commander and staff in framing problems and defining end state conditions.
- Challenge assumptions.
- Consider the perspectives of the adversary and other relevant actors as appropriate.
- Aid in identifying friendly and enemy vulnerabilities and opportunities.
- Assist in identifying areas for assessment as well as the assessment metrics.
- Anticipate the cultural perceptions of partners, adversaries, and other relevant actors.
- Conduct independent critical reviews and analyses of plans to identify potential weaknesses and vulnerabilities.

Red teams provide the commander and staff with an independent capability to challenge the organization's thinking.

The red team crosses staff functions and time horizons in JPP, which is different than a red cell, which is composed of members of the intelligence directorate of a joint staff (J-2) and performs threat emulation, or a joint intelligence operations center (JIOC) red team as an additive element on the J-2 staff to improve the intelligence analysis, products, and processes.

See p. 3-103 for an overview of red, white, and blue cells.

an operation or campaign. This requires assessment and reflection that challenge understanding of the existing problem and the relevance of actions addressing that problem. Due to complexity and constant change, commanders should be comfortable in the recognition that they will never know everything about the given OE and will never be able to fully define its problems. As such, many of the problems in the OE may not have solutions.

II. Operational Art

Commanders, skilled in the use of operational art, provide the vision that links strategic objectives to tactical tasks through their understanding of the strategic and operational environments (OEs) during both the planning and execution phases of an operation or campaign. More specifically, the interaction of operational art and operational design provides a bridge between strategy and tactics, linking national strategic aims to operations that must be executed to accomplish these aims and identifying how to assess the impact of the operations in achieving the strategic objectives. Likewise, operational art promotes unified action by helping JFCs and staffs understand how to facilitate the integration of other agencies and multinational partners toward achieving strategic and operational objectives.

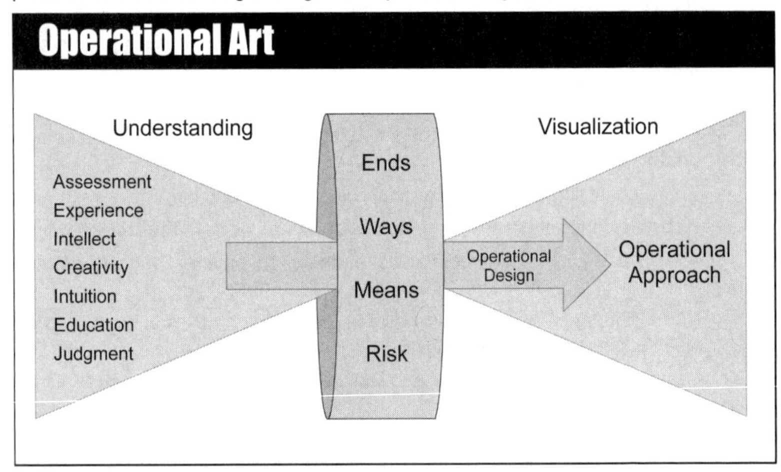

Ref: JP 5-0, fig. IV-2. Operational Art.

Through operational art, commanders link ends, ways, and means to attain the desired end state (see Figure IV-2). This requires commanders to answer the following questions:

- What is the current state of the OE?
- What are the military objectives that must be achieved, how are they related to the strategic objectives, and what objectives must be achieved to enable that strategic/national objective? How do these differ from the current conditions (state of the OE)? **(Ends)**
- What sequence of military actions, in conjunction with possible civilian actions, is most likely to achieve those objectives and attain the end state? How will I measure achievement of those objectives? **(Ways)**
- What military resources are required in concert with possible civilian resources to accomplish that sequence of actions within given or requested resources? **(Means)**
- What is the chance of failure or unacceptable consequences in performing that sequence of military actions? How will I identify if one or more of them occur? What is an acceptable level of "failure"? **(Risk)**

Role of Operational Art

Operational art enables commanders and staffs to take large amounts of data generated in the planning and analysis processes and distill it into useable information. During the plan development phase, detailed analysis may be required to determine feasible approaches and identify risk. Often during the decision-making process (and in IPRs), there is insufficient time to delve into the detail used to arrive at the proposed recommendation.

Operational art provides the ability to better understand the operational environment (OE), understand the decision-making process, and provide a concise and sufficiently detailed explanation without getting lost in the minutiae.

It also provides the commander the ability to make judgments and decisions with incomplete information. This is critical in crisis planning, time-constrained planning, and during execution, when there may not be the amount of time or analytic capability desired to conduct a full analysis of the OE.

Operational art also provides awareness of personal and organizational biases that could affect the analysis and decision processes. Although it is often difficult to completely ignore the biases, it enables an understanding of how they affect the decision process and risk associated with those decisions.

III. Operational Design

Operational design is a methodology to aid commanders and planners in organizing and understanding the OE.

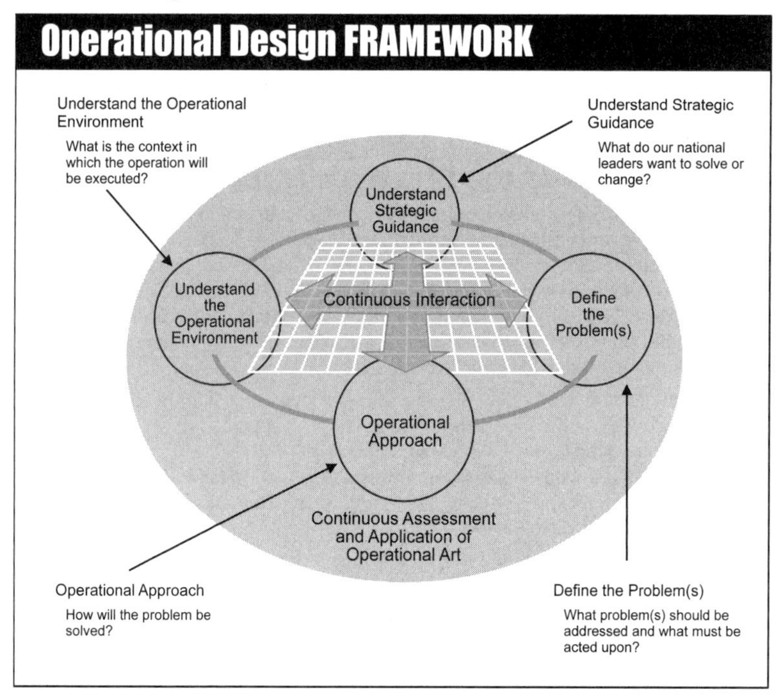

Ref: JP 5-0, fig. IV-3. Operational Design Framework.

There are four major components to operational design (see Figure IV-3). The components have characteristics that exist outside of each other and are not necessarily sequential. However, an understanding of the OE and problem must be established prior to developing operational approaches.

Operational design is one of several tools available to help the JFC and staff understand the broad solutions for mission accomplishment and to understand the uncertainty in a complex OE. Additionally, it supports a recursive and ongoing dialogue concerning the nature of the problem and an operational approach to achieve the desired objectives.

The process is continuous and cyclical in that it is conducted prior to, during, and for follow-on joint operations.

Operational Design Methodology

The general methodology in operational design is:

- Understand the strategic direction and guidance.
- Understand the strategic environment (policies, diplomacy, and politics).
- Understand the OE.
- Define the problem.
- Identify assumptions needed to continue planning (strategic and operational assumptions).
- Develop options (the operational approach).
- Identify decisions and decision points (external to the organization).
- Refine the operational approach(es).
- Develop planning guidance.

These steps are not necessarily sequential. Iteration and reexamination of earlier work is essential to identify how later decisions affect earlier assumptions and to fill in gaps identified during the process.

A. Understand the Strategic Direction and Guidance

Planning usually starts with the assignment of a planning task through a directive, order, or cyclical strategic guidance depending on how a situation develops. The commander and staff must analyze all available sources of guidance. These sources include written documents, such as the GEF and JSCP, written directives, oral instructions from higher headquarters, domestic and international laws, policies of other organizations that are interested in the situation, communication synchronization guidance, and higher headquarters' orders or estimates.

Strategic direction from strategic guidance documents can be vague, incomplete, outdated, or conflicting. This is due to the different times at which they may have been produced, changes in personnel that result in differing opinions or policies, and the staffing process where compromises are made to achieve agreement within the documents. During planning, commanders and staff must read the directives and synthesize the contents into a concise statement. Since strategic guidance documents can be problematic, the JFC and staff should obtain clear, updated direction through routine and sustained civilian-military dialogue throughout the planning process. When clarification does not occur, planners and commanders identify those areas as elements of risk.

Additionally, throughout the planning process, senior leaders will provide additional guidance. This can be through formal processes such as SGSs and IPRs, or through informal processes such as e-mails, conversations, and meetings. All of this needs to be disseminated to ensure the command has a common understanding of higher commander's intent, vision, and expectations.

In particular, commanders maintain dialogue with leadership at all levels to resolve differences of interpretation of higher-level objectives and the ways and means to accomplish these objectives. Understanding the OE, defining the problem, and devising a sound approach, are rarely achieved the first time. Strategic guidance addressing complex problems can initially be vague, requiring the commander to interpret and filter it for the staff. While CCDRs and national leaders may have a clear strategic perspective of the problem from their vantage point, operational-level commanders and subordinate leaders often have a better understanding of specific circumstances that comprise the operational situation and may have a completely different perspective on the causes and solutions.

Strategic guidance is essential to operational art and operational design. As discussed elsewhere,the President, SecDef, and CJCS all promulgate strategic guidance. In general, this guidance provides long-term as well as intermediate or ancillary objectives. It should define what constitutes victory or success (ends) and identify available forces, resources, and authorities (means) to achieve strategic objectives. The operational approach (ways) of employing military capabilities to achieve the ends is for the supported JFC to develop and propose, although policy or national positions may limit options available to the commander.

For situations that require the employment of military capabilities (particularly for anticipated large-scale combat), the President and SecDef may establish a set of operational objectives. However, in the absence of coherent guidance or direction, the CCDR/JFC may need to collaborate with policymakers in the development of these objectives.

B. Understand the Strategic Environment

After analyzing the strategic guidance, commanders and planners build an understanding of the strategic environment. This forms boundaries within which the operational approach must fit. Some considerations are:

- What actions or planning assumptions will be acceptable given the current US policies and the diplomatic and political environment?
- What impact will US activities have on third parties (focus on military impacts but identify possible political fallout)?
- What are the current national strategic objectives of the USG? Are the objectives expected to be long lasting or short-term only? Could they result in unintended consequences (e.g., if you provide weapons to a nation, is there sufficient time to develop strong controls so the weapons will not be used for unintended purposes)?

Strategic-Level Considerations

Strategic-level military activities affect national and multinational military objectives, develop CCMD campaign plans to achieve these objectives, sequence military operations, define limits and assess risks for use of the military instrument of national power, and provide military forces and capabilities in accordance with authorizing directives. Within the OE, there are strategic-level considerations that may include global aspects due to global factors such as international law, the capability of adversary/enemy information activities to influence world opinion, adversary and friendly organizations and institutions, and the capability and availability of national and commercial space-based systems and information technology. Strategic-level considerations of the OE are analyzed in terms of geopolitical regions, nations, and climate rather than local geography and weather. Nonmilitary aspects of the OE assume increased importance at the strategic level. For example, the industrial and technological capabilities of a nation or region will influence the type of military force it fields, and factors may influence the ability of a nation or region to endure a protracted conflict without outside assistance. In many situations, nonmilitary considerations may play a greater role than military factors in influencing adversary and relevant actor COAs. The JIPOE process analyzes all relevant aspects of the OE, including the adversary and other actors, and PMESII Systems and Subsystems.

C. Understand the Operational Environment (OE)

The OE is the composite of the conditions, circumstances, and influences that affect the employment of capabilities and bear on the decisions of the commander. It encompasses physical areas and factors of the air, land, maritime, and space domains; the electromagnetic spectrum; and the information environment (which includes cyberspace). Included within these areas are the adversary, friendly, and neutral actors that are relevant to a specific joint operation. Understanding the OE helps the JFC to better identify the problem; anticipate potential outcomes; and understand the results of various friendly, adversary, and neutral actions and how these actions affect attaining the military end state (see Figure IV-4).

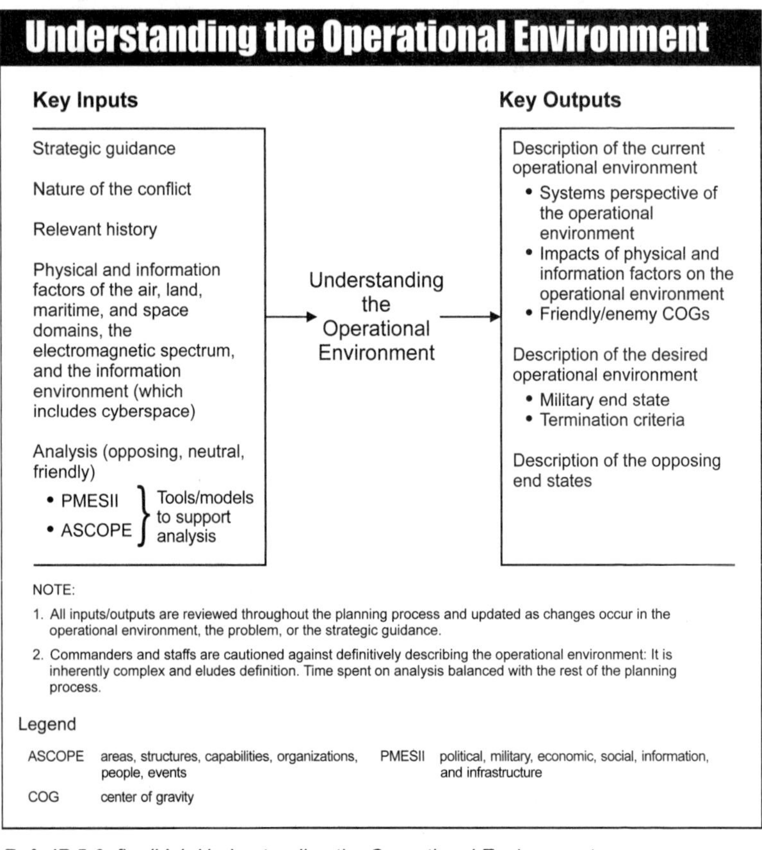

Ref: JP 5-0, fig. IV-4. Understanding the Operational Environment.

The commander must be able to describe both the current state of the OE and the desired state of the OE when operations conclude (desired military end state) to visualize an approach to solving the problem. Planners can compare the current conditions of the OE with the desired conditions. Identifying necessary objective conditions and termination criteria early in planning will help the commander and staff devise an operational approach with LOEs/LOOs that link each current condition to a desired end state condition.

See following pages (pp. 3-52 to 3-53) for an overview and further discussion of "Operational-Level Considerations" for understanding the current and future OE.

D. Define the Problem

Defining the problem is essential to addressing the problem. It involves understanding and isolating the root causes of the issue at hand—defining the essence of a complex, ill-defined problem. Defining the problem begins with a review of the tendencies and potentials of the relevant actors and identifying the relationships and interactions among their respective desired conditions and objectives. The problem statement articulates how the operational variables can be expected to resist or facilitate transformation and how inertia in the OE can be leveraged to ensure the desired conditions are achieved.

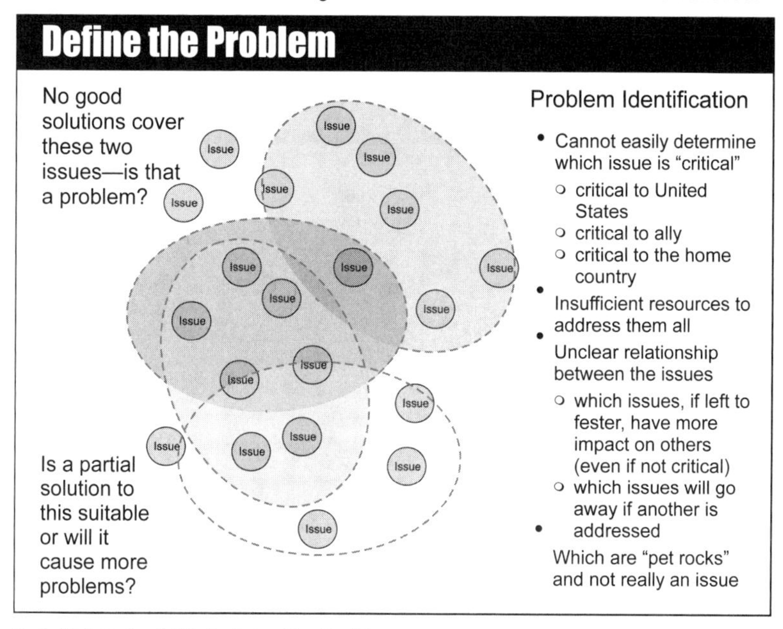

Ref: JP 5-0, fig. IV-6. Defining the Problem.

The problem statement identifies the areas for action that will transform existing conditions toward the desired end state. Defining the problem extends beyond analyzing interactions and relationships in the OE (see Figure IV-6). It identifies areas of tension and competition—as well as opportunities and challenges—that commanders must address to transform current conditions to attain the desired end state. Tension is the resistance or friction among and between actors. The commander and staff identify the tension by analyzing the context of the relevant actors' tendencies and potentials within the complex systems within the OE.

Critical to defining the problem is determining what needs to be acted on to reconcile the differences between existing and desired conditions. Some of the conditions are critical to success, some are not. Some may be achieved as a secondary or tertiary result of another condition. In identifying the problem, the planning team identifies the tensions between the desired conditions and identifies the areas of tension that merit further consideration as areas of possible intervention.

The JFC and staff must identify and articulate:

- Tensions between current conditions and desired conditions at the end state.
- Elements within the OE which must change or remain the same to attain desired end states.
- Opportunities and threats that either can be exploited or will impede the JFC from attaining the desired end state.
- Operational limitations.

Understand the Operational Environment (Operational-Level Considerations)

Ref: JP 5-0, Joint Planning (Jun '17), pp. IV-10 to IV-14.

The **JIPOE** process is a comprehensive analytic tool to describe all aspects of the OE relevant to the operation or campaign. In analyzing the current and future OE, the staff can use a **PMESII analytical framework** to determine relationships and interdependencies relevant to the specific operation or campaign (see Figure IV-5).

Holistic View of the Operational Environment

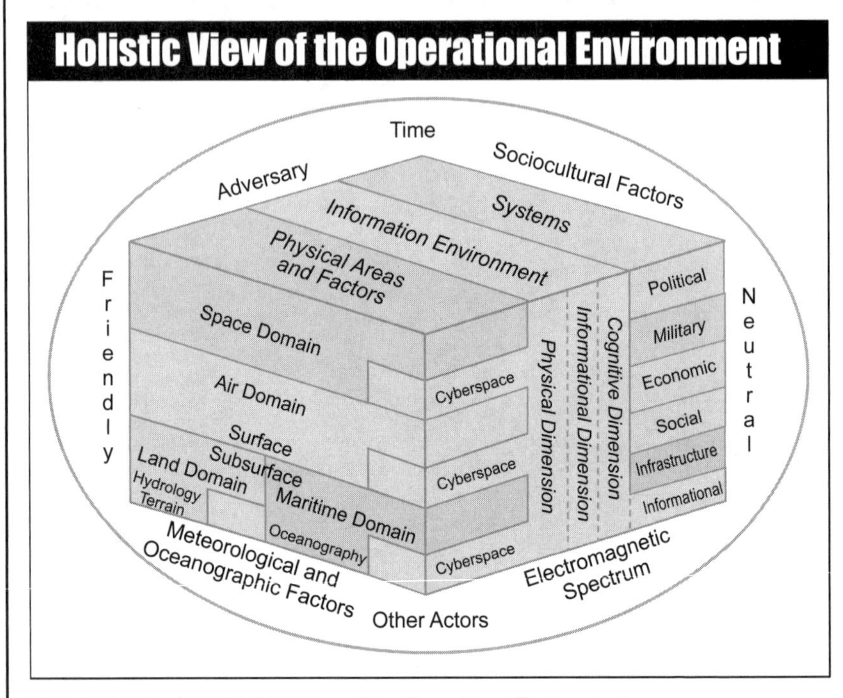

Ref: JP 5-0, fig. IV-5. Holistic View of the Operational Environment.

The size and scope of the analysis may also vary depending on particular aspects of the OE. For example, if a landlocked adversary has the capability to conduct space-based intelligence collection or cyberspace operations, then the relevant portions of space and the information environment would extend worldwide, while maritime considerations might be minimal. While most joint operations at the operational level may encompass many or all PMESII considerations and characteristics, the staff's balanced JIPOE efforts should vary according to the relevant OE aspects of the operation or campaign.

Additional factors that should be considered, include:

- Geographical features and meteorological and oceanographic characteristics.
- Population demographics (ethnic groups, tribes, ideological factions, religious groups and sects, language dialects, age distribution, income groups, public health issues).
- Social and cultural factors of adversaries, neutrals, and allies in the OE (beliefs, how and where they get their information, types and locations of media outlets).
- Political & socioeconomic factors (economic system, political factions, tribal factions).
- Infrastructure, such as transportation, energy, and information systems.

- Operational limitations such as rules of engagement (ROE), rules for the use of force (RUF), or legal restrictions on military operations as specified in US law, international law, or HN agreements.
- All friendly, adversary, and enemy conventional, irregular, and paramilitary forces and their general capabilities and strategic objectives (including all known and/or suspected chemical, biological, radiological, and nuclear threats and hazards).
- Environmental conditions (earthquakes, volcanic activity, pollution, naturally occurring diseases).
- Location of toxic industrial materials in the area of interest that may produce chemical, biological, radiological, or nuclear hazards.
- Psychological characteristics of adversary decision making.
- All locations of foreign embassies, international organizations, and NGOs.
- Friendly and adversary military and commercial capabilities provided by assets in space, their current or potential use, and critical vulnerabilities.
- Knowledge of the capabilities and intent, COGs, and critical vulnerabilities of forces, individuals, or organizations conducting cyberspace operations.
- Financial networks that could impact the adversary's ability to sustain operations.

To produce a holistic view of the relevant adversary, neutral, and friendly systems within a larger system that includes many external influences, analysis should define how these systems interrelate. Most important to this analysis is describing the relevant relationships within and between the various systems that directly or indirectly affect the problem at hand. Although the J-2 manages the JIPOE process, other directorates and agencies can contribute valuable expertise to develop and assess the complexities of the OE.

Tendencies and Potentials. In developing an understanding of the interactions and relationships of relevant actors in the OE, commanders and staffs consider observed tendencies and potentials in their analyses. Tendencies reflect the inclination to think or behave in a certain manner. Tendencies are not considered deterministic but rather model the thoughts or behaviors of relevant actors. Tendencies help identify the range of possibilities that relevant actors may develop with or without external influence. Once identified, commanders and staffs evaluate the potential of these tendencies to manifest within the OE. Potential is the inherent ability or capacity for the growth or development of a specific interaction or relationship. However, not all interactions and relationships support attaining the desired end state. The desired end state accounts for tendencies and potentials that exist among the relevant actors or other aspects of the OE. Early in JPP, pertinent lessons learned should be collected and reviewed as part of the analysis to allow previously learned lessons to make their way into the plan. The Joint Lessons Learned Information System provides a database of past lessons learned. However, people experienced in the mission, OE, and lessons learned functions should be sought for their knowledge and experience.

Describe the key conditions that must exist in the future OE to achieve the objectives. Planners should put a temporal aspect to this set of conditions in order to be able to conduct feasibility and acceptability analyses.

Determine the objectives of relevant actors affecting the OE. These actors will have different sets of conditions for achieving their respective objectives. Such opposition can be expected to take actions to thwart US and partner nations' objectives. Other actors, neutral or friendly, may not have an opposing mindset, but may have desired conditions (or unintended consequences of their actions) that oppose our desired end state conditions. The analysis of the OE should identify where the contradictions between the competing sides, allies, partners, and neutrals lie and recognize the conflicts of interests. In the course of developing the plan, planners should ask themselves if the COA being considered addresses these conflicts.

A concise problem statement is used to clearly define the problem or problem set to solve. It considers how tension and competition affect the OE by identifying how to transform the current conditions to the desired end state—before adversaries begin to transform current conditions to their desired end state. The statement broadly describes the requirements for transformation, anticipating changes in the OE while identifying critical transitions.

E. Identify Assumptions

Where there is insufficient information or guidance, the commander and staff identify assumptions to assist in framing solutions. At this stage, assumptions address strategic and operational gaps that enable the commander to develop the operational approach.

- Assumptions should be kept to the minimum required as each assumption adds to the probability of error in the plan and requires specific CCIRs to continuously check its validity.
- Assumptions address key and critical decisions required by senior leaders to enable the continuation of planning.

Commanders and staff should review strategic guidance and direction to see if any assumptions are imposed on the planning process. They should also regularly discuss planning assumptions with OSD and DOD leadership to see if there are changes in policy or guidance that affect the planning assumptions (examples could be basing or access permissions, allied or multinational contributions, alert and warning decision timelines, or anticipated threat actions and reactions). Assumptions should be phrased in terms of will or will not (rather than using "should" or "may") in order to establish specific conditions that enable planning to continue.

During JPP, the commander may develop additional assumptions to support detailed COA development.

F. Developing Operational Approaches

The operational approach is a commander's description of the broad actions the force can take to achieve an objective in support of the national objective or attain a military end state. It is the commander's visualization of how the operation should transform current conditions into the desired conditions—the way the commander envisions the OE at the conclusion of operations to support national objectives. The operational approach is based largely on an understanding of the OE and the problem facing the JFC. A discussion of operational approaches within and between options forms the basis of the IPRs between the CCDR and SecDef and staff (to ensure consistency with US policy and national objectives). Once SecDef approves the approach, it provides the basis for beginning, continuing, or completing detailed planning. The JFC and staff should continually review, update, and modify the approach as policy, the OE, end states, or the problem change. This requires frequent and continuing dialogue at all levels of command.

Commanders and their staffs can use operational design when planning any joint campaign or operation. Notwithstanding a commander's judgment, education, and experience, the OE often presents situations so complex that understanding them—let alone attempting to change them—exceeds individual capacity. Nor does such complexity lend itself to coherent planning. Bringing adequate order to complex problems to facilitate further detailed planning requires an iterative dialogue between commander, the planning staff, and policy staff. Rarely will members of either staff recognize an implicit operational approach during their initial analysis and synthesis of the OE. Successful development of the approach requires continuous analysis, learning, dialogue, and collaboration between commander and staff, as well as other subject matter experts. The challenge is even greater when the joint operation involves other agencies, the private sector, and multinational partners (which is typically the case), whose unique considerations can complicate the problem.

It is essential that commanders, through a dialogue with their staffs, planning teams, initiative groups, and any other relevant sources of information, first gain an understanding of the OE, to include the US policy perspective, and define the problem facing the joint force prior to conducting detailed planning. The problem as presented in guidance documents rarely includes all available guidance information and may identify the symptoms rather than the actual problem. From this understanding of the OE and definition of the problem, commanders develop their broad operational approach for transforming current conditions into desired conditions. The operational approach will underpin the operation and the detailed planning that follows. As detailed planning occurs, the JFC and staff continue discourse and refine their operational approach.

G. Identify Decisions and Decision Points

During planning, commanders inform leadership of the decisions that will need to be made, when they will have to be made, and the uncertainty and risk accompanying decisions and delay. This provides leaders, both military and civilian, a template and warning for the decisions in advance and provides them the opportunity to look across interagency partners and with allies to look for alternatives and opportunities short of escalation. The decision matrix also identifies the expected indicators needed in support of the intelligence collection plan.

Commanders are responsible to ensure senior leaders understand the risk and time lines associated with the decision points and the possible effects of delayed decisions.

H. Refine the Operational Approach

Throughout the planning processes, commanders and their staffs conduct formal and informal discussions at all levels of the chain of command. These discussions help refine assumptions, limitations, and decision points that could affect the operational approach and ensure the plan remains feasible, acceptable, and adequate.

The commander adjusts the operational approach based on feedback from the formal and informal discussions at all levels of command and other information.

I. Prepare Planning Guidance

Developing Commander's Planning Guidance

The commander provides a summary of the OE and the problem, along with a visualization of the operational approach, to the staff and to other partners through commander's planning guidance. As time permits, the commander may have been able to apply operational design to think through the campaign or operation before the staff begins JPP. In this case, the commander provides initial planning guidance to help focus the staff in mission analysis. Commanders should continue the analysis to further understand and visualize the OE as the staff conducts mission analysis. Upon completing analysis of the OE, the commander will issue planning guidance, as appropriate, to help focus the staff efforts. At a minimum, the commander issues planning guidance, either initial or refined, at the conclusion of mission analysis, and provides refined planning guidance as understanding of the OE, the problem, and visualization of the operational approach matures. It is critical for the commander to provide updated guidance as the campaign or operation develops in order to adapt the operational approach to a changing OE or changed problem.

See following page (p. 3-56) for discussion of the format and contents of the commander's planning guidance.

Commander's Planning Guidance

Ref: JP 5-0, Joint Planning (Jun '17), pp. IV-18 to IV-19.

The format for the commander's planning guidance varies based on the personality of the commander and the level of command, but should adequately describe the logic to the commander's understanding of the OE, the methodology for reaching the understanding of the problem, and a coherent description of the operational approach. It may include:

1. Describe the OE. Some combination of graphics showing key relationships and tensions and a narrative describing the OE will help convey the commander's understanding to the staff and other partners.

2. Define the problem to be solved. A narrative problem statement includes a timeframe to solve the problem will best convey the commander's understanding of the problem.

3. Describe the operational approach. A combination of a narrative describing objectives, decisive points, and potential LOEs and LOOs, with a summary of limitations (constraints and restraints) and risk (what can be accepted and what cannot be accepted) will help describe the operational approach.

4. Provide the commander's initial intent. The commander should also include the initial intent in planning guidance. The commander's initial intent describes the purpose of the operations, desired strategic end state, military end state, and operational risks associated with the campaign or operation. It also includes where the commander will and will not accept risk during the operation. It organizes (prioritizes) desired conditions and the combinations of potential actions in time, space, and purpose. The JFC should envision and articulate how military power and joint operations, integrated with other applicable instruments of national power, will achieve strategic success, and how the command intends to measure the progress and success of its military actions and activities. It should help staff and subordinate commanders understand the intent for unified action using interorganizational coordination among all partners and participants. Through commander's intent, the commander identifies the major unifying efforts during the campaign or operation, the points and events where operations must succeed to control or establish conditions in the OE, and where other instruments of national power will play a central role. The intent must allow for decentralized execution. It provides focus to the staff and helps subordinate and supporting commanders take actions to achieve the military objectives or attain the end state without further orders, even when operations do not unfold or result as planned.

Commander's Initial Intent (Format)

While there is no specified joint format for the commander's intent, a generally accepted construct includes the purpose, end state, and risk:

Purpose. Purpose delineates reason for the military action with respect to the mission of the next higher echelon. The purpose explains why the military action is being conducted. The purpose helps the force pursue the mission without further orders, even when actions do not unfold as planned. Thus, if an unanticipated situation arises, participating commanders understand the purpose of the forthcoming action well enough to act decisively and within the bounds of the higher commander's intent.

End State. An end state is the set of required conditions that defines achievement of the commander's objectives. This describes what the commander desires in military end state conditions that define mission success by friendly forces. It also describes the strategic objectives and higher command's military end state and describes how reaching the JFC's military end state supports higher headquarters' end state (or national objectives).

Risk. Defines aspects of the campaign or operation in which the commander will accept risk in lower or partial achievement or temporary conditions. It also describes areas in which it is not acceptable to accept such lower or intermediate conditions.

The intent may also include operational objectives, method, and effects guidance. The commander may provide additional planning guidance such as information management, resources, or specific effects that must be created or avoided.

IV. Elements of Operational Design

The elements of operational design can be used for all military planning. However, not all of the elements of operational design may be required for all plans.

Elements of Operational Design

- Termination
- Military end state
- Objectives
- Effects
- Center of gravity
- Decisive points
- Lines of operation and lines of effort
- Direct and indirect approach
- Anticipation
- Operational reach
- Culmination
- Arranging operations
- Forces and functions

See pp. 3-57 to 3-68 for an overview and further discussion.

A. Termination

Termination criteria are the specified standards approved by the President and/or SecDef that must be met before military operations can be concluded. Termination criteria are a key element in establishing a military end state. Termination criteria describe the conditions that must exist in the OE at the cessation of military operations. The conditions must be achievable and measurable so the commander can clearly identify the achievement of the military end state. Effective planning cannot occur without a clear understanding of the military end state and the conditions that must exist to end military operations. Knowing when to terminate military operations and how to preserve achieved advantages is key to attaining the national strategic end state. To plan effectively for termination, the supported JFC must know how the President and SecDef intend to terminate the joint operation and ensure that the conditions in the OE endure. CCMD campaign plans will not normally have termination criteria.

Termination criteria are developed first among the elements of operational design as they enable the development of the military end state and objectives. Commanders and their staffs must think through, in the early stages of planning, the conditions that must exist in order to terminate military operations on terms favorable to the US and its multinational partners. Termination criteria should account for a wide variety of operational tasks that the joint force may need to accomplish, to include disengagement, force protection, transition to post-conflict operations, reconstitution, and redeployment.

Military end states are briefed to SecDef as part of the IPR process to ensure the military end states support the termination criteria. Once approved, the criteria may change. It is important for commanders and staffs to keep an eye out for potential changes, as they may result in a modification to the military end state as well as the commander's operational approach. As such, it is essential for the military to keep a dialogue between the civilian national leadership, and the leadership of other agencies and partners involved.

B. Military End State

Military end state is the set of required conditions that defines achievement of all military objectives. It normally represents a point in time and/or circumstances beyond which the President does not require the military instrument of national power as the primary means to achieve remaining national objectives. As such, the military end state is often closely tied to termination. While it may mirror many of the conditions of the national strategic end state, the military end state typically will be more specific and contain other supporting conditions. These conditions contribute to developing termination criteria, the specified standards approved by the President and/or SecDef that must be met before a joint operation can be concluded. Aside from its obvious association with strategic or operational objectives, clearly defining the military end state promotes unity of effort, facilitates synchronization, and helps clarify (and may reduce) the risk associated with the campaign or operation. Commanders should include the military end state in their planning guidance and commander's intent statement.

C. Objectives

An objective is clearly defined, decisive, and attainable. Once the military end state is understood and termination criteria are established, operational design continues with development of strategic and operational military objectives. Joint planning integrates military actions and capabilities with those of other instruments of national power in time, space, and purpose in unified action to achieve the JFC's military objectives, which contribute to strategic national objectives. Objectives and their supporting effects provide the basis for identifying tasks to be accomplished. In GEF- and JSCP-directed campaign plans, objectives rather than an end state, define the path of the command's actions in contributing to national objectives.

Military missions are conducted to achieve objectives and are linked to national objectives. Military objectives are an important consideration in plan development. They specify what must be accomplished and provide the basis for describing desired effects.

A clear and concise end state allows planners to better examine objectives that must be met to attain the desired end state. Objectives describe what must be achieved to reach or attain the end state. These are usually expressed in military, diplomatic, economic, and informational terms and help define and clarify what military planners must do to support the national strategic end state. Objectives developed at the national-strategic and theater-strategic levels are the defined, decisive, and attainable goals toward which all military operations, activities, and investments are directed within the OA.

Achieving operational objectives ties execution of tactical tasks to reaching the military end state.

There are four primary considerations for an objective.

- An objective establishes a single desired result (a goal).
- An objective should link directly or indirectly to higher level objectives or to the end state.
- An objective is specific and unambiguous.
- An objective does not infer ways and/or means—it is not written as a task.

D. Effects

An effect is a physical and/or behavioral state of a system that results from an action, a set of actions, or another effect. A desired effect can also be thought of as a condition that can support achieving an associated objective, while an undesired effect is a condition that can inhibit progress toward an objective. In seeking unified action, a JFC synchronizes the military with the diplomatic, informational, and economic power of the US to affect the PMESII systems of relevant actors.

See facing page for further discussion of "End State, Objectives, Effects and Tasks."

End State, Objectives, Effects and Tasks

Ref: JP 5-0, Joint Planning (Jun '17), pp. IV-21 to IV-22 (fig. IV-8).

The CCDR plans joint operations based on analysis of national strategic objectives and development of theater strategic objectives supported by measurable strategic and operational desired effects and assessment indicators. At the operational level, a subordinate JFC develops supporting plans, which can include objectives supported by measurable operational-level desired effects and assessment indicators. This may increase operational- and tactical-level understanding of the purpose reflected in the higher-level commander's mission and intent. At the same time, commanders consider potential undesired effects and their impact on the tasks assigned to subordinate commands.

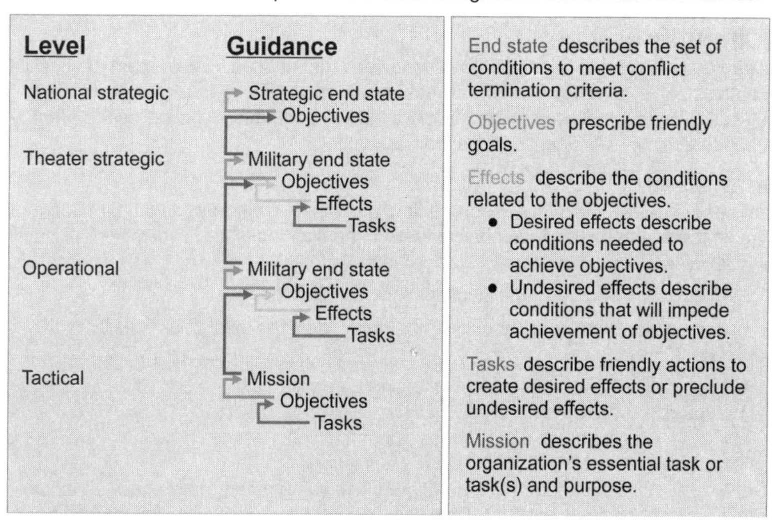

There are four primary considerations for writing a desired effect statement.

- Each desired effect should link directly to one or more objectives.
- The effect should be measurable.
- The statement should not specify ways and means for accomplishment.
- The effect should be distinguishable from the objective it supports as a condition for success, not as another objective or a task.

The proximate cause of effects in complex situations can be difficult to predict particularly when they relate to moral and cognitive issues. Further, there will always be gaps in our understanding of the OE. Commanders and their staffs must appreciate that unpredictable third-party actions, unintended consequences of friendly operations, subordinate initiative and creativity, and the fog and friction of conflict.

The use of effects in planning can help commanders and staff determine the tasks required to achieve objectives and use other elements of operational design more effectively by clarifying the relationships between COGs, LOOs, and/or LOEs; decisive points; and termination criteria. Once a systems perspective of the OE has been developed (and appropriate links and nodes have been identified), the linkage and relationship between COGs, LOOs, and decisive points can become more obvious throughout execution.

A mission is a task or set of tasks, together with the purpose, that clearly indicates the action to be taken and the reason for doing so. It is derived primarily from higher headquarters guidance.

Elements of Operational Design

Ref: JP 5-0, Joint Planning (Jun '17), executive summary and chap. IV.

Operational design employs various elements to develop and refine the commander's operational approach. These conceptual tools help commanders and their staffs think through the challenges of understanding the operational environment, defining the problem, and developing this approach, which guides planning and shapes the CONOPS.

Refer to JP 5-0, Joint Planning (Jun '17), pp. IV-19 to IV-40 for complete descriptions.

Termination

Termination criteria describe the conditions that must exist in the OE at the cessation of military operations.

Military End State

Military end state is the set of required conditions that defines achievement of all military objectives. It normally represents a point in time and/or circumstances beyond which the President does not require the military instrument of national power as the primary means to achieve remaining national objectives.

Objectives

An objective is clearly defined, decisive, and attainable. Objectives and their supporting effects provide the basis for identifying tasks to be accomplished. There are four primary considerations:

- An objective establishes a single desired result (a goal)
- An objective should link directly or indirectly to higher level objectives or to the end state
- An objective is prescriptive, specific, and unambiguous
- An objective does not infer ways and/or means—it is not written as a task

Effects

An effect is a physical and/or behavioral state of a system that results from an action, a set of actions, or another effect. There are four primary considerations for writing a desired effect statement:

- Each desired effect should link directly to one or more objectives
- The effect should be measurable
- The statement should not specify ways and means for accomplishment
- The effect should be distinguishable from the objective it supports as a condition for success, not as another objective or a task

Center of Gravity (COG)

A COG is a source of power that provides moral or physical strength, freedom of action, or will to act. An objective is always linked to a COG. In identifying COGs it is important to remember that irregular warfare focuses on legitimacy and influence over a population, unlike traditional warfare, which employs direct military confrontation to defeat an adversary's armed forces, destroy an adversary's war-making capacity, or seize or retain territory to force a change in an adversary's government or policies.

See following pages (pp. 3-62 to 3-63) for discussion of analyzing COGs.

Decisive Points (DPs)

A decisive point is a geographic place, specific key event, critical factor, or function that, when acted upon, allows a commander to gain a marked advantage over an enemy or contributes materially to achieving success.

Lines of Operation (LOO) and Lines of Effort (LOEs)

A **line of operations (LOO)** defines the interior or exterior orientation of the force in relation to the enemy or that connects actions on nodes and/or decisive points related in time and space to an objective(s). A **line of effort (LOE)** links multiple tasks and missions using the logic of purpose—cause and effect—to focus efforts toward establishing operational and strategic conditions. Major combat operations are typically designed using LOOs. These lines tie offensive, defensive, and stability tasks to the geographic and positional references in the OA. Commanders synchronize activities along complementary LOOs to achieve the end state.

- **Interior Lines.** A force operates on interior lines when its operations diverge from a central point. Interior lines usually represent central position, where a friendly force can reinforce or concentrate its elements faster than the enemy force can reposition.
- **Exterior Lines.** A force operates on exterior lines when its operations converge on the enemy. Operations on exterior lines offer opportunities to encircle and annihilate an enemy force.

Direct and Indirect Approach

The approach is the manner in which a commander contends with a COG. A direct approach attacks the enemy's COG or principal strength by applying combat power directly against it. An indirect approach attacks the enemy's COG by applying combat power against a series of decisive points that lead to the defeat of the COG while avoiding enemy strength.

Anticipation

During execution, JFCs should remain alert for the unexpected and for opportunities to exploit the situation.

Operational Reach

Operational reach is the distance and duration across which a joint force can successfully employ military capabilities.

Culmination

Culmination is that point in time and/or space at which the operation can no longer maintain momentum. In the offense, the culminating point is the point at which effectively continuing the attack is no longer possible and the force must consider reverting to a defensive posture or attempting an operational pause. A defender reaches culmination when the defending force no longer has the capability to go on the counteroffensive or defend successfully. Success in the defense is to draw the attacker to offensive culmination, then conduct an offensive to expedite the adversary's defensive culmination.

Arranging Operations

Commanders must determine the best arrangement of joint force and component operations to conduct the assigned tasks and joint force mission. This arrangement often will be a combination of simultaneous and sequential operations to reach the end state conditions with the least cost in personnel and other resources.

See pp. 3-66 to 3-67 for further discussion of the factors (simultaneity, depth, tempo and timing) and tools for arranging operations.

Operational Pause

Operational pauses may be required when a major operation may be reaching the end of its sustainability.

Forces and Functions

Typically, JFCs structure operations to attack both enemy forces and functions concurrently to create the greatest possible friction between friendly and enemy forces and capabilities.

E. Analysis of Friendly and Adversary COGs

Ref: JP 5-0, Joint Planning (Jun '17), pp. IV-23 to IV-26.

Analysis of friendly and adversary COGs is a key step in operational design. Joint force intelligence analysts identify adversary COGs, determining from which elements the adversary derives freedom of action, physical strength (means), and the will to fight. The J-2, in conjunction with other operational planners, then attempts to determine if the tentative or candidate COGs truly are critical to the adversary's strategy. This analysis is a linchpin in the planning effort. Others on the joint force staff conduct similar analysis to identify friendly COGs. Once COGs have been identified, JFCs and their staffs determine how to attack enemy COGs while protecting friendly COGs. The protection of friendly strategic COGs such as public opinion and US national capabilities typically requires efforts and capabilities beyond those of just the supported CCDR. An analysis of the identified COGs in terms of critical capabilities, requirements, and vulnerabilities is vital to this process. As the COG acts as the balance point or focal point that holds the system together, striking it should cause the system to collapse.

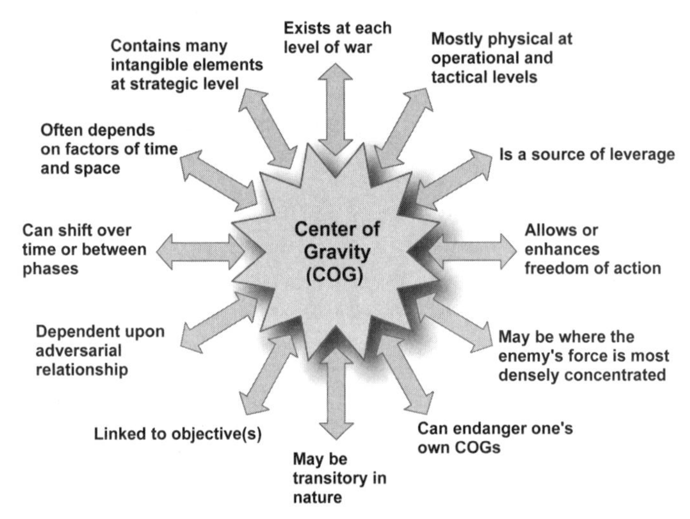

Ref: JP 5-0, Joint Planning, fig. IV-9, p IV-24.

Understanding the relationship among COGs not only permits but also compels greater precision in thought and expression in operational design. Planners should analyze COGs within a framework of three critical factors—capabilities, requirements, and vulnerabilities—to aid in this understanding.

- **Critical capabilities** are the primary abilities essential to the accomplishment of the objective.
- **Critical requirements** are essential conditions, resources, and means the COG requires to perform the critical capability.
- **Critical vulnerabilities** are those aspects or components of critical requirements that are deficient or vulnerable to direct or indirect attack in a manner achieving decisive or significant results.

In general, a JFC must possess sufficient operational reach and combat power or other relevant capabilities to take advantage of an adversary's critical vulnerabilities while protecting friendly critical capabilities within the operational reach of an adversary.

When identifying friendly and enemy critical vulnerabilities, the JFC and staff will understandably want to focus their efforts against the critical vulnerabilities that will do the most decisive damage to an enemy's COG. However, in selecting those critical vulnerabilities, planners must also compare their criticality with their accessibility, vulnerability, redundancy, resiliency, and impact on the civilian populace, and then balance those factors against friendly capabilities to affect those vulnerabilities. The JFC should seek opportunities aggressively to apply force against an adversary in as vulnerable an aspect as possible, and in as many dimensions as possible.

A proper analysis of adversary critical factors must be based on the best available knowledge of how adversaries organize, fight, think, and make decisions, and their physical and psychological strengths and weaknesses. JFCs and their staffs must develop an understanding of their adversaries' capabilities and vulnerabilities, as well as factors that might influence an adversary to abandon its strategic objectives. They must also envision how friendly forces and actions appear from the adversaries' viewpoints. Otherwise, the JFC and the staff may fall into the trap of ascribing to an adversary attitudes, values, and reactions that mirror their own.

Before solidifying COGs into the plan, planners should analyze and test the validity of the COGs. The defeat, destruction, neutralization, or substantial weakening of a valid COG should cause an adversary to change its COA or prevent an adversary from achieving its strategic objectives. If analysis and/or wargaming show this does not occur, then perhaps planners have misidentified the COG, and they must revise their COG and critical factors analysis. The conclusions, while critically important to the planning process itself, must be tempered with continuous evaluations because derived COGs and critical vulnerabilities are subject to change at any time during the campaign or operation.

Center of Gravity Analysis

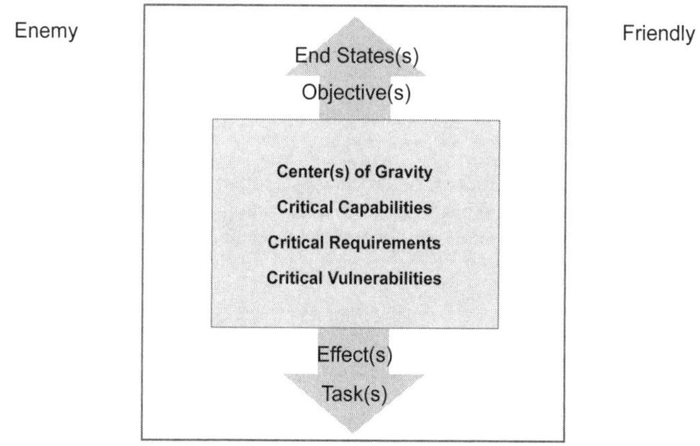

Enemy

Friendly

End States(s)

Objective(s)

Center(s) of Gravity

Critical Capabilities

Critical Requirements

Critical Vulnerabilities

Effect(s)

Task(s)

Center of Gravity Analysis supports achieving objective(s) and attaining end state(s) and helps determine the missions and tasks required to generate the desired affects.

Ref: JP 5-0, Joint Planning, fig. IV-10. Center of Gravity Analysis.

Commanders must also analyze friendly COGs and identify critical vulnerabilities (fig. IV-10). In conducting the analysis of friendly vulnerabilities, the supported commander must decide how, when, where, and why friendly military forces are (or might become) vulnerable to hostile actions and then plan accordingly. The supported commander must achieve a balance between prosecuting the main effort and protecting critical capabilities and vulnerabilities in the OA to protect friendly COGs.

For more information on COGs and the systems perspective, refer to JP 2-01.3, Joint Intelligence Preparation of the Operational Environment.

F. Defeat and Stability Mechanisms

Ref: JP 5-0, Joint Planning (Jun '17), pp. IV-31 to IV-33.

Defeat and stability mechanisms complement COG analysis. While COG analysis helps us understand a problem, defeat and stability mechanisms suggest means to solve it. They provide a useful tool for describing the main effects a commander wants to create along a line of operation (LOO) or line of effort (LOE).

Defeat Mechanisms

Defeat mechanisms primarily apply in combat operations against an active enemy force. Combat aims at defeating armed enemies—regular, irregular, or both, through the organized application of force to kill, destroy, or capture by all means available. There are two basic defeat mechanisms to accomplish this: attrition and disruption. The aim of disruption is to defeat an enemy's ability to fight as a cohesive and coordinated organization. The alternative is to destroy his material capabilities through attrition, which generally is more costly and time-consuming. Although acknowledging that all successful combat involves both mechanisms, joint doctrine conditionally favors disruption because it tends to be a more effective and efficient way of causing an enemy's defeat, and the increasing imperative for restraint in the application of violence may often preclude the alternative. The defeat mechanisms may include:

1. Destroy

To identify the most effective way to eliminate enemy capabilities; it may be attained by sequentially applying combat power over time or with a single, decisive attack.

2. Dislocate

To compel the enemy to expose forces by reacting to a specific action; it requires enemy commanders to either accept neutralization of part of their force or risk its destruction while repositioning.

3. Disintegrate

To exploit the effects of dislocation and destruction to shatter the enemy's coherence; it typically follows destruction and dislocation, coupled with the loss of capabilities that enemy commanders use to develop and maintain situational understanding.

4. Isolate

To limit the enemy's ability to conduct operations effectively by marginalizing critical capabilities or limiting the enemy's ability to influence events; it exposes the enemy to continued degradation through the massed effects of other defeat mechanisms.

Stability Mechanisms

A stability mechanism is the primary method through which friendly forces affect civilians in order to attain conditions that support establishing a lasting, stable peace. Combinations of stability mechanisms produce complementary and reinforcing effects that help to shape the human dimension of the OE more effectively and efficiently than a single mechanism applied in isolation. Stability mechanisms may include compel, control, influence, and support. Proper application of these stability mechanisms is key in irregular warfare where success is dependent on enabling a local partner to maintain or establish legitimacy and influence over relevant populations.

1. Compel

To maintain the threat—or actual use—of lethal or non-lethal force to establish control and dominance, effect behavioral change, or enforce cessation of hostilities, peace agreements, or other arrangements. Legitimacy and compliance are interrelated. While legitimacy is vital to achieving host-nation compliance, compliance depends on how the local populace perceives the force's ability to exercise force to accomplish the mission. The appropriate and discriminate use of force often forms a central component to success in stability operations; it closely ties to legitimacy. Depending on the circumstances, the threat or use of force can reinforce or complement efforts to stabilize a situation, gain consent, and ensure compliance with mandates and agreements. The misuse of force—or even the perceived threat of the misuse of force—can adversely affect the legitimacy of the mission or the military instrument of national power.

2. Control

To establish public order and safety, securing borders, routes, sensitive sites, population centers, and individuals and physically occupying key terrain and facilities. As a stability mechanism, control closely relates to the primary stability task, establish civil control. However, control is also fundamental to effective, enduring security. When combined with the stability mechanism compel, it is inherent to the activities that comprise disarmament, demobilization, and reintegration, as well as broader security sector reform programs. Without effective control, efforts to establish civil order—including efforts to establish both civil security and control over an area and its population—will not succeed. Establishing control requires time, patience, and coordinated, cooperative efforts across the OA.

3. Influence

To alter the opinions and attitudes of the HN population through IRCs, presence, and conduct. It applies nonlethal capabilities to complement and reinforce the compelling and controlling effects of stability mechanisms. Influence aims to effect behavioral change through nonlethal means. It is more a result of public perception than a measure of operational success. It reflects the ability of forces to operate successfully among the people of the HN, interacting with them consistently and positively while accomplishing the mission. Here, consistency of actions, words, and deeds is vital. Influence requires legitimacy. Military forces earn the trust and confidence of the people through the constructive capabilities inherent to combat power, not through lethal or coercive means. Positive influence is absolutely necessary to achieve lasting control and compliance. It contributes to success across the LOEs and engenders support among the people. Once attained, influence is best maintained by consistently exhibiting respect for, and operating within, the cultural and societal norms of the local populace.

4. Support

To establish, reinforce, or set the conditions necessary for the other instruments of national power to function effectively, coordinating and cooperating closely with HN civilian agencies and assisting aid organizations as necessary to secure humanitarian access to vulnerable populations. Support is vital to a comprehensive approach to stability activities. The military instrument of national power brings unique expeditionary capabilities to stabilization efforts. These capabilities enable the force to quickly address the immediate needs of the HN and local populace. In extreme circumstances, support may require committing considerable resources for a protracted period. However, easing the burden of support on military forces requires enabling civilian agencies and organizations to fulfill their respective roles. This is typically achieved by combining the effects of the stability mechanisms—compel, control, and influence—to reestablish security and control, restoring essential civil services to the local populace, and helping to secure humanitarian access necessary for aid organizations to function effectively.

G. Arranging Operations

Ref: JP 5-0, Joint Planning (Jun '17), pp. IV-36 to IV-39.

Commanders must determine the best arrangement of joint force and component operations to conduct the assigned tasks and joint force mission. This arrangement often will be a combination of simultaneous and sequential operations to reach the end state conditions with the least cost in personnel and other resources. Commanders consider a variety of factors when determining this arrangement, including geography of the OA, available strategic lift, changes in command structure, force protection, distribution and sustainment capabilities, adversary reinforcement capabilities, and public opinion. Thinking about the best arrangement helps determine the tempo of activities in time, space, and purpose. Planners should consider factors such as simultaneity, depth, timing, and tempo when arranging operations.

Simultaneity

Simultaneity refers to the simultaneous application of integrated military and nonmilitary power against the enemy's key capabilities and sources of strength. Simultaneity in joint force operations contributes directly to an enemy's collapse by placing more demands on enemy forces and functions than can be handled. This does not mean all elements of the joint force are employed with equal priority or that even all elements of the joint force will be employed. It refers specifically to the concept of attacking appropriate enemy forces and functions throughout the OE in such a manner as to damage their morale and physical cohesion.

Simultaneity also refers to the concurrent conduct of operations at the tactical, operational, and strategic levels. Tactical commanders fight engagements and battles, understanding their relevance to the contingency plan. JFCs set the conditions for battles within a major operation or campaign to achieve military strategic and operational objectives. GCCs integrate theater strategy and operational art. At the same time, they remain acutely aware of the impact of tactical events. Because of the inherent interrelationships between the various levels of warfare, commanders cannot be concerned only with events at their respective echelon, so commanders at all levels should understand how their actions contribute to the military end state.

Depth

The evolution of warfare and advances in technology have expanded the depth of operations. US joint forces can rapidly maneuver over great distances and strike with precision. Joint force operations should be conducted across the full breadth and depth of the OA, creating competing and simultaneous demands on enemy commanders and resources. The concept of depth seeks to overwhelm the enemy throughout the OA, creating competing and simultaneous demands on enemy commanders and resources and contributing to the enemy's speedy defeat. Depth applies to time as well as geography. Operations extended in depth shape future conditions and can disrupt an opponent's decision cycle. Global strike, interdiction, and the integration of IRCs with other capabilities are examples of the applications of depth in joint operations. Operations in depth contribute to protection of the force by destroying enemy potential before its capabilities can be realized or employed.

Tempo and Timing

The joint force should conduct operations at a tempo and point in time that maximizes the effectiveness of friendly capabilities and inhibits the adversary. With proper timing, JFCs can dominate the action, remain unpredictable, and operate beyond the enemy's ability to react.

The tempo of warfare has increased over time as technological advancements and innovative doctrines have been applied to military requirements. While in many situations

JFCs may elect to maintain an operational tempo that stretches the capabilities of both friendly and enemy forces, on other occasions JFCs may elect to conduct operations at a reduced pace. During selected phases of a campaign, JFCs could reduce the pace of operations, frustrating enemy commanders while buying time to build a decisive force or tend to other priorities in the OA such as relief to displaced persons. During other phases, JFCs could conduct high-tempo operations designed specifically to overwhelm enemy defensive capabilities. Assuring strategic mobility preserves the JFC's ability to control tempo by allowing freedom of theater access.

Several tools are available to planners to assist with arranging operations. Phases, branches and sequels, operational pauses, and the development of a notional TPFDD all improve the ability of the planner to arrange, manage, and execute complex operations

1. Phases

Phasing is a way to view and conduct a complex joint operation in manageable parts. The main purpose of phasing is to integrate and synchronize related activities, thereby enhancing flexibility and unity of effort during execution. Reaching the end state often requires arranging an operation or campaign in several phases. Phases in a contingency plan are sequential, but during execution there will often be some simultaneous and overlapping execution of the activities within the phases. In a campaign, each phase can represent a single major operation; while in a major operation, a phase normally consists of several subordinate operations or a series of related activities.

See following page (p. 3-68) for further discussion.

2. Branches and Sequels

Many plans require adjustment beyond the initial stages of the operation. Consequently, JFCs build flexibility into plans by developing branches and sequels to preserve freedom of action in rapidly changing conditions. They are primarily used for changing deployments or direction of movement and accepting or declining combat.

- **Branches** provide a range of alternatives often built into the basic plan. Branches add flexibility to plans by anticipating situations that could alter the basic plan. Such situations could be a result of adversary action, availability of friendly capabilities or resources, or even a change in the weather or season within the OA.
- **Sequels** anticipate and plan for subsequent operations based on the possible outcomes of the current operation—victory, defeat, or stalemate.

3. Operational Pause

The supported JFC should aggressively conduct operations to obtain and maintain the initiative. However, there may be certain circumstances when this is not feasible because of logistic constraints or force shortfalls. Therefore, operational pauses may be required when a major operation may be reaching the end of its sustainability.

As such, operational pauses can provide a safety valve to avoid potential culmination, while the JFC retains the initiative in other ways. However, if an operational pause is properly executed in relation to one's own culminating point, the enemy will not have sufficient combat power to threaten the joint force or regain the initiative during the pause.

4. Realistic Plans and an Accurate TPFDD

Realistic plans, branches, sequels, orders, and an accurate TPFDD are important to enable the proper sequencing of operations. Further, the dynamic nature of modern military operations requires adaptability concerning the arrangement of military capabilities in time, space, and purpose. For example, a rapidly changing enemy situation or other aspects of the OE may cause the commander to alter the planned arrangement of operations even as forces are deploying. Therefore, maintaining overall force visibility, to include both in-transit visibility and asset visibility, are critical to maintaining flexibility. The arrangement that the commander chooses should not foreclose future options.

V. Phasing

A phase can be characterized by the focus that is placed on it. Phases are distinct in time, space, and/or purpose from one another, but must be planned in support of each other and should represent a natural progression and subdivision of the campaign or operation. Each phase should have a set of starting conditions that define the start of the phase and ending conditions that define the end of the phase. The ending conditions of one phase are the starting conditions for the next phase.

Phases are necessarily linked and gain significance in the larger context of the campaign. As such, it is imperative the campaign or operation not be broken down into numerous arbitrary components that may inhibit tempo and lead to a plodding, incremental approach. Since a campaign is required whenever pursuit of a strategic objective is not attainable through a single major operation, the theater operational design includes provision for related phases that may or may not be executed.

Activities in phases may overlap. The commander's vision of how a campaign or operation should unfold drives subsequent decisions regarding phasing. Phasing, in turn, assists with synchronizing the CONOPS and aids in organizing the assignment of tasks to subordinate commanders. By arranging operations and activities into phases, the JFC can better integrate capabilities and synchronize subordinate operations in time, space, and purpose. Each phase should represent a natural subdivision of the campaign or operation's intermediate objectives. As such, a phase represents a definitive stage during which a large portion of the forces and joint/multinational capabilities are involved in similar or mutually supporting activities.

See pp. 2-50 to 2-51 for related discussion of phasing a joint operation from JP 3-0 (w/Chg 1), Joint Operations.

Number, Sequence, and Overlap

Working within the phasing construct, the actual phases used will vary (compressed, expanded, or omitted entirely) with the joint campaign or operation and will be determined by the JFC. During planning, the JFC establishes conditions, objectives, or events for transitioning from one phase to another and plans sequels and branches for potential contingencies. Phases are designed to be conducted sequentially, but some activities from a phase may begin in a previous phase and continue into subsequent phases. The JFC adjusts the phases to exploit opportunities presented by the adversary or operational situation or to react to unforeseen conditions. A joint campaign or operation may be conducted in multiple phases simultaneously if the OA has widely varying conditions. For instance, the commander may transition to stabilization efforts in some areas while still conducting combat operations in other areas where the enemy has not yet capitulated. Occasionally, operations may revert to a previous phase in an area where a resurgent or new enemy reengages friendly forces.

Transitions

Transitions between phases are planned as distinct shifts in focus by the joint force, often accompanied by changes in command or support relationships. The activities that predominate during a given phase, however, rarely align with neatly definable breakpoints. The need to move into another phase is normally identified by assessing that a set of objectives are achieved or that the enemy has acted in a manner that requires a major change in focus for the joint force and is therefore usually event driven, not time driven. Changing the focus of the operation takes time and may require changing commander's objectives, desired effects, measures of effectiveness (MOEs), measures of performance (MOPs), priorities, command relationships, force allocation, or even the approach. An example is the shift of focus from sustained combat operations to a preponderance of stability activities. Hostilities gradually lessen as the joint force facilitates reestablishing order, commerce, and local government and deters adversaries from resuming hostile actions while the US and international community take steps to establish or restore the conditions necessary for long-term stability.

V(a). Joint Planning Process (JPP)

Ref: JP 5-0, Joint Planning (Jun '17), chap. V.

The joint planning process (JPP) is an orderly, analytical set of logical steps to frame a problem; examine a mission; develop, analyze, and compare alternative COAs; select the best COA; and produce a plan or order.

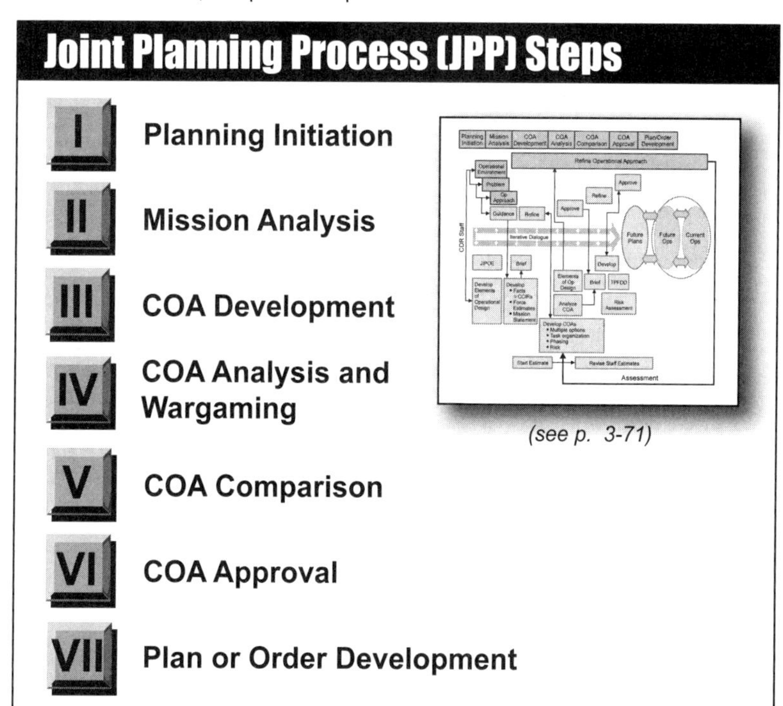

(see p. 3-71)

JPP provides a proven process to organize the work of the commander, staff, subordinate commanders, and other partners, to develop plans that will appropriately address the problem. It focuses on defining the military mission and development and synchronization of detailed plans to accomplish that mission. JPP helps commanders and their staffs organize their planning activities, share a common understanding of the mission and commander's intent, and develop effective plans and orders.

JPP is applicable for all planning. Like operational design, it is a logical process to approach a problem and determine a solution. It is a tool to be used by planners but is not prescriptive. Based on the nature of the problem, other tools available to the planner, expertise in the planning team, time, and other considerations, the process can be modified as required. Similarly, some JPP steps or tasks may be performed concurrently, truncated, or modified as necessary dependent upon the situation, subject, or time constraints of the planning effort.

In a crisis, the steps of JPP may be conducted simultaneously to speed the process. Supporting commands and organizations often conduct JPP simultaneously and iteratively with the supported CCMD.

I. Planning Initiation

Joint planning begins when an appropriate authority recognizes potential for military capability to be employed in support of national objectives or in response to a potential or actual crisis. At the strategic level, that authority—the President, SecDef, or CJCS—initiates planning by deciding to develop military options. Presidential directives, NSS, UCP, GEF, JSCP, and related strategic guidance documents (e.g., SGSs) serve as the primary guidance to begin planning.

CCDRs, subordinate commanders, and supporting commanders also initiate planning on their own authority when they identify a planning requirement not directed by higher authority. Additionally, analyses of the OE or developing or immediate crises may result in the President, SecDef, or CJCS directing military planning through a planning directive. CCDRs normally develop military options in combination with other nonmilitary options so that the President can respond with all the appropriate instruments of national power. Whether or not planning begins as described here, the commander may act within approved authorities and ROE/RUF in an immediate crisis.

The commander and staff will receive and analyze the planning guidance to determine the time available until mission execution; current status of strategic and staff estimates; and intelligence products, to include JIPOE, and other factors relevant to the specific planning situation. The commander will typically provide initial planning guidance based upon current understanding of the OE, the problem, and the initial operational approach for the campaign or operation. It could specify time constraints, outline initial coordination requirements, or authorize movement of key capabilities within the JFC's authority.

While planning is continuous once execution begins, it is particularly relevant when there is new strategic direction, significant changes to the current mission or planning assumptions, or the commander receives a mission for follow-on operations.

Planning for campaign plans is different from contingency plans in that contingency planning focuses on the anticipation of future events, while campaign planning assesses the current state of the OE and identifies how the command can shape the OE to deter crisis on a daily basis and support strategic objectives.

II. Mission Analysis

The CCDR and staff analyzes the strategic direction and derives the restated mission statement for the commander's approval, which allows subordinate and supporting commanders to begin their own estimates and planning efforts for higher headquarters' concurrence. The joint force's mission is the task or set of tasks, together with the purpose, that clearly indicates the action to be taken and the reason for doing so. Mission analysis is used to study the assigned tasks and to identify all other tasks necessary to accomplish the mission. Mission analysis is critical because it provides direction to the commander and the staff, enabling them to focus effectively on the problem at hand. When the commander receives a mission tasking, analysis begins with the following questions:

- What is the purpose of the mission received? (What problem is the commander being asked to solve or what change to the OE is desired?)
- What tasks must my command do for the mission to be accomplished?
- Will the mission achieve the desired results?
- What limitations have been placed on my own forces' actions?
- What forces/assets are needed to support my operation?
- How will I know when the mission is accomplished successfully?

The **primary inputs** to mission analysis are strategic guidance; the higher headquarters' planning directive; and the commander's initial planning guidance, which may include a description of the OE, a definition of the problem, the operational approach, initial intent, and the JIPOE (see following page).

Joint Planning Process (JPP) Overview

Ref: JP 5-0, Joint Planning (Jun '17), pp. V-2 to V-4 and fig. V-2.

Operational design and JPP are complementary tools of the overall planning process. Operational design provides an iterative process that allows for the commander's vision and mastery of operational art to help planners answer ends—ways—means—risk questions and appropriately structure campaigns and operations in a dynamic OE. The commander, supported by the staff, gains an understanding of the OE, defines the problem, and develops an operational approach for the campaign or operation through the application of operational design during the initiation step of JPP.

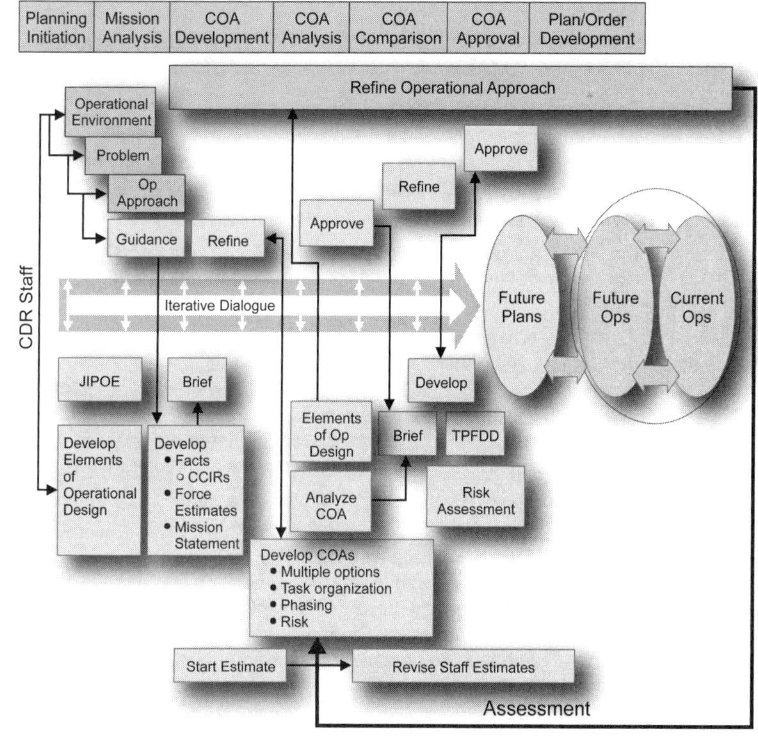

Commanders communicate their operational approach to their staff, subordinates, supporting commands, agencies, and multinational/nongovernmental entities as required in their initial planning guidance so that their approach can be translated into executable plans. As JPP is applied, commanders may receive updated guidance, learn more about the OE and the problem, and refine their operational approach. Commanders provide their updated approach to the staff to guide detailed planning. This iterative process facilitates the continuing development and refinement of possible COAs into a selected COA with an associated initial CONOPS and eventually into a resource informed executable plan or order.

The relationship between the application of operational art, operational design, and JPP continues throughout the planning and execution of the plan or order. By applying the operational design methodology in combination with the procedural rigor of JPP, the command can monitor the dynamics of the mission and OE while executing operations in accordance with the current approach and revising plans as needed. By combining these approaches, the friendly force can maintain the greatest possible flexibility.

Joint Intelligence Preparation of the Operational Environment (JIPOE)

Ref: Adapted from JP 2-0, Joint Intelligence (Jun '07 , pp. I-16 to I-17 and Oct '13, fig. I-5, p. I-17).

All planners need a basic familiarity with the JIPOE process in order to become critical consumers of the products produced by the intelligence community. Some steps in the JIPOE are conducted in parallel with the mission analysis and require input from other members of the maritime planning group. Although the specifics of the process vary depending on the situation and force involved, there is general agreement on the four major steps of JIPOE.

For a more detailed discussion of the JIPOE process, refer to JP 2-01.3, Joint Tactics, Techniques, and Procedures for Joint Intelligence Preparation of the Battlespace.

Step One: Define the Operational Environment

This first step is an initial survey of the geographic and non-geographic dimensions of the operational environment. It is used to bound the problem and to identify areas for further analysis. There are generally three tasks that must be accomplished.

- Identify the AO and the area of interest
- Determine the significant characteristics of the operational environment. This sub-step is an initial review of the factors of space, time, and forces and their interaction with one another.
- Evaluate existing databases and identify intelligence gaps and priorities. In this sub-step, intelligence personnel review the information found in various automated databases, Intelink sites (the classified version of the Internet), and other intelligence sources, both classified and unclassified. Intelligence requests and requirements may take the form of priority intelligence requirements (PIRs), requests for information (RFIs), production requests (PRs), and collection requirements.

Area of Operations: Defined by LAT/LONG or displayed on a map/chart for clarity and reference. The higher headquarters normally assigns this.

Area of Interest: Adjacent geographic area where political, military, economic, or other developments have an effect within a given theater; it might also extend to the areas enemy forces occupy that may endanger the accomplishment of one's mission; in practical terms, the area of interest determines the maximum scope of intelligence-gathering activities for the geographic combatant command; any theater (of war) also encompasses the pertinent parts of the cyberspace.

Step Two: Describe the Impact of the Operational Environment

The purpose of this step is to determine how the operational environment affects both friendly and enemy operations. It begins with an identification and analysis of all militarily significant environmental characteristics of each operational environment dimension. These factors are then analyzed to determine their effects on the capabilities and broad COAs of both enemy and friendly forces. Sub-steps include:

- Analyze the factor of space of the operational environment
- Analyze the factor of time of the operational environment
- Determine the operational environment effects on enemy and friendly capabilities and broad COAs

Summarize the Key Elements of the Factor of Space: military geography (area, position, distances, land use, environment, topography, vegetation, hydrography, oceanography, climate, and weather), politics, diplomacy, national resources, maritime infrastructure and positioning, economy, agriculture, transportation, telecommunications, culture, ideology, nationalism, sociology, science and technology.

Summarize the Key Elements of the Factor of Time: preparation, duration, warning, decision cycle, planning, mobilization, reaction, deployment, transit, concentration, maneuver, accomplish mission, rate of advance,reinforcements, commit reserves, regenerate combat power, redeployment, reconstruction

Step Three: Evaluate the Adversary & Other Relevant Actors

The third step is to identify and evaluate the adversary's forces and its capabilities; limitations; doctrine; and tactics, techniques, and procedures to be employed. In this step, analysts develop models that portray how the enemy normally operates and identifies capabilities in terms of broad ECOAs that the enemy might take. Analysts must take care not to evaluate enemy doctrine and concepts by mirror imaging U.S. doctrine. Sub-steps include:

- Identify adversary force capabilities
- Consider and describe general ECOAs in terms of DRAW-D (Defend, Reinforce, Attack, Withdraw, or Delay)
- Determine the current enemy situation (situation template)
- Identify broad COAs that would allow the enemy to achieve objectives

Summarize the Key Elements of the Factor of Forces (Enemy): defense system, armed forces, relative combat power of opposing forces (composition, reserves, reinforcements, location and disposition, strengths), logistics, combat efficiency (morale, leadership, doctrine, training, etc.)

Step Four: Determine Adversary & Other Relevant Actor COAs

Accurate identification of the full set of ECOAs requires the commander and his staff to think as the enemy thinks. From that perspective, it is necessary first to postulate possible enemy objectives and then to visualize specific actions within the capabilities of enemy forces that can be directed at these objectives and their impact upon potential friendly operations. From the enemy's perspective, appropriate physical objectives might include own-forces or their elements, own or friendly forces being supported or protected, facilities or lines of communication, and geographic areas or positions of tactical, operational, or strategic importance. The commander should not consider ECOAs based solely on factual or supposed knowledge of the enemy intentions.

The real COA by the enemy commander cannot be known with any confidence without knowing the enemy's mission and objective, and that information is rarely known. Even if such information were available, the enemy could change or feign the COA. Therefore, considering all the options the enemy could physically carry out is more prudent.

To develop an ECOA, one should ask the following three questions: Can the enemy do it? Will the enemy accomplish his objective? Would it materially affect the accomplishment of my mission? Each identified ECOA is examined to determine whether it meets the tests for suitability, feasibility, acceptability, uniqueness, and consistency with doctrine.

No ECOA should be dismissed or overlooked because it is considered as unlikely or uncommon, only if impossible. Once all ECOAs have been identified, the commander should eliminate any duplication and combine them when appropriate.

The **primary products** of mission analysis are staff estimates, the mission statement, a refined operational approach, the commander's intent statement, updated planning guidance, and initial CCIRs.

Mission analysis helps the JFC understand the problem and purpose of the operation and issue appropriate guidance to drive the rest of the planning process. The JFC and staff can accomplish mission analysis through a number of logical activities, such as those shown in Figure V-4 (facing page).

- Although some activities occur before others, mission analysis typically involves substantial concurrent processing of information by the commander and staff, particularly in a crisis situation.
- During mission analysis, it is essential the tasks (specified and implied) and their purposes are clearly stated to ensure planning encompasses all requirements, limitations (constraints—must do, or restraints—cannot do) on actions that the commander or subordinate forces may take are understood, and the correlation between the commander's mission and intent and those of higher and other commanders is understood. Resources and authorities must also be evaluated to ensure there is not a mission-resource¬authority mismatch and second, to enable the commander to prioritize missions and tasks against limited resources.
- Additionally, during mission analysis, specific information may need to be captured and tracked in order to improve the end products. This includes requests for information regarding forces, capabilities, and other resources; questions for the commander or special assistant (e.g., legal); and proposed battle rhythm for planning and execution. Recording this information during the mission analysis process will enable a more complete product and smoother mission analysis brief.

A. Analyze Higher Headquarters' Planning Directives and Strategic Guidance

Strategic Guidance

Strategic guidance is essential to joint planning and operational design. The President, SecDef, and CJCS promulgate strategic direction documents that cover a broad range of situations, and CCDRs provide guidance that covers a more narrow range of theater or functional situations. Documents such as the UCP, GEF, and JSCP provide near-term (0-2 years) strategic direction, and the CCDR's theater or functional strategy provide the mid- to long-term (greater than 3 years) GCC or FCC vision for the AOR or global employment of functional capabilities prepared in the context of SecDef's priorities. CCDR strategy links national strategic direction to joint planning.

Specific Guidance/Planning Directives

For a specific crisis, an order provides specific guidance, typically including a description of the situation, purpose of military operations, objectives, anticipated mission or tasks, and pertinent limitations. The GFMIG apportionment tables identify forces planners can reasonably expect to be available. Supported and supporting plans for the same military activity are constrained to the same resources. Planners should not expect to use additional forces beyond those listed in the apportionment tables without CJCS approval. The CJCS may amplify apportionment guidance for the specific crisis. This planning can confirm or modify the guidance for an existing contingency plan or order. This might simplify the analysis step, since consensus should already exist between the supported command and higher authority on the nature of the OE in the potential joint operations area (JOA)—such as the political, economic, social, and military circumstances—and potential US or multinational responses to various situations described in the existing plan. But even with a pre-existing contingency plan, planners need to confirm the actual situation matches the

Mission Analysis (Overview)

Ref: JP 5-0, Joint Planning (Jun '17), pp. V-5 to V-6 (fig. V-3 and V-4).

Mission analysis helps the JFC understand the problem and purpose of the operation and issue appropriate guidance to drive the rest of the planning process.

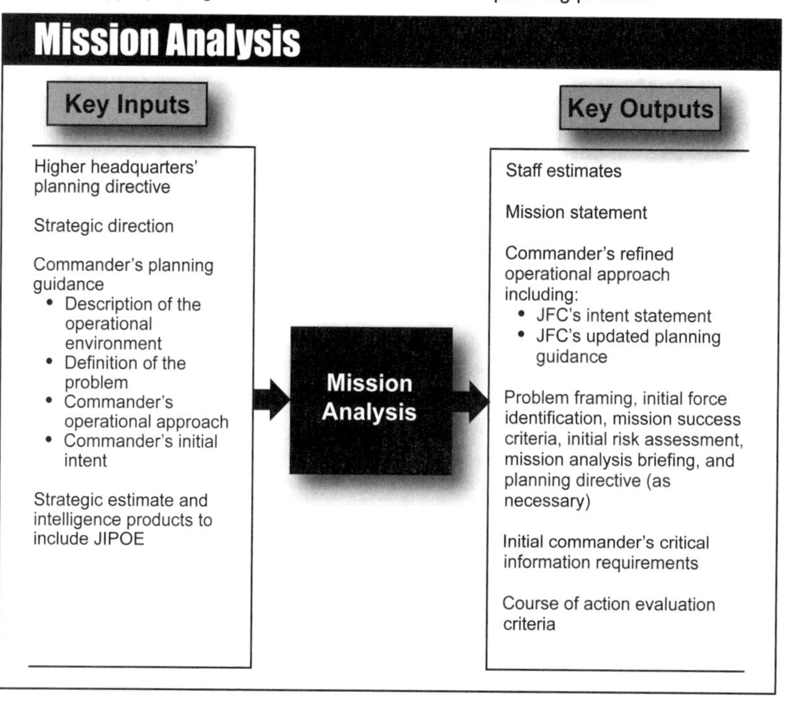

Mission Analysis Activities *(not necessarily sequential)*

- Begin logistics supportability analysis
- Analyze higher headquarters planning activities and strategic guidance
- Review commander's initial planning guidance, including his initial understanding of the operational environment, of the problem, and description of the operational approach
- Determine known facts and develop planning assumptions
- Determine and analyze operational limitations
- Determine specified, implied, and essential tasks
- Develop mission statement
- Conduct initial force allocation review
- Develop risk assessment
- Develop mission success criteria
- Develop commander's critical information requirements
- Prepare staff estimates
- Prepare and deliver mission analysis brief
- Publish commander's updated planning guidance, intent statement, and refined operational approach

hypothetical situation that the contingency plan was based on, as well as validating other assumptions. Significant changes may require refining or adapting the existing contingency plan. The dynamic nature of an emerging crisis can change many key aspects of the OE compared with earlier assumptions. These changes can greatly affect the plan's original operational approach upon which the commander and staff based decisions about COA alternatives and tasks to potential subordinate and supporting commands. In particular, planners must continuously monitor, assess, and adjust the strategic and operational objectives, planning assumptions, and criteria that comprise the military objectives. Differences between the commander's perspective and that of higher headquarters must be resolved at the earliest opportunity.

Higher Headquarters' Assessment

In time-compressed situations, especially with no preexisting plan, the higher headquarters' assessment of the OE and objectives may be the only guidance available. However, this circumstance is one that can benefit the most from the commander's and staff's independent assessment of circumstances to ensure they share a common understanding with higher headquarters assessment of the OE, strategic objectives, and the tasks or mission assigned to achieve these objectives. This is why CCMD JIPOE efforts should be continuous; these efforts maintain the intelligence portions of the CCDR's strategic estimate. Keeping the strategic estimate up to date greatly facilitates planning in a crisis as well as the transition of contingency plans to execution in crisis situations.

Multinational Strategic Guidance

CCDRs, JFCs, component and supporting commanders, and their staffs must clearly understand both US and partner nation strategic and military objectives and conditions that the national or multinational political leadership want the multinational military force to attain in terms of the internal and external balance of power, regional security, and geopolitics. To ensure unity of effort, planners should identify and attempt to resolve conflicts between participating nations' objectives and identify possible conflicts between different nations' national political and military objectives to ensure strategic planning accounts for these divergences. When multinational objectives are unclear, the senior US military commander must seek clarification and convey the positive or negative impact of continued ambiguity to the President and SecDef. For additional information on multinational operations, see JP 3¬16, Multinational Operations. For specific information on NATO operations, see Allied Joint Publication (AJP)-01, Allied Joint Doctrine; AJP-3, Allied Joint Doctrine for the Conduct of Operations; and AJP-5, Allied Joint Doctrine for Operational-Level Planning.

B. Review Commander's Initial Planning Guidance

Staff members and representatives from supporting organizations should maintain an open dialogue with the commander to better develop an appropriate solution to the problem and be able to adapt solutions to match the evolving OE and any potentially changing problems. Staffs should analyze the CCDR's initial planning guidance for the campaign or operation, which provides a basis for continued detailed analysis of the OE and of the tasks that may describe the mission and its parameters.

Commanders and planners must use caution in characterizing information as facts, as some items of information thought to be facts may be open to interpretation, based on the observer's perspective or incomplete information.

C. Determine Known Facts and Develop Planning Assumptions

The staff assembles both facts and assumptions to support the planning process and planning guidance.

Fact

A fact is a statement of information known to be true (such as verified locations of friendly and adversary force dispositions).

Assumption

An assumption provides a supposition about the current situation or future course of events, presumed to be true in the absence of facts. A valid assumption can be developed for both friendly and adversary situations and has three characteristics: logical, realistic, and essential for planning to continue. Commanders and staffs should never assume away adversary capabilities or assume that unrealistic friendly capabilities would be available. Assumptions address gaps in knowledge that are critical for the planning process to continue. Assumptions must be continually reviewed to ensure validity and challenged if they appear unrealistic. Subordinate commanders must not develop assumptions that contradict valid higher headquarters assumptions.

Commanders and staffs should anticipate changes to the plan if an assumption proves to be incorrect. Because of assumptions' influence on planning, planners must either validate the assumptions (treat as facts) or invalidate the assumptions (alter the plan accordingly) as quickly as possible.

During wargaming or red teaming, planners should review both the positive and negative aspect of all assumptions. They should review the plan from both the perspective that the assumption will prove true and from the perspective that the assumption will prove false. This can aid in preventing biases or tunnel vision during crisis action procedures.

Assumptions made in contingency planning should be addressed in the plan. Activities and operations in the plan can be used to validate, refute, or render unnecessary contingency plan assumptions.

Plans may contain assumptions that cannot be resolved until a potential crisis develops. As a crisis develops, assumptions should be replaced with facts as soon as possible. The staff accomplishes this by identifying the information needed to validate assumptions and submitting an information request to an appropriate agency as an information requirement. If the commander needs the information to make a key decision, the information requirement can be designated a CCIR. Although there may be exceptions, the staff should strive to resolve all assumptions before issuing the OPORD.

Planners should attempt to use as few assumptions as necessary to continue planning. By definition, assumptions introduce possibility for error. If the assumption is not necessary to continue planning, its only effect is to introduce error and add the likelihood of creating a bias in the commander's and planner's perspective. Since most plans require refinement, a simpler plan with fewer assumptions allows the commander and staff to act and react with other elements of the OE (including adversaries, allies, and the physical element). However, assumptions can be useful to identify those issues the commander and planners must validate on execution.

All assumptions should be identified in the plan or decision matrix to ensure they are reviewed and validated prior to execution.

D. Determine and Analyze Operational Limitations

Operational limitations are actions required or prohibited by higher authority and other restrictions that limit the commander's freedom of action, such as diplomatic agreements, political and economic conditions in affected countries, and partner nation and HN issues.

Constraint

A constraint is a requirement, "must do," placed on the command by a higher command that dictates an action, thus restricting freedom of action. For example, General Eisenhower was required to enter the continent of Europe instead of relying upon strategic bombing to defeat Germany.

Restraint

A restraint is a requirement, "cannot do," placed on the command by a higher command that prohibits an action, thus restricting freedom of action. For example, General MacArthur was prohibited from striking Chinese targets north of the Yalu River during the Korean War.

Many operational limitations are commonly expressed as ROE. Operational limitations may restrict or bind COA selection or may even impede implementation of the chosen COA. Commanders must examine the operational limitations imposed on them, understand their impacts, and develop options within these limitations to promote maximum freedom of action during execution.

Other operational limitations may arise from laws or authorities, such as the use of specific types of funds or training events. Commanders are responsible for ensuring they have the authority to execute operations and activities.

E. Determine Specified, Implied, and Essential Tasks

The commander and staff will typically review the planning directive's specified tasks and discuss implied tasks during planning initiation to resolve unclear or incorrectly assigned tasks with higher headquarters. If there are no issues, the commander and staff will confirm the tasks in mission analysis and then develop the initial mission statement.

Specified Tasks

Specified tasks are those that a commander assigns to a subordinate commander in a planning directive. These are tasks the commander wants the subordinate commander to accomplish, usually because they are important to the higher command's mission and/or objectives. One or more specified tasks often become essential tasks for the subordinate commander. Examples of specified tasks include:

- *Ensure freedom of navigation for US forces through the Strait of Gibraltar.*
- *Defend Country Green against attack from Country Red.*

Implied Tasks

Implied tasks are additional tasks the commander must accomplish, typically in order to accomplish the specified and essential tasks, support another command, or otherwise accomplish activities relevant to the operation or achieving the objective. In addition to the higher headquarters' planning directive, the commander and staff will review other sources of guidance for implied tasks, such as multinational planning documents and the GCC's TCP, FCPs, enemy and friendly COG analysis products, JIPOE products, relevant doctrinal publications, interviews with subject matter experts, and the commander's operational approach. The commander can also deduce implied tasks from knowledge of the OE, such as the enemy situation and political conditions in the assigned OA. However, implied tasks do not include routine tasks or SOPs that are inherent in most operations, such as conducting reconnaissance and protecting a flank. The following are examples of implied tasks:

- *Establish maritime superiority out to 50 miles from the Strait of Gibraltar*
- *Be prepared to conduct foreign internal defense and security force assistance operations to enhance the capacity and capability of Country Green security forces to provide stability and security if a regime change occurs in Country Red*

Essential Tasks

Essential tasks are those that the command must execute successfully to attain the desired end state defined in the planning directive. The commander and staff determine essential tasks from the lists of both specified and implied tasks. Depending on the scope of the operation and its purpose, the commander may synthesize certain specified and implied task statements into an essential task statement. See the example mission statement below for examples of essential tasks.

F. Develop Mission Statement

The mission statement describes the mission in terms of the elements of who, what, when, where, and why. The commander's operational approach informs the mission statement and helps form the basis for planning. The commander includes the mission statement in the planning guidance, planning directive, staff estimates, commander's estimate, CONOPS, and completed plan.

Example Mission Statement

When directed [when], United States X Command, in concert with coalition partners [who], deters Country Y from coercing its neighbors and proliferating weapons of mass destruction [what] in order to maintain security [why] in the region [where].

H. Conduct Initial Force Analysis

Initial Force Analysis

During mission analysis, the planning team begins to develop a rough-order of magnitude list of required forces and capabilities necessary to accomplish the specified and implied tasks. Planners consider the responsiveness of assigned and currently allocated forces. While more deliberate force requirement ID efforts continue during concept and plan development, initial ID of readily available forces during mission analysis may constrain the scope of the proposed operational approach.

Force requirements for a plan are initially documented in a force list developed from forces that are assigned, allocated, and apportioned. The force list may be an informal list (TPFDL) and later in the planning process entered into an information technology system such as JOPES as a baseline of forces to support subsequent time phasing. Planners should consider, at the onset of planning, that plan force requirements should be documented in a format and system that enables GFM allocation should the plan transition to execution.

In a crisis, assigned and allocated forces currently deployed to the geographic CCMD's AOR may be the most responsive during the early stages of an emergent crisis. Planners may consider assigned forces as likely to be available to conduct activities unless allocated to a higher priority. Re-missioning previously allocated forces may require SecDef approval and should be coordinated through the JS using procedures outlined CJCSM 3130.06, (U) Global Force Management Allocation Policies and Procedures.

Planners should also identify the status of reserve forces and identify the time required for call up and mobilization. Planners should evaluate appropriate requirements against existing or potential contracts or task orders to determine if the contracted support solution could meet the requirements. Planners must take into consideration force requirements for supported and supporting plans are drawing from the same quantity of apportioned forces and will compete with requirements for military activities and ongoing operations when the plan is executed.

Finally, planners compare the specified and implied tasks to the forces and resources available and identify shortfalls.

Identify Non-Force Resources Available for Planning

In many types of operations, the commander (and planners) may have access to non-force resources, such as commander's initiative funds, other funding sources (such as train and equip funding, support to foreign security forces funding, etc.), or can work with other security assistance programs (foreign military sales, excess defense article transfers, etc.). Planners and commanders can weave together resources and authorities from several different programs to create successful operations.

Refer to JP 3-20, Security Cooperation, for additional information on integrating multiple resources. See the GFMIG, for more information on the GFM processes and CJCSM 3130.06, (U) Global Force Management Allocation Policies and Procedures, for additional guidance on GFM allocation.

I. Develop Mission Success Criteria

Mission success criteria describe the standards for determining mission accomplishment. The JFC includes these criteria in the initial planning guidance so the joint force staff and components better understand what constitutes mission success. Mission success criteria apply to all joint operations. Specific success criteria can be utilized for development of supporting objectives, effects, and tasks and therefore become the basis for operation assessment. These also help the JFC determine if and when to move to the next phase. The initial set of criteria determined during mission analysis becomes the basis for operation assessment.

If the mission is unambiguous and limited in time and scope, mission success criteria can derive directly from the mission statement. For example, if the JFC's mission is to "evacuate all US personnel from the US Embassy in Grayland," then mission analysis could identify two primary success criteria: all US personnel are evacuated and established ROE are not violated.

However, more complex operations will require more complex assessments with MOEs and MOPs for each task, effect, and phase of the operation. These measures must evaluate not only the success of the specific task or mission, but that the desired objective was achieved (the conditions in the OE are those in support of US objectives or interests).

Campaigns and complex operations will often require multiple phases or steps to accomplish the mission. Planners can use a variety of methods through a developed operational approach to identify progress toward the desired objective or end state. Attainment of objectives is one method to assess progress. Commanders review MOEs and MOPs as two additional methods to measure success.

Measuring the status of tasks, effects, and objectives becomes the basis for reports to senior commanders and civilian leaders on the progress of the operation. The CCDR can then advise the President and SecDef accordingly and adjust operations as required. Whether in a supported or supporting role, JFCs at all levels must develop their mission success criteria with a clear understanding of termination criteria established by the CJCS and SecDef. Commanders and staffs should be aware that successful accomplishment of the task or objective might not produce the desired results—and be ready to make recommendations to the President or SecDef on changes to the campaign or operation.

Develop COA Evaluation Criteria

Evaluation criteria are standards the commander and staff will later use to measure the relative effectiveness and efficiency of one COA relative to other COAs. Developing these criteria during mission analysis or as part of commander's planning guidance helps to eliminate a source of bias prior to COA analysis and comparison. Evaluation criteria address factors that affect success and those that can cause failure. Criteria change from mission to mission and must be clearly defined and understood by all staff members before starting the wargame to test the proposed COAs. Normally, the chief of staff (COS) (or executive officer) initially determines each proposed criterion with weights based on its relative importance and the commander's guidance. Commanders adjust criterion selection and weighting according to their own experience and vision. The staff member responsible for a functional area scores each COA using those criteria. The staff presents the proposed evaluation criteria to the commander at the mission analysis brief for approval.

Planners conducting a preliminary risk assessment must identify the obstacles or actions that may preclude mission accomplishment and then assess the impact of these impediments to the mission. Once planners identify the obstacles or actions, they assess the probability of achieving objectives and severity of loss linked to an obstacle or action, and characterize the military risk. Based on judgment, military risk assessment is an integration of probability and consequence of an identified impediment.

J. Develop Risk Assessment

The probability of the impediment occurring may be ranked as very likely: occurs often, continuously experienced; likely: occurs several times; questionable: unlikely, but could occur at some time; or unlikely: can assume it will not occur. Based on probabilities, military risk (consequence) may be high: critical objectives cannot be achieved; significant: only the most critical objectives can be achieved; moderate: can partially achieve all objectives; or low: can fully achieve all objectives.

Determining military risk is more an art than a science. Planners use historical data, intuitive analysis, and judgment. Military risk characterization is based on an evaluation of the probability that the commander's objectives will be accomplished. The level of risk is high if achieving objectives or obtaining end states is unlikely, significant if achieving objectives or obtaining end states is questionable, moderate if achieving objectives or obtaining end states is likely, and low if achieving objectives or obtaining end states is very likely.

Planners and commanders need to be able to explain military risk to civilian leadership who may not be as familiar with military operations as they are. Additionally, since military risk is often a matter of perspective and personal experience, they must be able to help decision makers understand how they evaluated the probability of accomplishing objectives, how they characterized the resultant military risk, and the sources or causes of that risk.

During decision briefs, risks must be explained using standard terms that support the decision-making process, such as mission success (which missions will and which will not be accomplished), time (how much longer will a mission take to achieve success), and forces (casualties, future readiness, etc.), and political implications.

See p. 2-17 for discussion of risk management from JP 3-0, and p. 3-41 for discussion of the types of risk from JP 5-0.

K. Determine Commander's Critical Information Requirements (CCIRs)

Ref: JP 5-0, Joint Planning (Jun '17), pp. V-14 to V-16.

CCIRs are elements of information the commander identifies as being critical to timely decision making. CCIRs help focus information management and help the commander assess the OE, validate (or refute) assumptions, identify accomplishment of intermediate objectives, and identify decision points during operations. CCIRs belong exclusively to the commander. They are situation-dependent, focused on predictable events or activities, time-sensitive, and always established by an order or plan. The CCIR list is normally short so that the staff can focus its efforts and allocate scarce resources. The CCIR list is not static; JFCs add, delete, adjust, and update CCIRs throughout plan development, assessment, and execution based on the information they need for decision making. PIRs and FFIRs constitute the total list of CCIRs.

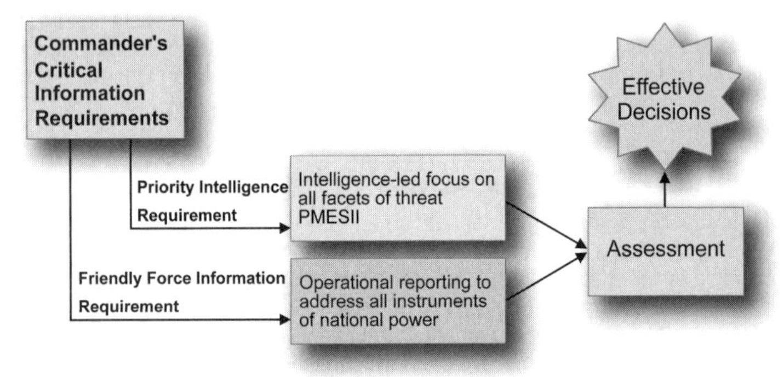

Priority Intelligence Requirements (PIRs)

PIRs focus on the adversary and the OE and are tied to commander's decision points. They drive the collection of information by all elements of a command requests for national-level intelligence support and requirements for additional intelligence capabilities. All staff sections can recommend potential PIRs they believe meet the commander's guidance. However, the joint force J-2 has overall staff responsibility for consolidating PIR nominations and for providing the staff recommendation to the commander. JFC-approved PIRs are automatically CCIRs.

Friendly Force Intelligence Requirements (FFIRs)

FFIRs focus on information the JFC must have to assess the status of the friendly force and supporting capabilities. All staff sections can recommend potential FFIRs they believe meet the commander's guidance. Commander-approved FFIRs are automatically CCIRs.

A CCIR must be a decision required of the commander, not of the staff, and responding to a CCIR must be critical to the success of the mission.

CCIRs support the commander's future decision requirements and are often related to MOEs and MOPs. PIRs are often expressed in terms of the elements of PMESII while FFIRs are often expressed in terms of the diplomatic,

See p. 2-16 for discussion of CCIRs from JP 3-0.

L. Prepare Staff Estimates

A staff estimate is an evaluation of how factors in a staff section's functional area support and impact the mission. The purpose of the staff estimate is to inform the commander, staff, and subordinate commands how the functional area supports mission accomplishment and to support COA development and selection.

Staff estimates are initiated during mission analysis, at which point functional planners are focused on collecting information from their functional areas to help the commander and staff understand the situation and conduct mission analysis. Later, during COA development and selection, functional planners fully develop their estimates providing functional analysis of the COAs, as well as recommendations on which COAs are supportable. They should also identify critical shortfalls or obstacles that impact mission accomplishment. Staff estimates are continually updated based on changes in the situation. Operation assessment provides the means to maintain running staff estimates for each functional area.

Not every situation will require or permit a lengthy and formal staff estimate process. In a crisis, staff estimates may be given orally to support the rapid development of plans. However, with sufficient time, planning will demand a more formal and thorough process. Staff estimates should be shared with subordinate and supporting commanders to help them prepare their supporting estimates, plans, and orders. This will improve parallel planning and collaboration efforts of subordinate and supporting elements and help reduce the planning times for the entire process.

See following pages (pp. 3-84 to 3-85) for further discussion of staff estimates.

Intelligence Support to Joint Operation Planning

Intelligence support to joint planning includes Defense Intelligence Agency-produced dynamic threat assessments (DTAs) for top-priority contingency plans and theater intelligence assessments with a 2-5 year outlook to support CCDR campaign plan development and assessment. Additionally, CCMD JIOCs and subordinate JFC joint intelligence support elements produce intelligence assessments and estimates resulting from the JIPOE process. The intelligence estimate constitutes the intelligence portion of the commander's estimate and is typically published as Appendix 11 to Annex B (Intelligence) to a plan or an order. These are baseline information and finished intelligence products that inform the four continuous operational activities of situational awareness, planning, execution, and assessment within APEX.

During mission analysis, intelligence planners lead the development of PIRs to close critical knowledge gaps in initial estimative intelligence products or to validate threat and OE-related planning assumptions. Throughout JPP, additional PIRs may be nominated to support critical decisions needed throughout all phases of the operation. The intelligence planner then prepares a J-2 staff estimate, which is an appraisal of available capabilities within the intelligence joint function to satisfy commanders' PIRs. This estimate drives development of Annex B (Intelligence) to a plan or an order.

See pp. 2-28 to 2-29 for discussion of intelligence as a joint function.

Logistics Staff Estimate

The commander's logistics staff and Service component logisticians should develop a logistics overview, which includes but is not restricted to critical logistics facts, assumptions, and information requirements that must be incorporated into the CCIRs; current or anticipated HNS and status; ID of existing contracts and task orders available for use; identifying aerial and sea ports of debarkation; any other distribution infrastructure and associated capacity; inventory (e.g., on-hand, prepositioned, theater reserve); combat support and combat service support capabilities; known or potential capability shortfalls; and contractor support required to replace or augment unavailable military capabilities. From this TLO, a logistics estimate can identify known and anticipated factors that may influence the logistics support.

See p. 4-37 for further discussion of logistics estimates.

Sample Staff Estimate Format

Ref: JP 5-0, Joint Planning (Jun '17), app. c.

Staff estimates are central to formulating and updating military action to meet the requirements of any situation. Staff estimates should start with the strategic estimate and be comprehensive and continuous and visualize the future, while optimizing the limited time available to not become overly time-consuming. Comprehensive estimates consider both the quantifiable and the intangible aspects of military operations. They translate friendly and enemy strengths, weapons systems, training, morale, and leadership into combat capabilities. The estimate process requires the ability to visualize the battle or crisis situations requiring military forces.

The following is a sample format that can be used as a guide when developing an estimate. The exact format and level of detail may vary somewhat among joint commands and primary staff sections based on theater-specific requirements and other factors. Refer to the CJCSM 3130.03, Adaptive Planning and Execution (APEX) Planning Formats and Guidance.

1. Mission

a. Mission Analysis

(1) Determine the higher command's purpose. Analyze national security and national military strategic direction, as well as appropriate guidance in partner nations' directions, including long- and short-term objectives. Determine if a clearly defined military end state and related termination criteria are warranted.

(2) Determine specified, implied, and essential tasks and their priorities.

(3) Determine objectives and consider desired and undesired effects.

(4) Reassess if strategic direction & guidance support the desired objectives or end state

b. Mission Statement

(1) Express in terms of who, what, when, where, and why (purpose).

(2) Frame as a clear, concise statement of the essential tasks to be accomplished and the purpose to be achieved.

2. Situation and Courses of Action

a. Situation Analysis

(1) Geostrategic Context

(a) Domestic and international context: political and/or diplomatic long- and short-term causes of conflict; domestic influences, including public will, competing demands for resources and political, economic, legal, and moral constraints; and international interests (reinforcing or conflicting with US interests, including positions of parties neutral to the conflict), international law, positions of international organizations, and other competing or distracting international situations. Similar factors must be considered for theater and functional campaigns and noncombat operations.

(b) A systems perspective of the operational environment: all relevant political, military, economic, social, infrastructure, informational, and other aspects.

(2) Analysis of the Adversary/Competitors. Scrutiny of the opponent situation, including capabilities and vulnerabilities (at the theater level, commanders normally will have available a formal intelligence estimate), should include the following:

(a) Political and military intentions and objectives (to extent known).

(b) Broad military COAs being taken and available in the future.

(c) Military strategic and operational advantages and limitations.

(d) Possible external military support.

(e) COGs (strategic and operational) and decisive points.

(f) Specific operational characteristics such as strength, composition, location, and disposition; reinforcements; logistics; time, and space factors (including basing utilized and available); and combat/noncombat efficiency and proficiency.

(g) Reactions of third parties/competitors in theater and functional campaigns.

(3) **Friendly Situation.** Should follow the same pattern used for the analysis of the adversary. At the theater level, CCDRs normally will have available specific supporting estimates, including personnel, logistics, and communications estimates. Multinational operations require specific analysis of partner nations' objectives, capabilities, and vulnerabilities. Interagency coordination required for the achievement of objectives should also be considered.

(4) **Operational Limitations.** Actions either required or prohibited by higher authority, such as constraints or restraints, and other restrictions that limit the commander's freedom of action, such as diplomatic agreements, political or economic conditions in affected countries, and HN issues.

(5) **Assumptions.** Assumptions are intrinsically important factors upon which the conduct of the operation is based and must be noted as such.

(6) **Deductions.** Deductions from the analysis should yield estimates of relative combat power, including enemy capabilities that affect mission accomplishment.

b. Course of Action Development and Analysis. COAs are based on the above analysis and a creative determination of how the mission will be accomplished. Each COA must be adequate, feasible, and acceptable. State all practical COAs open to the commander that, if successful, will accomplish the mission. For a CCDR's strategic estimate, each COA typically will constitute an alternative theater strategic or operational concept and should outline:

(1) Major strategic & operational tasks (in the order in which they are to be accomplished).

(2) Major forces or capabilities required (joint, interagency, and multinational).

(3) C2 concept.

(4) Sustainment concept.

(5) Deployment concept.

(6) Estimate of time required to achieve the termination criteria.

(7) Concept for establishing and maintaining a theater reserve.

3. Analysis of Adversary/Competitor Capabilities & Intentions

a. Determine the probable effect of possible adversary capabilities and intentions on the success of each friendly COA.

b. Conduct this analysis in an orderly manner by time phasing, geographic location, and functional event. Consider:

(1) The potential actions of subordinates two echelons down.

(2) Conflict termination issues; think through own action, opponent reaction, and counteraction.

(3) The potential impact on friendly desired effects and the likelihood that the adversary's actions will cause specific undesired effects.

c. Conclude with revalidation of friendly COAs. Determine additional requirements, make required modifications, and list advantages and disadvantages of each adversary capability.

4. Comparison of Own Courses of Action

a. Evaluate the advantages and disadvantages of each COA.

b. Compare with respect to evaluation criteria.

(1) Fixed values for joint operations (the principles of joint operations, the fundamentals of joint warfare, and the elements of operational design).

(2) Other factors (for example, political constraints).

(3) Mission accomplishment.

c. If appropriate, merge elements of different COAs into one.

d. Identify risk specifically associated with the assumptions (i.e., what happens if each assumptions prove false).

5. Recommendation

Provide an assessment of which COAs are supportable, an analysis of the risk for each, and a concise statement of the recommended COA with its requirements.

M. Prepare and Deliver Mission Analysis Brief

Upon conclusion of the mission analysis, the staff will present a mission analysis brief to the commander. This brief provides the commander with the results of the staff's analysis of the mission, offers a forum to discuss issues that have been identified, and ensures the commander and staff share a common understanding of the mission. The results inform the commander's development of the mission statement. The commander provides refined planning guidance and intent to guide subsequent planning. Figure V-6, below, shows an example mission analysis briefing.

The mission analysis briefing may be the only time the entire staff is present and the only opportunity to make certain all staff members start from a common reference point. The briefing focuses on relevant conclusions reached as a result of the mission analysis.

Immediately after the mission analysis briefing, the commander approves a restated mission. This can be the staff's recommended mission statement, a modified version of the staff's recommendation, or one that the commander has developed personally. **Once approved, the restated mission becomes the unit mission.**

At the mission analysis brief, the commander will likely describe an updated understanding of the OE, the problem, and the vision of the operational approach to the entire assemblage, which should include representatives from subordinate commands and other partner organizations. This provides the ideal venue for facilitating unity of understanding and vision, which is essential to unity of effort.

Example Mission Analysis Briefing

Introduction

Situation overview
- Operational environment (including joint operations area) and threat overview
- Political, military, economic, social, information, and infrastructure strengths and weaknesses
- Enemy (including center[s] of gravity) and objectives

Friendly assessment
- Facts and assumptions
- Limitations—constraints/restraints
- Capabilities allocated
- Legal considerations

Communications synchronization

Objectives, effects, and task analysis
- United States Government interagency objectives
- Higher commander's objectives/mission/guidance
- Objectives and effects
- Specified/implied/essential tasks
- Centers of gravity

Operational protection
- Operational risk
- Mitigation

Proposed initial commander's critical information requirements

Mission
- Proposed mission statement
- Proposed commander's intent

Command relationships

Conclusion—potential resource shortfalls

Mission analysis approval and commander's COA planning guidance

N. Commander's Refined Planning Guidance

Ref: JP 5-0, Joint Planning (Jun '17), pp. V-19 to V-20.

After approving the mission statement and issuing the intent, the commander provides the staff (and subordinates in a collaborative environment) with enough additional guidance (including preliminary decisions) to focus the staff and subordinate planning activities during COA development.

At a minimum, this refined planning guidance should include the following elements:

- An approved mission statement
- Key elements of the OE (operational environment)
- A clear statement of the problem
- Key assumptions
- Key operational limitations
- National strategic objectives with a description of how the operation will support them
- Termination criteria (if appropriate, CCMD-level campaign plans will not have termination criteria and many operations will have transitions rather than termination)
- Military objectives or end state and their relation to the national strategic end state
- The JFC's initial thoughts on the conditions necessary to achieve objectives
- Acceptable or unacceptable levels of risk in key areas
- The JFCs visualization of the operational approach to achieve the objectives in broad terms. This operational approach sets the basis for development of COAs. The commander should provide as much detail as appropriate to provide the right level of freedom to the staff in developing COAs. Planning guidance should also address the role of interorganizational and multinational partners in the pending operation and any related special considerations as required.

Commanders describe their visualization of the forthcoming campaign or operations to help build a shared understanding among the staff. Enough guidance (preliminary decisions) must be provided to allow the subordinates to plan the action necessary to accomplish the mission consistent with commander's intent. The commander's guidance must focus on the essential tasks and associated objectives that support the accomplishment of the assigned national objectives. It emphasizes in broad terms when, where, and how the commander intends to employ military capabilities integrated with other instruments of national power to accomplish the mission within the higher JFC's intent.

The JFC may provide the planning guidance to the entire staff and/or subordinate JFCs or meet each staff officer or subordinate unit individually as the situation and information dictates. The guidance can be given in a written form or orally. No format for the planning guidance is prescribed. However, the guidance should be sufficiently detailed to provide a clear direction and to avoid unnecessary efforts by the staff or subordinate and supporting commands.

Planning guidance can be very explicit and detailed, or it can be very broad, allowing the staff and/or subordinate commands wide latitude in developing subsequent COAs. However, no matter its scope, the content of planning guidance must be arranged in a logical sequence to reduce the chances of misunderstanding and to enhance clarity. Moreover, one must recognize that all the elements of planning guidance are tentative only. The JFC may issue successive planning guidance during the decision-making process; yet the focus of the JFC's staff should remain upon the framework provided in the initial planning guidance. The commander should continue to provide refined planning guidance during the rest of the planning process while understanding of the problem continues to develop.

III. Course of Action Development

A COA is a potential way (solution, method) to accomplish the assigned mission. The staff develops COAs to provide unique options to the commander, all oriented on accomplishing the military end state. A good COA accomplishes the mission within the commander's guidance, provides flexibility to meet unforeseen events during execution, and positions the joint force for future operations. It also gives components the maximum latitude for initiative.

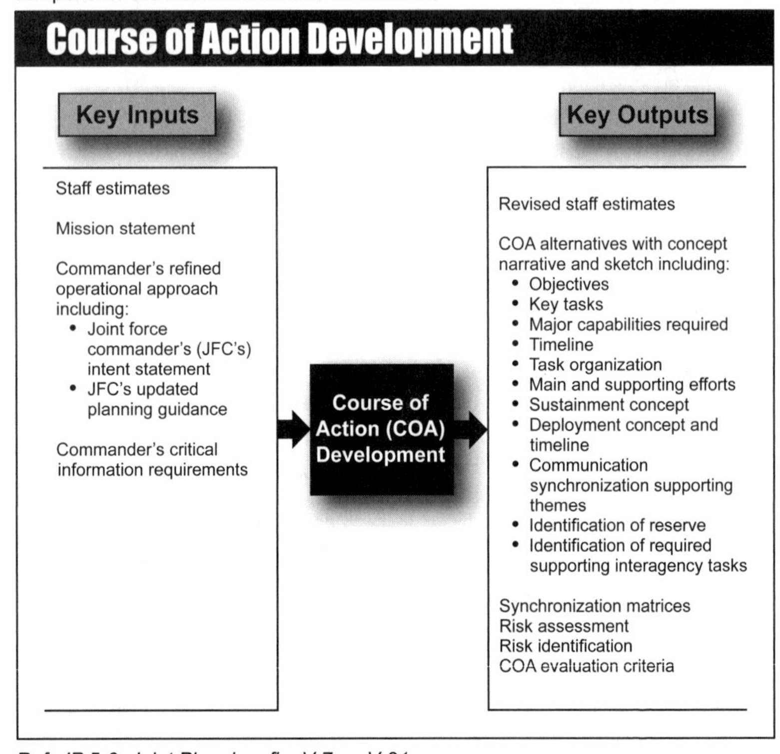

Ref: JP 5-0, Joint Planning, fig. V-7, p. V-21.

COA Elements

Each COA typically has an associated initial CONOPS with a narrative and sketch and includes the following:

- Operational Environment (OE)
- Objectives, key tasks and purpose
- Forces and capabilities required, to include anticipated interagency roles, actions, and supporting tasks
- Integrated timeline
- Task organization
- Operational concept
- Sustainment concept
- Communication synchronization
- Risk
- Required decisions and decision timeline (e.g., mobilization, DEPORD)
- Deployment concept
- Main and supporting efforts

A. COA Development Considerations

Ref: JP 5-0, Joint Planning (Jun '17), pp. V-21 to V-22.

The products of COA development are potential COA alternatives, with a sketch for each if possible. Each COA describes, in broad but clear terms, what is to be done throughout the campaign or operation, the size of forces deemed necessary, time in which joint force capabilities need to be brought to bear, and the risks associated with the COA. These COAs will undergo additional validity testing, analysis, wargaming, and comparison, and they could be eliminated at any point during this process. These COAs provide conceptualization and broad descriptions of potential CONOPS for the conduct of operations that will accomplish the desired end state.

Available Planning Time

Available planning time is always a key consideration, particularly in a crisis. The JFC gives the staff additional considerations early in COA development to focus the staff's efforts, helping the staff concentrate on developing COAs that are the most appropriate. There should always be more than one way to accomplish the mission, which suggests that commanders and planners should give due consideration to the pros and cons of valid COA alternatives. However, developing several COAs could violate time constraints. Usually, the staff develops two or three COAs to focus their efforts and concentrate valuable resources on the most likely scenarios. However, COAs must be substantially distinguishable from each other. Commanders should not overburden staffs by developing similar solutions to the problem. The commander's involvement in the early operational design process can help ensure only value-added options are considered. If time and personnel resources permit, different COAs could be developed by different teams to ensure they are unique.

Employment of all Participants in the Operation

For each COA, the commander must envision the employment of all participants in the operation as a whole—US military forces, MNFs, and interagency and multinational partners—taking into account operational limitations, political considerations, the OA, existing FDOs, and the conclusions previously drawn during the mission analysis, the commander's guidance and informal dialogue and formal IPRs with DOD leadership held to date.

Adversary COAs

During COA development, the commander and staff consider all feasible adversary COAs. Other actors may also create difficult conditions that must be considered during COA development. It is best to consider all opposing actors' actions likely to challenge the attainment of the desired end states when exploring adversary COAs.

Alternatives

An alternative is an activity within a COA that may be executed to enable achieving an objective. Alternatives, and groups of alternatives comprising branches, allow the commander to act rapidly and transition as conditions change through the campaign or operation. Alternatives, and more broadly branches, should enable the commander to progress sequentially or skip ahead based on success or other changes to the conditions or strategic direction from dialogue with higher commanders, SecDef, and/or the President. They should also enable the commander to transition rapidly, exploit success, and control escalation and tempo while denying the same to the enemy. The development of alternatives within COAs empowers the commander and translates up and down the chain of command and enables strategic flexibility for SecDef and the President. COAs should be simple and brief, yet complete. Individual COAs should have descriptive titles. Distinguishing factors of the COA may suggest titles that are descriptive in nature.

The products of mission analysis drive COA development. Since the operational approach contains the JFC's broad approach to solve the problem at hand, each COA will expand this concept with the additional details that describe who will take the action, what type of military action will occur, when the action will begin, where the action will occur, why the action is required (purpose), and how the action will occur (method of employment of forces). Likewise, the essential tasks identified during mission analysis (and embedded in the draft mission statement) must be common to all potential COAs.

Planners can vary COAs by adjusting the use of joint force capabilities throughout the OE by employing the capabilities in combination for effectiveness making use of the information environment (including cyberspace) and the electromagnetic spectrum.

B. COA Development Techniques and Procedures

1. Review Information

Review information contained in the mission analysis and commander's operational approach, planning guidance, and intent statement. All staff members must understand the mission and the tasks that must be accomplished within the commander's intent to achieve mission success.

2. Determine the COA Development Technique

A critical first decision in COA development is whether to conduct simultaneous or sequential development of the COAs. Each approach has distinct advantages and disadvantages. The advantage of simultaneous development of COAs is potential time savings. Separate groups are simultaneously working on different COAs. The disadvantage of this approach is that the synergy of the JPG may be disrupted by breaking up the team. The approach is manpower intensive and requires component and directorate representation in each COA group, and there is an increased likelihood that the COAs will not be distinctive. While there is potential time to be saved, experience has demonstrated that it is not an automatic result. The simultaneous COA development approach can work, but its inherent disadvantages must be addressed and some risk accepted up front. The recommended approach if time and resources allows is the sequential method.

See facing page for further discussion.

3. Review Operational Objectives and Tasks and Develop Ways to Accomplish Tasks

Planners must review and refine theater and supporting operational objectives from the initial work done during the development of the operational approach. These objectives establish the conditions necessary to help accomplish the national strategic objectives. Tasks are shaped by the CONOPS—intended sequencing and integration of air, land, maritime, special operations, cyberspace, and space forces. Tasks are prioritized while considering the enemy's objectives and the need to gain advantage.

Regardless of the eventual COA, the staff should plan to accomplish the higher commander's intent by understanding its essential task(s) and purpose and the intended contribution to the higher commander's mission success.

The staff must ensure all the COAs developed will fulfill the command mission and the purpose of the operation by conducting a review of all essential tasks developed during mission analysis. They should then consider ways to accomplish the other tasks.

C. Step-by-Step Approach to Course of Action Development

Ref: JP 5-0, Joint Planning (Jun '17), p. V-24 (fig. V-8).

There are several planning sequence techniques available to facilitate COA development. One option is the step-by-step approach, which uses the backward-planning technique (also known as reverse planning):

Step 1
Determine how much force will be needed in the theater at the end of the operation or campaign, what those forces will be doing, and how those forces will be postured geographically. Use troop-to-task analysis. Draw a sketch to help visualize the forces and their locations.

Step 2
Looking at the sketch and working backwards, determine the best way to get the forces postured in Step 1 from their ultimate positions at the end of the operation or campaign to a base in friendly territory. This will help formulate the desired basing plan.

Step 3
Using the mission statement as a guide, determine the tasks the force must accomplish en route to their locations/positions at the end of the operation or campaign. Draw a sketch of the maneuver plan. Make sure the force does everything the Secretary of Defense (SecDef) has directed the commander to do (refer to specified tasks from the mission analysis).

Step 4
Determine the basing required to posture the force in friendly territory, and the tasks the force must accomplish to get to those bases. Sketch this as part of the deployment plan.

Step 5
Determine if the planned force is enough to accomplish all the tasks SecDef has given the commander.

Step 6
Given the tasks to be performed, determine in what order the forces should be deployed into theater. Consider the force categories such as combat, protection, sustainment, theater enablers, and theater opening.

Step 7
The information developed should now allow determination of force employment, major tasks and their sequencing, sustainment, and command relationships.

4. Integrate & Synchronize Requirements (Joint Functions)

Once the staff has begun to visualize COA alternatives, it should see how it can best synchronize (arrange in terms of time, space, and purpose) the actions of all the elements of the force. The staff should estimate the anticipated duration of the operation. One method of synchronizing actions is the use of phasing as discussed earlier. Phasing assists the commander and staff to visualize and think through the entire operation or campaign and to define requirements in terms of forces, resources, time, space, and purpose. Planners should then integrate and synchronize these requirements by using the joint functions of C2, intelligence, fires, movement and maneuver, protection, sustainment, and information. Additionally, planners should consider IRCs as additional tools to create desired effects. At a minimum, planners should make certain the synchronized actions answer the following questions:

- How do land, maritime, air, space, cyberspace, and special operations forces integrate across the joint functions to accomplish their assigned tasks?
- How can the joint forces synchronize their actions and messages (words and deeds) and integrate IRCs with lethal fires?

5. Tentative COAs Should Focus on COGs and Decisive Points

The COAs should focus on COGs and decisive points or areas of influence for CCMD-level campaigns. The commander and the staff review and refine their COG analysis begun during mission analysis based on updated intelligence, JIPOE products, and initial staff estimates. The refined enemy and friendly COG analysis, particularly the critical vulnerabilities, is considered in the development of the initial COAs. The COG analysis helps the commander become oriented to the enemy and compare friendly strengths and weakness with those of the enemy. By looking at friendly COGs and vulnerabilities, the staff understands the capabilities of their own force and critical vulnerabilities that will require protection. Protection resource limitations will probably mean the staff cannot plan to protect every capability, but rather will look at prioritizing protection for critical capabilities and developing overlapping protection techniques. The strength of one asset or capability may provide protection from the weakness of another.

6. Identify the Sequencing

Identify the sequencing (simultaneous, sequential, or a combination) of the actions for each COA. Understand when and what resources become available during the operation or campaign. Resource availability will significantly affect sequencing operations and activities.

7. Identify Main and Supporting Efforts

Identify main and supporting efforts by phase, the purposes of these efforts, and key supporting/supported relationships within phases.

8. Identify Decision Points and Assessment Process

The commander will need to know when a critical decision has to be made and how to know specific objectives have been achieved. This requires integration of decision points and assessment criteria into the COA as these processes anticipate a potential need for decisions from outside the command (SecDef, the President, or a functional or adjacent command).

9. Identify Component-Level Missions and Tasks

Identify component-level missions/tasks (who, what, and where) that will accomplish the stated purposes of main and supporting efforts. Think of component and joint function tasks such as movement and maneuver, intelligence, fires, protection, sustainment, C2, and information. Display them with graphic control measures as much as possible. A designated LOO will help identify these tasks.

10. Integrate Information-Related Capabilities (IRCs)

Some IRCs help to create effects and influence adversary decision making. Planners should consider how IRCs can influence positioning of adversary units, disrupt adversary C2, and decrease adversary morale when developing COAs.

11. Determine Task Organization

The staff should develop an outline task organization to execute the COA. The commander and staff determine appropriate command relationships and appropriate missions and tasks.

Determine command relationships and organizational options. Joint force organization and command relationships are based on the operation or campaign CONOPS, complexity, and degree of control required. Establishing command relationships includes determining the types of subordinate commands and the degree of authority to be delegated to each. Clear definition of command relationships further clarifies the intent of the commander and contributes to decentralized execution and unity of effort. The commander has the authority to determine the types of subordinate commands from several doctrinal options, including Service components, functional components, and subordinate joint commands. Regardless of the command relationships selected, it is the JFC's responsibility to ensure these relationships are understood and clear to all subordinate, adjacent, and supporting headquarters. The following are considerations for establishing joint force organizations:

- Joint forces will normally be organized with a combination of Service and functional components with operational responsibilities.
- Functional component staffs should be joint with Service representation in approximate proportion to the mix of subordinate forces. These staffs should be organized and trained prior to employment in order to be efficient and effective, which will require advanced planning.
- Commanders may establish support relationships between components to facilitate operations.
- Commanders define the authority and responsibilities of functional component commanders based on the strategic CONOPS and may alter their authority and responsibility during the course of an operation.
- Commanders must balance the need for centralized direction with decentralized execution.
- Major changes in the joint force organization are normally conducted at phase changes.

12. Sustainment Concept

No COA is complete without a plan to sustain it properly. The sustainment concept is more than just gathering information on various logistic and personnel services. It entails identifying the requirements for all classes of supply, creating distribution, transportation, OCS, and disposition plans to support the commander's execution, and organizing capabilities and resources into an overall theater campaign or operation sustainment concept. It concentrates forces and material resources strategically so the right force is available at the designated times and places to conduct decisive operations. It requires thinking through a cohesive sustainment for joint, single Service and supporting forces relationships in conjunction with CSAs, multinational, interagency, nongovernmental, private sector, or international organizations.

13. Deployment Concept

A COA must consider the deployment concept in order to describe the general flow of forces into theater. There is no way to determine the feasibility of the COA without including the deployment concept. While the detailed deployment concept will be developed during plan synchronization, enough of the concept must be described in the COA to visualize force buildup, sustainment requirements, and military-political considerations.

14. Define the Operational Area (OA)

The OA is an overarching term that can encompass more descriptive terms for geographic areas. It will provide flexibility/options and/or limitations to the commander. The OA must be precisely defined because the specific geographic area will impact planning factors such as basing, overflight, and sustainment. OAs include, but are not limited to, such descriptors as AOR, theater of war, theater of operations, JOA, amphibious objective area, joint special operations area, and area of operations. Except for AOR, which is assigned in the UCP, GCCs and their subordinate JFCs designate smaller OAs on a temporary basis. OAs have physical dimensions composed of some combination of air, land, maritime, and space domains.

15. Develop Initial COA Sketches and Statements

Each COA should answer the following questions:

- Who (type of forces) will execute the tasks?
- What are the tasks?
- When will the tasks begin?
- What are key/critical decision points?
- How (but do not usurp the components' prerogatives) the commander should provide "operational direction" so the components can accomplish "tactical actions."
- Why (for what purpose) will each force conduct its part of the operation?
- How will the commander identify successful accomplishment of the mission?
- Develop an initial intelligence support concept

16. Test the Validity of Each Tentative COA

All COAs selected for analysis must be valid, and the staff should reject COA alternatives that do not meet all five of the validity criteria.

See facing page for listing and discussion of the five validity criteria.

17. Conduct COA Development Brief to Commander

See following page (p. 3-96) for example COA development briefing.

18. JFC Provides Guidance on COAs

- Review and approve COA(s) for further analysis
- Direct revisions to COA(s), combinations of COAs, or development of additional COA(s)
- Direct priority for which enemy COA(s) will be used during war-gaming of friendly COA(s)

19. Continue the Staff Estimate Process

The staff must continue to conduct their staff estimates of supportability for each COA.

20. Conduct Vertical and Horizontal Parallel Planning

- Discuss the planning status of staff counterparts with both commander's and JFC components' staffs
- Coordinate planning with staff counterparts from other functional areas
- Permit adjustments in planning as additional details are learned from higher and adjacent echelons, and permit lower echelons to begin planning efforts and generate questions (e.g., requests for information)

16. Test the Validity of Each Tentative COA

Ref: JP 5-0, Joint Planning (Jun '17), pp. V-28 to V-29.

All COAs selected for analysis must be valid, and the staff should reject tentative COAs that do not meet all five of the following validity criteria:

Adequate. Can accomplish the mission within the commander's guidance. Preliminary tests include:

1. Does it accomplish the mission?
2. Does it meet the commander's intent?
3. Does it accomplish all the essential tasks?
4. Does it meet the conditions for the end state?
5. Does it take into consideration the enemy and friendly COGs?

Feasible. Can accomplish the mission within the established time, space, and resource limitations.

1. Does the commander have the force structure and lift assets (means) to execute it? The COA is feasible if it can be executed with the forces, support, and technology available within the constraints of the physical environment and against expected enemy opposition.
2. Although this process occurs during COA analysis and the test at this time is preliminary, it may be possible to declare a COA infeasible (for example, resources are obviously insufficient). However, it may be possible to fill shortfalls by requesting support from the commander or other means.

Acceptable. Must balance cost and risk with the advantage gained.

1. Does it contain unacceptable risks? (Is it worth the possible cost?) A COA is considered acceptable if the estimated results justify the risks. The basis of this test consists of an estimation of friendly losses in forces, time, position, and opportunity.
2. Does it take into account the limitations placed on the commander (must do, cannot do, other physical limitations)?
3. Acceptability is considered from the perspective of the commander by reviewing the strategic objectives.
4. Are COAs reconciled with external constraints, particularly ROE? This requires visualization of execution of the COA against each enemy capability. Although this process occurs during COA analysis and the test at this time is preliminary, it may be possible to declare a COA unacceptable if it violates the commander's definition of acceptable risk.

Distinguishable. Must be sufficiently different from other COAs in the following:

1. The focus or direction of main effort
2. The scheme of maneuver (land, air, maritime, and special operation)
3. Sequential versus simultaneous maneuvers
4. The primary mechanism for mission accomplishment
5. Task organization
6. The use of reserves

Complete. Does it answer the questions who, what, where, when, how, and why? Must incorporate:

1. Objectives, desired effects to be created, and tasks to be performed
2. Major forces required
3. Concepts for deployment, employment, and sustainment
4. Time estimates for achieving objectives
5. Military end state and mission success criteria (including the assessment: how the commander will know they have achieved success).

COA Development Briefing (Example)

Ref: JP 5-0, Joint Planning (Jun '17), p. V-30 (fig. V-9).

Operations Directorate of a Joint Staff (J-3)/ Plans Directorate of a Joint Staff (J-5)

- Context/background (i.e., road to war)
- Initiation—review guidance for initiation
- Strategic guidance—planning tasks assigned to supported commander, forces/ resources apportioned, planning guidance, updates, defense agreements, theater campaign plan(s), Guidance for Employment of the Force/Joint Strategic Capabilities Plan
- Forces apportioned/assigned

Intelligence Directorate of a Joint Staff (J-2)

- Joint Intelligence Preparation of the Operational Environment
- Enemy objectives
- Enemy courses of action (COAs)—most dangerous, most likely; strengths and weaknesses

J-3/J-5

- Update facts and assumptions
- Mission statement
- Commander's intent (purpose, method, end state)
- End state: political/military
  - termination criteria
- Center of gravity analysis results: critical factors; strategic/operational
- Joint operations area/theater of operations/communications zone sketch
- Phase 0 shaping activities recommended (for current theater campaign plan)
- Flexible deterrent options with desired effect
- For each COA, sketch and statement by phase
  - task organization
  - component tasking
  - timeline
  - recommended command and control by phase
  - lines of operation/lines of effort
  - logistics estimates and feasibility
  - COA risks
  - synchronization matrices
- COA summarized distinctions
- COA priority for analysis
- Update Course of Action Development Briefing to Include:
  - Red objectives

Commander's Guidance

Joint Planning

D. The Planning Directive

The planning directive identifies planning responsibilities for developing joint force plans. It provides guidance and requirements to the staff and subordinate commands concerning coordinated planning actions for plan development. The JFC normally communicates initial planning guidance to the staff, subordinate commanders, and supporting commanders by publishing a planning directive to ensure everyone understands the commander's intent and to achieve unity of effort.

Generally, the plans directorate of a joint staff (J-5) coordinates staff action for planning for the CCMD campaign and contingencies, and the operations directorate of a joint staff (J-3) coordinates staff action in a crisis situation. The J-5 staff receives the JFC's initial guidance and combines it with the information gained from the initial staff estimates. The JFC, through the J-5, may convene a preliminary planning conference for members of the JPEC who will be involved with the plan. This is the opportunity for representatives to meet face-to-face. At the conference, the JFC and selected members of the staff brief the attendees on important aspects of the plan and may solicit their initial reactions. Many potential conflicts can be avoided by this early exchange of information.

IV. Course of Action Analysis and Wargaming

COA analysis is the process of closely examining potential COAs to reveal details that will allow the commander and staff to tentatively identify COAs that are valid and identify the advantages and disadvantages of each proposed friendly COA. The commander and staff analyze each COA separately according to the commander's guidance. While time-consuming, COA analysis should reaffirm the validity of the COA while answering 'is the COA feasible, and is it acceptable?'

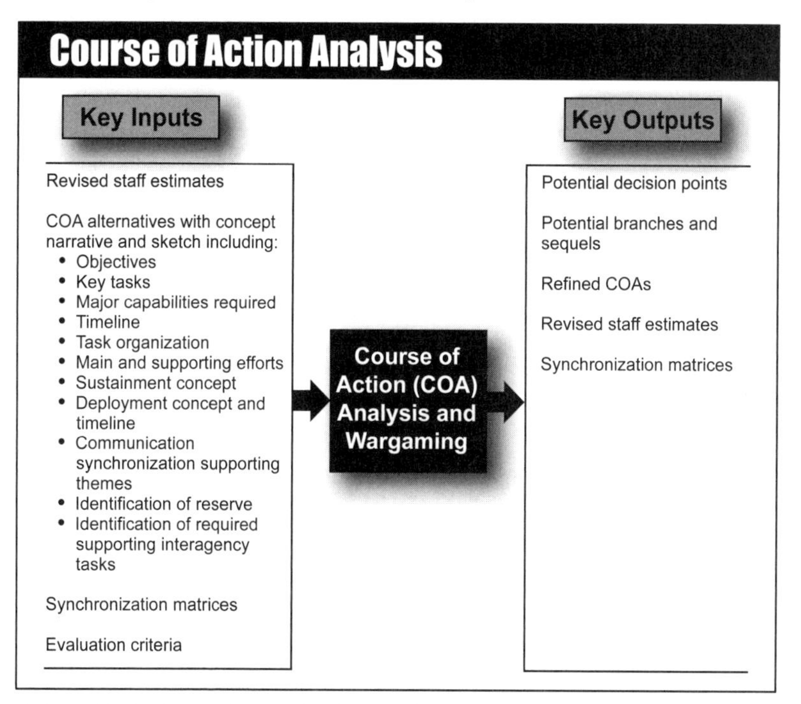

Ref: JP 5-0, Joint Planning, fig. V-10, p. V-33.

Wargaming is a primary means to conduct this analysis. Wargames are representations of conflict or competition in a synthetic environment, in which people make decisions and respond to the consequences of those decisions. COA wargaming is a conscious attempt to visualize the flow of the operation, given joint force strengths and dispositions, adversary capabilities and possible COAs, the OA, and other aspects of the OE. Each critical event within a proposed COA should be wargamed based upon time available using the action, reaction, and counteraction method of friendly and/or opposing force interaction. The basic COA wargaming method can be modified to fit the specific mission and OE, and be applied to noncombat, CCMD campaign activities, and combat operations. Wargaming is most effective when it contains the following elements:

- People making decisions
- A fair competitive environment (i.e., the game should have no rules or procedures designed to tilt the playing field toward one side or another)
- Adjudication
- Consequences of actions
- Iterative (i.e., new insights will be gained as games are iterated)

COA wargaming allows the commander, staff, and subordinate commanders and their staffs to gain a common understanding of friendly and enemy COAs, and other actor actions that may (intentionally or otherwise) work in opposition to achieving the objectives or attaining desired end state conditions. This common understanding allows them to determine the advantages and disadvantages of each COA and forms the basis for the commander's comparison and approval. COA wargaming involves a detailed evaluation of each COA as it pertains to the enemy and the OE. Each of the selected friendly COAs is then wargamed against selected enemy or OE COAs, as well as other actor actions as applicable (for example, wargaming theater campaign or functional campaign activities can identify how a HN or third party might react/respond to US campaign activities). The commander will select the COAs he wants wargamed and provide wargaming guidance along with refined evaluation criteria.

Wargaming stimulates thought about the operation so the staff can obtain ideas and insights that otherwise might not have emerged. An objective, comprehensive analysis of COA alternatives is difficult even without time constraints. Based upon time available, the commander should wargame each COA alternative against the most probable and the most dangerous adversary COAs (or most difficult objectives in noncombat and campaign operations) identified through the JIPOE process.

A. Analysis and Wargaming Process

The analysis and wargaming process can be as simple as a detailed narrative effort that describes the action, probable reaction, counteraction, assets, and time used. A more comprehensive version is the "sketch-note" technique, which adds operational sketches and notes to the narrative process in order to gain a clearer picture. Sophisticated wargames employ more extensive means to depict the range of actions by competitors and the consequences of the synthesis of those actions. The most sophisticated form of wargaming is one where all competitors in a conflict are represented (and emulated to the best degree possible) and have equal decision space to enable a full exploration of the competition within the OE. Modeling and simulation are distinct and separate analytic tools and not the same as wargames. Modeling and simulation can be complementary and assist wargaming through bookkeeping, visualization, and adjudication for well understood actions. The heart of the commander's estimate process is analysis of multiple COAs. The items selected for wargaming and COA comparison will depend on the nature of the mission. For plans or orders involving combat operations, the staff considers opposing COAs based on enemy capabilities, objectives, an estimate of the enemy's intent, and integrated actions by other actors (neutral, other adversaries, and even friendly ac-

COA Analysis Considerations

Ref: JP 5-0, Joint Planning (Jun '17), pp. V-35 to V-37.

Evaluation criteria and known critical events are two of the many important considerations as COA analysis begins.

Evaluation Criteria

The commander and staff use evaluation criteria during follow-on COA comparison (JOPP step 5) for the purpose of selecting the best COA. The commander and staff consider various potential evaluation criteria during wargaming, and select those that the staff will use during COA comparison to assess the effectiveness and efficiency of one COA relative to others following the war game. These evaluation criteria help focus the wargaming effort and provide the framework for data collection by the staff. These criteria are those aspects of the situation (or externally imposed factors) that the commander deems critical to mission accomplishment.

Evaluation criteria change from mission to mission. It will be helpful during future wargaming steps for all participants to be familiar with the criteria so any insights that influence a criterion are recorded for later comparison. The criteria may include anything the commander desires. If they are not received directly, the staff can derive them from the commander's intent statement. Evaluation criteria do not stand alone. Each must be clearly defined. Precisely defining criteria reduces subjectivity and ensures consistent evaluation. The following sources provide a good starting point for developing a list of potential evaluation criteria.

- Commander's guidance and commander's intent
- Mission accomplishment at an acceptable cost
- The principles of joint operations
- Doctrinal fundamentals for the type of operation being conducted
- The level of residual risk in the COA
- Implicit significant factors relating to the operation (e.g., need for speed, security)
- Factors relating to specific staff functions
- Elements of operational design
- Other factors to consider: diplomatic or political constraints, residual risks, financial costs, flexibility, simplicity, surprise, speed, mass, sustainability, C2, and infrastructure survivability.

Critical Events

These are essential tasks, or a series of critical tasks, conducted over a period of time that require detailed analysis (such as the series of component tasks to be performed on D-day). This may be expanded to review component tasks over a phase(s) of an operation or over a period of time (C-day through D-day). The planning staff may wish at this point to also identify decision points (those decisions in time and space that the commander must make to ensure timely execution and synchronization of resources). These decision points are most likely linked to a critical event (e.g., commitment of the reserve force).

For CCMD campaigns, this includes identifying linked events and activities: the staff must identify if campaign activities are sensitive to the sequence in which they are executed and if subsequent activities are dependent on the success of earlier ones. If resources are cut for an activity early in the campaign, the staff must identify to the commander the impact of the loss of that event (or if the results were different from those anticipated), a decision point to continue subsequent events, and alternates if the planned events were dependent on earlier ones.

tions that would not be favorable) that would challenge achievement of the objective. For noncombat operations or CCMD campaign plans, the staff may analyze COAs based on partner capabilities, partner and US objectives, criticality, and risk. In the analysis and wargaming step, the staff analyzes the probable effect each opposing COA has on the chances of success of each friendly COA. The aim is to develop a sound basis for determining the feasibility and acceptability of the COAs. Analysis also provides the planning staff with a greatly improved understanding of their COAs and the relationship between them. COA analysis identifies which COA best accomplishes the mission while best positioning the force for future operations. It also helps the commander and staff to:

- Determine how to maximize combat power against the enemy while protecting the friendly forces and minimizing collateral damage in combat or maximize the effect of available resources toward achieving CCMD and national objectives in noncombat operations and campaigns
- Have as near an identical visualization of the operation as possible
- Anticipate events in the OE and potential reaction options
- Determine conditions and resources required for success while also identifying gaps and seams
- Determine when and where to apply the force's capabilities
- Plan for and coordinate authorities to integrate IRCs early
- Focus intelligence collection requirements
- Determine the most flexible COA
- Identify potential decision points
- Determine task organization options
- Develop data for use in a synchronization matrix or related tool
- Identify potential plan branches and sequels
- Identify high-value targets
- Assess risk
- Determine COA advantages and disadvantages
- Recommend CCIRs
- Validate end states and objectives
- Identify contradictions between friendly COAs and expected enemy end states.

Wargaming is a disciplined process, with rules and steps that attempt to visualize the flow of the operation. The process considers friendly dispositions, strengths, and weaknesses; enemy assets and probable COAs; and characteristics of the physical environment. It relies heavily on joint doctrinal foundation, tactical judgment, and operational and regional/area experience. It focuses the staff's attention on each phase of the operation in a logical sequence. It is an iterative process of action, reaction, and counteraction. Wargaming stimulates ideas and provides insights that might not otherwise be discovered. It highlights critical tasks and provides familiarity with operational possibilities otherwise difficult to achieve. Wargaming is a critical portion of the planning process and should be allocated significant time.

Each retained COA should, at a minimum, be wargamed against both the most likely and most dangerous enemy COAs.

Wargaming Steps

Ref: JP 5-0, Joint Planning (Jun '17), p. V-38 (fig. V-12).

The primary steps for wargaming are prepare for the war game, conduct the war game and assess the results, and prepare products. Figure V-12 shows sample wargaming steps:

1. Prepare for the War Game

- Gather tools
- List and review opposing forces and capabilities
- List known critical events
- Determine participants
- Determine opposing course of action (COA) to war game
- Select wargaming method (manual or computer-assisted)
- Select a method to record and display wargaming results
    - narrative
    - sketch and note
    - war game worksheets
    - synchronization matrix

2. Conduct War Game and Assess Results

- Purpose of war game (identify gaps, visualization, etc.)
- Basic methodology (e.g., action, reaction, counteraction)
- Record results

3. Prepare Products

- Results of the war game brief
    - potential decision points
    - evaluation criteria
    - potential branches and sequels
- Revised staff estimates
- Refined COAs
- Time-phased force and deployment data refinement and transportation feasibility
- Feedback through the COA decision brief

B. Prepare for the War Game

There are two key decisions to make before COA analysis (wargaming) begins. The first decision is to decide what type of wargame will be used. This decision should be based on commander's guidance, time and resources available, staff expertise, and availability of simulation models. The second decision is to prioritize the enemy COAs or the partner capabilities, partner and US objectives for noncombat operations, and the wargame that it is to be analyzed against. In time-constrained situations, it may not be possible to wargame against all COAs.

The two forms of wargames are computer-assisted and manual. There are many forms of computer-assisted wargames; most require a significant amount of preparation to develop and load scenarios and then to train users. However, the potential to utilize the computer model for multiple scenarios or blended scenarios makes it valuable. For both types, consider how to organize the participants in a logical manner.

For manual wargaming, three distinct methods are available to run the event:

1. Deliberate Timeline Analysis

Consider actions day-by-day or in other discrete blocks of time. This is the most thorough method for detailed analysis when time permits.

2. Phasing

Used as a framework for COA analysis. Identify significant actions and requirements by functional area and/or joint task force (JTF) component.

3. Critical Events/Sequence of Essential Tasks

The sequence of essential tasks, also known as the critical events method, highlights the initial actions necessary to establish the conditions for future actions, such as a sustainment capability and engage enemy units in the deep battle area. At the same time, it enables the planners to adapt if the enemy or other actor in the OE reacts in such a way that necessitates reordering of the essential tasks. This technique also allows wargamers to analyze concurrently the essential tasks required to execute the CONOPS. Focus on specific critical events that encompass the essence of the COA. If necessary, different MOEs should be developed for assessing different types of critical events (e.g., destruction, blockade, air control, neutralization, ensure defense). As with the focus on phasing, the critical events discussion identifies significant actions and requirements by functional area and/or by JTF component and enables a discussion of possible or expected reactions to execution of critical tasks.

C. Conduct the Wargame and Evaluate the Results

The facilitator and the red cell chief get together to agree on the rules of the wargame. The wargame begins with an event designated by the facilitator. It could be an enemy offensive or defensive action, a friendly offensive or defensive action, or some other activity such as a request for support or campaign activity. They decide where (in the OA) and when (H-hour or L-hour) it will begin. They review the initial array of forces and the OE. Of note, they must come to an agreement on the effectiveness of capabilities and previous actions by both sides prior to the wargame. The facilitator must ensure all members of the wargame know what events will be wargamed and what techniques will be used. This coordination within the friendly team and between the friendly and the red team should be done well in advance.

Each COA wargame has a number of turns, each consisting of three total moves: action, reaction, and counteraction. If necessary, each turn of the wargame may be extended beyond the three basic moves. The facilitator, based on JFC guidance, decides how many total turns are made in the wargame.

Red/White/Blue Cells

Ref: JP 5-0, Joint Planning (Jun '17), pp. V-38 to V-39.

Red Cell

The J-2 staff will provide a red cell to role-play and model the enemies and other relevant actors in the OE during planning and specifically during wargaming.

A robust, well-trained, imaginative, and skilled red cell that aggressively pursues the enemy's point of view during wargaming is essential. By accurately portraying the full range of realistic capabilities and options available to the enemy, they help the staff address friendly responses for each enemy COA. For campaign and noncombat operation planning, the red cell provides expected responses to US actions based on their knowledge and analysis of the OE.

The red cell is normally composed of personnel from the joint force J-2 staff and augmented by other subject matter experts. The red cell develops critical decision points, projects enemy and other actor's OE reactions to friendly actions, and estimates impacts and implications on the enemy forces and objectives. By trying to win the wargame, the red cell helps the staff identify weaknesses and vulnerabilities before a real enemy does.

Given time constraints, as a minimum, the most dangerous and most likely COAs should be wargamed and role-played by the red cell during the wargame.

White Cell

A small cell of arbitrators normally composed of senior individuals familiar with the plan is a smart investment to ensure the wargame does not get bogged down in unnecessary disagreement or arguing. The white cell will provide overall oversight to the wargame and any adjudication required between participants. The white cell may also include the facilitator and/or highly qualified experts as required.

Blue Cell

In addition to a red cell and a white cell, there should also be a blue cell that represents friendly forces, and a green cell represents transnational groups, NGOs, and neutral regional populations.

See related discussion of red teaming on p. 3-45.

During the wargame, the participants must continually evaluate the COA's feasibility. Can it be supported? Can this be done? Will it achieve the desired results? Are more forces, resources, intelligence collection capabilities, or time needed? Are necessary logistics and communications available? Is the OA large enough? Has the threat successfully impacted key enablers like logistics or communications, or countered a certain phase or stage of a friendly COA? Based on the answers to the above questions, revisions to the friendly COA may be required. Major revisions to a COA are not made in the midst of a wargame. Instead, stop the wargame, make the revisions, and start over at the beginning.

The wargame is for comparing and contrasting friendly COAs with the enemy COAs. Planners compare and contrast friendly COAs with each other in the fifth step of JPP, COA comparison. Planners avoid becoming emotionally attached to a friendly COA and avoid comparing one friendly COA with another friendly COA during the wargame so they can remain unbiased. The facilitator ensures adherence to the timeline. A wargame for one COA at the JTF level may take six to eight hours. The facilitator must allocate enough time to ensure the wargame will thoroughly test a COA.

Synchronization Matrix

A synchronization matrix is a decision-making tool and a method of recording the results of wargaming. Key results that should be recorded include decision points, potential evaluation criteria, CCIRs, COA adjustments, branches, and sequels. Using a synchronization matrix helps the staff visually synchronize the COA across time and space in relation to the enemy's possible COAs and (or) other actor's activities within the OA. The wargame and synchronization matrix efforts will be particularly useful in identifying cross-component support resource requirements.

The wargame considers friendly dispositions, strengths, and weaknesses; enemy assets and probable COAs; and characteristics of the OA. Through a logical sequence, it focuses the participants on essential tasks to be accomplished.

When the wargame is complete and the worksheet and synchronization matrix are filled out, there should be enough detail to flesh out the bones of the COA and begin orders development (once the COA has been selected by the commander in a later JPP step).

Additionally, the wargame will produce a refined event template and the initial decision support template (DST), decision points (and the CCIR related to them), or other decision support tools. These are similar to a football coach's game plan. The tools can help predict what the threat will do and how partner nations or other actors will react to US actions. The tools also provide the commander options for employing forces to counter an adversary action. The tools will prepare the commander (coach) and the staff (team) for a wide range of possibilities and a choice of immediate solutions.

The wargame relies heavily on doctrinal foundation, tactical and operational judgment, and experience. It generates new ideas and provides insights that might have been overlooked. The dynamics of the wargame require the red cell to be aggressive, but realistic, in the execution of threat activities. The wargame:

- Records advantages and disadvantages of each COA as they become evident.
- Creates decision support tools (a game plan).
- Focuses the planning team on the threat and commander's evaluation criteria.

D. COA Analysis/Wargaming Products

Ref: JP 5-0, Joint Planning (Jun '17), pp. V-41 to V-42.

Certain products should result from the war game in addition to wargamed COAs.

Event Template

Planners enter the wargame with a rough event template and must complete the wargame with a refined, more accurate event template. The event template with its named areas of interest (NAIs) and time-phase lines will help the J-2 focus the intelligence collection effort. An event matrix can be used as a "script" for intelligence reporting during the wargame. It can also tell planners if they are relying too much on one or two collection platforms and if assets have been overextended.

Decision Support Template and Matrix (DST/DSM)

A first draft of a DST should also come out of the COA wargame. As more information about friendly forces and threat forces becomes available, the DST may change.

The critical events are associated with the essential tasks identified in mission analysis. The decision points are tied to points in time and space when and where the commander must make a critical decision.

Priority Intelligence Requirements (PIRs) and Friendly Force Information Requirements (FFIRs)

Decision points should be tied to the CCIRs. CCIRs generate two types of information requirements: PIRs and FFIRs. The commander approves CCIRs. From a threat perspective, PIRs tied to a decision point will require an intelligence collection plan that prioritizes and tasks collection assets to gather information about the threat. JIPOE ties PIRs to NAIs, which are linked to adversary COAs. The synchronization matrix is a tool that will help determine if adequate resources are available.

Primary Outputs of the Wargame

Primary outputs of the COA analysis/wargame are:

1. Wargamed COAs with graphic and narrative. Branches and sequels identified
2. Information on commander's evaluation criteria
3. Initial task organization
4. Critical events and decision points
5. Newly identified resource shortfalls to include force augmentation
6. Refined/new CCIRs and event template/matrix
7. Initial DST/DSM
8. Refined synchronization matrix
9. Refined staff estimates
10. Assessment plan and criteria

The outputs of the COA wargame will be used in the JPP steps COA comparison, COA approval, and plan or order development. The results of the wargame are an understanding of the strengths and weaknesses of each friendly COA, the core of the back brief to the commander.

The commander and staff normally will compare advantages and disadvantages of each COA during COA comparison. However, if the suitability, feasibility, or acceptability of any COA becomes questionable during the analysis step, the commander should modify or discard it and concentrate on other COAs. The need to create additional combinations of COAs may also be identified.

V. Course of Action Comparison

COA comparison is a subjective process whereby COAs are considered independently and evaluated/compared against a set of criteria that are established by the staff and commander. The objective is to identify and recommend the COA that has the highest probability of accomplishing the mission.

The figure below depicts inputs and outputs for COA comparison. Other products not graphically shown in the chart include updated JIPOE products, updated CCIRs, staff estimates, and commander's ID of branches for further planning.

COA comparison facilitates the commander's decision-making process by balancing the ends, means, ways, and risk of each COA. The end product of this task is a briefing to the commander on a COA recommendation and a decision by the commander. COA comparison helps the commander answer the following questions:

- What are the differences between each COA?
- What are the advantages and disadvantages?
- What are the risks?

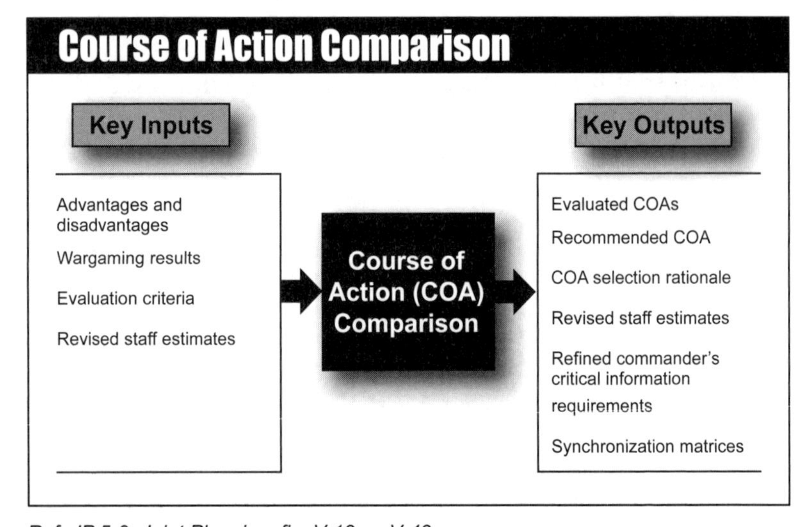

Ref: JP 5-0, Joint Planning, fig. V-13, p. V-42.

In COA comparison, the staff determines which COA performs best against the established evaluation criteria. The commander reviews the criteria list and adds or deletes, as required. The number of evaluation criteria will vary, but there should be enough to differentiate COAs. COAs are not compared with each other within any one criterion, but rather they are individually evaluated against the criteria that are established by the staff and commander. Their individual performances are then compared to enable the staff to recommend a preferred COA to the commander.

Staff officers may each use their own matrix to compare COAs with respect to their functional areas. Matrices use the evaluation criteria developed before the wargame. Decision matrices alone cannot provide decision solutions. Their greatest value is providing a method to compare COAs against criteria that the commander and staff believe will produce mission success. They are analytical tools that staff officers use to prepare recommendations. Commanders provide the solution by applying their judgment to staff recommendations and making a decision.

A. Prepare for COA Comparison

Ref: JP 5-0, Joint Planning (Jun '17), pp.V-44 to V-45.

The staff helps the commander identify and select the COA that best accomplishes the mission. The staff supports the commander's decision-making process by clearly portraying the commander's options and recording the results of the process.

The staff evaluates feasible COAs to identify the one that performs best within the evaluation criteria against the enemy's most likely and most dangerous COAs.

The commander and staff use the evaluation criteria developed during mission analysis to identify the advantages and disadvantages of each COA. Comparing the strengths and weaknesses of the COAs identifies their advantages and disadvantages relative to each other.

1. Determine/Define Comparison/Evaluation Criteria

As discussed earlier, criteria are based on the particular circumstances and should be relative to the situation. There is no standard list of criteria, although the commander may prescribe several core criteria that all staff directors will use. Individual staff sections, based on their estimate process, select the remainder of the criteria.

- Criteria are based on the particular circumstances and should be relative to the situation
- Review commander's guidance for relevant criteria
- Identify implicit significant factors relating to the operation
- Each staff identifies criteria relating to that staff function
- Other criteria might include:
    - Political, social, and safety constraints; requirements for coordination with embassy/interagency personnel
    - Fundamentals of joint warfare
    - Elements of operational art
    - Mission accomplishment
    - Risks
    - Costs
    - Time

2. Define and Determine the Standard for Each Criterion

- Establish standard definitions for each evaluation criterion. Define the criteria in precise terms to reduce subjectivity and ensure the interpretation of each evaluation criterion remains constant between the various COAs.
- Establish definitions prior to commencing COA comparison to avoid compromising the outcome.
- Apply standards for each criterion to each COA.

3. Evaluation Criteria

The staff evaluates COAs using those evaluation criteria most important to the commander to identify the one COA with the highest probability of success. The selected COA should also:

- Place the force in the best posture for future operations.
- Provide maximum latitude for initiative by subordinates.
- Provide the most flexibility to meet unexpected threats and opportunities.

B. Determine the Comparison Method and Record

Ref: JP 5-0, Joint Planning (Jun '17), pp. V-45 and app. G.

Actual comparison of COAs is critical. The staff may use any technique that facilitates reaching the best recommendation and the commander making the best decision. There are a number of techniques for comparing COAs. COA comparison remains a subjective process and should not be turned into a mathematical equation. The key element in this process is the ability to articulate to the commander why one COA is preferred over another.

Decision Matrixes

The most common technique for COA comparison is the weighted numerical comparison, which uses evaluation criteria to determine the preferred COA based upon the wargame. COAs are not compared to each other directly until each COA is considered independently against the evaluation criteria. The CCDR may direct some of these criteria, but most criteria are developed by the JPG.

Below are examples of common methods.

1. Weighted Numerical Comparison Technique

The example below provides a numerical aid for differentiating COAs. Values reflect the relative advantages or disadvantages of each COA for each criterion selected. Certain criteria have been weighted to reflect greater value.

Determine the weight of each criterion based on its relative importance and the commander's guidance. The commander may give guidance that results in weighting certain criteria. The staff member responsible for a functional area scores each COA using those criteria. Multiplying the score by the weight yields the criterion's value. The staff member then totals all values. However, the staff member must be careful not to portray subjective conclusions as the results of quantifiable analysis. Comparing COAs by category is more accurate than comparing total scores.

Evaluation Criterion	Weight	COA 1		COA 2		COA 3	
		Score	Weighted	Score	Weighted	Score	Weighted
Surprise	2	3	6	1.5	3	1.5	3
Risk	2	3	6	1	2	2	4
Flexibility	1	3	3	1.5	1.5	1.5	1.5
Retaliation	1	1.5	1.5	3	3	1.5	1.5
Damage to alliance	1	3	3	1.5	1.5	1.5	1.5
Legal basis	1	2	2	3	3	1	1
External support	1	3	3	2	2	1	1
Force protection	1	2.5	2.5	2.5	2.5	1	1
OPSEC	1	3	3	1.5	1.5	1.5	1.5
Total			30		20		16

Ref: JP 5-0, fig. G-2. Example #2 Course of Action Comparison Matrix Format.

(1) Criteria are those selected through the process described earlier.

(2) The criteria can be rated (or weighted). The most important criteria are rated with the highest numbers. Lesser criteria are weighted with lower numbers.

(3) The highest number is best.

(4) Each staff section does this separately, perhaps using different criteria on which to base the COA comparison. The staff then assembles and arrives at a consensus for the criterion and weights.

2. Non-Weighted Numerical Comparison Technique

The same as the previous method except the criteria are not weighted. Again, the highest number is best for each of the criteria.

3. Narrative or Bulletized Descriptive Comparison of Strengths & Weaknesses (or Advantages & Disadvantages)

Summarize comparison of all COAs by analyzing strengths and weaknesses or advantages and disadvantages for each criterion.

	Criteria 1		Criteria 2		Criteria 3	
COA 1	Strengths • •	Weaknesses • •	Strengths • •	Weaknesses • •	Strengths • •	Weaknesses • •
COA 2	Strengths • •	Weaknesses • •	Strengths • •	Weaknesses • •	Strengths • •	Weaknesses • •
COA 3	Strengths • •	Weaknesses • •	Strengths • •	Weaknesses • •	Strengths • •	Weaknesses • •

Ref: JP 5-0, Joint Operations, fig. G-3. Criteria for Strengths and Weaknesses Example.

4. Plus/Minus/Neutral Comparison

Base this comparison on the broad degree to which selected criteria support or are reflected in the COA. This is typically organized as a table showing (+) for a positive influence, (0) for a neutral influence, and (–) for a negative influence.

Criteria	COA 1	COA 2	COA 3
Casualty estimate	+	–	–
Casualty evacuation routes	–	+	+
Suitable medical facilities	0	0	0
Flexibility	+	–	–

Ref: JP 5-0, Joint Operations, fig. G-5. Plus/Minus/Neutral Comparison Example.

5. Descriptive Comparison

This is simply a description of advantages and disadvantages of each COA.

VI. Course of Action Approval

In this JOPP step, the staff briefs the commander on the COA comparison and the analysis and wargaming results, including a review of important supporting information. The staff determines the best COA to recommend to the commander.

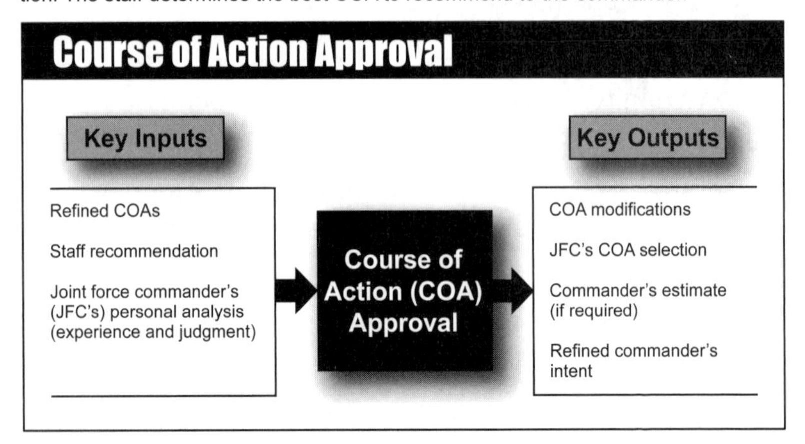

Ref: JP 5-0, Joint Planning, fig. V-15, p. V-45.

The nature of the OE or contingency may make it difficult to determine the desired end state until the crisis actually occurs. In these cases, the JFC may choose to present two or more valid COAs for approval by higher authority. A single COA can then be approved when specific circumstances become clear. However, in a crisis, the desired end state should be based on the set of objectives approved by the President.

A. Prepare and Present the COA Decision Briefing

The staff briefs the commander on the COA comparison, COA analysis, and wargaming results. The briefing should include a review of important supporting information such as the current status of the joint force, the current JIPOE, and assumptions used in COA development. All principal staff directors and the component commanders should attend this briefing (physically or virtually). Figure V-16 shows a sample COA briefing guide.

B. Commander Selects/Modifies the COA

COA selection is the end result of the COA comparison process. Throughout the COA development process, the commander conducts an independent analysis of the mission, possible COAs, and relative merits and risks associated with each COA. The commander, upon receiving the staff's recommendation, combines personal analysis with the staff recommendation, resulting in a selected COA. It gives the staff a concise statement of how the commander intends to accomplish the mission, and provides the necessary focus for planning and plan development. During this step, the commander should:

- Review staff recommendations
- Apply results of own COA analysis and comparison
- Consider any separate recommendations from supporting and subordinate commanders
- Review guidance from the higher headquarters/strategic guidance

Sample Course of Action Briefing Guide

Ref: JP 5-0, Joint Planning (Jun '17), pp. V-47 (fig. V-16).

1. **Purpose of the briefing**
2. **Opposing situation**
   - **Strength**. A review of opposing forces, both committed and available for reinforcement
   - **Composition**. Order of battle, major weapons systems, and operational characteristics
   - **Location and disposition.** Ground combat and fire support forces; air, naval, and missile forces; logistics forces and nodes; command and control facilities; and other combat power
   - **Reinforcements**. Land; air; naval; missile; chemical, biological, radiological, and nuclear; other advanced weapons systems; capacity for movement of these forces
   - **Logistics**. Summary of opposing forces ability to support combat operations
   - **Time and Space Factors.** The capacity to move and reinforce positions
   - **Combat Efficiency.** The state of training, readiness, battle experience, physical condition, morale, leadership, motivation, tactical doctrine, discipline, and significant strengths and weaknesses
3. **Friendly situation (similar elements as opposing situation)**
4. **Mission statements**
5. **Commander's intent statement**
6. **Operational concepts and courses of action (COAs)**
   - Any changes from the mission analysis briefing in the following areas:
     – assumptions
     – limitations
     – adversary and friendly centers of gravity (COGs)
     – phasing of the operation (if phased)
     – lines of operation/lines of effort
   - Present courses of action. As a minimum, discuss:
     – COA# _____ (short name, e.g., "Simultaneous Assault")
     – COA statement (brief concept of operations)
     – COA sketch
     – COA architecture (task organization, command relationships, organization of the operational area)
     – major differences between each COA
     – summaries of COAs
   - COA analysis
     – review of the joint planning group's wargaming efforts
     – add considerations from own experiences
   - COA comparisons
     – description of comparison criteria (e.g., evaluation criteria) and comparison methodology
     – weigh strengths and weaknesses with respect to comparison criteria COA recommendations
   - COA Recommendations
     – staff
     – components

- The commander may:
  - Concur with staff/component recommendations, as presented
  - Concur with recommended COAs, but with modifications
  - Select a different COA from the staff/component recommendation
  - Combine COAs to create a new COA.
  - Reject all and start over with COA development or mission analysis.
  - Defer the decision and consult with selected staff/commanders prior to making a final decision.

C. Refine Selected COA

Once the commander selects a COA, the staff will begin the refinement process of that COA into a clear decision statement to be used in the commander's estimate. At the same time, the staff will apply a final "acceptability" check.

1. Staff Refines Commander's COA Selection into Clear Decision Statement

- Develop a brief statement that clearly and concisely sets forth the COA selected and provides only whatever information is necessary to develop a plan for the operation (no defined format)
- Describe what the force is to do as a whole, and as much of the elements of when, where, and how as may be appropriate
- Express decision in terms of what is to be accomplished, if possible
- Use simple language so the meaning is unmistakable
- Include statement of what is acceptable risk

2. Apply Final "Acceptability" Check

- Apply experience and an understanding of situation
- Consider factors of acceptable risk versus desired outcome consistent with higher commander's intent and concept. Determine if gains are worth expenditures.

D. Prepare the Commander's Estimate

Once the commander selects the COA, provides guidance, and updates intent, the staff then completes the commander's estimate. The commander's estimate provides a concise narrative statement of how the commander intends to accomplish the mission and provides the necessary focus for campaign planning and contingency plan development. Further, it responds to the establishing authority's requirement to develop a plan for execution. The commander's estimate provides a continuously updated source of information from the perspective of the commander. Commanders at various levels use estimates during JPP to support COA determination and plan or order development.

See facing page for further discussion and a sample format.

E. CJCS Estimate Review

The estimate review determines whether the scope and concept of planned operations satisfy the tasking and will accomplish the mission, determines whether the assigned tasks can be accomplished using available resources in the timeframes contemplated by the plan, and ensures the plan is proportional and worth the expected costs. As planning is approved by SecDef (or designated representative) during an IPR, the commander's estimate informs the refinement of the initial CONOPS for the plan.

Commander's Estimate

Ref: JP 5-0, Joint Planning (Jun '17), pp. V-48 to V-49.

A commander uses a commander's estimate as the situation dictates. The commander's initial intent statement and planning guidance to the staff can provide sufficient information to guide the planning process. The commander will tailor the content of the commander's estimate based on the situation and ongoing analysis.

A typical format for a commander's estimate is in CJCSM 3130.03, Adaptive Planning and Execution (APEX) Planning Formats and Guidance.

Contents may vary, depending on the nature of the plan or contingency, time available, and the applicability of prior planning. In a rapidly developing situation, the formal commander's estimate may be impractical, and the entire estimate process may be reduced to a commanders' conference.

With appropriate horizontal and vertical coordination, the commander's COA selection may be briefed to and approved by SecDef. In the strategic context, where military operations are strategically significant, even a commander's selected COA is normally briefed to and approved by the President or SecDef. The commander's estimate then becomes a matter of formal record keeping and guidance for component and supporting forces.

The supported commander may use simulation and analysis tools in the collaborative environment to evaluate a variety of options, and may also choose to convene a concept development conference involving representatives of subordinate and supporting commands, the Services, JS, and other interested parties. Review of the resulting commander's estimate requires collaboration and coordination among all planning participants. The supported commander may highlight issues for future interagency consultation, review, or resolution to be presented to SecDef during the IPR.

Sample Commander's Estimate Format

Operational Description

- Purpose of the operation
- References
- Description of military operations

Narrative—Five Paragraphs

- Mission
- Situation and courses of action
- Analysis of opposing courses of action (adversary capabilities and intentions)
- Comparison of friendly courses of action
- Recommendation or decision

Remarks

- Remarks—Cite plan identification number of the file where detailed requirements have been loaded into the Joint Operation Planning and Execution System

Note: This sample format is adapted from the Aug '11 edition of JP 5-0.

Joint Planning

VII. Plan or Order Development

A. Concept of Operations (CONOPS)

The CONOPS clearly and concisely expresses what the JFC intends to accomplish and how it will be done using available resources. It describes how the actions of the joint force components and supporting organizations will be integrated, synchronized, and phased to accomplish the mission, including potential branches and sequels.

See facing page for further discussion of CONOPS.

Planning results in a plan that is documented in the format of a plan or an order. If execution is imminent or in progress, the plan is typically documented in the format of an order. During plan or order development, the commander and staff, in collaboration with subordinate and supporting components and organizations, expand the approved COA into a detailed plan or OPORD by refining the initial CONOPS associated with the approved COA. The CONOPS is the centerpiece of the plan or OPORD.

CJCSM 3130.03, Adaptive Planning and Execution (APEX) Planning Formats and Guidance, provides detailed guidance on CONOPS content and format.

B. Format of Military Plans and Orders

Plans and orders can come in many varieties from very detailed campaign plans and contingency plans to simple verbal orders. They may also include orders and directives such as OPORDs, WARNORDs, PLANORDs, ALERTORDs, EXORDs, and FRAGORDs, as well as PTDOs, DEPORDs, and the GFMAP. The more complex directives will contain much of the amplifying information in appropriate annexes and appendices. However, the directive should always contain the essential information in the main body. The information contained may depend on the time available, the complexity of the operation, and the levels of command involved. \

In most cases, the directive will be standardized in the five-paragraph format that is described in CJCSM 3130.03, Adaptive Planning and Execution (APEX) Planning Formats and Guidance. *See pp. 3-119 to 3-124 for related discussion.*

C. Plan or Order Development

For most plans and orders, the CJCS monitors planning activities, resolves shortfalls when required, and reviews the supported commander's plan for adequacy, feasibility, acceptability, completeness, and compliance with policy and joint doctrine. When required, the commander will conduct one or more IPRs with SecDef (or designated representative) to confirm the plan's strategic guidance, assumptions (including timing and national-level decisions required), any limitations (restrictions and constraints), the mission statement, the operational approach, key capability shortfalls, areas of risk, and acceptable levels of risk; and any further guidance required for plan refinement. During the IPRs, the CJCS and the USD(P) will separately address issues arising from, or resolved during, plan review (e.g., key risks, decision points). Commanders should show how the plan supports the objectives identified in the GEF and JSCP and identify the links to other plans, both within the AOR (or functional area) and with those of other CCMDs. The result of an IPR should include an endorsement of the planning to date or acknowledgement of friction points and guidance to shape continued planning. All four APEX operational activities (situational awareness, planning, execution, and assessment) continue in a complementary and iterative process.

See following pages (pp. 3-116 to 3-117) for an overview and further discussion of plan development activities. *CJCSI 3141.01, Management and Review of Joint Strategic Capabilities Plan (JSCP)-Tasked Plans, provides further details on the IPR process.*

Concept of Operations (CONOPS)

Ref: JP 5-0, Joint Planning (Jun '17), pp. V-49 to V-50.

The CONOPS clearly and concisely expresses what the JFC intends to accomplish and how it will be done using available resources. It describes how the actions of the joint force components and supporting organizations will be integrated, synchronized, and phased to accomplish the mission, including potential branches and sequels.

- States the commander's intent
- Describes the central approach the JFC intends to take to accomplish the mission
- Provides for the application, sequencing, synchronization, and integration of forces and capabilities in time, space, and purpose (including those of multinational and interagency organizations as appropriate)
- Describes when, where, and under what conditions the supported commander intends to give or refuse battle, if required
- Focuses on friendly and adversary COGs and their associated critical vulnerabilities
- Avoids discernible patterns and makes full use of ambiguity and deception
- Provides for controlling the tempo of the operation
- Visualizes the campaign in terms of the forces and functions involved
- Relates the joint force's objectives and desired effects to those of the next higher command and other organizations as necessary. This enables assignment of tasks to subordinate and supporting commanders.

The staff writes (or graphically portrays) the CONOPS in sufficient detail so subordinate and supporting commanders understand their mission, tasks, and other requirements and can develop their supporting plans. During CONOPS development, the commander determines the best arrangement of simultaneous and sequential actions and activities to accomplish the assigned mission consistent with the approved COA, and resources and authorities available. This arrangement of actions dictates the sequencing of activities or forces into the OA, providing the link between the CONOPS and force planning. The link between the CONOPS and force planning is preserved and perpetuated through the sequencing of forces into the OA via a TPFDD. The structure must ensure unit integrity, force mobility, and force visibility as well as the ability to transition to branches or sequels rapidly as operational conditions dictate. Planners ensure the CONOPS, force plan, deployment plans, and supporting plans provide the flexibility to adapt to changing conditions, and are consistent with the JFC's intent.

If the scope, complexity, and duration of the military action contemplated to accomplish the assigned mission warrants execution via a series of related operations, then the staff outlines the CONOPS as a campaign. They develop the preliminary part of the operational campaign in sufficient detail to impart a clear understanding of the commander's concept of how the assigned mission will be accomplished.

During CONOPS development, the JFC must assimilate many variables under conditions of uncertainty to determine the essential military conditions, sequence of actions, and application of capabilities and associated forces to create effects and achieve objectives. JFCs and their staffs must be continually aware of the higher-level objectives and associated desired and undesired effects that influence planning at every juncture. If operational objectives are not linked to strategic objectives, the inherent linkage or "nesting" is broken and eventually tactical considerations can begin to drive the overall strategy at cross-purposes.

D. Plan Development Activities

Ref: JP 5-0, Joint Planning (Jun '17), pp. V-52 to V-60.

The JFC guides plan development by issuing a PLANORD or similar planning directive to coordinate the activities of the commands and agencies involved. A number of activities are associated with plan development, as Figure V-17 shows. These planning activities typically will be accomplished in a parallel, collaborative, and iterative fashion rather than sequentially, depending largely on the planning time available. The same flexibility displayed in COA development is seen here again, as planners discover and eliminate shortfalls and conflicts within their command and with the other CCMDs.

The CJCS APEX family of documents referenced in CJCS Guide 3130, Adaptive Planning and Execution Overview and Policy Framework, provides policy, procedures, and guidance on these activities for organizations required to prepare a plan or order. These are typical types of activities that supported and supporting commands and Services accomplish collaboratively as they plan for joint operations.

Application of Forces and Capabilities

When planning forces and capabilities, the commander is constrained by the total quantity of forces in the force apportionment tables. If additional resources are deemed necessary to reduce risk, CJCS approval is required. The supported commander should address the additional force requirement as early as possible in the IPR process, justify the requirement, and identify the risk associated if the forces are not made available. Risk assessments will include results using both apportioned capabilities and augmentation capabilities.

The supported commander should designate the main effort and supporting efforts as soon as possible and identify interdependent missions (especially subsequent tasks dependent on the successful completion of earlier tasks). This action is necessary for economy of effort. The main effort is based on the supported JFC's prioritized objectives. It identifies where the supported JFC will concentrate capabilities or prioritize efforts to achieve specific objectives. Designation of the main effort can be addressed in geographical (area) or functional terms

Force Planning

The primary purposes of force planning are to identify all forces needed to accomplish the CONOPS and effectively phase the forces into the OA. Force planning consists of determining the force requirements by operation phase, mission, mission priority, mission sequence, and operating area. It includes force requirements review, major force phasing, integration planning, and force list refinement. Force planning is the responsibility of the supported CCDR, supported by component commanders in coordination with the JS, JFPs, and FPs. Force planning begins early during plan development and focuses on applying the right force to the mission at the right time, while ensuring force visibility, force mobility, and adaptability. The commander determines force requirements and as necessary, develops a TPFDD letter of instruction specific to the OA and plans force modules to align and time-phase the forces in accordance with the CONOPS.

Support Planning

Support planning is conducted concurrently with force planning to determine and sequence logistics and personnel support in accordance with the plan CONOPS. Support planning includes all core logistics functions: deployment and distribution, supply, maintenance, logistic services, OCS, health services, and engineering.

See pp. 4-23 to 4-32 for further discussion of support planning.

Nuclear Strike Planning

Commanders must assess the military as well as strategic impact a nuclear strike would have on conventional operations. Nuclear planning guidance is provided in Presidential policy documents and further clarified in DOD documents such as the GEF Annex B, and the Nuclear Supplement to the JSCP. Guidance issued to the CCDR is based on national-level considerations and supports the accomplishment of US objectives. United States Strategic Command (USSTRATCOM) is the lead organization for nuclear planning and coordination with appropriate allied commanders.

Deployment and Redeployment Planning

Deployment and redeployment planning is conducted on a continuous basis for all approved contingency plans and as required for specific crisis action plans. Planning for redeployment should be considered throughout the operation and is best accomplished in the same time-phased process in which deployment was accomplished. In all cases, mission requirements of a specific operation define the scope, duration, and scale of both deployment and redeployment operation planning.

Shortfall Identification

Along with hazard and threat analysis, shortfall ID is conducted throughout the plan development process. The supported commander continuously identifies limiting factors, capability shortfalls, and associated risks as plan development progresses. Where possible, the supported commander resolves the shortfalls and required controls and countermeasures through planning adjustments and coordination with supporting and subordinate commanders. If the shortfalls and necessary controls and countermeasures cannot be reconciled or the resources provided are inadequate to perform the assigned task, the supported commander reports these limiting factors and assessment of the associated risk to the CJCS.

Feasibility Analysis

This step in plan or order development is similar to determining the feasibility of a COA, except it typically does not involve simulation-based wargaming. The focus in this step is on ensuring the assigned mission can be accomplished using available resources within the time contemplated by the plan. The results of force planning, support planning, deployment and redeployment planning, and shortfall ID will affect feasibility. The primary factors to consider are the capacity of lift and throughput constraints of transit points and JRSOI infrastructure that can support the plan. The primary factors analyzed for feasibility include forces, resources, and transportation

Documentation

When the TPFDD is complete and end-to-end transportation feasibility has been achieved and is acceptable to the supported CCDR, the supported CCDR completes the documentation of the plan or OPORD and coordinates access with respective JPEC stakeholders to the TPFDD as appropriate.

Movement Plan Review and Approval

When the plan or OPORD is complete, JS J-5 coordinates with the JPEC for review. The JPEC reviews the plan or OPORD and provides the results of the review to the supported CCDR and CJCS. The CJCS reviews and provides recommendations to SecDef, if necessary. The JCS provides a copy of the plan to OSD to facilitate their parallel review of the plan and to inform USD(P)'s recommendation of approval/disapproval to SecDef. After the CJCS's and USD(P)'s review, SecDef or the President will review, approve, or modify the plan. The President or SecDef is the final approval authority for OPORDs, depending upon the subject matter.

See CJCSI 3141.01, Management and Review of Joint Strategic Capabilities Plan (JSCP)-Tasked Plans, for more information on plan review and approval.

E. Transition

Ref: JP 5-0, Joint Planning (Jun '17), pp. V-60 to V-62.

Transition is an orderly turnover of a plan or order as it is passed to those tasked with execution of the operation. It provides information, direction, and guidance relative to the plan or order that will help to facilitate situational awareness. Additionally, it provides an understanding of the rationale for key decisions necessary to ensure there is a coherent shift from planning to execution. These factors coupled together are intended to maintain the intent of the CONOPS, promote unity of effort, and generate tempo. Successful transition ensures those charged with executing an order have a full understanding of the plan. Regardless of the level of command, such a transition ensures those who execute the order understand the commander's intent and CONOPS. Transition may be internal or external in the form of briefs or drills. Internally, transition occurs between future plans and future/current operations. Externally, transition occurs between the commander and subordinate commands.

See pp. 3-137 to 3-141 for related discussion of "transition to execution."

Transition Brief

At higher levels of command, transition may include a formal transition brief to subordinate or adjacent commanders and to the staff supervising execution of the order. At lower levels, it might be less formal. The transition brief provides an overview of the mission, commander's intent, task organization, and enemy and friendly situation. It is given to ensure all actions necessary to implement the order are known and understood by those executing the order. The brief may include items such as:

1. Higher headquarters' mission and commander's intent
2. Mission
3. Commander's intent
4. CCIRs
5. Task organization
6. Situation (friendly and enemy)
7. CONOPS
8. Execution (including branches and potential sequels)
9. Planning support tools (such as a synchronization matrix)

Confirmation Brief

A confirmation brief is given by a subordinate commander after receiving the order or plan. Subordinate commanders brief the higher commander on their understanding of commander's intent, their specific tasks and purpose, and the relationship between their unit's missions and the other units in the operation. The confirmation brief allows the higher commander to identify potential gaps in the plan, as well as discrepancies with subordinate plans. It also gives the commander insights into how subordinate commanders intend to accomplish their missions.

Transition Drills

Transition drills increase the situational awareness of subordinate commanders and the staff and instill confidence and familiarity with the plan. Sand tables, map exercises, and rehearsals are examples of transition drills.

Plan Implementation

Military plans and orders should be prepared to facilitate implementation and transition to execution. For a plan to be implemented, the following products and activities must occur: confirm assumptions, model the TPFDD to confirm the sourcing and transportation feasibility assessment, establish execution timings, confirm authorities for execution, Conduct execution sourcing from assigned forces or request allocation of required forces, and issue necessary orders for execution.

V(b). Joint Operation Plan (OPLAN) Format

Ref: JP 5-0, Joint Operation Planning (Aug '11), app. A.

Below is a sample format that a joint force staff can use as a guide when developing a joint OPLAN. The exact format and level of detail may vary somewhat among joint commands, based on theater-specific requirements and other factors. However, joint OPLANs/CONPLANs will always coSntain the basic five paragraphs (such as paragraph 3, "Execution") and their primary subparagraphs (such as paragraph 3a, "Concept of Operations"). The JPEC typically refers to a joint contingency plans that encompasses more than one major operation as a campaign plan, but JFCs prepare a plan for a campaign in joint contingency plan format.

The CJCSM 3130 series volumes describe joint planning interaction among the President, SecDef, CJCS, the supported joint commander, and other JPEC members, and provides models of planning messages and estimates. CJCSM 3130.03, Adaptive Planning and Execution (APEX) Planning Formats and Guidance, provides the formats for joint plans.

Notional Operation Plan Format

a. Copy No. ____________________

b. Issuing Headquarters

c. Place of Issue

d. Effective Date/Time Group

e. OPERATION PLAN: (Number or Code Name)

f. USXXXXCOM OPERATIONS TO . . .

g. References: (List any maps, charts, and other relevant documents deemed essential to comprehension of the plan.)

1. Situation

(This section briefly describes the composite conditions, circumstances, and influences of the theater strategic situation that the plan addresses [see national intelligence estimate, any multinational sources, and strategic and commanders' estimates].)

a. General. (This section describes the general politico-military variables that would establish the probable preconditions for execution of the contingency plans. It should summarize the competing political goals that could lead to conflict, identify primary antagonists, state US policy objectives and the estimated objectives of other parties, and outline strategic decisions needed from other countries to achieve US policy objectives and conduct effective US military operations to achieve US military objectives. Specific items can be listed separately for clarity as depicted below.)

(1) Assessment of the Conflict. (Provide a summary of the national and/or multinational strategic context [JSCP, UCP].)

Continued on next page

Continued from previous page

(2) Policy Goals. (This section relates the strategic guidance, end state, and termination criteria to the theater situation and requirements in its global, regional, and space dimensions, interests, and intentions.)

(a) US/Multinational Policy Goals. (Identify the national security, multinational or military objectives, and strategic tasks assigned to or coordinated by the CCMD.)

(b) End State. (Describe the national strategic end state and relate the military end state to the national strategic end state.)

(3) Non-US National Political Decisions.

(4) Operational Limitations. (List actions that are prohibited or required by higher or multinational authority [ROE, law of armed conflict, termination criteria, etc.].)

b. Area of Concern

(1) OA. (Describe the JFC's operational area. A map may be used as an attachment to graphically depict the area.)

(2) Area of Interest. (Describe the area of concern to the commander, including the area of influence, areas adjacent thereto, and extending into enemy territory to the objectives of current or planned operations. This area also includes areas occupied by enemy forces who could jeopardize the accomplishment of the mission.)

c. Deterrent Options. (Delineate FDOs and FROs desired to include those categories specified in the current JSCP. Specific units and resources must be prioritized in terms of latest arrival date relative to C-day. Include possible diplomatic, informational, or economic deterrent options accomplished by non-DOD agencies that would support US mission accomplishment.) *See pp. 2-64 to 2-65 for discussion and examples of FDOs and FROs.)*

d. Risk. (Risk is the probability and severity of loss linked to hazards. List the specific hazards that the joint force may encounter during the mission. List risk mitigation measures.)

e. Enemy Forces. (Identify the opposing forces expected upon execution and appraise their general capabilities. Refer readers to Annex B [Intelligence] for details. However, this section should provide the information essential to a clear understanding of the magnitude of the hostile threat. Identify the adversary's strategic and operational COGs and critical vulnerabilities as depicted below.)

(1) Enemy COGs.

(a) Strategic.

(b) Operational.

(2) Enemy Critical Factors.

(a) Strategic.

(b) Operational.

(3) Enemy COAs (most likely and most dangerous to friendly mission accomplishment).

(a) General.

(b) Enemy End State.

(c) Enemy's Strategic Objectives.

(d) Enemy's Operational Objectives.

(e) Adversary CONOPs.

(4) Enemy Logistics and Sustainment.

(5) Other Enemy Forces/Capabilities.

(6) Enemy Reserve Mobilization.

Continued from previous page

f. Friendly Forces

(1) Friendly COGs. (This section should identify friendly COGs, both strategic and operational; this provides focus to force protection efforts.)

(a) Strategic.

(b) Operational.

(2) Friendly Critical Factors.

(a) Strategic.

(b) Operational.

(3) MNF.

(4) Supporting Commands and Agencies. (Describe the operations of unassigned forces, other than those tasked to support this contingency plan that could have a direct and significant influence on the operations in the plan. Also list the specific tasks of friendly forces, commands, or government departments and agencies that would directly support execution of the contingency plan, for example, USTRANSCOM, USSTRATCOM, Defense Intelligence Agency, and so forth.)

g. Assumptions. (List all reasonable assumptions for all participants contained in the JSCP or other tasking on which the contingency plan is based. State expected conditions over which the JFC has no control. Include assumptions that are directly relevant to the development of the plan and supporting plans and assumptions to the plan as a whole. Include both specified and implied assumptions that, if they do not occur as expected, would invalidate the plan or its CONOPS. Specify the mobility [air and sea lift], the degree of mobilization assumed [i.e., total, full, partial, selective, or none].)

(1) Threat Warning/Timeline.

(2) Pre-positioning and Regional Access (including international support and assistance).

(3) In-Place Forces.

(4) Strategic Assumptions (including those pertaining to nuclear weapons employment).

h. Legal Considerations. (List those significant legal considerations on which the plan is based.)

(1) ROE.

(2) International law, including the law of war.

(3) US law.

(4) HN and partner nation policies.

(5) Status-of-forces agreements.

(6) Other bilateral treaties and agreements.

(7) HN agreements to include HNS agreements.

2. Mission

(State concisely the essential task(s) the JFC has to accomplish. This statement should address: who, what, when, where, and why.)

3. Execution

a. CONOPs. (For a CCDR's contingency plan, the appropriate commander's estimate can be taken from the campaign plan and developed into a strategic concept of operation for a theater campaign or OPLAN. Otherwise, the CONOPS will be developed as a result of the COA selected by the JFC during COA development. The concept should be stated in terms of who, what, where, when, why, and how. It also contains the JFC's strategic vision, intent, and guidance for force projection operations, including mobilization, deployment, employment, sustainment, and redeployment of all participating forces, activities, and agencies.) (Refer to Annex C.)

Continued on next page

Continued on next page

Continued from previous page

(1) Commander's Intent. (This should describe the JFC's intent [purpose and end state], overall and by phase. This statement deals primarily with the military conditions that lead to mission accomplishment, so the commander may highlight selected objectives and their supporting effects. It may also include how the posture of forces at the end state facilitates transition to future operations. It may also include the JFC's assessment of the enemy commander's intent and an assessment of where and how much risk is acceptable during the operation. The commander's intent, though, is not a summary of the CONOPS.)

(a) Purpose and End State.

(b) Objectives.

(c) Effects, if discussed.

(2) General. (Base the CONOPS on the JFC's selected COA. The CONOPS states how the commander plans to accomplish the mission, including the forces involved, the phasing of operations, the general nature and purpose of operations to be conducted, and the interrelated or cross-Service support. For a CCDR's contingency plan, the CONOPS should include a statement concerning the perceived need for Reserve Component mobilization based on plan force deployment timing and Reserve Component force size requirements. The CONOPS should be sufficiently developed to include an estimate of the level and duration of conflict to provide supporting and subordinate commanders a basis for preparing adequate supporting plans. To the extent possible, the CONOPS should incorporate the following:

(a) JFC's military objectives, supporting desired effects, and operational focus.

(b) Orientation on the adversary's strategic and operational COGs.

(c) Protection of friendly strategic and operational COGs.

(d) Phasing of operations, to include the commander's intent for each phase.

1. Phase I:

a. JFC's Intent.

b. Timing.

c. Objectives and desired effects.

d. Risk.

e. Execution.

f. Employment.

(1) Land Forces.

(2) Air Forces.

(3) Maritime Forces.

(4) Space Forces.

(5) Cyberspace Forces.

(6) SOF.

g. Operational Fires. List those significant fires considerations on which the plan is based. The fires discussion should reflect the JFC's concept for application of available fires assets. Guidance for joint fires may address the following:

(1) Joint force policies, procedures, and planning cycles.

(2) Joint fire support assets for planning purposes.

(3) Priorities for employing target acquisition assets.

(4) Areas that require joint fires to support operational maneuver.

(5) Anticipated joint fire support requirements.

(6) Fire support coordinating measures (if required). *Refer to JP 3-09, Joint Fire Support, for a detailed discussion.*

Continued from previous page

2. **Phases II through XX.** *(Cite information as stated in subparagraph 3a(2)(d)1 above for each subsequent phase based on expected sequencing, changes, or new opportunities.)*

b. Tasks. (List the tasks assigned to each element of the supported and supporting commands in separate subparagraphs. Each task should be a concise statement of a mission to be performed either in future planning for the operation or on execution of the OPORD. The task assignment should encompass all key actions that subordinate and supporting elements must perform to fulfill the CONOPS, including operational and tactical deception. If the actions cannot stand alone without exposing the deception, they must be published separately to receive special handling.)

c. Coordinating Instructions. (Provide instructions necessary for coordination and synchronization of the joint operation that apply to two or more elements of the command. Explain terms pertaining to the timing of execution and deployments. Coordinating instructions should also include CCIRs and associated reporting procedures that may be expanded upon in Annex B [Intelligence], Annex C [Operations], and Annex R [Reports].)

Continued on next page

4. Administration and Logistics

a. Concept of Sustainment. (This should provide broad guidance for the theater strategic sustainment concept for the campaign or operation, with information and instructions broken down by phases. It should cover functional areas of logistics, transportation, personnel policies, and administration.)

b. Logistics. (This paragraph addresses the CCDR's logistics priorities and intent: basing, combat, general, and geospatial engineering requirements, HNS, required contracted support, environmental considerations, mortuary affairs, and Service responsibilities. Identify the priority and movement of logistic support for each option and phase of the concept.)

c. Personnel. (Identify detailed planning requirements and subordinate taskings. Assign tasks for establishing and operating joint personnel facilities, managing accurate and timely personnel accountability and strength reporting, and making provisions for staffing them. Discuss the administrative management of participating personnel, the reconstitution of forces, command replacement and rotation policies, and required joint individual augmentation [JIA] to command headquarters and other operational requirements.) Refer to Annex E (if published).

d. Public Affairs. Refer to Annex F.

e. Civil-Military Operations. Refer to Annex G.

f. Meteorological and Oceanographic Services. Refer to Annex H.

g. Environmental Considerations. Refer to Annex L. *See JP 3-34, Joint Engineer Operations.*

h. Geospatial Information and Services. Refer to Annex M.

i. Health Service Support. Refer to Annex Q. (Identify planning requirements and subordinate taskings for joint health services functional areas. Address critical medical supplies and resources. Assign tasks for establishing joint medical assumptions and include them in a subparagraph.)

5. Command and Control

a. Command

(1) Command Relationships. (State the organizational structure expected to exist during plan implementation. Indicate any changes to major C2 organizations and the time of expected shift. Identify all command arrangement agreements and memorandums of understanding used and those that require development.)

Continued on next page

(2) **Command Posts.** (List the designations and locations of each major headquarters involved in execution. When headquarters are to be deployed or the plan provides for the relocation of headquarters to an alternate command post, indicate the location and time of opening and closing each headquarters.)

(3) **Succession to Command.** (Designate in order of succession the commanders responsible for assuming command of the operation in specific circumstances.)

b. Joint Communications System Support. (Provide a general statement concerning the scope of communications systems and procedures required to support the operation. Highlight any communications systems or procedures requiring special emphasis.) Refer to Annex K.

[Signature]

[Name]

[Rank/Service]

Commander

Annexes:

A—Task Organization

B—Intelligence

C—Operations

D—Logistics

E—Personnel

F—Public Affairs

G—Civil-Military Operations

H—Meteorological and Oceanographic Operations

J—Command Relationships

K—Communications Systems

L—Environmental Considerations

M—Not currently used

N—Not currently used

P—Host-Nation Support

Q—Medical Services

R—Reports

S—Special Technical Operations

T—Consequence Management

U—Notional Counterproliferation Decision Guide

V—Interagency Coordination

W—Operational Contract Support

X—Execution Checklist

Y—Communication Synchronization

Z—Distribution

Note: Annexes A—D, K, and Y are required annexes for a crisis OPORD per APEX. All others may either be required by the JSCP or deemed necessary by the supported commander.

Continued from previous page

Continued from previous page

VI. Operation Assessment

Ref: JP 5-0, Joint Planning (Jun '17), chap. VI.

Operation assessments are an integral part of planning and execution of any operation, fulfilling the requirement to identify and analyze changes in the OE and to determine the progress of the operation. Assessments involve the entire staff and other sources such as higher and subordinate headquarters, interagency and multinational partners, and other stakeholders. They provide perspective, insight, and the opportunity to correct, adapt, and refine planning and execution to make military operations more effective. Operation assessment applies to all levels of warfare and during all military operations.

Assessment

A continuous activity that supports decision making by ascertaining progress toward accomplishing a task, creating an effect,achieving an objective, or attaining an end state for the purpose of developing, adapting, and refining plans and for making campaigns and operations more effective.

Commanders maintain a personal sense of the progress of the operation or campaign, shaped by conversations with senior and subordinate commanders, key leader engagements (KLEs), and battlefield circulation. Operation assessment complements the commander's awareness by methodically identifying changes in the OE, identifying and analyzing risks and opportunities, and formally providing recommendations to improve progress towards mission accomplishment. Assessment should be integrated into the organization's planning (beginning in the plan initiation step) and operations battle rhythm to best support the commander's decision cycle.

The starting point for operation assessment activities coincides with the initiation of joint planning. Integrating assessments into the planning cycle helps the commander ensure the operational approach remains feasible and acceptable in the context of higher policy, guidance, and orders. This integrated approach optimizes the feedback senior leadership needs to appropriately refine, adapt, or terminate planning and execution to be effective in the OE.

CCMDs, subordinate Service, joint functional components, and JTFs devote significant effort and resources to plan and execute operations. They apply appropriate rigor to determine whether an operation is being effectively planned and executed as needed to accomplish specified objectives and end states. Assessment complements that rigor by analyzing the OE objectively and comprehensively to estimate the effectiveness of planned tasks and measure the effectiveness of completed tasks with respect to desired conditions in the OE.

Background

TCPs and country-specific security cooperation sections/country plans are continuously in some stage of implementation. Accordingly, during implementation CCMD planners should annually extend their planning horizon into the future year. The simultaneity of planning for the future while implementing a plan requires a CCMD to continually assess its implementation in order to appropriately revise, adapt, or terminate elements of the evolving (future) plan. This synergism makes operation assessment a prerequisite to plan adaptation. Operation assessment is thus fundamental to revising implementation documents ahead of resource allocation processes.

- Events can arise external to the CCMD's control that affect both plan execution and future planning. Some of these events can impede achievement of one or more objectives while others may present opportunities to advance the plan more rapidly than anticipated.
- External events generally fall into two categories. The first are those that change the strategic or OE in which a CCMD implements a plan (typically a J-2 focus). The second category involves those events that change the resource picture with respect to funding, forces, and time available (typically a force structure, resource, and assessment directorate of a joint staff [J-8] focus). This document treats these two types of external events as separate considerations because they can influence plan implementation independent of each other.

The overall purpose of operation assessment is to provide recommendations to make operations more effective. As it relates to campaigns, where strategic objectives frame the CCMD's mission, operation assessment helps the CCDR and supporting organizations refine or adapt the campaign plan and supporting plans to achieve the campaign objectives or, with SecDef, to adapt the GEF-directed objectives to changes in the strategic objectives and OEs.

The assessment process serves as part of the CCMD's feedback mechanism throughout campaign planning and execution. It also feeds external requirements such as the CCDR's inputs to the CJCS AJA. Assessment analysis and products should identify where the CCMD's ways and means are sufficient to attain their ends, where they are not and why not, and support recommendations to adapt or modify the campaign plan or its components. The analyses might provide insight into basic questions such as:

- Are the objectives (strategic and intermediate) achievable given changes in the OE and emerging diplomatic/political issues?
- Is the current plan still suitable to achieve the objectives?
- Do changes in the OE impose additional risks or provide additional opportunities to the command?
- To what degree are the resources employed making a difference in the OE?

I. Campaign Assessments

Campaign assessments determine whether progress towards achieving CCMD campaign objectives is being made by evaluating whether progress towards intermediate objectives is being made. Intermediate objectives are desired conditions in the OE the CCDR views as critical for successfully executing the campaign plan and achieving CCMD campaign objectives. Essentially, intermediate objectives are multiple objectives that are between initiation of the campaign and achievement of campaign objectives. Accordingly, at the strategic assessment level, intermediate objectives are criteria used to observe and measure progress toward campaign desired conditions and evaluate why the current status of progress exists.

Functional campaign assessments assist the FCCs in evaluating progress toward or regression from achieving their global functional objectives. FCCs provide unique support to GCCs in their respective specialties and are required to assess progress towards their intermediate objectives in support of their global functional objectives or DOD-wide activities.

The JSCP, GEF, and other strategic guidance provide CCMDs with strategic objectives. CCMDs translate and refine those long-range objectives into near-term (achievable in 2-5 years) intermediate objectives. Intermediate objectives represent unique military contributions to the achievement of strategic objectives. In some cases, the CCMD's actions alone may not achieve strategic objectives; additional intermediate objectives may be required to achieve them. Consequently, other instruments of national power may be required, with the CCMD operating in a supported or supporting role.

See following pages (pp. 3-128 to 3-129) for an overview and further discussion.

Purpose/Tenets of Operation Assessments

Ref: JP 5-0, Joint Planning (Jun '17), pp. VI-7 to VI-8.

Operation assessments help the commander and staff determine progress toward mission accomplishment. Assessment results enhance the commander's decision making enable more effective operations and help the commander and the staff to keep pace with a constantly evolving OE. A secondary purpose is to inform senior civil-military leadership dialogue to support geopolitical and resource decision making throughout planning and execution. Integrating assessment during planning and execution can help commanders and staffs to:

- Develop mission success criteria.
- Compare observed OE conditions to desired objectives and/or end state conditions.
- Determine validity of key planning facts and assumptions.
- Determine whether or not the desired effects have been created and whether the objectives are being achieved.
- During execution, determine the effectiveness of allocated resources against specific task and mission performance and effects, and test the validity of intermediate objectives.
- Determine whether an increase, decrease, or change to resources is required.
- Identify the risks and barriers to mission accomplishment.
- Identify opportunities to accelerate mission accomplishment.

Tenets of Operation Assessment

The following tenets should guide the commander and the staff:

Commander Centricity. The commander's involvement in operation assessment is essential. The assessment plan should focus on the information and intelligence that directly support the commander's decision making.

Subordinate Commander Involvement. Assessments are more effective when used to support conversations between commanders at different echelons.

Integration. Staff integration is crucial to planning and executing effective assessments. Operation assessment is the responsibility of commanders, planners, and operators at every level and not the sole work of an individual advisor, committee, or assessment entity.

Integration into the Planning Process and Battle Rhythm. To deliver information at the right time, the operation assessment should be synchronized with the commander's decision cycle. The assessment planning steps occur concurrently with the steps of JPP. The resulting assessment plan should support the command's battle rhythm.

Integration of External Sources of Information. Operation assessment should allow the commander and staff to integrate information that updates the understanding of the OE in order to plan more effective operations.

Credibility and Transparency. Assessment reports should cite all sources of information used to build the report.

Continuous Operation Assessment. While an operation assessment product may be developed on a specific schedule, assessment is continuous in any operation. The information collected and analyzed can be used to inform planning, execution, and assessment of operations.

Campaign Assessment Process

Ref: JP 5-0, Joint Planning (Jun '17), pp. VI-3 to VI-7.

The basic process for campaign assessment is similar to that used for contingency and crisis applications but the scale and scope are generally much larger. While operational level activities such as a JTF typically focus on a single end state with multiple desired conditions, the campaign plan must integrate products from a larger range of strategic objectives, each encompassing its own set of intermediate objectives and desired conditions, subordinate operations, and subordinate plans (i.e., country-specific security cooperation sections/country plans, contingency plans not in execution, on-going operations, directed missions).

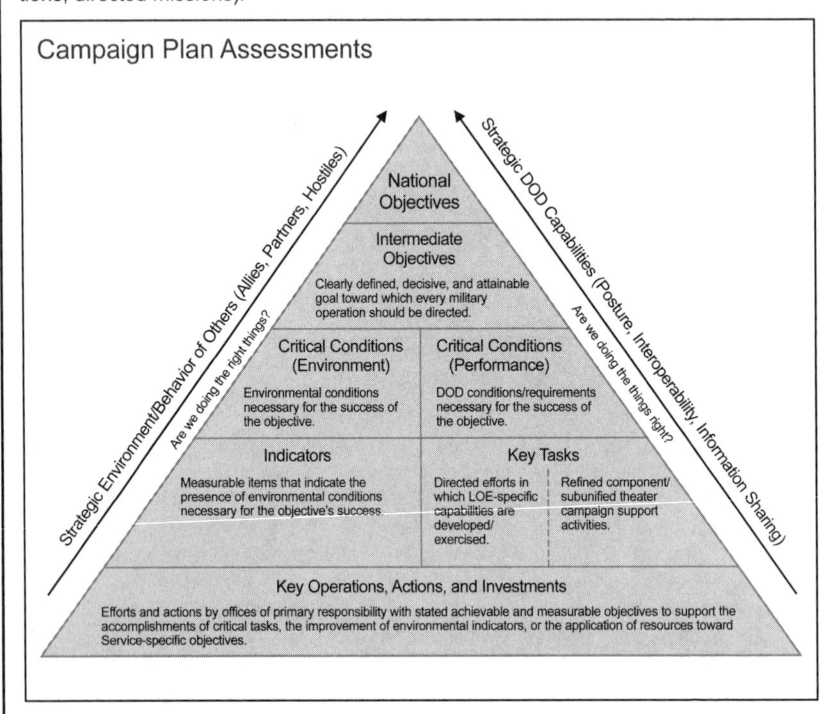

Ref: JP 5-0, fig. VI-1. Campaign Plan Assessments.

One common method to establish more manageable campaign plans is for CCMDs to establish LOEs with associated intermediate objectives for each campaign objective. This method allows the CCMD to simultaneously assess each LOE and then assess the overall effort using products from the LOE assessments. The following discussion uses several boards, cells, and working groups. The names merely provide context for the process and are not intended to be a requirement for organizations to follow.

The assessment needs to nest with and support the campaign and national objectives and cannot rely on accomplishment of specific tasks. Commanders and staffs should make certain the established intermediate objectives will change the OE in the manner desired.

See fig. VI-2 and facing page for further discussion.

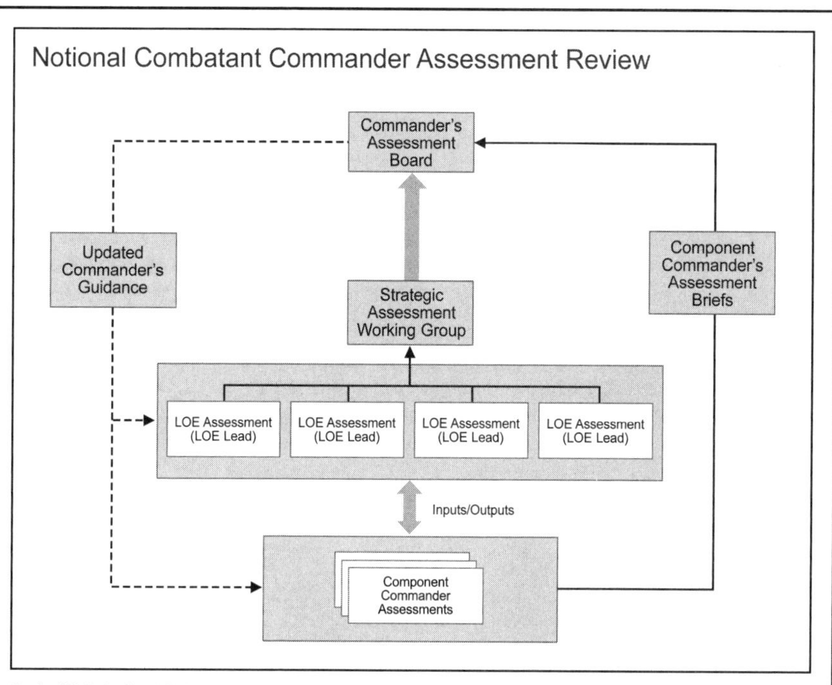

Ref: JP 5-0, fig. VI-2. Notional Combatant Commander Assessment Review.

Line of Effort (LOE) Assessment

1. Leads. LOE leads should guide the development and assessment of LOE intermediate objectives, critical conditions, indicators, tasks, and associated metrics and recommendations through the LOE working groups.
2. Output. The LOE assessment produces updated findings, insights, and recommendations by LOE. These are consolidated for presentation and validation during the strategic assessment working group (SAWG).

Strategic Assessment Working Group (SAWG)

1. Leads. Designated lead (typically from a J-3, J-5, or J-8 element) chairs this O-6 level review working group. LOE assessors and leads brief their subcampaign assessments, findings, insights, and recommendations to this group.
2. Output. The SAWG produces an assessment brief and recommendations for presentation and approval during the commander's assessment board (CAB).

Commander's Assessment Board (CAB)

1. Leads. CCDRs chair this board. LOE leads present a consolidated assessment brief with SAWG-validated, command-level recommendations for the commander's decision. As a note, this board may occur as part of the commander's council or the commander's update brief.
2. Outputs. The CAB validates recommendations for staff action and higher level coordination and produces refined commander's guidance.

Component Command Assessment

If required by the CCDR, component and subordinate commands will provide an annual assessment briefing to CCDR detailing their progress toward key LOE objectives and conduct of key operations and activities.

II. Operation Assessment Process

Ref: JP 5-0, Joint Planning (Jun '17), pp. VI-12 to VI-26 (and fig. VI-3).

The assessment process is continuous. Throughout JPP, assessment provides support to and is supported by operational design and operational art. The assessment process complements and is concurrent with JPP in developing specific and measurable task-based end states, objectives, and effects during operational design. These help the staff identify the information and intelligence requirements (including CCIRs). During execution, assessment provides information on progress toward creating effects, achieving objectives, and attaining desired end states. Assessment reports are based on continuous situational awareness and OE analysis from internal and external sources and address changes in the OE and their proximate causes, opportunities to exploit and risks to mitigate, and recommendations to inform decision making throughout planning and execution.

Activity	Primarily in Planning or Execution	Personnel Involved	Input	Associated Staff Activity	Output
Develop Assessment Approach	Planning • Operational Design • JPP Steps 1-6	• Commander • Planners • Primary staff • Special staff • Assessment element	Strategic guidance CIPG Description of OE Problem to be solved Operational approach Commander's intent (purpose, end state, risk)	• Conduct JIPOE • Develop operational approach • Support development and refinement of end states, objectives, effects, and tasks • Conduct joint planning (JPP and operational design) • Determine and develop how to assess tasks, effects, objectives, and end state progress for each course of action: • Identify indicators	Assessment approach which includes: assessment framework and construct Specific outcomes (end state, objectives, effects) Commander's estimate/CONOPS (from JPP)
Develop Assessment Plan	Planning • JPP Step 7	• Commander • Planners • Primary staff • Special staff • Assessment element • Operations planners • Intelligence planners • Subordinate commanders • Interagency and multinational partners • Others, as required	Assessment approach which includes: assessment framework and construct Specific outcomes (end state, objectives, effects) Commander's estimate/ CONOPS (from JPP)	• Document assessment framework and construct ○ Finalize the data collection plan ○ Coordinate and assign responsibilities for monitoring, collection, and analysis ○ Identify how the assessment is integrated into battle rhythm/ feedback mechanism • Vet and staff the draft assessment plan	Approved assessment plan Data collection plan Approved contingency plan/operation order
Collect Information and Intelligence	Execution	• Intelligence analysts • Current operations • Assessment element • Subordinate commanders • Interagency and multinational partners • Others, as required	Approved assessment plan Data collection plan Approved contingency plan/operation order	• JIPOE • Staff estimates • IR management • ISR planning and optimization	Data collected and organized, relevant to joint force actions, current and desired conditions
Analyze Information and Intelligence	Execution	• Primary staff • Special staff • Assessment element	Data collected and organized, relevant to joint force actions, current and desired conditions	• Assessment working group • Staff estimates • Vet and validate recommendations	Draft assessment products Vetted and validated recommendations
Communicate Feedback and Recommend-ations	Execution	• Commander • Subordinate commanders (periodically) • Primary staff • Special staff • Assessment element	Draft assessment products Vetted and validated recommendations	• Provide timely recommendations to appropriate decision makers	Approved assessment products, decisions, and recommendations to higher headquarters
Adapt Plans or Operations/ Campaigns	Execution Planning	• Commander • Planners • Primary staff • Special staff • Assessment element	Approved assessment products, decisions, and recommendations to higher headquarters	• Develop branches and sequels • Modify operational approach/plan • Modify objectives, effects, tasks • Modify assessment approach/plan)	Revised plans or fragmentary orders Updated assessment plan Updated data collection plan
Repeat Steps 3-6 until operation terminated/replaced/transitioned. (Adjust using steps 1 and 2 as required during execution.)					

There is no single way to conduct assessment. Every mission and OE has its own unique challenges, making every assessment unique. The following steps can help guide the development of an effective assessment plan and assessment performance during execution. Organizations should consider these steps as necessary to fit their needs.

Step 1—Develop the Operation Assessment Approach

Operation assessment begins during the initiation step of JPP when the command identifies possible operational approaches and their associated objectives, tasks, effects, and desired conditions in the OE. Concurrently, the staff begins to develop the operation assessment approach by identifying and integrating the appropriate assessment plan framework and structure needed to assess planning and execution effectiveness. The assessment approach identifies the specific information needed to monitor and analyze effects and conditions associated with achieving operation or campaign objectives. The assessment approach becomes the framework for the assessment plan and will continue to mature through plan development, refinement, adaptation, and execution in order to understand the OE and measure whether anticipated and executed operations are having the desired impact on the OE.

Step 2—Develop Operation Assessment Plan

Developing, refining, and adapting the assessment plan is concurrent and complementary throughout joint planning and execution. This step overlaps with the previous step during ID of the objectives and effects. Developing the assessment plan is a whole of staff effort and should include other key stakeholders to better shape the assessment effort. The assessment plan should identify staff or subordinate organizations to monitor, collect, analyze information, and develop recommendations and assessment products as required. Requirements for staff coordination and presentation to the commander should also be included in the plan and integrated into the command's battle rhythm to support the commander's decision cycle.

Step 3—Collect Information and Intelligence

Commands should collect relevant information throughout planning and execution. Throughout planning and execution the joint force refines and adapts information collection requirements to gather information about the OE and the joint force's anticipated and completed actions as part of normal C2 activities. Typically, staffs and subordinate commands provide information about planning and execution on a regular cycle through specified battle rhythm events. Intelligence staffs continually provide intelligence about the OE and operational impact to support the collective staff assessment effort. In accordance with the assessment plan, assessment considerations may help the staff determine the presence of decision point triggers and other mission impacts.

Step 4—Analyze Information and Intelligence

Accurate, unbiased analysis seeks to identify operationally significant trends and changes to the OE and their impact on the operation or campaign. Based on analysis, the staff can estimate the effects of force employment and resource allocation, determine whether objectives are being achieved, or determine if a decision point has been reached. Using these determinations, the staff also may identify additional risks and challenges to mission accomplishment or identify opportunities to accelerate mission accomplishment.

Step 5—Communicate Feedback and Recommendations

The staff may be required to develop assessment products (which may include summary reports and briefings) containing recommendations for the commander based upon the guidelines set forth in the assessment plan. The commander's guidance is the most critical step in developing assessment products. Regardless of quality and effort, the assessment process is useless if the communication of its results is deficient or inconsistent with the commander's personal style of digesting information and making decisions.

Step 6—Adapt Plans or Operations/Campaigns

Once feedback and recommendations have been provided, commanders typically direct changes or provide additional guidance that dictate updates or modifications to operation or campaign plan. The commander's guidance may also induce modifications to the assessment plan. Even without significant changes to the plan or order, changes to the assessment plan may be necessary to reflect changes in the OE or adjustments to the information or intelligence requirements.

III. Linking Effects, Objectives, and End States to Tasks Through Indicators

An operation's desired effects, objectives, and end states should help focus the staff's assessment efforts by identifying and analyzing a subset of the overall changes within the overall OE. As the staff develops the desired effects, objectives, and end states during planning, they should concurrently identify the specific pieces of information needed to infer changes in the OE supporting them. These pieces of information are commonly referred to as indicators.

Indicator

In the context of operation assessment, a specific piece of information that infers the condition, state, or existence of something, and provides a reliable means to ascertain performance or effectiveness.

Joint Planning

Indicators share common characteristics with carefully selected MOPs and MOEs and link tasks to effects, objectives, and end states (see Figure VI-12). Commanders and staffs should develop an approach that best fits their organization, operation, and requirements.

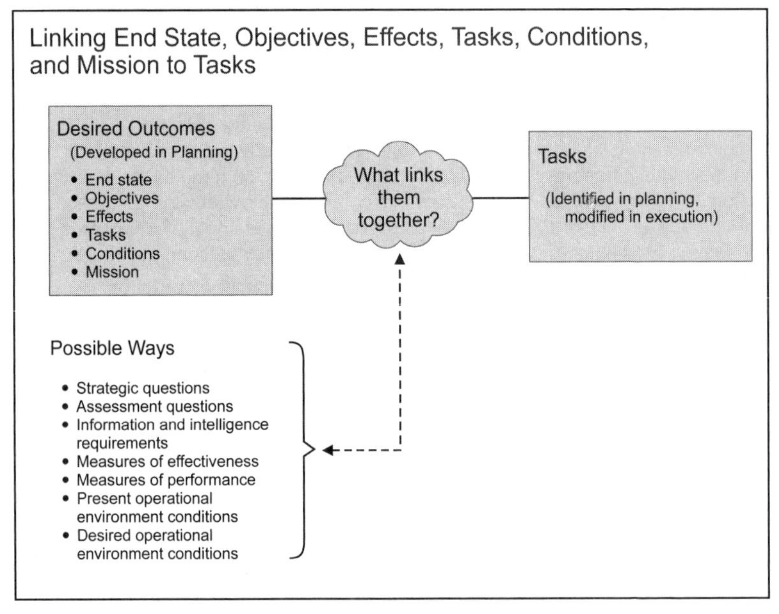

Ref: JP 5-0, fig. VI-12. Linking End State, Objectives, Effects, Tasks, Conditions, and Mission to Tasks.

Ensuring effects, objectives, and end states are linked to tasks through carefully selected MOPs and MOEs is essential to the analytical rigor of an assessment framework. Establishing strong, cogent links between tasks and effects, objectives, and end states through MOPs and MOEs facilitates the transparency and clarity of the assessment approach. Additionally, links between tasks and effects, objectives, and end states assist in mapping the plan's strategy to actual activities and conditions in the OE and subsequently to desired effects, objectives, and end states.

See following pages (pp. 3-134 to 3-135) for an overview and further discussion.

Guidelines for Indicator Development

Ref: JP 5-0, Joint Planning (Jun '17), pp. VI-24 to VI-26.

Indicators should be relevant, observable or collectable, responsive, and resourced.

Relevant

Indicators should be relevant to a desired effect, objective, or end state within the plan or order. A valid indicator bears a direct relationship to the desired effect, objective, or end state and accurately signifies the anticipated or actual status of something about the effect, objective, or end state that must be known. This criterion helps avoid collecting and analyzing information that is of no value to a specific operation. It also helps ensure efficiency by eliminating redundant efforts.

Observable and Collectable

Indicators must be observable (and therefore collectable) such that changes can be detected and measured or evaluated. The staff should make note of indicators that are relevant but not collectable and report them to the commander. Collection shortfalls can often put the analysis quality at risk. The commander must decide whether to accept this risk, realign resources to collect required information, or modify the plan or order.

Responsive

Indicators should signify changes in the OE timely enough to enable effective response by the staff and timely decisions by the commander. Assessors must consider an indicator's responsiveness to stimulus in the OE. If it reacts too slowly, opportunities for response are likely to be missed; if too quickly, it exposes the staff and commander to false alarms. The JFC and staff should consider the time required for a task or mission to produce desired results within the OE and develop indicators that can respond accordingly. Many actions directed by the JFC require time to implement and may take even longer to produce a measurable result.

Resourced

The collection of indicators should be adequately resourced so the command and subordinate units can obtain the required information without excessive effort or cost. Indicator information should be derived from other staff processes whenever possible. Assessors should avoid indicators that require development of an additional collection system. Staffs should ensure resource requirements for indicator collection efforts and analysis are included in plans and monitored. Data collection and analysis requirements associated with the threat and the OE should be embodied in the commander's PIRs with relevant tasks specified through Annex B (Intelligence) to a plan or an order. Given the focus of PIRs, the collection and analysis they drive provides the commander with insights on changes associated with MOEs. On the other hand, FFIRs provide insights to the commander on the ability of major force elements and other critical capabilities to execute their assigned tasks. Thus, they are associated with MOPs and should be published in Annex C (Operations) with reporting requirements and procedures specified in Annex R (Reports). Effective assessment planning can help avoid duplicating tasks and unnecessary actions, which in turn can help preserve combat power.

Collection plans must clearly articulate why an indicator is necessary for the accurate assessment of an action. Collection may draw on subordinate unit operations, KLEs, joint functions and functional estimates, and battle damage assessment. Staffs need to understand the fidelity of the available information, choose appropriate information, and prioritize use of scarce collection resources.

Some assessment indicators must compete for prioritization and collection assets. Assessors should coordinate with intelligence planners throughout planning and execution to identify collection efforts already gathering indicator information, alternative indicator information that might be available, and coordinate and synchronize assessment-related collection requirements with the command's integrated collection plan.

Linking Effects, Objectives, and End States to Tasks Through Indicators (Overview)

Ref: JP 5-0, Joint Planning (Jun '17), pp. VI-12 to VI-26 (and fig. VI-3).

Approach 1—Using Assessment Questions and Information and Intelligence Requirements

This approach uses the model shown in Figure VI-14 to guide the development of assessment questions and information and intelligence requirements in order to identify indicators.

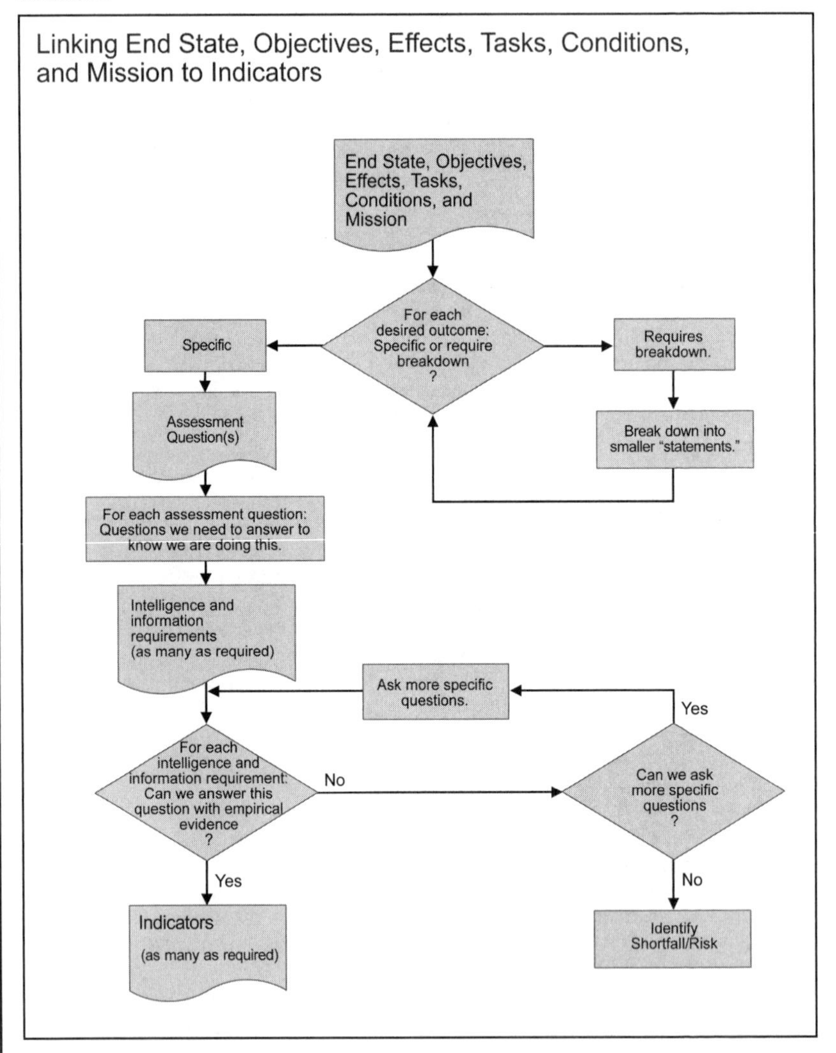

Ref: JP 5-0, fig. VI-14. Linking End State, Objectives, Effects, Tasks, Conditions, and Mission to Indicators.

Approach 2—Develop Indicators To Assess Operations

This approach facilitates the development of MOPs and MOEs (see Figure VI-20). During planning, the OPT, as supported by assessors, determines a hierarchy of increasingly specific or more refined statements. For example, these may be the objectives to be achieved, the effects to be created in the OE to achieve those objectives, and perhaps the tasks intended to create those effects.

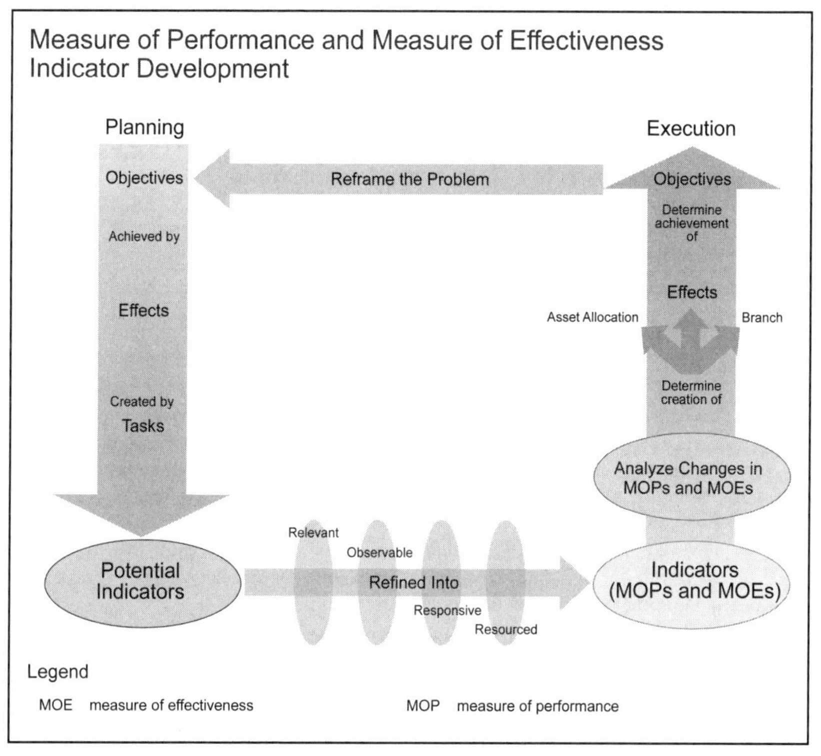

Ref: JP 5-0, fig. VI-20. Measure of Performance and Measure of Effectiveness Indicator Development.

Functional experts, supported by assessors, then develop potential indicators for each effect. Potential indicators should answer the questions "What happened?" and "How do we know we are creating the desired effects?" The answers to these questions are indicators that may inform MOPs and MOEs. The following steps present a logical process the staff can use to develop measures and indicators (either MOPs or MOEs) for each desired effect.

- Analyze the desired effects and tasks.
- Identify candidate MOPs and MOEs for subsequent refinement. Consider developing MOPs, (and MOP indicators, if used) that reflect progress in achieving key tasks as the approach to performance assessment.
- Refine MOEs and MOPs. They should be relevant to the desired effect (MOEs) or associated task (MOPs), observable, responsive, and resourced.
- Identify collection requirements for MOPs and MOEs.
- Incorporate indicators into the DCP and assessment plan.
- Monitor and modify indicators as necessary during execution.

IV. Selecting Indicators

The two types of indicators commonly used by the joint forces are MOPs and MOEs.

Measures of Performance (MOPs)

MOPs are indicators used to assess friendly (i.e., multinational) actions tied to measuring task accomplishment. MOPs commonly reside in task execution matrices and confirm or deny proper task performance. MOPs help answer the question, "Are we doing things right?" or "Was the action taken?" or "Was the task completed to standard?"

Measures of Effectiveness (MOEs)

MOEs are indicators used to help measure a current system state, with change indicated by comparing multiple observations over time to gauge the achievement of objectives and attainment of end states. MOEs help answer the question, "Are we doing the right things to create the effects or changes in the conditions of the OE that we desire?"

Choose Distinct Indicators

Using indicators that are too similar to each other can result in the repetitious evaluation of change in a particular condition. In this way, similar indicators skew analyses by overestimating, or 'double-counting,' change in one item in the OE.

Include indicators from different causal chains. When indicators have a cause and effect relationship with each other, either directly or indirectly, it decreases their value in measuring a particular condition. Measuring progress toward a desired condition by multiple means adds rigor to the analyses.

Use the same indicator for more than one end state, objective, effect, task, condition, or mission when appropriate. This sort of duplication in organizing OE information does not introduce significant bias unless carried to an extreme.

Avoid or minimize additional reporting requirements for subordinate units. In many cases, commanders may use information generated by other staff elements as indicators in the assessment plan. Units collect many assessment indicators as part of routine operational and intelligence reporting. With careful consideration, commanders and staffs can often find viable alternative indicators without creating new reporting requirements. Excessive reporting requirements can render an otherwise valid assessment plan untenable.

Maximize Clarity

An indicator describes the sought-after information, including specifics on time, information, geography, or unit, as necessary. Any staff member should be able to read the indicator and precisely understand the information it describes.

V. Information Categories and Data Types

Information Categories

The specific type of information that is expressed in indicators can typically be categorized as quantitative or qualitative, and subjective or objective.

Information Types

Assessment information is used to calculate, analyze, and recommend. Whenever possible, information should be empirical—originating in or based on observation or experience. Generally, there are four information types: nominal, ordinal, interval, and ratio. Knowing the type is essential to understand the type of analysis that can be performed, and whether the information can be interpreted to draw conclusions, such as the quantity and speed of change in an OE condition over time.

VII. Transition to Execution

Ref: JP 5-0, Joint Planning (Jun '17), chap. VII.

> ***"A good plan, violently executed now, is better than a perfect plan next week."***
> **- George S. Patton**

Plans are rarely executed as written. Regardless of how much time and effort went into the planning process, commanders and their staffs should accept that the plan, as written, will likely need changes on execution. Often, the decision to deploy the military will be in conditions significantly different from the original planning guidance or the conditions planned. Planning provides a significant head start when called to deploy the military. Assessments and reframing the problem, if required, inform the applicability of, or necessary modifications to the plan in response to changes in the operational environment (OE).

Effective planning enables transition. Integrated staff effort during planning ensures the plan is a team effort and the knowledge gained across the staff in the planning process is shared and retained. This staff work assists in identifying changes in the OE and guidance, speeding transition to execution.

Detailed planning provides the analysis of the adversary and the OE. The knowledge and understanding gained enables a well-trained staff to quickly identify what is different between their plan and current conditions and make recommendations based on their prior work.

Detailed OPLANs (levels 3 or 4) may require more significant changes due to their specificity. Forces identified in the plan may not be available, assumptions may not be validated, and policy and strategic decisions (and the decision timeline) may have changed or not support the original concept. However, the extra time spent on analysis provides a deeper understanding of the OE, adversaries, and the technical issues with projecting forces.

Less detailed plans (levels 1-2) may be more readily adaptable to execution due to their generality. However, they will require significantly more analysis (e.g., forces, transportation, logistics) to provide the detail required to enable decisions at the strategic level and ensure the plan's executability and suitability for the problem at hand.

The decision to execute will often be presented as an examination of options in response to a developing crisis or action by a competitor state or adversary (state or non-state) rather than a specific directive to execute a specific CONPLAN or OPLAN.

- If an existing plan is appropriate, the commander and staff should review and update the plan.
- If no existing plan meets the guidance, the commander and staff conduct crisis planning (planning in reduced timeline). More often than not, the commander and staff have conducted some previous analysis of the OE which will speed the planning process.

See p. 3-118 for discussion of transition as related to plan or order development within the joint planning process (including transition brief).

I. Transition Process

Ref: JP 5-0, Joint Planning (Jun '17), pp. VII-3 to VII-6.

The transition from plan to execution should consider the following points. These are not meant to be exclusive and may be conducted simultaneously.

A. Transition Requirements

1. Update environmental frame and intelligence analysis. Identify what has changed since plan development and how that affects the plan.

2. Identify any changes to strategic direction or guidance. This will require dialogue with senior civilian leadership to ensure the military objectives remain synchronized with policy and strategic objectives.

- Confirm and update strategic objectives or end states.
- Confirm and update operational limitations (constraints and restraints).
- Validate assumptions.
- Review and validate assessment criteria.
  - External (strategic) assumptions, especially those dealing with policy, diplomacy, and multinational partners, should be validated as part of the plan review with senior civilian leadership. These are usually the assumptions dictated to the command through strategic directives (GEF, JSCP, SGSs) or previous planning IPRs.
  - Internal (operational) assumptions should be validated by the staff through their update of the OE.
- Identify partners and allies.
- Identify interagency participation, actions, and responsibilities.

3. Identify forces and resources, to include transportation. The forces assumed in planning are for planning purposes only; execution sourced forces may or may not match those assumed in planning. Execution sourcing requires a dialogue between the supported CCDR, the JS, JFPs, Services, and USTRANSCOM.

4. Identify decision points and CCIRs to aid in decision making. Ensure consideration is taken to include lead times, to include notification and mobilization for reserve forces, transportation timelines, and JRSOI requirements. These decision points are critical for senior DOD leadership to understand when decisions should be made to enable operations and reduce risk. During this discussion, commanders and planners should identify alternative COAs and the cost and risk associated with them should decisions be delayed or deferred. Decision points should specifically address how the US might use the military in:

- **Flexible Deterrent Options (FDOs)**. When and what FDOs should be deployed and the expected impact. The discussion should identify indicators that the FDOs are creating the desired effect. *See p. 2-74.*
- **Flexible Response Options (FROs).** FROs, usually used in response to terrorism, can also be employed in response to aggression by a competitor or adversary. Like FDOs, the discussion should include indicators of their effectiveness and probability of consequences, desired and undesired. *See p. 2-75*
- **De-Escalation.** During transition to execution, commanders should identify a means for de-escalation and steps that could be taken to enable de-escalation without endangering US forces or interests. .
- **Escalation.** Similarly, commanders need to identify decision points at which senior leaders must make decisions to escalate in order to ensure strategic advantage, to include the expected risk associated should the adversary gain the advantage prior to US commitment.

5. Confirm Authorities for Execution. Request and receive President or SecDef authority to conduct military operations. Authorities granted may be for execution of an approved plan or for limited execution of select phases of an approved plan.

6. Direct Execution. The JS, on behalf of the CJCS, prepares orders for the President or SecDef to authorize the execution of a plan or order. The authorities for execution, force allocation, and deployment are often provided separately vice in a comprehensive order. Upon approval, CCDRs and Services pass orders down the chain of command directing action ordered by higher headquarters. The following orders are some of those that may be used in the process of transitioning from planning to execution: WARNORD, PLANORD, ALERTORD, OPORD, PTDO, DEPORD, EXORD, and FRAGORD.

- **Contingency Plans.** The authority to execute a contingency plan may be provided incrementally. Initial execution authority may be limited to early phase activities and CCDRs should be prepared to request additional or modified execution authorities as an operation develops.
- **CCMD Campaign Plans.** CCMD campaign plans are in constant execution. While they are reviewed by SecDef, the authorization to execute a campaign plan does not provide complete authority for the CCDR to execute all of the individual military activities that comprise the plan. Additional CCMD coordination is required to execute the discrete military activities within a campaign plan to include: posture, force allocation, and country team coordination.

Refer to CJCSM 3130.03, Adaptive Planning and Execution (APEX) Planning Format and Guidance, for more information on the content and format of orders.

B. Impact on Other Operations

As the plan transitions to execution, the commander and staff synchronize that operation with the rest of the CCMD's theater (or functional) campaign.

The commander identifies how the additional operation will affect the campaign.

- **Resources**. Resources may be diverted from lower priority operations and activities to support the new operation. This may require modifying the campaign or adjusting objectives.
- **Secondary effects**. Adding new operations, especially combat operations, will impact the perception and effects of other operations within the AOR (and likely in other CCMD's AORs as well). Both the new operation and existing ones may need to be adjusted to reflect the symbiotic effect of simultaneous operations.

The commander may require support from other CCMDs. In addition to support within the plan transitioning to execution, the CCDR may require external support to ensure continued progress toward theater or functional objectives. By using a pre¬established capability (force) sharing agreement, a CCDR can gain the support needed without requiring additional JS or OSD coordination. Support from other CCMDs often require shared battle rhythm activities. Balancing the benefit of improved awareness without overburdening commanders and their staffs remains a challenge. Informal cross-CCMD, directorate-level coordination has proven beneficial and can expand when security conditions necessitate deeper coordination and synchronization. However, identifying standardized staff organizations provides additional structure when planning and scheduling across organizational boundaries.

Depending on the significance of the new operation, the CCDR may need to update the theater or functional campaign objectives. This will require a conversation with senior civilian leaders to see if the US national objectives should be adjusted given the change in the strategic landscape.

II. Types of Transition

There are three possible conditions for transitioning planning to execution.

A. Contingency Plan Execution

Contingency plans are planned in advance to typically address an anticipated crisis. If there is an approved contingency plan that closely resembles the emergent scenario, that plan can be refined or adapted as necessary and executed. The APEX execution functions are used for all plans.

Members of the planning team may not be the same as those responsible for execution. They may have rotated out or be in the planning sections of the staff rather than the operations. This is the most likely situation where the conditions used in developing the plan will have changed, due to the time lag between plan development and execution. Staff from the planning team need to provide as much background information as possible to the operations team.

The planning team should be a key participant, if not the lead, in updating the plan for the current (given) conditions. This enables the command to make effective use of the understanding gained by the staff during the planning process. The operations team should be the co-lead for the plan update to ensure they understand the decision processes and reasoning used in development of the operational approach and COAs.

B. Crisis Planning to Execution

Crisis planning is conducted when an emergent situation arises. The planning team will analyze approved contingency plans with like scenarios to determine if an existing plan applies. If a contingency plan is appropriate to the situation, it may be executed through an OPORD or FRAGORD. In a crisis, planning usually transitions rapidly to execution, so there is limited deviation between the plan and initial execution. Planners from the command J-5 can assist in the planning process through their planning expertise and knowledge gained of the OE during similar planning efforts.

C. Campaign Plan Execution

Activities within campaign plans are in constant execution.

Planning is conducted based upon assumed forces and resources. Upon a decision to execute, these assumptions are replaced by the facts of actual available forces and resources. Disparities between planning assumptions and the actual OE conditions at execution will drive refinement or adaptation of the plan or order. Resource informed planning during plan development allows planners to make more realistic force and resource planning assumptions.

During execution, the commander will likely have reason to consider updating the operational approach. It could be triggered by significant changes to understanding of the OE and/or problem, validation or invalidation of assumptions made during planning, identifying (through continuous assessment process) that the tactical actions are not resulting in the expected effects, changes in the conditions of the OE, or the end state. The commander may determine one of three ways ahead:

- The current OPLAN is adequate, with either no change or minor change (such as execution of a branch)—the current operational approach remains feasible.
- The OPLAN's mission and objectives are sound, but the operational approach is no longer feasible or acceptable—a new operational approach is required.
- The mission and/or objectives are no longer valid, thus a new OPLAN is required—a new operational approach is required to support the further detailed planning.

Assessment could cause the JFC to shift the focus of the operation, which the JFC would initiate with a new visualization manifested through new planning guidance for an adjusted operation or campaign plan.

Chap 4

Joint Logistics

Ref: JP 4-0, Joint Logistics (Feb '19), chap. I.

Sustainment

Sustainment—one of the seven joint functions (command and control [C2], information, intelligence, fires, movement and maneuver, protection, and sustainment)— is the provision of logistics and personnel services to maintain operations until mission accomplishment and redeployment of the force. Joint force commanders (JFCs) are called upon to maintain persistent military engagement in an uncertain, complex, and rapidly changing environment to advance and defend US values and interests, achieve objectives consistent with national strategy, and conclude operations on terms favorable to the US. Effective sustainment provides the JFC the means to enable freedom of action and endurance and to extend operational reach. Sustainment determines the depth to which the joint force can conduct decisive operations, allowing the JFC to seize, retain, and exploit the initiative. Joint logistics supports sustained readiness for joint forces.

I. Joint Logistics

The relative combat power that military forces can generate against a threat is constrained by their capability to plan for, gain access to, and deliver forces and materiel to points of application. Joint logistics is the coordinated use, synchronization, and often sharing of two or more combatant commands (CCMDs) or Military Departments' logistics resources to support the joint force. To meet the wide variety of global challenges, combatant commanders (CCDRs), subordinate commanders, and their staffs must develop a clear understanding of joint logistics, to include the relationship between logistic organizations, personnel, core functions, principles, imperatives, and the operational environment (OE). This publication provides logistics guidance essential to the operational capability and success of the joint force. It focuses on the integration of strategic, operational, and tactical support efforts while leveraging the global joint logistics enterprise (JLEnt) to affect the mobilization and movement of forces and materiel to sustain a JFC's concept of operations (CONOPS). Additionally, it provides guidance for joint logistics; describes core logistics functions essential to success; and offers a framework for CCDRs and subordinate commanders to integrate capabilities from national, multinational, Services, and combat support agencies (CSAs) to provide forces properly equipped and trained, when and where required. The identification of established coordination frameworks, agreements, treaties, theater distribution, and posture plans creates an efficient and effective logistics network to support the JFC's mission.

Joint logistics planning must account for the adversary's threat to logistics. It must also identify and reduce logistics and operational risks. The challenge for future joint logistics is to adequately support globally integrated operations given the combination of five ongoing trends:

- Increasing logistics requirements caused by global demand for US joint forces and operations.
- Constrained and degraded resources, both overall and within the logistics force structure.
- The growing complexity of logistics operations.
- The proliferation of advanced antiaccess/area denial capabilities by adversaries that would degrade logistics capabilities and capacities.
- The increase of cyberspace threats to joint and partner logistics networks and mission systems.

Logistics integrates strategic, operational, and tactical support efforts to project and sustain military power across the globe at a chosen time and place, and represents a comparative advantage that provides multiple options to leadership and multiple dilemmas to potential adversaries. A relevant and resilient JLEnt remains essential to the pursuit of national interests through assurance, deterrence, and responding to a full range of contingencies.

II. Joint Logistics Environment (JLE)

Military leaders conduct globally integrated logistics operations in a complicated, interconnected, transregional environment (see Figure I-1). These operations involve the total force, which consists of the Active Component and the Reserve Component and Department of Defense (DOD) civilians and contracted support. Additional capabilities in the area of responsibility (AOR) or joint operations area (JOA) could also include a variety of military forces, other governmental organizations, nongovernmental organizations (NGOs), and multinational forces (MNFs). The joint logistics environment is the sum of conditions and circumstances that affect logistics. The joint logistics environment exists at the strategic, operational, and tactical levels. Globalization, technology advancements, antiaccess/area denial, and flexible threats create a complex, ever-changing OE. The essential challenge is to support unified action by meeting increasingly demanding logistics requirements with constrained resources in a potentially contested environment. Globally integrated logistics is the capability to allocate and adjudicate joint logistics support on a global scale to maximize effectiveness and responsiveness, and to reconcile competing demands for limited logistics resources based on strategic priorities. Understanding the global environment is essential to plan, execute, synchronize, assess, and coordinate logistics operations.

Joint logistics takes place throughout the OE. Service components and CSAs provide the forces, materials, and capabilities while the JFC's staff focuses on integrating the capabilities with operations. Access to secure networks is necessary to sustain joint force readiness. Effective networks are used to find and access relevant information, facilitate collaboration, distribute data to forward deployed areas, increase performance and reliability, ensure the enterprise infrastructure for evolving DOD systems is resilient, and leverage partner nations' (PNs') capabilities.

See facing page for further discussion of the JLE operating framework.

Building Partnership Capacity (BPC)

Complicated supply lines, finite resources, the challenges of providing robust logistics in austere environments, and shared lines of communications (LOCs) require the ability to establish and foster nontraditional partnerships. For some operations, logistics forces may be employed in quantities disproportionate to their normal military roles and in nonstandard tasks. Further, logistics forces may precede other military

Joint Logistics Environment (JLE) Operating Framework

Ref: JP 4-0, Joint Logistics (Feb '19), fig. I-1, p. I-3.

Military leaders conduct globally integrated logistics operations in a complicated, interconnected, transregional environment (see Figure I-1). These operations involve the total force, which consists of the Active Component and the Reserve Component and Department of Defense (DOD) civilians and contracted support. Additional capabilities in the area of responsibility (AOR) or joint operations area (JOA) could also include a variety of military forces, other governmental organizations, nongovernmental organizations (NGOs), and multinational forces (MNFs). The joint logistics environment is the sum of conditions and circumstances that affect logistics. The joint logistics environment exists at the strategic, operational, and tactical levels.

Strategic Level Campaign Quality	Operational Level Coordinate, Integrate, and Synchronize	Tactical Level Effectiveness
• Industrial base capacity enables sustained operations • End-to-end processes drive efficiencies across Services, agencies, and industry • Effectiveness dependent upon optimizing processes against required outcomes	• Combatant commander integrates joint requirements with national systems • Must optimize component, agency, and other partner nation capabilities to meet requirements • Most significant impact for joint logistics and the joint force	• Outcome is measured • Operational readiness enables "freedom of action" • Desired outcomes should drive optimization–from strategic to tactical

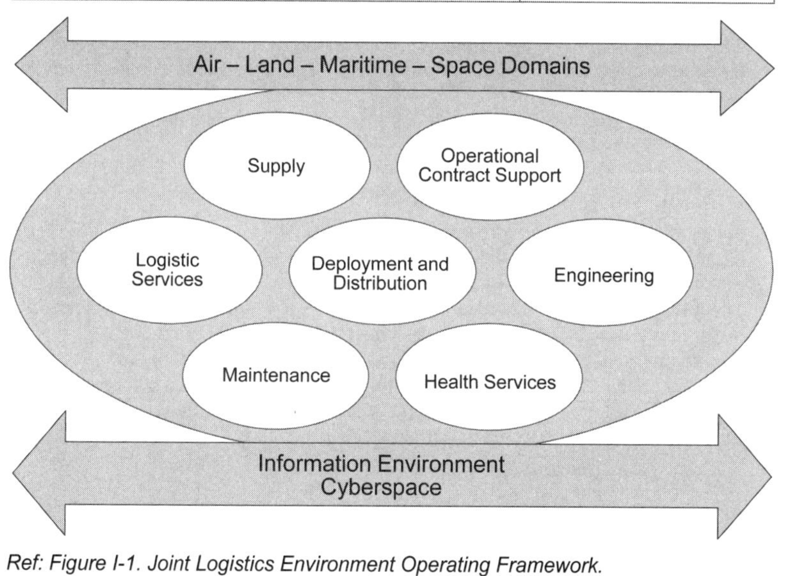

Ref: Figure I-1. Joint Logistics Environment Operating Framework.

forces or may be the only forces deployed. Logistics forces also may continue to support other military personnel and civilians after the departure of combat forces. BPC is important for sharing the costs and responsibilities, improving information flow, and establishing PN agreements. BPC includes coordination of resources with multinational partners, international organizations, and NGOs. BPC improves unity of effort within the entire JLEnt. BPC is an essential component of joint operations because the Services seldom have sufficient capability to support a joint force independently. BPC is an ongoing, long-term relationship development process that may not yield immediate results. The earlier the BPC efforts begin, the better the chance of success for securing partner logistics support when needed. By combining capabilities, commanders can provide maximum effectiveness and flexibility to the joint force focused on objectives that deliver sustained logistics support.

III. Joint Logistics Enterprise (JLEnt)

The purpose of the JLEnt is to protect and sustain military power across the globe at a time and place of our choosing and represents a US comparative advantage that provides multiple options to our nation's leadership and multiple dilemmas to potential adversaries. The JLEnt is a multitiered matrix of key global logistics providers cooperatively structured through an assortment of collaborative agreements, contracts, policy, legislation, or treaties utilized to provide the best possible support to the JFC or other supported organization. The key DOD organizations in the JLEnt include the Services, CCMDs, Defense Logistics Agency (DLA), Joint Staff J-3 [Operations Directorate], and Joint Staff J-4 [Logistics Directorate]. Other US Government departments and agencies, NGOs, and commercial partners also play a vital role in virtually all aspects of the JLEnt and function on a global scale, providing comprehensive, end-to¬end capabilities. The JLEnt may also include multinational partners and international organizations. Participants operate across the strategic, operational, and tactical levels— many are affiliated with either supported or supporting commands and operate under a variety of command relationships.

The JLEnt is interconnected among global logistics providers, supporting and supported organizations and units, and other entities. Knowing the roles, responsibilities, relationships, and authorities of JLEnt partners is essential to planning, executing, controlling, and assessing logistics operations. JLEnt partners must collaborate to ensure the coordinated employment and sharing of capabilities and resources. Global logistics providers manage end-to-end processes that provide capabilities to the supported CCDR to fulfill requirements. The transregional, multi-domain, and multifunctional nature of future threats, combined with budgetary pressures, require enterprise-wide tradeoffs; these tradeoffs can be mitigated through persistent responsiveness.

IV. Joint Logistics Imperatives

Joint logistics focuses on three imperatives to influence mission success: unity of effort, JLEnt visibility, and rapid and precise response. These imperatives define the desired attributes of a federation of systems, processes, and organizations that effectively adapt within a constantly changing OE to meet the emerging needs of the supported JFC. The joint logistics imperatives support operations, which is the primary purpose of logistics. Successfully meeting the needs of operational forces will build trust in the logistics process and between organizations during joint operations. These imperatives guide joint logisticians in the performance of the integrating functions needed for successful joint logistics operations.

See facing page for further discussion.

Joint Logistics Imperatives

Ref: JP 4-0, Joint Logistics (Feb '19, pp. I-5 to I-6.

Joint logistics focuses on three imperatives to influence mission success: unity of effort, JLEnt visibility, and rapid and precise response.

1. Unity of Effort

Unity of effort is the coordination and cooperation toward common objectives, even if the participants are not necessarily part of the same command or organization. Unity of effort is the product of successful unified action. For joint logisticians, unified action synchronizes and integrates logistics capabilities focused on the commander's intent. Unified action is critical to joint logistics objectives. To achieve unity of effort, joint logisticians must develop a clear understanding of how joint and multinational logistics (MNL) processes work, know the roles and responsibilities of the providers executing tasks in those processes, build agreement around common measures of performance, and ensure appropriate members of the JLEnt have visibility into the processes.

2. JLEnt Visibility

JLEnt visibility is access to logistics processes, resources, and requirements data to provide the information necessary to make effective decisions. JLEnt visibility is inclusive of the sub-components:

- **In-Transit Visibility (ITV)** is the ability to track the identity, status, and location of DOD units, non-unit cargo or supplies, passengers, patients, and personal property from origin to consignee or destination.
- **Defense Transportation System (DTS)** is that portion of the worldwide transportation infrastructure that supports DOD transportation needs in peace and war. It consists of three major sources of transportation resources and capabilities: military (organic), commercial (nonorganic), and host nation (HN). Resources include inland surface transportation (rail, road, and inland waterway), sea transportation (coastal and ocean), air transportation, and pipelines.

3. Rapid and Precise Response

Rapid and precise response is the ability of the core logistics functions, military and commercial, to meet the constantly changing needs of the joint force. The effectiveness of joint logistics can be measured by assessing the following attributes or key performance indicators.

- **Velocity** is at the core of responsiveness. Velocity does not mean everything moves at the same rate or fastest rate, but everything moves according to priority at the rate that produces a balance between efficiency and effectiveness to fully meet the CCDR's operational needs.
- **Reliability** is reflected in the dependability of the global providers and the development of a resilient distribution network able to deliver required support when promised. Reliability is characterized by a high degree of predictability or time-definite delivery of support. Time-definite delivery is the consistent delivery of requested logistics support at a time and destination specified by the requiring activity.
- **Efficiency** is related directly to the amount of resources required to achieve a specific objective. In the tactical and operational environments, inefficiency increases the logistics footprint, force protection requirements, and risk. At the strategic level, inefficiency increases the cost and risk for the operation.
- **Effectiveness** is the ability of the JLEnt to fully meet the CCDR's operational requirements within acceptable risk. Effectiveness is providing the right logistics solutions at the right time and place.

V. Joint Logistics Focus Areas

Ref: JP 4-0, Joint Logistics (Feb '19, p. I-7.

The joint logistics community must focus on the following five areas to influence mission success: warfighting readiness, competition below armed conflict, global integration, innovation, and the strengthening of alliance and partner networks. These areas discuss the desired attributes of a federation of systems, processes, and organizations that effectively adapt within a constantly changing OE to meet the emerging needs of the supported JFC. These focus areas will guide joint logisticians in the performance of the integrating functions needed for successful joint operations.

1. Warfighting Readiness

The Joint Staff will champion efforts to enhance and protect the JLEnt's capability, capacity, and comprehensive readiness to project/sustain military power globally at a chosen time and place. In particular, JFCs will assess and mitigate risk within operational planning activities to ensure the joint force is logistically positioned to support the range of military operations.

2. Competition Below Armed Conflict

Adversaries understand that strategic logistics is a comparative advantage of the US and will attempt to undermine its ability to project/sustain military power. These "supply chain wars" cut across all instruments of national power and include infiltration of traditional business systems. The joint force must understand this competition space, and our adversaries' capabilities/intentions in targeting the JLEnt, and pursue actions to protect mission assurance.

3. Global Integration

The JLEnt must be able to effectively allocate scarce resources to meet global priorities. Joint logisticians must have access to strategic logistics information and institute flexible processes that provide an accurate picture of the logistics environment to facilitate timely resource-informed decision making and enables operational success.

4. Innovation

A "data culture" improves the understanding of potential concepts like big data, artificial intelligence, machine learning, and modern computing power with regard to revolutionary improvements across the JLEnt. Adversaries will focus efforts on eroding the comparative competitive advantage in technology. Success in future conflicts may depend on the ability to expeditiously adopt and field new technologies that assure the continued ability to project and sustain power.

5. Strengthen Alliance and Partner Networks

Relationships with like-minded partners are essential to advancing US interests. It is critical to continue to advocate and support JLEnt efforts that increase joint force lethality, global agility, interoperability, and operational effectiveness through expanded access, visibility, and cooperation. Joint logisticians must understand the ability to project and sustain power is inextricably linked to the JLEnt, its array of partners and allies within DOD and broader US Government organizations, the industrial base, and aligned nations.

VI. Principles of Logistics

Ref: JP 4-0, Joint Logistics (Feb '19), pp. I-8 to I-9.

1. Responsiveness

Responsiveness is providing the right support when and where it is needed. Responsiveness is characterized by the reliability of support and the speed of response to the needs of the joint force. Clearly understood processes and well-developed decision support tools are key elements enabling responsiveness to emerging requirements. By monitoring the battle rhythm, the joint logistician can anticipate logistic issues.

2. Simplicity

Simplicity fosters efficiency in planning and execution, and allows for more effective control over logistic operations. Clarity of tasks, standardized and interoperable procedures, and clearly defined command relationships contribute to simplicity. Simplicity is a way to reduce the "fog of war" or the friction caused by combat. Clear objectives, relevant processes, and documented procedures assist unity of effort.

3. Flexibility

Flexibility is the ability to improvise and adapt logistic structures and procedures to changing situations, missions, and operational requirements. Flexibility is how well logistics responds in a dynamic environment. Where responsiveness is a commander's view of logistic support, flexibility is a logistician's view of being responsive. The logistician's ability to anticipate requirements in an operational environment allows for the development of viable options able to support operational needs.

4. Economy

Economy is the minimum amount of resources required to bring about or create a specific outcome. Economy is achieved when support is provided using the fewest resources within acceptable levels of risk. At the tactical and operational levels, economy is reflected in the number of personnel, units and equipment required to deliver support. Among the key elements of the logistic principle of economy is the identification and elimination of redundancy.

5. Attainability

Attainability is the assurance that the essential supplies and services available to execute operations will achieve mission success. Attainability is the point at which the CCDR or subordinate JFC judges that sufficient supplies, support, distribution capabilities, and LOC capacity exist to initiate operations at an acceptable level of risk. Some examples of minimal requirements are inventory on hand (days of supply), critical support and Service capabilities, theater distribution assets (surge capability), combat service support (CSS) sufficiency, and force reception throughput capabilities.

6. Sustainability

Sustainability is the ability to maintain the necessary level and duration of logistics support to achieve military objectives. Sustainability is a function of providing for and maintaining those levels of ready forces, materiel, and consumables necessary to support military action. Sustainability is focused on the long-term objectives and requirements of the supported forces. Sustainability provides the JFC with the means to enable freedom of action and extend operational reach.

7. Survivability

Survivability is the capacity of an organization to prevail in spite of adverse impacts or potential threats. To provide continuity of support, critical logistic infrastructure must be identified and plans developed for its protection. Survivability is directly affected by dispersion, design of operational logistic processes, and the allocation of forces to protect critical logistic infrastructure. Examples of critical logistic infrastructure include industrial centers, airfields, seaports, railheads, supply points, depots, LOCs, bridges, intersections, logistic centers, and military installations.

VII. Logistics Integration

Commanders and staffs apply basic principles, control resources, and manage capabilities to provide sustained joint logistics. Logisticians can use the principles of logistics as a guideline to assess how effective logistics are integrated into plans and execution. To achieve full integration, commanders and their logisticians coordinate, synchronize, plan, execute, and assess logistics support to joint forces during all phases of the operation.

Coordinating and Synchronizing

Effective coordination of joint logistics includes choosing organizational options to execute effective joint logistics operations.

See pp. 4-11 to 4-26 for further discussion.

Planning

Logistics planners at every level should set conditions for subordinate success. Timely, accurate, and responsive planning enables trade-offs, alternate courses of action (COAs), and, therefore, freedom of action for JFCs. Joint logistics planning links the mission and commander's intent to core logistics functions, procedures, and organizations. This establishes the JFC's ability to meet requirements in terms of forces, capabilities, movement, projection, sustainment, duration of operations, redeployment, and retrograde. Joint logistics operations overseas should be planned and conducted with appropriate consideration of their effect on the environment in accordance with applicable US and HN agreements, environmental laws, policies, and regulations. Joint logistics operations planned and conducted within the US and territories will be conducted in compliance with applicable federal, state, or local environmental laws and regulations. Early planning is essential to ensure all appropriate environmental reviews have been completed in accordance with Department of Defense Instruction (DODI) 4715.06, Environmental Compliance in the United States, and for installations outside the continental United States (CONUS), see DODI 4715.05, Environmental Compliance at Installations Outside the United States

See pp. 4-27 to 4-38 for further discussion.

Executing

Executing joint logistics involves the employment of capabilities and resources to support joint and multinational operations.

See pp. 4-39 to 4-42 for further discussion.

Assessing

Assessing joint logistics facilitates future success through plan refinement and adaptation. The joint logistician must be able to assess and respond to requirements by monitoring dynamic situations and providing accurate feedback to subordinates and decision makers.

Chap 4

I. Core Logistics Functions

Ref: JP 4-0, Joint Logistics (Feb '19), chap. II and pp. I-10 to I-12.

Core logistics functions provide a framework to facilitate integrated decision making, enable effective synchronization and allocation of resources, and optimize joint logistics processes. The challenges associated with support cut across all core logistics functions, especially when multiple joint task forces (JTFs) or multinational partners are involved.

Core Logistic Functions

Core Functions	Functional Capabilities
Deployment and Distribution	• Move the force • Sustain the force • Operate the joint deployment and distribution enterprise
Supply	• Manage supplies and equipment • Inventory management • Manage global supplier networks
Maintenance	• Depot maintenance operations • Field maintenance operations • Equipment reset
Logistics Services	• Food service • Water and ice service • Contingency base services • Hygiene services • Mortuary affairs
Operational Contract Support	• Contract support integration • Contracting support • Contractor management
Engineering	• General engineering • Combat engineering • Geospatial engineering
Joint Health Services	• Force health protection • Health service support

Ref: Figure II-1. Core Logistics Functions

The core logistics functions are: deployment and distribution, supply, maintenance, logistics services, operational contract support (OCS), engineering, and joint health services. The core logistics functions are considered during the employment of US military forces in coordinated action toward a common objective and provide global force projection and sustainment.

Joint Logistics

Core Logistics Functions

Ref: JP 4-0, Joint Logistics (Feb '19), pp. II-1 to II-13.

A. Deployment and Distribution

The global dispersion of the threats, coupled with the necessity to rapidly deploy, execute, and sustain operations worldwide, makes the deployment and distribution capability the cornerstone of joint logistics. These operational factors necessitate a shift from a supply-based system to a system that is primarily distribution-based with beginning-to-end synchronization to meet JFC requirements. Through sharing critical information, it is possible to create unity of effort among diverse distribution organizations to satisfy deployment, execution, and sustainment operations.

B. Supply

The joint logistician must understand the complexities of supply operations, the functions and processes that define them, and the organizations and personnel responsible for executing tasks to meet the JFC's requirements. The Services and DLA are primarily responsible for DOD supply chain operations and manage the supply processes to provide common commodities and services to joint forces. Planning for supply operations requires a collaborative environment to fully consider all major components of the JLEnt, to include the return and retrograde of equipment and supplies.

C. Maintenance

Maintenance supports system readiness for the JFC. The Services, as part of their Title 10, United States Code (USC), responsibilities, execute maintenance as a core logistics function. The Services employ a maintenance structure of depot- and field-level maintenance to improve the JFC's freedom of action and sustain the readiness and capabilities of assigned units.

D. Logistics Services

Logistics services comprise the support capabilities that collectively enable the US to rapidly provide global sustainment for our military forces. Logistics services include many scalable and disparate capabilities. Included in this area are food service, water and ice service, contingency base services, hygiene services, and mortuary affairs (MA).

E. Operational Contract Support (OCS)

OCS is a core logistics function and a critical component of total force readiness. DOD relies on contractors to perform many tasks. OCS provides the CCDR flexibility and options to employ commercially sourced logistics solutions from JLEnt partners such as BOS intra-theater transportation, logistics services, maintenance, storage, construction, security operations, and common-user commodities

F. Engineering

Engineer capabilities enable joint operations by facilitating freedom of action necessary for the JFC to meet mission objectives. Engineer operations integrate combat, general, and geospatial engineering to meet national and JFC requirements. Joint engineer operations facilitate the mobility and survivability of friendly forces; counter the mobility of enemy forces; provide infrastructure to position, project, protect, and sustain the joint force; contribute to a clear understanding of the physical environment; and provide support to civilian authorities and other nations.

G. Health Services

Joint health care services are conducted as part of an interrelated health system that shares medical services, capabilities, and specialists among the Service components and partners with multiple agencies and nations to implement a seamless unified health care effort in support of a joint force. Joint medical capabilities encompass both health service support (HSS) and force health protection (FHP) functions and are employed across the full range of military operations. These capabilities span the OA from prevention to point of injury/illness to definitive care.

II. Coordinating & Synchronizing Joint Logistics

Ref: JP 4-0, Joint Logistics (Feb '19), chap. III.

This section describes the authorities, organizations, and controls that synchronize logistics in support of the JFC. *JP 3-0, Joint Operations,* identifies C2 as a joint function. Command includes both the authority and responsibility for effectively using available resources and the art of motivating and directing people and organizations to accomplish missions. Control is inherent in command. However, logistics assets will rarely fall under one command, which makes control, coordination, collaboration, synchronization, and management of joint logistics more challenging. To control joint logistics, commanders direct forces and functions consistent with a commander's command authority. It involves organizing the joint logistics staff, operational-level logistics elements, CSAs, and their capabilities to assist in planning and executing joint logistics. Designating lead Service, assigning agency responsibilities, and developing procedures to execute the CCDR's directive authority for logistics (DAFL) will assist in planning, integrating, synchronizing, and executing joint logistics support operations. While logistics remains a Service responsibility, there are other logistics organizations, processes, and tasks to consider when developing a concept of logistics support (COLS) to optimize joint logistics objectives.

I. Logistics Authority

Directive Authority for Logistics (DAFL)

CCDRs exercise authoritative direction over logistics, in accordance with Title 10, USC, Section 164. DAFL cannot be delegated or transferred. However, the CCDR may delegate the responsibility for the planning, execution, and/or management of common support capabilities to a subordinate JFC or Service component commander to accomplish the subordinate JFC's or Service component commander's mission. For some commodities or support services common to two or more Services, the Secretary of Defense (SecDef) or the Deputy Secretary of Defense may designate one provider as the EA (see Appendix D, "Logistic-Related Executive Agents"). Other control measures to assist in developing common user logistics are joint tasks or inter-Service support agreements. However, the CCDR must formally delineate this delegated authority by function and scope to the subordinate JFC or Service component commander. The exercise of DAFL by a CCDR includes the authority to issue directives to subordinate commanders, including peacetime measures necessary for the execution of military operations in support of the following: execution of approved OPLANs; effectiveness and economy of operation; and prevention or elimination of unnecessary duplication of facilities and overlapping of functions among the Service component commands.

DAFL of a GCC applies to the entire AOR and affects all subordinate components, commands, and direct reporting units in the AOR. Some CCDR responsibilities include:

- Issuing directives to subordinate commanders, including peacetime measures necessary for the execution of military operations, in support of the following: execution of approved OPLANs, effectiveness and economy of operation, and prevention or elimination of unnecessary duplication of facilities and overlapping of functions among the Service component commands.

- Coordinating with USTRANSCOM to identify transportation-related requirements and initiatives (e.g., establishment of aerial port of debarkation [APOD]/seaport of debarkation [SPOD], determining transportation routes and infrastructure to support).
- Coordinating with DLA to identify logistics requirements and initiatives (e.g., establishing storage locations, identifying pre-positioned material and equipment, determining fuel requirements, providing contingency contracting solutions).
- Establishing host-nation support (HNS) (e.g., acquisition and cross-servicing agreements [ACSAs]/mutual logistics support agreements, status-of-forces agreements, cost-sharing agreements).

Unless otherwise directed by SecDef, the Military Departments and Services continue to have responsibility for the logistics support of their forces assigned or attached to joint commands, subject to the following guidance:

- Under peacetime conditions, the scope of the logistics authority exercised by the CCDR will be consistent with the peacetime limitations imposed by legislation, DOD policy or regulations, budgetary considerations, local conditions, and other specific conditions prescribed by SecDef or the Chairman of the Joint Chiefs of Staff (CJCS). Where these factors preclude execution of a CCDR's directive by component commanders, the comments and recommendations of the CCDR, together with the comments of the component commander concerned, normally will be referred to the appropriate Military Department for consideration. If the matter is not resolved in a timely manner with the appropriate Military Department, it will be referred by the CCDR, through the CJCS, to SecDef.
- Under crisis, wartime conditions, or where critical situations make diversion of the normal logistics process necessary, the logistics authority of CCDRs enables them to use all facilities and supplies of all forces assigned to their commands for the accomplishment of their missions. The President or SecDef may extend this authority to attached forces when transferring those forces for a specific mission and should specify this authority in the establishing directive or order. Joint logistics doctrine and policy developed by the CJCS establishes wartime logistics support guidance to assist the CCDR in conducting successful joint operations.

A CCDR's DAFL does not:

- Discontinue Service responsibility for logistics support.
- Discourage coordination by consultation and agreement.
- Disrupt effective procedures or efficient use of facilities or organizations.
- Include the ability to provide contracting authority or make binding contracts for the US Government.

In exercising DAFL, CCDRs have an inherent obligation to ensure accountability of resources. This obligation is an acknowledgement of the Military Departments' Title 10, USC, responsibilities and recognizes that the Military Departments, with rare exceptions, do not resource their forces to support other DOD forces. In that regard, CCDRs will coordinate with appropriate Service components before exercising DAFL or delegating authority for subordinate commanders to exercise common support capabilities to one of their components. In keeping with the Title 10, USC, roles of the Military Departments, CCDRs should maintain an accounting of resources taken from one Service component and provided to another. This accounting can be used to reimburse the losing Service component in kind over time within the AOR when possible, or can be used to pass back a requirement to DOD for resource actions to rebalance Military Department resource accounts.

II. Joint Logistics Roles and Responsibilities

Clearly articulating responsibilities is the first step in fully synchronized and coordinated logistics support during joint operations.

A. Secretary of Defense (SecDef)

SecDef is the principal advisor to the President on defense matters and serves as the leader and chief executive officer of DOD. The offices of SecDef most concerned with logistics matters are the Under Secretary of Defense for Policy (USD[P]), Under Secretary of Defense for Acquisition and Sustainment (USD[A&S]) (formerly Under Secretary of Defense for Acquisition, Technology, and Logistics), and Assistant Secretary of Defense for Sustainment (ASD[S]) (formerly Assistant Secretary of Defense for Logistics and Materiel Readiness).

- **USD(P)** is SecDef's principal staff assistant (PSA) and advisor for all matters on the formulation of national security and defense policy and the integration and oversight of DOD policy and plans to achieve national security objectives.
- **USD(A&S)** is the PSA and advisor to SecDef and Deputy Secretary of Defense (DepSecDef) for all matters relating to logistics; installation management; military construction; procurement; environment, safety, and occupational health management; utilities and energy management; and nuclear, chemical, and biological defense programs.
- **ASD(S)** is the principal advisor to USD(A&S), SecDef, and DepSecDef on logistics and materiel readiness in DOD and is the principal logistics official within senior management. ASD(S) is the principal advisor to USD(A&S), SecDef, and DepSecDef for energy policy, plans, and programs, and advises the CJCS regarding the role of energy in the DOD planning process.

B. Chairman of the Joint Chiefs of Staff (CJCS)

The CJCS is the principal military adviser to the President and the National Security Staff (which consists of the National Security Council and the Homeland Security Council) and SecDef. The CJCS prepares joint logistics and mobility plans to support strategic and contingency plans and recommends the assignment of logistics and mobility responsibilities to the Armed Forces of the United States. The CJCS also advises SecDef on critical deficiencies in force capabilities (including manpower, logistics, intelligence, and mobility support).

C. Military Departments

The Military Departments exercise authority to conduct all affairs of their departments, including to recruit, organize, supply, equip, train, service, mobilize, demobilize, administer, and maintain forces; construct, outfit, and repair military equipment; adhere to environmental compliance; construct, maintain, and repair buildings, structures, and utilities; and acquire, manage, and dispose of real property or natural resources.

D. Services

In accordance with Title 10, USC, the Services are responsible for preparing for employment of Service forces. They recruit, supply, organize, train, equip, service, mobilize, demobilize, provide administrative support, and maintain ready forces. Services are the center of a collaborative network, and their logistics organizations form the foundation of the JLEnt. The Services are the primary force providers and executors of joint logistics, as well as the primary providers of logistics in support of their own Service organizations supporting the CCDR. They are responsible for operational logistics support systems, platforms, and their execution to support the force. They are responsible for maintaining systems' life-cycle readiness.

See pp. 4-20 to 4-21 for an overview of the Services' logistics execution.

E. Defense Logistics Agency (DLA)

As the nation's combat logistics support agency, DLA manages the global supply chain and in collaboration with JLEnt partners sustains the readiness and lethality of the Armed Forces of the United States. As a statutory CSA, DLA provides logistics advice, advocacy, and assistance to the Office of the Secretary of Defense, Joint Chiefs of Staff, the CCDRs, Military Departments, DOD components, and interagency partners.

F. The Joint Staff J-3

The Joint Staff J-3 is responsible for maintaining the global capability for rapid and decisive military force power projection. The Joint Staff J-3 is also responsible for leading the collaborative efforts of the joint planning and execution community to improve the joint deployment and redeployment processes, while maintaining the overall effectiveness of these processes so that all supported JFCs and supporting DOD components can execute military force power projection more effectively and efficiently. Additionally, the Joint Staff J-3 serves as the joint force coordinator and is responsible for coordinating the staffing of all force requirements among the joint force providers (JFPs), consolidating all execution and contingency sourcing recommendations, and performing the duties of a JFP for all conventional force requirements.

G. The Joint Staff J-4

The Joint Staff J-4 leads the DOD efforts in the JLEnt and coordinates policy and makes recommendations to improve the preparedness of the DOD global logistics force. Additionally, the Joint Staff J-4 advises the CJCS on the readiness assessments of the CCMDs and Services.

H. The Joint Staff J-5 [Directorate for Strategy, Plans, and Policy]

The Joint Staff J-5 collaborates with the Joint Staff J-4 to ensure contingency plans are resource-informed as the coordinating authority for global logistics.

I. Combatant Commands (CCMDs)

Unless otherwise directed by the President or SecDef, the CCDR exercises authority, direction, and control over the commands and forces assigned to that command through combatant command (command authority) (COCOM). CCDRs coordinate and approve the administration, support (including control of resources and equipment, internal organization, and training), and discipline necessary to carry out missions assigned to the command.

J. Executive Agent (EA)

A DOD EA is the head of a DOD component to whom SecDef or the DepSecDef has assigned specific responsibilities, functions, and authorities to provide defined levels of support for operational missions, or administrative or other designated activities that involve two or more of the DOD components. The DOD EA may delegate to a subordinate designee, within that official's component, the authority to act on that official's behalf for any or all of those DOD EA responsibilities, functions, and authorities assigned by SecDef or DepSecDef.

Refer to DODD 5101.1, DOD Executive Agent, and JP 4-0, Appendix D, "Logistics-Related Executive Agents," for details.

K. Combat Support Agencies (CSAs)

CSAs designated under Title 10, USC, Section 193, fulfill combat support (CS) or CSS functions for joint operating forces across the range of military operations and in support of CCDRs executing military operations. CSAs perform support functions or provide supporting operational capabilities, consistent with their establishing directives and pertinent DOD planning guidance.

For more information on CSAs, refer to DODD 3000.06, Combat Support Agencies.

L. US Transportation Command (USTRANSCOM)

USTRANSCOM is responsible for providing air, land, and sea transportation, terminal management, and aerial refueling to support the global deployment, employment, sustainment, and redeployment of US forces. USTRANSCOM serves as DOD's mobility JFP, DOD's single manager for defense transportation, and DOD's single manager for PM.

M. General Services Administration (GSA)

GSA provides logistics support for the functions and missions of the US Government, including DOD.

N. Defense Health Agency (DHA)

DHA is a CSA that enables the Army, Navy, and Air Force medical services to provide a medically ready force and ready medical force to CCMDs.

O. Lead Service

A Service or Service Component responsible for the programming and execution of common-user items, logistics functions, and/or service support. A CCDR may choose to assign specific CUL functions, to include both planning and execution to a lead Service. These assignments can be for single or multiple common logistics functions and may also be based on phases or OAs within the CCDR's AOR. In circumstances where one Service is the predominant provider of forces, or the owner of the preponderance of logistics capability, it may be prudent to designate that Service as the joint logistics lead for BOS-I. The CCDR may augment the lead Service logistics organization with capabilities from another component's logistics organizations, as appropriate. Key lead Service functions at operating areas typically include, but are not limited to, BOS-I, communications synchronization, and senior airfield authority (SAA) synchronization; budget programming; real property management; and provision (provide and fund) of common-user items or service support. The lead Service may consider a commercially contracted solution to meet the requirements in addition to, or in place of, organic support.

P. Base Operating Support-Integrator (BOS-I)

BOS-I is a sub-function of lead Service. The BOS-I is responsible for planning and synchronizing the efficient application of resources and contracting to facilitate unity of effort in the coordination of sustainment functions at designated contingency locations. When multiple Service components share a common base of operations, a GCC may designate a Service component or JTF as the BOS-I at each contingency location. The GCC, commensurate with special operations forces' (SOF's) capacity and capability, may assign SOF the synchronization of BOS functions in specific instances where SOF and their enablers are the only forces at a contingency location. The designated BOS-I is responsible for coordinating common user contract support, as well as the efficient use of other support resources, for all joint forces at the contingency location. Additional BOS-I responsibilities may include, but are not limited to: coordinating the issuance of war reserve materiel assets, collecting and prioritizing construction requirements, seeking infrastructure funding support, environmental management, emergency management, emergency services, force protection, and hazardous waste management. The BOS-I must closely coordinate with the SAA or single port or terminal manager. If no SAA or single port or terminal manager is assigned, the BOS-I is responsible for their functions.

III. Combatant Commander's Logistics Directorate

The logistics directorate of a joint staff (J-4) at the CCMD conducts logistics planning and execution in support of joint operations. They integrate, coordinate, and synchronize Service component and CSA logistics capabilities to support the joint force. The J-4 also advises the JFC on logistics support to optimize available resources. Although the organizational considerations outlined below could apply to a CCDR's J-4 staff, they will most frequently be applied to subordinate joint force J-4 organizations. The J-4 staff supports the operations directorate of a joint staff (J-3) in the planning and executing of requirements for the joint reception, staging, onward movement, and integration (JRSOI) process, as well as contingency base planning and sustainment. The J-4 coordinates, synchronizes, plans, and executes core logistics functions in joint and multinational environments.

- **Planning**. The J-4 provides logistics expertise as part of the joint planning process (JPP). In accordance with JP 5-0, Joint Planning, the J-4 establishes a logistics planning cell in coordination with the plans directorate of a joint staff to fulfill this responsibility. Planning occurs at every level of warfare in a networked, collaborative environment, which requires dialogue among senior leaders, concurrent and parallel plan development, and collaboration across multiple planning levels.
- **Execution**. The GCC's J-4 coordinates and synchronizes joint theater logistics. This includes communicating the logistics priorities of the GCC to the Services responsible for executing joint logistics operations. The J-4s organize their logistics staff functions to respond to anticipated or ongoing operations.

A. Joint Logistics Operations Center (JLOC)

The J-4 establishes a JLOC to monitor and control the execution of logistics in support of on-going operations. The JLOC is an integral part of the CCDR's operations element and provides joint logistics expertise to the J-3 operations cell. The JLOC is tailored to the operation and staffed primarily by the J-4 staff.

B. Joint Deployment Distribution Operations Center (JDDOC)

At time of need, a supported GCC can create a JDDOC and incorporate its capabilities into the staff functions. The GCC can place the JDDOC at any location required or under the operational control (OPCON) of other commanders. The JDDOC can reach back to the national partners to address and solve deployment and distribution issues for the GCC. The JDDOC develops deployment and distribution plans; integrates multinational and/or interagency deployment and distribution; and coordinates and synchronizes supply, transportation, and related distribution activities. The JDDOC synchronizes the strategic to operational movement of forces and sustainment into theater by providing advance notice to the GCC's air and surface theater movement C2 elements. In concert with the GCC's overall priorities, and on behalf of the GCC, the JDDOC coordinates common user and theater distribution operations above the tactical level. A joint movement center (JMC) may be established at a subordinate unified or JTF level to coordinate the employment of all means of transportation (including that provided by allies or HNs) to support the CONOPS. This coordination is accomplished through establishment of theater and JTF transportation policies within the assigned OA, consistent with relative urgency of need, port and terminal capabilities, transportation asset availability, and priorities set by a JFC. The JTF JMC will work closely with the JDDOC.

For more information, refer to JP 4-09, Distribution Operations.

C. Joint Logistic Boards, Centers, Offices, and Cells

Ref: JP 4-0, Joint Logistics (Feb '19), app. B.

The CCDR may also establish boards, centers, offices, and cells (e.g., subarea petroleum office [SAPO], joint facilities utilization board [JFUB], joint mortuary affairs office [JMAO], operational contract support integration cell [OCSIC]) to meet increased requirements and to coordinate the logistics effort.

Strategic-level Boards, Offices, and Centers

Strategic-level joint logistic boards, offices, and centers provide advice or allocation recommendations to the CJCS concerning prioritizations, allocations, policy modifications or procedural changes.

- Joint Logistics Board (JLB)
- Joint Materiel Priorities and Allocation Board (JMPAB)
- Joint Transportation Board (JTB)
- Joint Logistics Operations Center (JLOC)
- Deployment and Distribution Operations Center (DDOC)
- Defense Health Agency (DHA)
- Contingency Basing Executive Council (CBEC)
- Global Posture Executive Council (GPEC)
- Medical Logistics Division
- United States Transportation Command, Office of the Command Surgeon (TCSG)
- Armed Services Blood Program (ASBP)

Operational Joint Logistic Boards, Centers, and Cells

Operational-level joint logisticians must provide advice and recommendations to the supported CCDR concerning prioritizations, allocations, or procedural changes based upon the constantly changing operational environment.

- Joint Logistics Operations Center (JLOC)
- Joint Deployment and Distribution Operations Center (JDDOC)
- Combatant Commander Logistic Procurement Support Board (CLPSB)
- Joint Requirements Review Board (JRRB)
- Joint Contracting Support Board (JCSB)
- Joint Environmental Management Board (JEMB)
- Joint Facilities Utilization Board (JFUB)
- Logistics Coordination Board
- Joint Movement Center (JMC)
- Theater Patient Movement Requirements Center (TPMRC)
- Joint Patient Movement Requirements Center (JPMRC)
- Joint Blood Program Office (JBPO)
- Joint Petroleum Office (JPO)
- Sub-area Petroleum Office (SAPO)
- Joint Mortuary Affairs Office (JMAO)
- Explosive Hazards Coordination Cell (EHCC)
- Joint Munitions Office (JMO)
- Operational Contract Support Integration Cell (OCSIC)

IV. Logistics Execution Organizations

The fundamental role of joint logistics is to integrate and coordinate logistics capabilities from Service, agency, and other providers of logistics support and to facilitate execution of the Services' Title 10, USC, responsibilities while supporting the ever-changing needs of the JFC. Logistics may also be called upon to support the National Guard in Title 32, USC, status. It may also include special assignment airlift missions in addition to channel airlift, surface, and sealift movements. Joint logisticians should understand how each of the Services conducts logistics at the operational level.

See following pages (pp. 4-20 to 4-21) for discussion of the Services' logistics execution, and facing page for discussion of TRANSCOM, DLA, DCMA, and DSCA.

V. Logistics Control Options

The CCDR's logistics authority enables use of all logistics capabilities of the forces assigned as necessary for the accomplishment of the mission. The President or SecDef may extend this authority to attached forces when transferring those forces for a specific mission and should specify this authority in the establishing directive or order. The CCDR may elect to control logistics through the J-4 staff tailored and augmented as discussed previously in "Combatant Commander's Logistics Directorate." The CCDR may also decide to control joint logistics by designating a subordinate logistics organization. In these instances, the CCDR will delineate the authorities and command relationships that will be used by the subordinate commander to control logistics. In both cases, the CCDR exercises effective control of joint force logistics by fusing procedures and processes to provide visibility and control over the logistics environment and integrating joint logistics planning with operations planning. Control of joint logistics is enhanced by how effectively the logistician combines the capabilities of the global providers and the Services' logistics elements with the JFC's requirements in a way that achieves unity of effort.

A. Staff Control

The J-4 staff may be used to support a wide range of operations, including campaigns; complex or long-duration major operations; or complex operations involving multiagency, international organizations, NGOs, or MNFs, if properly augmented. For example, the staff may be sized and tasked to provide increased movement control or material management capabilities; it could be augmented with a robust OCS planning and integration capability; the J-4 could receive augmented capability to coordinate multinational support operations or execute JOA-wide infrastructure repair/restoration missions. J-4 staff augmentation can come from a combination of military, civilian emergency workforce, and contractor personnel. When exercising this option, the CCDR will specify the control authorities delegated to the J-4 over the components logistics elements. Taskings to Service component logistics elements in this case must come from formal tasking orders issued through the CCDR's J-3. The logistics taskings, which could come in the form of a fragmentary order (FRAGORD), formalizes the authorities given the J-4 by the JFC and enables the rapid response to operational logistics requirements.

B. Organizational Control

As another alternative for controlling the major operations outlined above, the CCDR may elect to assign responsibility to establish a joint command for logistics to a subordinate Service component. The senior logistics headquarters of the designated Service component will normally serve as the basis for this command, an organization joint by mission (e.g., campaigns, major operations, humanitarian missions), but not by design. When exercising this option, the CCDR retains DAFL and must specify the control and tasking authorities being bestowed upon the subordinate joint

Logistics Execution Organizations

Ref: JP 4-0, Joint Logistics (Feb '19), p. III-10 to III-11.

The fundamental role of joint logistics is to integrate and coordinate logistic capabilities from Service, agency and other providers of logistic support, and to facilitate execution of the Services' Title 10, USC responsibilities while supporting the ever changing needs of the JFC.

US Transportation Command (USTRANSCOM)

Serves as the global distribution synchronizer, as outlined in the Unified Command Plan, on behalf of and in coordination with the JDDE COI to establish processes to plan, apportion, allocate, route, schedule, validate priorities, track movements, and redirect forces and supplies per the supported commander's intent. This coordination and synchronization will not infringe upon either the supported CCDRs, or Services Title 10, USC, designated responsibilities, but serves to facilitate a unity of effort throughout the JDDE to support the CCDR or subordinate JFC. The supported GCC is responsible to plan, identify requirements, set priorities, and redirect forces and sustainment as needed to support operations within the respective AOR. USTRANSCOM, as the Mobility Joint Force Provider, exercises responsibility for planning, resourcing, and operating a worldwide defense transportation system in support of distribution operations, to include reviewing taskings and analyzing supported CCDR's requirements for transportation feasibility, and advising on changes required to produce a sustainable force deployment. During the deployment, sustainment, and redeployment phases of a joint operation, CCDRs coordinate their movement requirements and required delivery dates with USTRANSCOM, and supported GCCs are responsible for deployment and distribution operations executed with assigned/attached force in their respective AORs.

Defense Logistics Agency (DLA)

DLA manages, integrates, and synchronizes suppliers and supply chains to support the Armed Forces of the US, allies, and multinational partners. The ASD(L&MR), under the USD(AT&L), exercises authority, direction, and control over DLA. As a statutory CSA, DLA provides logistics advice and assistance to the Office of the Secretary of Defense, the CJCS, Joint Chiefs of Staff, the CCDRs, Military Departments, DOD components, and interagency partners, as appropriate. Additionally, DLA operates as part of the JLEnt in providing humanitarian assistance. DLA manages nine diverse supply chains and serves as the DOD EA for four supply chains: subsistence, construction and barrier materiel, bulk petroleum, and medical materiel. DLA also directs a network of distribution depots located throughout the US, Europe, Pacific, and South West Asia.

Defense Contract Management Agency (DCMA)

DCMA is the CSA responsible for ensuring major DOD acquisition programs (systems, supplies, and services) are delivered on time, within projected cost or price, and meet performance requirements. DCMA's major role and responsibility in contingency operations is to provide contingency contract administration services for external and theater support contracts. DCMA is also responsible for selected weapons system support contracts with place of performance in the operational area and theater support contracts when contract administration services are delegated by the procuring contracting officers.

Defense Security Cooperation Agency (DSCA)

DSCA arranges DOD funded and space available transportation for NGOs for delivery of humanitarian goods to countries in need; coordinates foreign disaster relief missions; and, in concert with DLA, procures, manages, and arranges for delivery of humanitarian daily rations and other humanitarian materiel in support of US policy objectives.

Service Logistics Execution

Ref: JP 4-0, Joint Logistics (Feb '19), p. III-10 to III-14.

Army

The overarching theater-level headquarters is the theater Army/Army Service component command (ASCC), which provides support to Army forces and other Services as directed. It is important for the ASCC and theater special operations command (TSOC) J-4 to enhance conventional forces and SOF synchronization of sustainment. The theater sustainment command (TSC) is the logistics C2 element assigned to the ASCC and is the senior Army logistics headquarters within a theater of operations.

The TSC is responsible for executing port opening, theater opening, theater surface distribution, and sustainment functions in support of Army forces and provides lead Service and EA support for designated CUL to other government departments and agencies, MNFs, and NGOs as directed. The TSC is also responsible for establishing and synchronizing the intratheater segment of the surface distribution system in coordination with the JDDOC with the strategic-to-theater segment of the global distribution network.

The TSC establishes C2 of operational-level logistics in a specified area of operations by employing one or more expeditionary sustainment commands (ESCs), which provide a rapidly deployable, regionally focused, forward-based C2 capability until a TSC can assume that function. When the Army is the predominant land force operating within an OA, the TSC or ESC, at the discretion of the JFC, has the capability to become a joint logistics headquarters providing logistics support to all joint forces within the OA. This is contingent upon the other Services, DOD agencies, and CCMDs providing the appropriate augmentation of personnel and capabilities to support this joint mission. Though the TSC can be sourced from any component of the Army, the preponderance of the Army's logistic capability is in the Reserve Component, either Army Reserve or Army National Guard.

Marine Corps

The Marine expeditionary force (MEF) is the principle warfighting organization in the Marine Corps, capable of conducting and sustaining expeditionary operations in any geographic environment. The Marine logistics group is responsible for providing tactical logistics above the organic capability of supported units to all elements of the MEF. It is a permanently organized command structured with functional and multi-functional units which are organized to support a MEF possessing one Marine division and one Marine aircraft wing. Integration with strategic- and operational-level logistics support is coordinated through the Marine Corps component commander.

Navy

For numbered fleets, the senior logistician is the assistant chief of staff for logistics. The assistant chief of staff for logistics is normally the logistics readiness center (LRC) director. Coordination and unity of effort between the LRC and logistics supporting staffs and commands providing logistics resources and support is key to effectively controlling and executing logistics support.

The logistics forces of each numbered fleet are organized into standing task forces, and the commanders of these task forces are the principal logistics agents for the fleet commander. The logistics task force commander is responsible to the fleet commander for management of logistics support forces for maritime sustainment of Navy, United States Coast Guard (USCG), and Marine Corps units. The logistics task force commander has tactical control of Military Sealift Command Combat Logistics Force ships, plans resupply for all classes of supply, and plans and manages theater ship repairs in military and commercial yards outside the CONUS.

Fleet operational forces are normally organized into task forces under the command of a task force commander. The task force commander exercises control of logistics through a fleet logistics coordinator, task force logistics coordinator, or task group logistics coordinator and coordinates the replenishment of forces at sea.

Air Force

The air expeditionary task force (AETF) is the organizational structure for deployed US Air Force forces. AETF presents a scalable, tailorable organization with three elements: a single commander, embodied in the commander, Air Force forces (COMAFFOR); appropriate C2 mechanisms; and tailored and fully supported forces. The Air Force forces staff is the vehicle through which the COMAFFOR fulfills operational and administrative responsibilities for assigned and attached forces, and is responsible for long-range planning that occurs outside the air tasking cycle. The PSAs to the COMAFFOR for JOA-wide integration of agile CS capabilities and processes are the director of manpower, personnel, and services (A1); the director of logistics, engineering, and force protection (A4); and the surgeon general (SG).

A1 is responsible for the functions of billeting; MA assistance; and food service, to include bottled water to support planned meals. Responsibility for planning daily consumable water outside of planned meals resides with A4, civil engineering. Contracting is the responsible agent to procure bottled water (when the requirements have been established) from approved sources that are coordinated with bioenvironmental engineers and public health. A4 controls logistics planning; distribution; material management; fuels; maintenance; and munitions; civil engineering; fire emergency services; explosive ordnance disposal; chemical, biological, radiological, and nuclear (CBRN) defense and response elements of emergency management; and force protection. The SG advises on FHP and HSS.

USCG

USCG maritime patrol and deployable specialized forces are capable of supporting joint military operations worldwide. In order to accomplish the many missions, deployable units and assets consist of high-endurance cutters, patrol boats, buoy tenders, aircraft, port security units, maritime safety and security teams, maritime security response teams, tactical law enforcement teams, and the National Strike Force. Logistics support for the USCG is provided by the Deputy Commandant for Mission Support and its subordinate elements. When USCG forces operate as part of a JTF, they may draw upon the logistics support infrastructure established by/for the JTF. These general support functions normally include, but are not limited to, the following: berthing, subsistence, ammunition, fuel, and accessibility to the naval supply systems. The Navy logistics task force commander coordinates the replenishment, intratheater organic airlift, towing, salvage, ship maintenance, and material control, as well as commodity management for the task force group.

Special Operations (SOF)

Commander, United States Special Operations Command (CDRUSSOCOM), exercises COCOM over all SOF and the TSOCs unless otherwise directed by SecDef. SOF are dependent on Service and joint logistics support as the primary means of support. As directed, GCCs exercise OPCON over assigned TSOCs and SOF.

When a GCC establishes and employs subordinate JTFs and task forces, the GCC or commander, theater special operations command (CDRTSOC), may establish and employ a special operations joint task force (SOJTF), joint force special operations component, special operations command-forward, or joint special operations air component to control SOF assets and accommodate special operations requirements. Accordingly, the GCC establishes command relationships between SOF commanders and other JTF/task force commanders. CDRUSSOCOM can establish and employ a SOJTF or a joint special operations task force as a JFC in coordination with GCCs for special operations in their AOR.

For more information regarding special operations task organizations, see JP 3-05, Special Operations. *See also pp. 6-27 to 6-30.*

command for logistics, as well as the command relationships it will have with the Service components. This command would control logistics taskings as directed by the CCDR and must not infringe on the authorities and responsibilities as specified in "Joint Logistics Roles and Responsibilities." The CCDR, through the global force management process, would request augmentation with joint, agency, and other Service capabilities to effectively integrate and control logistics requirements, processes, and systems and with forces made available.

C. CUL Control

Planners should consider areas where CUL organizational options are best suited. CCMD and subordinate logistics planners must keep in mind that while CUL support can be very efficient, it may not always be the most effective method of support. By its very nature, CUL support will normally take place outside routine support channels, which may lead to reduced responsiveness if not properly planned, coordinated, and executed. CCDRs, along with their subordinate commanders, must review, coordinate, and direct CUL requirements with DLA; functional CCDRs, which include their supporting contracting activities; and Service component commanders to provide an integrated joint logistics system from the strategic to tactical levels. All parties must ensure that the advantages and disadvantages of each CUL-related COA are properly considered, to include the extent of reliance on commercially sourced, contracted support. However, the GCC has overall responsibility for deciding the amount and type of CUL support for a particular joint operation. The CCDR's decision to use DAFL to direct CUL support within a subordinate joint force must be deliberate and coordinated to ensure proper CUL execution.

Cross-Leveling CUL Assets

It must be clearly understood that only the CCDR has the authority to direct the cross-leveling of supplies within a joint force. Cross-leveling of a supply for one Service component will be only for common items, should be accomplished in a very prudent and deliberate manner, and consider reimbursement between Services. CUL suitability for commodities is displayed in Figure III-1, as well as other potential CUL areas that should be considered in reducing redundancy, risks, and costs.

Organizational Control Options

Based on the operational situation, the CCDRs can modify or mix two major control options: single-Service logistic support or lead Service/agency support.

- **Single-Service Logistic Support.** . In this organizational option, each Service retains primary responsibility for providing support to their subordinate organizations. CUL would be limited to existing support relationships between Services as identified in inter-Service support agreements. If delegated by the CCDR, the J-4 may coordinate limited CUL support to other Services or agencies in certain situations. This method would most likely be used in major operations where the operational situation allows for, and calls for, the deployment of the requisite Service component logistics assets in a timely manner and where logistics effectiveness is paramount.
- **Lead Service or Agency CUL Support.** The CCDR may designate a lead Service or DOD agency to provide selected CUL support to one or more Service components, governmental organizations, and/or NGOs in a joint or multinational operation. This CUL option is normally based on the dominant user and/or most capable Service concepts and may or may not involve OPCON or tactical control of one Service component logistics units to the lead Service.

D. Control Option Selection Considerations

Ref: JP 4-0, Joint Logistics (Feb '19), p. III-15.

After determining what commodities and functions will be joint, the CCDR must decide how to control those logistics operations. The selection of a control option should benefit from a careful analysis to include the following considerations. These considerations are not designed to stand alone. They should be considered comprehensively to properly inform the commander's decision.

Mission

The mission is the foremost consideration from the commander when selecting the option that will be used to control joint logistics. Mission analysis helps identify the complexity and scale of the joint logistics requirements the command will face during execution. Generally, the more complex operations have greater need for an organizational control option.

The Most Capable Service Component

This consideration aligns with the most prevalent Service capabilities in the OA. It is one of the most important considerations to analyze because no Service component's logistics organization or supporting contracting activity is staffed or equipped to plan and execute joint logistics or joint contracting support. To some degree, the most capable Service component organization will have to be augmented to provide common-user support responsibilities. Without adequate Service component logistics C2 capability available, the staff control option would be the most appropriate.

The Geographic and Physical Infrastructure in the Operational Area

This consideration is related to the most capable Service component consideration. The geographic and physical infrastructure in the OA usually dictates the nature of the LOCs needed to support the joint force and the need for contingency basing. The LOCs will influence the distribution system, to include the location of distribution points and the challenges brought on by the ITV technology need to support the operation. Additionally, the condition of the LOCs may force CUL, common-user land transport, and intratheater plans. The GCC should coordinate with USTRANSCOM, DOD agencies, and other stakeholders when analyzing the geography and physical infrastructure in the OA and when selecting the control option.

Geographic Combatant Commander (GCC) Option Selection and Design

Figure III-2 (following page) details a logical sequence that can be used by GCCs when evaluating, selecting, and designing the option they will use to control joint logistics. For more amplifying information detailing the joint logistics factors and enablers with regard to the staff and organization control

See following pages (pp. 4-24 to 4-25) for an overview and further discussion.

Joint Logistics

Geographic Combatant Commander (GCC) Option Selection and Design

Ref: JP 4-0, Joint Logistics (Feb '19), p. III-19 to III-21 and app. D.

JP 4-0, appendix E provides amplifying information detailing the joint logistics factors and enablers with regard to the staff and organization control options.

GCCs require visibility over the JLEnt to meet the command priorities. The factors below should be considered when the GCC is establishing the logistics control required by the JFC. These factors are not absolute nor all inclusive; but they do reflect the best practices observed in the field. These factors are applicable regardless of the control option selected by the GCC.

Centralized Joint Logistics Planning

This factor implies a capability to match joint logistics planning with the planning done during the execution of a mission.

Maintenance of Situational Awareness

This factor represents more than using radio signals and internet-based application data to track cargo movement (ITV). It involves elements such as the design and use of logistics situation reports and the building of ground truth in logistics input to the JFC's COP.

Adjudication of Conflicting Priorities

This factor is to have processes in place to identify conflicts when following the commander's priorities. For example, a reliable logistics input to the JFC's COP may provide the means to identify conflicts, and a fusion cell may provide the capability to adjudicate.

Timely Identification of Factors and Shortfalls

To meet this factor a process that links the logistics portion of the battle rhythm with the planning windows must exist.

Clear Understanding of Component Capabilities

This factor involves the building of databases that reflect current Service component and support agencies logistics capabilities. Fulfilling this factor may require liaison and physical presence of logisticians representing all appropriate Service components within the selected joint logistics control option.

Ability to Synchronize Components Capabilities

This factor matches the best capability, regardless of Service component, to the joint logistics need.

Integrated Logistics Processes

This factor is founded on the notion that the joint logistic staff comprehends the Service components logistic processes and uses this understanding to build the visibility required by the JFC to control joint logistics.

Integrated Distribution

This factor deals with the establishment of the JDDOC and its integration within the joint theater logistics construct. It maximizes the capabilities of the JDDOC to fill the seams between strategic and operational level deployment and distribution tasks. The JDDOC also strives to maximize and synchronize the use of common user land transportation and intratheater lift.

Cross Component Supply

This factor involves the establishment of CUL responsibilities and the processes required to achieve their execution.

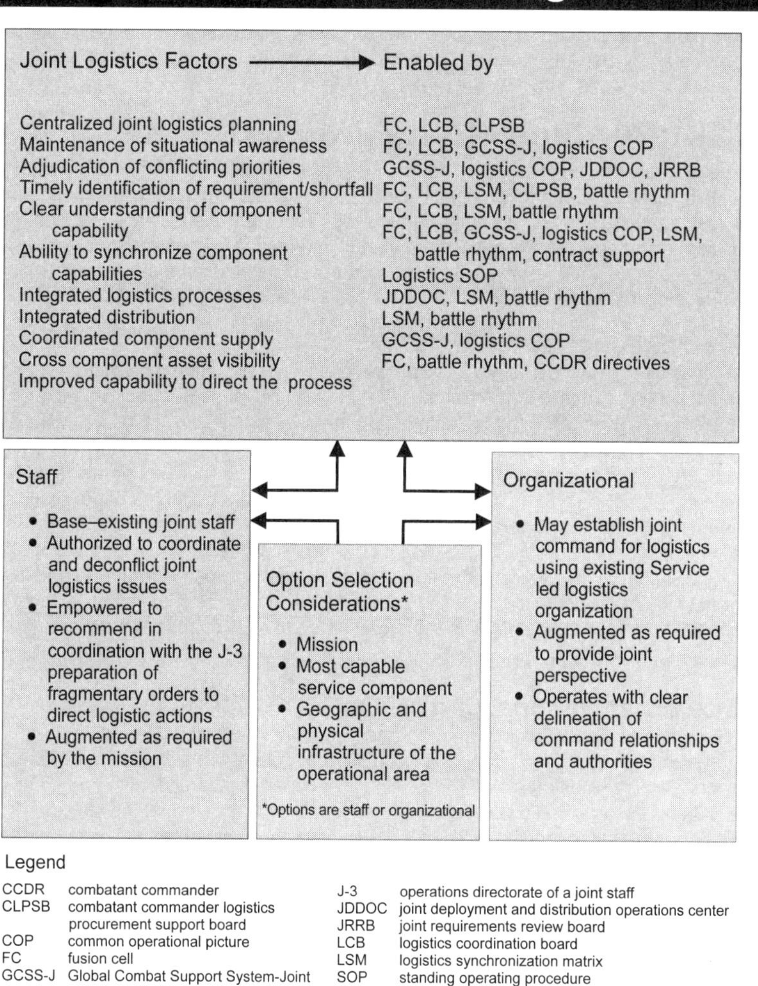

Ref: Figure III-2. Geographic Combatant Commander Option Selection and Design.

Cross Component Visibility

This factor refers to the ability for the Service components to see and understand assets available from other components.

Improved Capability to Direct the Process

This factor proposes the establishment of a decision-making process to direct logistics actions. These actions usually are directed in the form of further guidance to enhance the planning or assessment processes, or the publication of a FRAGORD to direct an action.

Designation of Contracting Construct

It is imperative that a detailed analysis of the OCS aspects of the OE be prepared to help shape COA development and determine the possible intended and unintended outcomes of OCS.

VI. Technology

The rapid advance of technology, if leveraged effectively, can enable the JFC to effectively control logistics within the OA. Technology, in the form of information systems, decision support tools, and communications capabilities, can improve visibility of logistics processes, resources, and requirements and provide the information necessary to make effective decisions.

VII. Interorganizational Cooperation

Interorganizational cooperation that results in operational arrangements regarding joint logistics are bound together by a web of relationships among global providers. These relationships are critical to joint logistics success because logistics capabilities, resources, and processes are vested in a myriad of organizations, which interact across multiple physical domains, the information environment, and span the range of military operations.

Multinational Operations

In today's OE, logisticians will likely be working with multinational partners. While the US maintains the capability to act unilaterally, it is likely that the requirement, and the desire, to operate with multinational partners will continue to increase. MNL is a challenge. However, leveraging MNL capabilities increases the CCDR's freedom of action. Additionally, many multinational challenges can be resolved or mitigated by having a thorough understanding of the capabilities and procedures of our multinational partners before operations begin. Integrating and synchronizing logistics in a multinational environment requires multinational information sharing, developing interoperable logistics concepts and doctrine, as well as clearly identifying and integrating the appropriate logistics processes, organizations, and C2 options. Careful consideration should be given to the broad range of MNL support structures.

See chap. 7 for further discussion.

Interorganizational and NGOs

Integration and coordination among military forces, NGOs, and international organizations are different from the coordination requirements of a purely military operation. These differences present significant challenges to coordination. First, NGO and international organization culture is different from that of the military. Their operating procedures will undoubtedly differ from one organization to another and with DOD. However, their similar needs (e.g., distribution, materials handling equipment, shelter, water, and power) in a contingency environment will add another requirement for resources that must be addressed early in any operation. Ultimately, some NGOs and international organizations may even have policies not in consonance with those of DOD. In the absence of a formal command structure, the joint logistician will need to collaborate and elicit cooperation to accomplish the mission. NGOs and international organizations possess unique skills and capabilities that can assist in providing the joint warfighter more robust logistics.

See chap. 8 for further discussion.

Logistics Support of US Government Organizations and Agencies

Logistics must be integrated at lowest echelons and are complicated by the creation of more support relationships of greater variety across Service lines and at lower levels. Likewise, multinational operations and support to interagency partners can complicate logistics by introducing a wider variety of potential partners. This complication is both the challenge and the solution, as it demands working with partners with a variety of requirements while also providing access to external resources and expertise. Partner logistics capabilities vary, as do their specific materiel resources, procedures, and information systems.

Chap 4

III. Planning Joint Logistics

Ref: JP 4-0, Joint Logistics (Feb '19), chap. IV.

Joint logistics planning provides the process and the means to integrate, synchronize, and prioritize joint logistics capabilities toward achieving the supported commander's operational objectives during all phases of plan development. This section is applicable to combatant command campaign plans (CCPs), subordinate campaign plans, campaign support plans, and contingency plans tasked in Chairman of the Joint Chiefs of Staff Instruction (CJCSI) 3110.01, (U) 2015 Joint Strategic Campaign Plan (JSCP) (commonly referred to as the JSCP), or as directed by the CCDR. This section also addresses planning considerations, input and output products used by joint logisticians to create OPLANs and operation orders (OPORDs) that enable transition from peacetime activities to execution of orders. Focus is on the JPP in development of the theater logistics overview (TLO) as a segment of the CCP.

Joint logistics planning is conducted under the construct of joint planning and the JPP addressed in JP 5-0, Joint Planning. Joint planning consists of planning activities associated with joint military operations by CCDRs and their subordinate commanders in response to contingencies and crises. It transforms national strategic objectives into activities by development of operational products that include planning for the mobilization, deployment, employment, sustainment, redeployment, and demobilization of joint forces and supporting contractors. Joint planning occurs at multiple strategic national and operation levels using process, procedures, tactics, techniques, and facilitating information technology tools/applications/systems aligned to the Joint Operation Planning and Execution System (JOPES) and the Adaptive Planning and Execution (APEX) enterprise.

The theater logistics overview (TLO) segment of the CCP articulates the overarching logistic architecture of the GCC's AOR. It is the start point of subsequent JPP logistics planning for regional OPLAN development and other contingencies.

I. Planning Functions

Joint planning encompasses a number of elements, including four planning functions: strategic guidance, concept development, plan development, and plan assessment. Depending upon the type of planning and time available, these functions can be sequential or concurrent. Joint planning features detailed planning guidance and frequent dialogue between senior leaders and commanders to promote a common understanding of planning assumptions, considerations, risks, COA, implementing actions, and other key factors. Plans may be rapidly modified throughout their development and execution. This process involves expeditious plan reviews and feedback, which can occur at any time, from SecDef and the CJCS. The intent is to give SecDef and the CCDR a mechanism for adapting plans rapidly as the situation dictates.

Integrated planning coordinates resources, timelines, decision points, and authorities across CCMD functional areas and AORs to attain strategic end states. Integrated planning produces a shared understanding of the OE, required decisions, resource prioritization, and risk across the CCMDs. JFCs and component commanders need to involve all associated commands and agencies within DOD in their plans and planning efforts. Moreover, planning efforts must be coordinated with other US Government department and agency stakeholders in the execution of the plan to assure unity of effort across the whole-of-government. The integrated planning process is the way the joint force will address complex challenges that span multiple CCMD

AORs and functional responsibilities. Integrated planning also synchronizes resources and integrates timelines, decision points, and authorities across multiple GCCs to achieve GEF-directed campaign objectives and attain contingency end states.

Supported CCDR

The supported CCDRs lead integrated logistics planning for their problem sets, inclusive of all associated plans related to the logistics problem both intertheater and intratheater. As such, supported CCDRs have coordinating authority for logistics planning. They lead the logistics planning process with all supporting CCMDs to develop a common understanding of logistics requirements, synchronize logistics planning activities, identify problem set logistics resource requirements, and provide logistics supportability analyses (quantitative and qualitative), as well as risk and supportability assessments associated with the plans. The supported commander designates and prioritizes objectives, timing, and duration of the supporting action. The supported commander ensures supporting commanders understand the operational approach and the support requirements of the plan. If required, SecDef will adjudicate competing demands for resources when there are simultaneous requirements amongst multiple supported CCDRs.

Supporting Commander

Supporting commanders will ensure their logistics planning is sufficiently integrated and synchronized across the problem set. They assist the supported CCMDs' efforts to develop a unified view of the logistics environment and synchronize resources, timelines, logistics C2, decision points, and authorities. The supporting commander determines the forces, tactics, methods, procedures, and communications to be employed in providing support. The supporting commander advises and coordinates with the supported commander on matters concerning the employment and limitations (e.g., logistics) of required support, assists in planning for the integration of support into the supported commander's effort, and ensures support requirements are appropriately communicated throughout the supporting commander's organization.

A. Strategic Guidance

The primary end product of the strategic guidance function and an in-progress review (IPR) is an approved CCDR's mission statement for contingency planning and a commander's assessment (operational report-3 pinnacle command assessment) or commander's estimate for crisis planning.

B. Concept Development

During concept development, if an IPR is required, the CCDR outlines COAs and makes recommendations to higher authority for approval and further development. Products from concept development include an approved mission statement, preliminary COAs, and prepared staff estimates. The CCDR recommends a COA for SecDef approval in the commander's estimate. The SecDef's approved COA from a concept development IPR is the basis for CONOPS.

C. Plan Development

This function is used to develop a feasible plan or order that is ready to transition into execution. This function fully integrates mobilization, deployment, employment, sustainment, conflict termination, redeployment, and demobilization activities through all phases of the plan. When the CCDR believes the plan is sufficiently developed, the CCDR briefs the final plan to SecDef (or a designated representative) for approval. Plan development solidifies the CONOPS and the OPLAN, concept plan (CONPLAN), or OPORD and required supporting documents are prepared.

Logistics Planning Integration (Strategic Guidance, Plans, & Operations)

Ref: JP 4-0, Joint Logistics (Feb '19), fig. IV-1, p. IV-5.

Using the JPP framework for planning, Figure IV-1 reflects the cascading relationship from strategic guidance and tasking to planning and developing OPORDs with a focus on CCP and associated key logistics area products. These key logistics area products, TLO, logistics estimate, and COLS support the CCP and provide the basis for plan and OPORD development. These products are key to the GCC's conduct of missions throughout the AOR. Figures IV-2 and IV-3 (following pages) reflect the joint logistics planning process combined with elements of the joint planning activities, functions, and products depicted in Figure IV-1 (below).

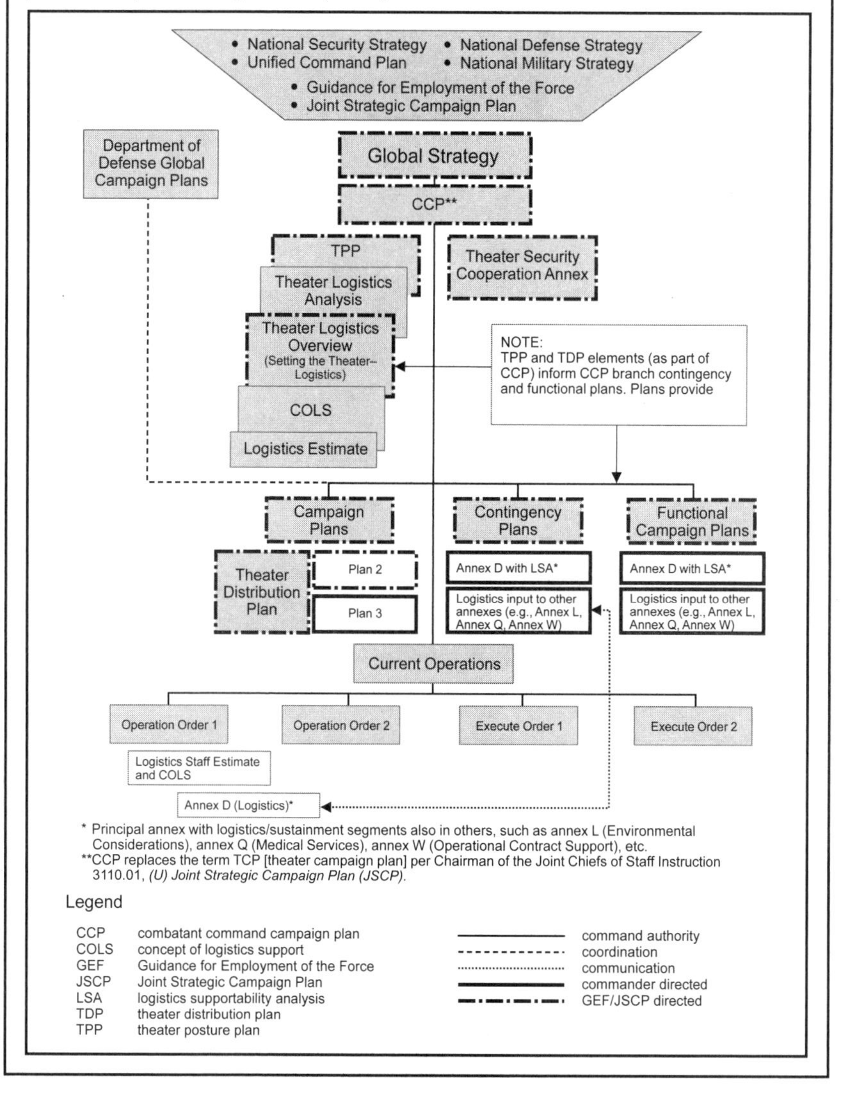

Joint Logistics

II. Joint Logistics Planning Considerations

Ref: JP 4-0, Joint Logistics (Feb '19), pp. IV-3 to IV-6 (figs. IV-2 and IV-3).

Figures IV-2 and IV-3 reflect the joint logistics planning process combined with elements of the joint planning activities, functions, and products depicted in Figure IV-1 (pervious page).

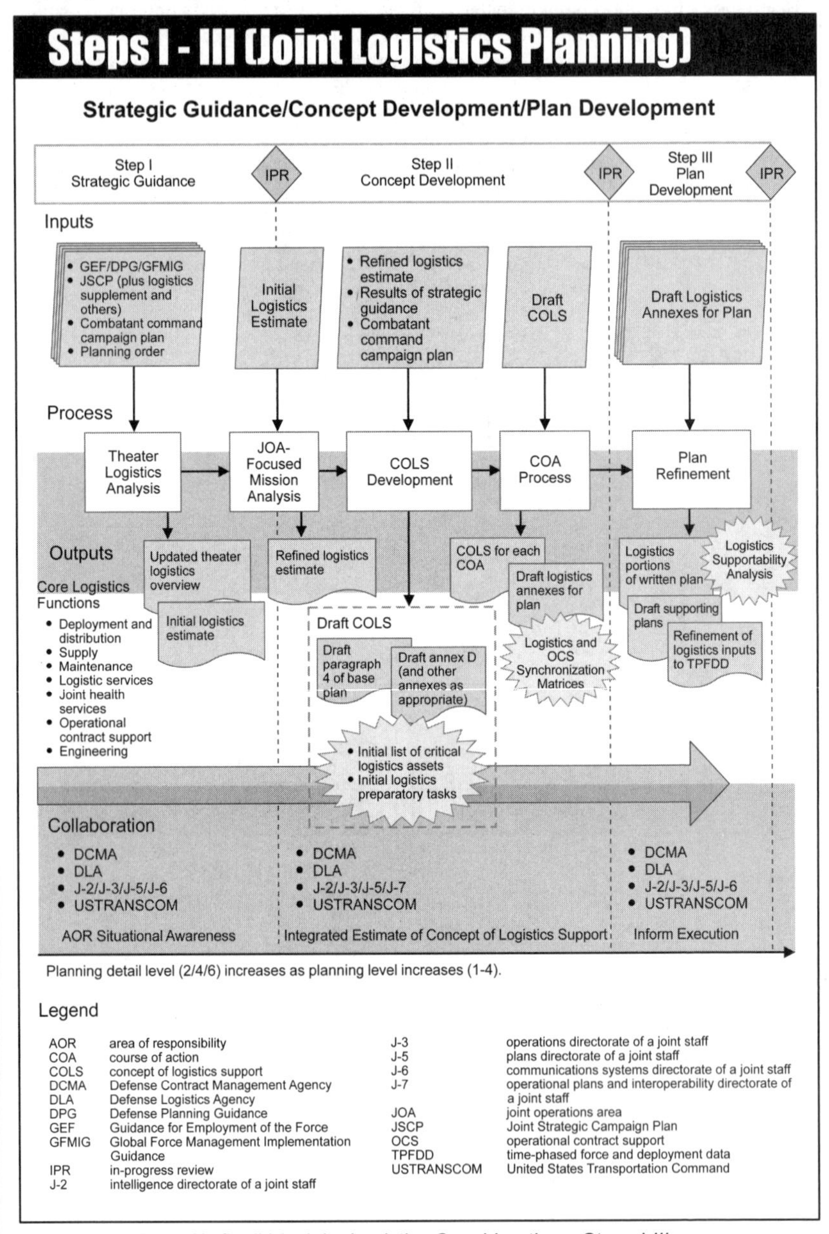

Ref: JP 4-0 (Feb '19), fig. IV-2. Joint Logistics Considerations, Steps I-III.

A means of anticipating future requirements is through the theater logistics analysis (TLA) process supporting TLO development and codification, logistics estimate, and logistics planning process Anticipating requirements is essential to ensuring responsiveness and determining adequacy of support. The purpose of the logistics planning process is to ensure the logistics facts, assumptions, information, and considerations are properly analyzed and effectively synthesized within an integrated plan that supports the CONOPS.

Steps III - IV (Joint Logistics Planning)

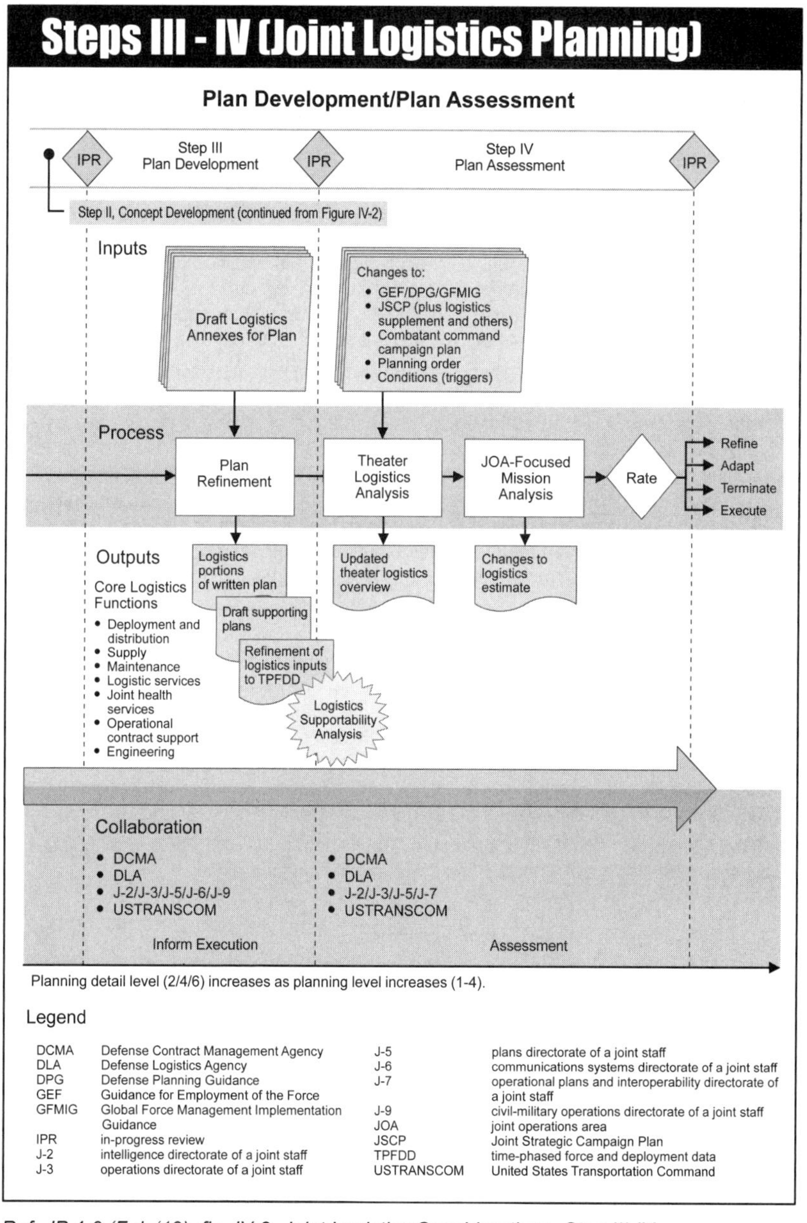

Ref: JP 4-0 (Feb '19), fig. IV-3. Joint Logistics Considerations, Step III-IV.

D. Plan Assessment

The joint planning and execution community continually reviews and evaluates the plan; determines one of four possible outcomes: refines, adapts, terminates, or executes; and then acts accordingly. Commanders and the joint planning and execution community continue to evaluate the situation for any changes that would require changes in the plan. The CCDR will brief SecDef during routine plan update IPRs of modifications and updates to the plan based on the CCDR's assessment of the situation, changes in resources or guidance, and the plan's ability to achieve the objectives and attain the end states.

E. Concept Development

This planning function includes the following JPP steps: COA development, COA analysis and wargaming, COA comparison, and COA approval. The staff, in coordination with supporting commands, Services, and agencies, develops, analyzes, and compares valid COAs and prepares staff estimates. The output is an approved COA. Critical elements include a common understanding of the situation, interagency coordination requirements, multinational involvement (if applicable), and capability requirements. Logistics planners must integrate planning efforts with operation planners, as deployment, redeployment, distribution, contracted support requirements, and sustainment requirements are an integral part of COA development.

The logistician must also identify requirements, critical logistics assets (CLAs), and services needed. A CLA is a logistics asset that is essential to completing key tasks that ensure mission accomplishment; if nonoperational or absent, it would have a seriously debilitating effect on the ability of a CCMD to execute their mission. The logistician must be aware of force structure planning, TPFDD development, and existing contracts and task orders, as well as the limitations of the OCS and JRSOI requirements. The logistician also uses this planning data during concept of support development to meet sustainment requirements from theater entry and operations to redeployment and reset. Logistics planners address all the core joint logistics functions.

During COA refinement, phasing of joint operations is done to ensure joint capabilities are available in the proper sequence to meet the operational requirements. Events drive phase changes, not time. Phasing helps the planning community visualize the entire operation to define requirements in terms of forces, resources, time, space, and purpose. The CCDR determines the number and nature of the phases during the operational design. Transitions between phases are designed to be distinct shifts in joint force focus and may be accompanied by changes in command relationships. Phase transition often changes priorities, command relationships, force allocation, or even the design of the OA, thereby creating new support challenges.

CCMD campaigns focus on shaping the OE to support the CCDR's overall objectives. Questions to ask when setting the theater can include:

- Do we have the right C2 and communications systems?
- Do we have access to critical infrastructure?
- Do we have a good theater distribution plan?
- Have we coordinated at the higher levels with the strategic partners (e.g., DLA and Army Materiel Command)?
- Have we properly positioned logistics assets at the tactical level?

Military engagement, security cooperation, and deterrence activities occur during shaping operations, seeking to improve cooperation with allies and other partners. These activities complement broad diplomacy and economic development in support of a friendly government's own security activities and setting the theater for major combat operations. Military engagement, security cooperation, and deterrence activities may also occur with new emerging governments and those previously considered as non-friendly to US national interests.

SOF are normally highly engaged during CCMD operations. Joint logistics planners must be aware of SOF requirements during day-to-day operations. SOF logistics support includes the sustainment and replenishment of all classes of supply, maintenance, transportation, joint health services, facilities, BOS, and services. Logistics support of SOF units is the responsibility of the parent Service, except where otherwise provided for by support agreements and/or directives. This may include Service support, joint in-theater support, nonstandard support, special operations-peculiar support.

Shaping offers logisticians the opportunity to expand knowledge of and access to additional capabilities in anticipation of future events. If it becomes apparent that an event will occur, the logistician can begin preliminary actions, such as pre-positioning (PREPO) of materiel, preparing organic or commercial JLEnt partners to surge capabilities, coordinating award of contracts (e.g., external and theater), and readying the assets to move on short notice. Shaping is a critical period to identify potential risks in terms of access, capabilities, and capacities so alternatives and mitigating measures can be developed. Planners must identify and assess critical infrastructure and installation needs and compare the results to current and programmed military construction requirements and authorities. Analysis of required logistics support for deployment and sustainment of flexible deterrent options may occur during this phase. This analysis must carefully balance joint logistics capabilities currently assigned, projected early joint deployers, and changes to OCS requirements.

F. Plan Development

During the plan development function, the CCDR's staff creates a detailed OPLAN, OPORD, or CONPLAN, with required annexes. The supported CCDR, subordinate commanders, supporting commanders, CSAs, and staff conduct a number of different planning activities, to include force planning, support planning, deployment planning, redeployment or unit rotation planning, shortfall identification, feasibility analysis, refinement, documentation, plan review and approval, and supporting plan development. Planning activities culminate in training and wargaming exercises to provide feedback on the planned concept of support. The joint logistics concept of support specifies how capabilities will be delivered over time, identifies who is responsible for delivering a capability, and defines the critical logistical tasks necessary to achieve objectives during all phases of the operation. Annex W (Operational Contract Support) is closely tied to the COLS since contracted support may fill critical operational and logistics capability gaps. The COLS encompasses joint capabilities of all force capabilities, to include multinational, HN, interagency partners, international organizations, NGOs, DOD OCS, plus Active Component and Reserve Component forces.

G. Plan Assessment

The supported commander extends and refines planning, while supporting and subordinate commanders and CSAs complete their support plans. Branch plans and other options may be developed. The CCDR and staff continually evaluate the situation for changes which trigger plan refinement, adaptation, termination, or execution. Additional means of assessing joint logistics planning are LSAs completed as appendix 4 (Logistics Supportability Analysis), annex D (Logistics) during plan development, Service component analysis, joint combat capability assessments-plans assessment, Global Logistics Readiness Dashboard, and Defense Readiness Reporting System assessments.

Guidance for development of an LSA is available in CJCSI 3110.03, (U) Logistics Supplement (LOGSUP) to the Joint Strategic Capabilities Plan (JSCP), and Chairman of the Joint Chiefs of Staff Manual (CJCSM) 3130.03, Adaptive Planning and Execution (APEX) Planning Formats and Guidance, provides the LSA format.

III. Joint Planning Process (JPP)

Joint planning is the overarching process that guides CCDRs in developing plans for the employment of military power within the context of national strategic objectives and national military strategy to shape events, meet contingencies, and respond to unforeseen crises. Logisticians provide key inputs, analysis, and assessments throughout the process. Logistics input is derived from mission analysis; COA development, analysis, and selection; and plan development, to include preparation and submission of LSA. Previously completed TLA, TLO (setting the theater-logistics), and COLS prepared for the CCP provide a foundational basis for complementary sections for tasked contingency plans. This foundation can also assist with transition to OPORD preparation for crisis execution under a plan and/or no-plan scenario.

See pp. 3-69 to 3-118 for complete discussion of the JPP from JP 5-0.

IV. Planning Levels

JP 5-0, Joint Planning, identifies four levels of planning detail and establishes a minimum level of effort for each.

See facing page (fig. IV-4) and following pages (pp. 4-36 to 4-37) for discussion of key logistics planning process outputs supporting and/or included in CCP development and execution planning.

A. Level 1 Planning Detail—Commander's Estimate

This level of planning involves the least amount of detail and focuses on producing multiple COAs to address a contingency. The product for this level can be a COA briefing, command directive, a commander's estimate, or a memorandum with a required force list.

B. Level 2 Planning Detail—Base Plan (BPLAN)

A BPLAN describes the CONOPS, major forces, concepts of support, and anticipated timelines for completing the mission. It does not normally contain annexes. Unless the CCDR opts to produce an annex D or the JSCP requires an annex D, there will be a paragraph 4 (Administrative and Logistics) only within the BPLAN summary. A BPLAN may contain alternatives, including flexible deterrent options, to provide flexibility in addressing a contingency as it develops or aid in developing the situation. Command logisticians should develop a Logistics Estimate (paragraph 4).

C. Level 3 Planning Detail—CONPLAN

This level is an abbreviated OPLAN with selected annexes and a CCDR's estimate of the plan's feasibility with respect to forces, logistics, and transportation. It will produce, if applicable, a COLS to include a "gross¬transportation-feasible" TPFDD, thus, the further delineation of a 3T plan (i.e., a CONPLAN or TPFDD). The COLS for CONPLANs or 3T plans will mirror the level of detail contained in the supported annex D. Appendix 4 to annex D provides the LSA for the plan. Level 3T plans (level three plan with TPFDD) and above require an annex W.

D. Level 4 Planning Detail—OPLAN

This plan requires a full description of the CONOPS, a complete set of annexes, and a TPFDD. Figure IV-4 (facing page) depicts logistics planning products by level of plan. Within the JPP, key logistics outputs are OPORD TLO, logistics estimate supporting development of the commander's estimate, and COLS. The COLS further supports annex D plans and OPORDs. Appendix 4 to annex D provides the LSA for the plan. In terms of operations execution, logistics supportability is addressed and status update reported in the JFC's situation report (SITREP) per CJCSM 3150.05, Joint Reporting System Situation Monitoring Manual. Logistics input to the SITREP provides shared

Likely Expected Logistics Outputs to JPP

Ref: JP 4-0, Joint Logistics (Feb '19), pp. IV-13 to IV-14.

Figure IV-4, below, depicts logistics planning products by level of plan. Within JPP, key logistics outputs are OPORD TLO; logistics estimate supporting development of the commander's estimate and COLS. The COLS further supports annex D for deliberate plans and OPORDs. Appendix 4 to annex D for deliberate plans provides the LSA for the plan. In terms of operations execution, logistics supportability is addressed and status update reported in the JFC's situation report (SITREP) per CJCSM 3150.5, Joint Reporting System Situation Monitoring Manual.

Likely Expected Logistics Outputs

Plan Level	Strategic Guidance	Concept Development	Plan Approval	Plan Review
Level 1 "Commander's Estimate"	TLO, ILE, and RLE (briefing)			Δs to TLO and RLE (briefing)
Level 2 Base Plan	TLO, ILE, and RLE (briefing)	Paragraph 4 (written and briefing)		Δs to TLO and RLE (briefing)
Level 3 Base Plan with Select Annexes	TLO, ILE, and RLE (briefing)	Paragraph 4, Annex D, logistics enablers, preparation tasks, COLS, and LSM (written and briefing) Annex Q		Δs to TLO and RLE (briefing)
Level 3 with TPFDD		Transportation feasible TPFDD, Annex W, CSSM		
Level 4 Base Plan with Annexes and Detailed TPFDD	TLO, ILE, and RLE (briefing)	Paragraph 4, Annex D, Annex W, logistics enablers, preparation tasks, COLS, LSM, and CSSM (written and briefing) Annex Q	Logistics portions of plan, draft supporting plans, logistics inputs to TPFDD, and LSA (written and briefing)	Δs to TLO, RLE, COLS, and LSA; status of supporting plans (briefing)

Legend

Δ	change
COLS	concept of logistic support
CSSM	contracted support synchronization matrix
ILE	initial logistics estimate
LSA	logistics supportability analysis
LSM	logistics synchronization matrix
RLE	refined logistics estimate
TLO	theater logistics overview
TPFDD	time-phased force and deployment data

Ref: JP 4-0 (Feb '19), fig. IV-4. Likely Expected Logistics Outputs.

Logistics input to the SITREP assists in providing shared situational awareness and visibility within and across echelons of command to address the core logistics functions, force and sustainment tracking, JRSOI supporting declaration of force closure for operational employment, and other conditions that increase, or materially detract from, the adaptability and readiness of forces.

V. Key Logistics Planning Process Outputs

Ref: JP 4-0, Joint Logistics (Feb '19), pp. IV-13 to IV-16.

A. Theater Logistics Analysis (TLA)

The TLA is a supporting process facilitating development of the TLO through examination, assessment, and codification of an understanding of current conditions of the OE. Analysis determines infrastructure, logistics assets/resources, and environmental factors in the OE that will optimize or adversely impact means for supporting and sustaining operations within the theater. To facilitate developing the TLA, logistics planners leverage all interactions with PN logistics professional counterparts (e.g., during multinational exercises logistics planning and execution) to capture insights into their capabilities, processes, and policies by writing and distributing detailed after action reports (AARs). To effectively share best practices and lessons learned from logistics operations across DOD, observations, insights, and AARs should be entered into the Joint Lessons Learned Information System (JLLIS). Entry into JLLIS facilitates awareness of issues and may provide solutions to logistics planning issues. The TLA provides a detailed country-by-country analysis of key infrastructure by location or installation (main operating base/forward operating site/cooperative security location); footprint projections (including contingency locations); HN agreements; and available contracted support capabilities, existing contracts, and task orders to logistically support the theater during peacetime through contingency operations. Work completed supports TLO development as a segment of the CCP and development of directed plans and OPORDs. Information and data collected and codified during the TLA process are the basis for analysis which assists in identifying, resolving, and/or mitigating risk associated with theater shaping operations. Additionally, the TLA provides the framework for conceptual planning, which involves understanding the OE and the problem, determining the operation's end state, and visualizing an operational approach. Using the TLA, the operational approach is initially addressed in a logistics estimate and transitions to culminate in the TLO. Detailed planning works out the scheduling, coordination, or technical problems involved with moving, sustaining, and synchronizing the actions of force as a whole to achieve objectives. Effective planning requires the integration of both the conceptual and detailed components of planning. The TLA assists in improving the JFC's situational awareness and understanding of theater logistics support capabilities and readiness to support/execute theater operations.

B. Theater Logistics Overview (TLO)

The TLO is a segment of the iterative planning process which addresses identification, understanding, and framing the theater's mission at the campaign level, not for a specific operation. The TLO uses TLA information to inform decisions about the approaches to be used for sourcing and distribution of logistics support for theater operations. Having captured influencing elements in the TLA as a frame, the JFC's logistics staff elements develop and codify an overarching approach to theater operations in the TLO. The TLO then serves as an important link between conceptual planning and the detailed planning tasked in the GEF/JSCP. Additionally, the TLO helps the JFC and operations and logistics staff segments measure the overall effectiveness of employing forces, force sustainability, and logistics capability readiness to ensure that the operational approach remains feasible and acceptable. As such, the TLO is key to help identify and address capability gaps, risk mitigations, and residual risk. If risk cannot be resolved or mitigated to an acceptable level then the operational concept may be reframed. Reframing involves revisiting earlier COAs, conclusions, and decisions that underpin the current operational CONOPS. Reframing can lead to a modification of the current CONOPS or result in preparation of a branch plan or entirely new plan. In developing the TLO, logistics planners, in coordination with intelligence and operations staff segments, identify opportunities/initiatives by anticipating events. This allows them to identify decision points to operate inside the threat's decision cycle or to react promptly to deteriorating situation advancing beyond shaping operations. Time to complete the TLA and resulting TLO assists in optimizing available planning time for associated detailed plans. Based on their understanding and learning gained during TLO development, the JFC and senior logistics staff representative issue logistics planning guidance to support and enable the operational approach expressed in the CONOPS and to guide more detailed planning.

Refer to JP 4-0, Appendix A, for an example of a TLO format.

C. Logistics Estimate

Logistics estimate supports the commander's estimate, COLS, OPORD development, and execution. Execution planning may involve abbreviated and compressed timelines from situational awareness/initiating event and reporting to potential JFC planning guidance or CJCS planning order to OPORD and execution. The TLA and TLO provide a foundation for rapid review and response development. Due to accelerated timelines, availability, and incorporation of TLA information and TLO segments, preparation of the logistics estimate may be compressed supporting the commander's estimate and initial work for COA development, analysis, and selection. Updating the TLA/TLO baseline, the logistics estimate supporting the commander's estimate informs the COLS and OCS concept prepared for OPORD annex D and annex W development and iterative planning during operations execution. The logistics estimate is an analysis of how CSS factors can affect mission accomplishment. It contains the logistics staff's comparison of requirements and capabilities, conclusions, and recommendations about the feasibility of supporting a specified COA. This estimate includes how the core logistics functions affect various COAs. Preparation of the logistics estimate provides a coordinated and formalized means for the staff to identify and consider logistics shaping in support of the operational CONOPS. Planners should evaluate the feasibility of OPLANs in light of strategic lift capabilities and limitations. The logistics effort and development of the logistics estimate refined as COLS for OPORD annex D must be integrated into the JPP and OPORD development upfront. Using the TLA/TLO baseline, logistics staff segments will be able to identify if specific operational actions to augment or expand theater logistics capabilities to support the operational CONOPS must be taken. The previously developed TLA/TLO assists the logistics planners in providing logistics characteristics of the AOR and area of operations/area of interest for the specified operations. The TLA/TLO aids planners in identification of logistics infrastructure of the OE (what exists in the OA that may be put to use).

D. Concept of Logistic Support (COLS)

In support of the CCDR and preparation of plans/OPORDs, the logistics staff elements prepare a logistics estimate which is further refined and developed into a COLS. The COLS provides a foundational basis in preparation of annex D for assigned contingency plans and/or OPORD development tasks. The COLS establishes priorities of support across all phases of operations to support the JFC's CONOPS. Logistics staff elements' active participation within and across JPP activities at all echelons facilitates CONOPS and associated COLS development. A COLS addresses the sustainment of forces, to include identification and status of contingency basing. Through exercising DAFL, the CCDR may assign a component commander with the responsibility for conducting various theater logistics functions, as well as base support at designated theater locations. Logistics functions may include management of afloat assets; identification and status of theater sustainment elements, to include identification and/or forecast of required augmentation; priority of sustainment by class of supply with guidance on days of supply to be maintained (minimum and maximum); movement priorities for airlift and sealift aligned to JFC's CONOPS; guidance for employment of sea-air interfaces to facilitate JRSOI; controlling CUL; JFC's declaration of force closure; actions by phase; logistics assets required; and designation of contracting construct (e.g., lead Service for contracting [LSC], joint theater support contracting command [JTSCC]).

For more information on the COLS, refer to CJCSI 3110.03, (U) Logistics Supplement (LOGSUP) for the 2015 Joint Strategic Capabilities Plan (JSCP).

situational awareness and visibility within and across echelons of command to address the core logistics functions, force, and sustainment tracking; JRSOI supporting declaration of force closure for operational employment; and other conditions that increase, or materially detract from, the adaptability and readiness of forces.

VI. Transition to Execution

Planning does not cease with development, submission, and approval of a plan or OPORD. Planning is iterative and continues throughout as actions and assessments evolve in a dynamic manner across command echelons from the strategic national to operational to tactical levels. Strategic guidance for plans, as well as plan segments and resulting OPORDs, is refined as situational awareness and understanding evolves. Through assessment, guidance and/or plans may be reframed. Assessment is a determination of the progress toward accomplishing a task, creating an effect, or achieving an objective. Assessment is a continuous activity to support the operation process and associated planning and execution activities. During planning, assessment focuses on understanding current conditions of an OE and assumptions to address mission, enemy, terrain and weather, troops and support available, time available, and civil considerations. During preparation, assessment focuses on determining force readiness to execute the operation and verifying the assumptions on which the plan is based. During execution, assessment focuses on evaluating progress of the operation. Based on their assessments, commanders at various echelons direct adjustment to the plan/OPORD ensuring the plan/OPORD/operation stays focused on mission.

VII. Sustainment Distribution Planning and Management Process

USTRANSCOM's sustainment distribution planning and management process supports its JDDC role by enhancing the JDDE ability to ensure an agile, scalable, and resilient distribution network. Sustainment distribution planning and management provides the JDDE with a suite of five capabilities: distribution lane validation, distribution workload forecast/demand planning, advanced air route planning, strategic surface route plan, and sustainment distribution plans. Sustainment distribution plans are codified in USTRANSCOM's Campaign Plan for Global Distribution 9033.

For more information on distribution planning, refer to JP 4-09, Distribution Operations.

Chap 4

IV. Executing Joint Logistics

Ref: JP 4-0, Joint Logistics (Feb '19), chap. V.

The term "executing joint logistics" is used to describe actions and operations conducted by joint logistics forces in support of the JFC mission. Force reception, theater distribution, and MA are examples of joint logistics operations. Since joint logistics operations span the strategic, operational, and tactical levels, the transition from planning to execution is critical.

I. Essential Elements for Joint Logistics Execution

A. Organizing for Execution

The CCMD J-4 monitors, assesses, plans, synchronizes, and directs logistics operations throughout the theater. This transition may occur through the directed expansion of the JLOC and/or the CCDR's JDDOC. The CCDR's or JFC's staff is augmented (either physically or virtually) with representatives from Service components, USTRANSCOM, other supporting CCDRs, CSAs, and other national partners or agencies outside the command's staff. For example, each GCC has established a JDDOC to synchronize and optimize the flow of arriving forces and materiel between the intertheater and intratheater transportation. As the operating tempo increases during a contingency or crisis, additional joint logisticians and selected subject matter experts (e.g., maintenance, ordnance, supply) can augment JDDOCs and use established networks and command relationships instead of creating new staffs with inherent startup delays and inefficiencies. This expanded organization must be organized and situated to ensure increased coordination and synchronization of requirements in the deployment and distribution process. This organization must have clear roles and responsibilities between the various elements and clearly understood relationships between the logistics elements and the CCMD staff.

B. Expeditionary Capabilities

The joint logistician should understand the expeditionary theater opening capability options available to the commander. Expeditionary theater opening capabilities provide GCCs critical initial actions for rapid insertion/expansion of force capabilities into an OA that directly affects the JFC's ability to expand and adjust force flow to allow flexible, agile response to asymmetric and dynamic operational requirements. Expeditionary theater opening capabilities support the first critical OA entry missions with the eventual transition of theater port of debarkation (POD) operations to a JFC-designated Service component and establish conditions to facilitate the arrival of larger Service theater distribution and sustainment forces where/when appropriate.

C. Technology and Communications

Logisticians use a variety of automated tools to assist in planning and execution. Effective execution of logistics plans requires a robust data communications architecture. Planning should anticipate communications in degraded environments at all levels and phases of operations and include considerations for alternate routing, redundant systems, use of other systems, protocols, and message standards. These degradations may be imposed by the threat, the environment, by the JFC as part of operational security, or a combination of all of them. Sustained impaired/ inadequate information exchange capability must be anticipated and incorporated into risk management considerations during logistics operations planning and execution.

D. Achieving Situational Awareness

A role of the joint logistician is to provide situational awareness of the current logistics posture to support the JFC in making decisions and disseminating and executing directives. Maintaining situational awareness requires visibility of the status and location of resources. This includes status of existing contracts and task orders over the current and future requirements of the force and over the joint and component processes that deliver support to the joint force. In order to provide this visibility, timely, and accurate data and information are required for all equipment, sustaining supplies, repair parts, munitions, fuel and etc., moving into, within, exiting, or being stored in the GCC's AOR. This kind of visibility is the key to continuously monitoring progress and is enabled by operational inputs which serve to inform joint logisticians about the current situation. Service reports, operational summaries, logistics SITREPs, and HN reports all serve to expand the joint logistician's awareness of the JOA.

E. Battle Rhythm

The JFC will establish a battle rhythm for the operation along with mechanisms to establish and maintain visibility for all functional areas, to include logistics. The joint logistician must develop a supporting battle rhythm for the sustainment staff that supports the JFC's battle rhythm and is designed to provide proactive logistics options. Synchronizing logistics reporting with operational updates, ensuring that the operational planning cycle is part of the logistics battle rhythm, and minimizing shift changes at critical points in the battle rhythm will enable more effective execution. Additionally, tying the component logistics elements to the JFC's battle rhythm will provide more accurate and timely situational awareness and promote better integrated support to the joint force.

F. Joint Logistics Boards, Offices, Centers, Cells, and Groups

The joint logistician will often use boards, centers, or other organizations to assist the J-4 staff in executing joint logistics operations, by prioritizing and/or allocating resources, controlling functions, or prioritizing requirements.

See p. 4-17 for an overview and further discussion.

G. Execution Synchronization

A synchronization matrix or decision support tool/template can establish common reference points to help assess the progress of an operation. Joint logisticians may use a matrix to display progress against actual execution and recommend adjustments as needed. A logistics synchronization matrix is built around the concept of the operation and normally contains the phasing of the operation over time along the horizontal axis. The vertical axis normally contains the functions the joint logistician integrates into a concept of support. The body of the matrix contains the critical tasks, arrayed in time and linked to responsible elements for execution. This decision support tool enables logisticians to graphically display the logistics concept of support, see potential gaps, develop options to mitigate those gaps, and respond to a changing OE.

H. Commander's Critical Information Requirements (CCIRs)

CCIRs are elements of friendly and enemy information the commander identifies as critical to timely decision making. Joint logisticians update the critical information requirements related to logistics. Joint logisticians will most often use friendly forces information requirements to guide decision making. Those requirements are often a direct reflection of resources (force availability, unit readiness, or materiel availability).

II. Joint Logistics Execution

Ref: JP 4-0, Joint Logistics (Feb '19), pp. IV-1 to IV-3.

JFCs adapt to evolving mission requirements and operate effectively across a range of military operations. These operations differ in complexity and duration. The joint logistician must be aware of the characteristics and focus of these operations and tailor logistics support appropriately. This range of military operations extends from shaping activities to major operations and campaigns. US and multinational partners collaborate to expand mutual support and leverage capabilities to quickly respond to future contingencies.

Military Engagement, Security Cooperation, and Deterrence

The GEF directs development of CCPs focused on current operations, military engagement, security cooperation, deterrence, and other shaping or preventative activities. Specific issues that can be addressed in the CCMD campaigns include securing interagency approvals; addressing PN and regional sensitivities, changing politics, and overall stability; determining optimal presence and posture; BPC; and developing formal agreements/permissions between the US and PNs. Developing mutually supportive relationships to enhance coordination is an important enabler for joint logistics operations. ACSAs are bilateral international agreements that allow for the provision of cooperative logistics support under the authority granted in Title 10, USC, Sections 2341-2350. They are governed by DODD 2010.9, Acquisition and Cross-Servicing Agreements, and implemented by CJCSI 2120.01, Acquisition and Cross-Servicing Agreements.

Crisis Response and Limited Contingency Operations

US military history indicates crisis response and limited contingency operations are typically single, small-scale, limited-duration operations. Many of these operations involve a combination of military forces, the private sector, and capabilities in close cooperation with other US Government departments and agencies, international organizations, and NGOs. Logisticians must understand multinational, private-sector, and interagency logistics capabilities and coordinate mutual support, integrating them into the joint operation when appropriate. Many crisis response missions, such as foreign humanitarian assistance and disaster relief operations, require time-sensitive sourcing of critical commodities and capabilities, and rapid delivery to the point of need. In these operations, joint logistics is often the main effort, often operating in support of the Department of State.

Major Operations or Campaigns

Major operations or campaigns typically involve the deployment, sustainment, redeployment, and retrograde of large combat forces. Joint logistics can be executed by an appointed lead Service or agency for CUL. Joint logisticians develop support plans for the duration of the operation, as well as the return of personnel and equipment to CONUS or other locations. These plans often leverage contractor support to augment Service logistics capabilities. The primary challenges for logisticians during these types of operations are identifying the requirements, ensuring logistics issues are considered among competing priorities and adjusting to the situation to ensure sustained readiness and synchronized timelines as the operation transitions across phases. Logistics plans must account for and have the flexibility to mitigate the impact of CBRN-contaminated APODs and SPODs on force flow. This includes identifying locations for transload and exchange zone operations. A critical planning requirement during any operation is to plan for the transition to the final phase, where logisticians will have competing requirements to support stability activities, provide basic services while conducting contract closeout and changes to the contractor management plan, support foreign humanitarian assistance, and assist with reconstruction efforts all while conducting movement of redeploying forces and equipment. The retrograde of contaminated materiel will require special handling to control contamination and protect the force and mission resources.

III. Joint Logistics Assessment

Assessment is an integral part of planning for and execution of any operation, fulfilling the necessary requirement for analyzing changes in the OE and determining progress of an operation. Assessment activity involves the entire staff and other sources such as higher and subordinate headquarters, interagency and multinational partners, and other stakeholders. Logisticians not only feed assessment data to the commander to determine progress towards objectives but also assess the adequacy of logistics support, making adjustments to the logistics plan as required. Logisticians collect information from both the end-user and service providers to adjust and improve logistics support. Assessment is a continuous process throughout a campaign or operation that measures the overall effectiveness of employing joint force logistics capabilities. It involves monitoring and evaluating the current situation and progress of logistics support toward mission completion and requires input from not just the logistician but the end-user and JLEnt membership as well.

IV. Terminating Joint Operations

Terminating joint operations is an aspect of the CCDR's strategy that links to achievement of national strategic objectives. The supported CCDR can develop and propose specified conditions approved by the President or SecDef that must be met before a joint operation can be concluded. These termination criteria help define the desired military end state, which normally represents a period in time or set of conditions beyond which the President does not require the military instrument of national power as the primary means to achieve remaining national objectives.

A. Concluding Joint Logistics Operations

Joint logistics operations are always ongoing, but it is possible that some aspects of logistics operations could be completed before the operation has concluded. For example, force reception operations could be completed when forces have been placed under the control of the commander for integration and employment, and no other forces are flowing into the JOA. Joint logisticians monitor transitional activities and ensure resources are fully utilized or redeployed. Withdrawal and redeployment from an operation are challenging and require a synchronized and holistic effort by joint logisticians. Maintenance support planning should address the process for determining equipment disposition and the requirements for preparing equipment for shipment. In addition, maintenance support planning should ensure that equipment is available for movement when required while minimizing the impact on readiness. In accordance with DOD policies, logisticians plan for the disposition of materiel, such as retrograde and demilitarization, scrap removal, and disposal of hazardous waste, and, when required, clearance decontamination of supplies and equipment.

B. Theater Closure

When it has been determined that joint operations should be terminated, joint logistics operations focus on tasks that include redeploying personnel and materiel from the JOA to a new OA or home station/demobilization station, departure of contractor personnel, disposal of equipment, transitioning materiel and facilities to HN, foreign military sales, or disposal of materiel. Joint logistics operations also play a major role in closing ports to military operations and terminating operational contracts and agreements. Plans should be developed to monitor or assist the retrograde of contractor equipment and personnel. DOD must receive back any government-furnished property loaned to contractors as part of their mission. Operational contracts and agreements are not considered closed out until the force has confirmation of receipt of all goods and services and full payment has been made. Contracting and payment officials should not redeploy until all contracts and agreements are closed out.

Chap 5

Joint Task Forces (JTF)

Ref: JP 3-33, Joint Task Force Headquarters (Jan '18), chap. I.

A joint task force (JTF) is one of several command and control (C2) options for conducting joint operations. A JTF may be established when the scope, complexity, or other factors of the operation require capabilities of Services from at least two Military Departments operating under a single joint force commander (JFC). The size, composition, capabilities, and other attributes will vary significantly among JTFs based on the mission and various factors of the operational environment (OE), such as the threat, the geography of the joint operations area (JOA), the nature of the crisis (e.g., flood, earthquake), and the time available to accomplish the mission. CJTFs typically function at the operational level and employ their capabilities throughout the JOA.

See pp. 1-23 to 1-30 for discussion of considerations for establishing joint forces.

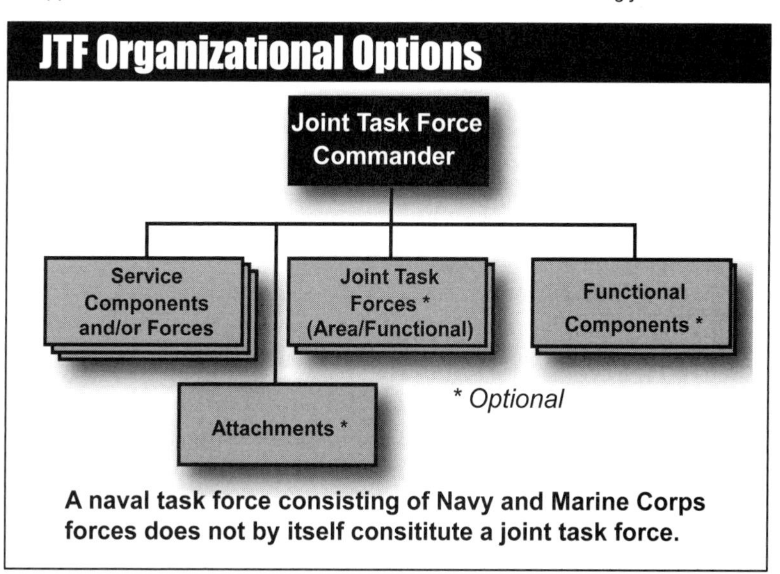

Ref: JP 3-33, fig. I-2. Joint Task Force Organizational Options.

Operational JTFs are the most common type of JTF and is established in response to a Secretary of Defense (SecDef)-approved military operation or crisis.

Contingency JTFs are identified and designated to support operation plan execution or specific on-call missions, such as national special security events, on a contingency basis.

A **Standing JTF** is a JTF originally established as an operational JTF, but that has an enduring mission that is projected to continue indefinitely.

A **combined JTF** is a multinational JTF the commander commands from a multinational and joint headquarters (HQ).

A National Guard (NG) **joint force headquarters-state** (NG JFHQ-State) liaison officer (LNO) provides C2 of all NG forces in the state for the governor and can act as a joint HQ for national-level response efforts during contingency operations

Joint Task Force Establishing Authority Responsibilities

Ref: JP 3-33, Joint Task Force Headquarters (Jan '18), fig. I-1, pp. I-1 to I-2.

A CJTF has authority to assign missions, redirect efforts, and require coordination among subordinate commanders. Unity of command, centralized planning and direction, and decentralized execution are key considerations. Generally, a CJTF should allow Service tactical and operational groupings to function as they were designed. The intent is to meet the CJTF's mission requirements while maintaining the functional integrity of Service components. A CJTF may elect to centralize selected functions within the joint force, but should strive to avoid reducing the versatility, responsiveness, and initiative of subordinate forces.

- Appoint the commander, joint task force (CJTF), assign the mission and forces, and exercise command and control of the joint task force (JTF)
  - In coordination with the CJTF, determine the military forces and other national means required to accomplish the mission
  - Allocate or request forces required
- Provide the overall mission, purpose, and objectives for the directed military operations
- Define the joint operations area (JOA) in terms of geography or time. (Note: The JOA should be assigned through the appropriate combatant commander and activated at the date and time specified)
  - Provide or coordinate communications, personnel recovery, and security for forces moving into or positioned outside the JOA thus facilitating the commander's freedom of action.
- Develop, modify as required, and promulgate to all concerned rules of engagement and rules for the use of force tailored to the situation.
- Monitor the operational situation and keep superiors informed through periodic reports
- Provide guidance (e.g., planning guidelines with a recognizable end state, situation, concepts, tasks, execute orders, administration, logistics, media releases, and organizational requirements)
- Promulgate changes in plans and modify mission and forces as necessary
- Provide or coordinate administrative and sustainment support
- Recommend to higher authority which organizations should be responsible for funding various aspects of the JTF
- Establish or assist in establishing liaison with US embassies and foreign governments involved in the operation
- Determine supporting force requirements
  - Prepare a directive that indicates the purpose, in terms of desired effect, and the scope of action required. The directive establishes the support relationships with amplifying instructions (e.g., strength to be allocated to the supporting mission; time, place, and duration of the supporting effort; priority of the supporting mission; and authority for the cessation of support).
- Approve CJTF plans
- Delegate directive authority for common support capabilities (if required)

I. Forming & Organizing the Joint Task Force HQ

Ref: JP 3-33, Joint Task Force Headquarters (Jan '18), chap. II.

JTFs are established by SecDef, a CCDR, subordinate unified commander, existing JTF commander, or a state governor (for JTF-State) in accordance with the National Security Act of 1947, Title 10, USC, and Title 32, USC. Establishing authorities for subordinate unified commands and JTFs normally direct the delegation of OPCON over forces attached to those subordinate commands. A JFC has the authority to organize assigned or attached forces with specification of OPCON to best accomplish the assigned mission based on commander's intent, the CONOPS, and consideration of Service organizations.

I. Forming the Joint Task Force Headquarters

The preferred approach to forming a JTF HQ is around an existing C2 structure. Typically, this is a CCMD's Service component HQ or a subordinate Service component HQ. The establishing authority (typically the CCDR) determines the appropriate level based on the scope and scale of the operation and nature of the mission.

A. Building Upon an Existing Service HQ

DOD relies primarily on Service component HQ to adapt with little or no notice into a JTF HQ, often under crisis planning conditions. However, the newly designated JTF HQ typically requires additional resources that are not organic to the core Service HQ. Examples include joint C2 equipment and training, regional language and cultural experts, an operational contract support integration cell (OCSIC), augmentation from the JTF's Service components, and CSA elements. The CJTF and staff should plan for the time required to integrate new personnel and capabilities and accommodate other military and nonmilitary liaison personnel and the private sector expected to be involved in the operation. Once the JTF HQ is established, it takes time to receive, train, and integrate new members and then to begin functioning as a cohesive HQ with common processes, standards, and procedures. To mitigate these challenges, CCDRs may designate one or more Service HQ from their assigned forces to become "JTF-capable" HQ and integrate them into CCMD planning efforts and joint exercise programs. A designated Service HQ can be considered a JTF-capable HQ if it has achieved and can sustain a level of readiness to transition, organize, establish, and operate as a JTF HQ and is acceptable to the establishing authority (CCDR).

B. Tasking an Existing JTF HQ

In some situations, the establishing authority could task an existing JTF with the mission rather than establish a new joint force. Many of the organizational factors mentioned before apply to this option.

C. Build a New JTF HQ

The optimal sourcing method for a new JMD is the identification of a core unit around which to build the JTF HQ, submitting RFFs as necessary for skills/capabilities resident in the core unit (or units), and finally augmenting the JTF HQ with JIAs for subject matter expertise or unique experience not inherently resident in the core unit or the enabling units contributing to the JTF HQ's JMD. Sourcing a JTF HQ from a new JMD would only be used in a narrow set of circumstances where an existing Service HQ is not available, nor is there an existing JTF HQ

II. Options for Augmenting the Headquarters

Augmentation of the JTF HQ is a function of both the CJTF's mission and the JTF's force composition. The CJTF's mission is the most important factor in determining the required type of core staff augmentation. Mission analysis should consider the JTF HQ-required capabilities and other related functions.

The following factors are among many that the new CJTF and staff must consider as they determine augmentation requirements and sources.

- Current staffing level of the designated HQ
- Linguist and Interpreter Support
- Interagency Requirements
- Multinational Involvement
- Sustainment
- Liaison Requirements

The JTF composition is a key factor that affects the type of augmentation the core staff should receive. Generally, the JTF staff should be representative of the force composition as to numbers, experience, and influence of position and rank of members among the JTF's Service and functional components. The CJTF should also consider whether and how to represent supporting commands and MNFs in the HQ. However, HQ composition is more an issue of having relevant expertise in the right positions to ensure the most effective employment of the JTF's capabilities in the context of the mission than of having equal component representation.

A. Joint Enabling Capabilities Command (JECC)

JECC is a subordinate command of USTRANSCOM and provides global, rapidly deployable, temporary joint expeditionary capabilities across the range of military operations to assist in the initial establishment, organization, and operation of joint force HQs; fulfill global response force execution; and bridge joint operational requirements. Its joint capability packages are mission-tailored plans, operations, logistics, knowledge sharing, intelligence, communications, and public affairs (PA) capabilities:

- **Communication capabilities** include deployable, en route, early entry, and scalable command, control, communications, computer, and combat systems services to CCDRs and other agencies as directed to facilitate the rapid establishment of a joint force HQ, bridge system requirements, and provide reliable and interoperable communications that link the CJTF and staff to the President and SecDef, geographic combatant commanders (GCCs), their component HQ, and multinational partners. The tactical communication packages vary in capability from small initial-entry and early-entry teams to a significantly larger deployable joint C2 system. Packages can support operations worldwide, as well as in homeland defense (HD) and DSCA missions.
- **Public Affairs (PA) capabilities** provide ready, rapidly deployable, combat ready joint PA capability to CCDRs to facilitate the rapid establishment of a joint force HQ, bridge joint PA requirements, and advise and assist commanders to address current and emerging challenges in the information environment. Early-entry PA capabilities enable the CJTF to gain and maintain the initiative in the information environment.
- **Planning support capabilities** provide rapidly deployable, tailored, ready joint planners, operations planners, logisticians, knowledge sharing managers, and intelligence specialists to accelerate the formation, and increase the effectiveness, of newly formed joint force HQs. These planners are specialists in the JPP.

(Discussion continued on p. 5-8.)

Joint Task Forces (JTFs)

III. Typical Joint Task Force Organization

Ref: JP 3-33, Joint Task Force Headquarters (Jan '18), fig. II-6, p. II-22.

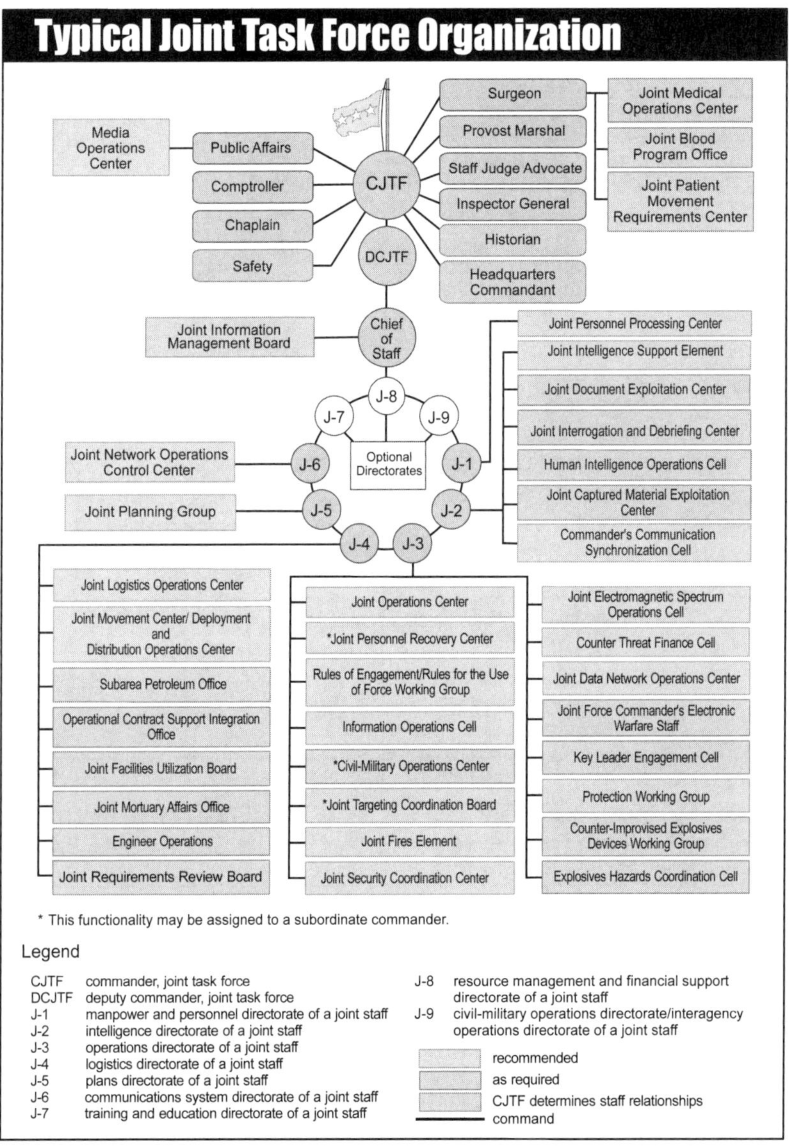

Figure II-6 is not meant to be directive or all-inclusive, depicts an example of a JTF HQ's staff alignment of cross-functional organizations. The figure shows the most common proponent (by staff directorate or special staff group) for each cross-functional organization. As a practical matter, the CJTF and staff establish and maintain only those cross-functional organizations that enhance planning and decision making within the HQ. They establish, modify, and dissolve these entities as the needs of the HQ evolve.

Joint Task Forces (JTFs)

IV. JTF HQ Organization Options

Ref: JP 3-33, Joint Task Force Headquarters (Jan '18), II-12 to II-18 and chaps. V - X.

Mission requirements drive organization and manning since each joint force's mission is different. The CJTF provides early guidance that affects how the JTF HQ will organize and function.

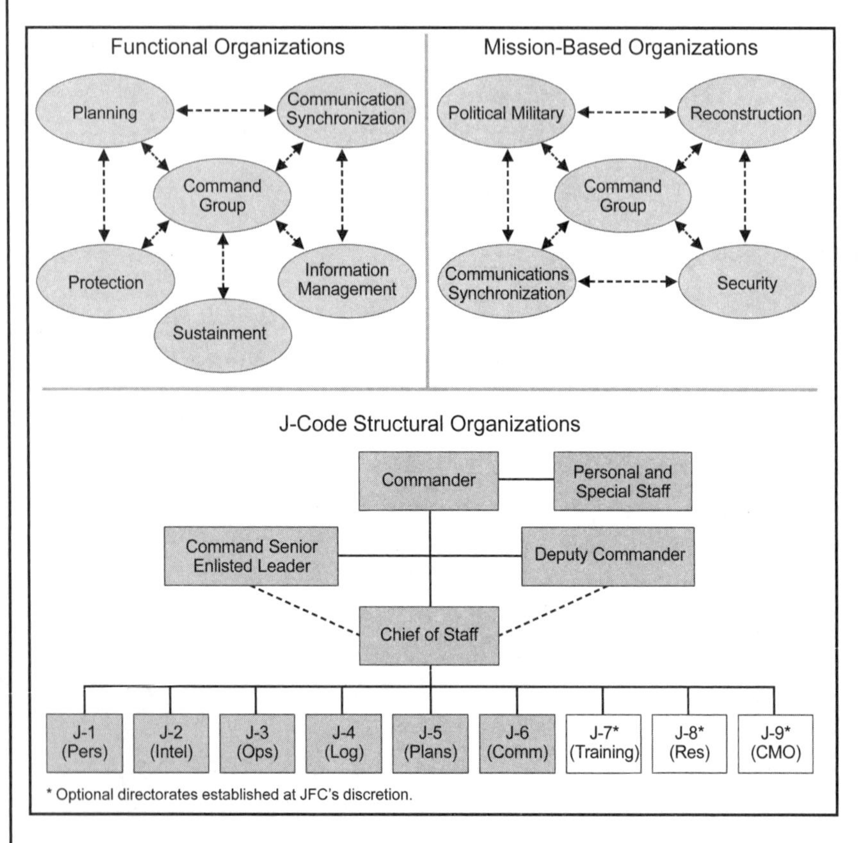

Experience continues to support the J-code structure as the best HQ organizational option for most missions. Figure II-2 depicts a traditional J-code structure on the lower portion and other organizational options on top.

J-1

JTF manpower and personnel support requirements must be determined early in the planning process and continue through the execution of operations and redeployment. To ensure optimal JTF personnel support, the JTF J-1 synchronizes and coordinates personnel support efforts with Service components and functional components and establishes and publishes policies, procedures, and standards to accomplish personnel visibility in the JOA.

J-2

The JTF J-2 informs the commander; describes the OE; identifies, defines, and nominates objectives; supports planning and execution of operations; counters adversary deception and surprise; supports friendly deception efforts; and assesses the effectiveness of operations. Both DOD and non-DOD intelligence agencies and organizations provide assistance to the J-2 in support of activities and operations within the JOA.

The CJTF may establish a JTF-level intelligence element, such as a joint intelligence support element (JISE) or a joint intelligence operations center (JIOC), under a JTF J-2 director, to manage intelligence collection, production, and dissemination. The decision as to the type of intelligence element required will be based on the scope and breadth of the mission assigned to the JTF. If the CJTF requires a JIOC, the decision to establish a fully staffed JIOC at the JTF level may require augmentation and should be approved by the CCDR. This publication uses "JISE" to refer to the JTF-level intelligence element.

J-3

The JTF J-3 helps the commander direct and control current operations. The J-3, typically in concert with the higher HQ and JTF components' operations directorates, plans, coordinates, and integrates operations. The J-3 accomplishes this, in part, by working closely with the rest of the staff. The staff makes recommendations to be included in the commander's intent, informs the commander's decision-making process, and contributes to the execution and assessment of current operations. JTF planning and execution typically consider supported or supporting activities and the integration of interagency partners, multinational participants, NGOs, and international organizations.

The CJTF should establish a joint operations center (JOC), under the director JTF J-3, to manage all matters related to planning and executing current operations.

J-4

The J-4 helps the JFC manage the provision of logistics to the joint force. The ultimate goal is for logistics planners to develop a feasible, supportable, and efficient concept of logistic support that is able to identify risks to the execution of the CONOPS.

The JTF J-4's concept of logistic support often involves interorganizational coordination with the HN, private contractors, and other participants or stakeholders. The quality and quantity of potential HNS and acquisition and cross-servicing agreements (ACSAs) can significantly affect the logistics concept and the JFC's CONOPS.

The JTF J-4 should establish a joint movement center (JMC). The CJTF or GCC, respectively, approve forming a joint logistics operations center (JLOC) or a joint deployment and distribution operations center (JDDOC), if the director JTF J-4 determines one is necessary. Because logistic limitations affect all planning and execution, it is essential that logistic planners are integral members of the joint planning group (JPG) and appropriately integrated throughout the staff.

J-5

Before execution, the JTF J-5 develops, updates, reviews, and coordinates joint plans required for successful accomplishment of JTF mission(s). During execution of current operations, the J-5 focuses on future plans, which are typically for the next phase of operations or sequels to the current operation. The J-5 also supports the future operations planning effort, which normally occurs in the J-3.

The J-5 typically establishes a JPG to facilitate integrated planning across the staff. For HD or DSCA operations, the J-5 typically includes National Guard Bureau (NGB) and/or NG JFHQ-State elements in the JPG.

J-6

The JTF J-6 provides a protected, interoperable joint communications system to enable a CJTF to have C2 of the JTF throughout the OE. This includes development and integration of communications systems, architecture, and plans that support the command's operational-and strategic-level requirements, as well as policy and guidance for implementation and integration of interoperable communications system support to exercise command in the execution of the mission covered under DODIN operations. The JTF J-6 also manages activities described in JP 3-12, Cyberspace Operations.

Optional directorates include Training and Education Directorate of a Joint Staff (J-7), Resource Management and Financial Support Directorate of a Joint Staff (J-8), Civil-Military Operations Directorate/Interagency Operations Directorate of a Joint Staff (J-9), and Personal and Special Staff Groups.

JP 3-33, chapters V to XIV, describe these directorates in detail.

B. Joint Individual Augmentation (JIA)

Individual augmentation is an important mechanism for providing personnel to a JTF HQ. JIA requirements are identified by the HQ designated to be the JTF or the establishing authority (usually a CCDR) and documented in a joint manning document (JMD).

C. Joint Organization Augmentation

Following is a list of joint organizations other than the JECC that may provide JTF augmentation. This list is not all-inclusive, but it should provide insight into the type of augmentation a JTF can receive and the purpose behind that augmentation.

- **National Intelligence Support Team (NIST)**. The NIST is a nationally sourced team of intelligence and communications experts from Defense Intelligence Agency (DIA), Central Intelligence Agency (CIA), National Security Agency (NSA), National Geospatial-Intelligence Agency (NGA), and other agencies.
- **Defense Threat Reduction Agency (DTRA).** DTRA's mission is to safeguard America and its allies from weapons of mass destruction (WMD) (chemical, biological, radiological, or nuclear [CBRN]) by providing capabilities to reduce, eliminate, and counter the threat and mitigate its effects.
- **Joint Information Operations Warfare Center (JIOWC).** The JIOWC is a CJCS-controlled activity and the principal field agency for joint information operations (IO) support of CCMDs.
- **Joint Communications Security Monitoring Activity (JCMA).** JCMA can provide information security monitoring and analysis support to JTFs.
- **Joint Personnel Recovery Agency (JPRA).** JPRA is the principal DOD agency for coordinating and advancing personnel recovery (PR).
- **Joint Warfare Analysis Center (JWAC).** JWAC assists in preparation and analysis of joint OPLANs and Service chiefs' analysis of weapons effectiveness. JWAC normally provides this support to JTFs through the supported CCMD.
- **JTF-PO** is a subordinate command of USTRANSCOM that provides rapid opening and establishment of ports of debarkation and facilitates distribution operations in response to emergencies, incidents, and global contingencies.
- **Defense Logistics Agency (DLA).** DLA supports the JTF using various capabilities, which include, but are not limited to, robust planning experience, surge and sustainment expertise, expeditionary organizations, and personnel embedded physically and virtually with warfighting and support organizations. DLA is the executive agent for the following classes of supply: I (subsistence), IIIB (bulk petroleum), IV (construction and barrier materiel), and VIII (medical materiel). DLA also exercises item manager duties for supply support across the other classes of supply, except class V. DLA can access and use a variety of IM tools to monitor supplies and equipment. GCCs can also request DLA's Joint Contingency Acquisition Support Office (JCASO) as temporary augmentation to the CCMD or a JTF to synchronize and integrate OCS.
- **United States Cyber Command (USCYBERCOM).** USCYBERCOM plans, coordinates, integrates, synchronizes, and conducts activities to direct the operations and defense of specified DOD information networks and prepares to, when directed, conduct full-spectrum military cyberspace operations.
- **Joint Electronic Warfare Center (JEWC)**. The JEWC provides CCMD support along multiple lines of effort (LOEs) to provide a deployable electronic warfare (EW) planning and coordination cell to assist in the initial establishment of a joint electromagnetic spectrum (EMS) cell within the JTF, assist in the preparation and analysis of joint OPLANs from an EW and EMS perspective, provide reachback support to CCMD EW officers, and perform EMS wave propagation analysis supporting CCMD requests.

V. Cross-Functional Organizations and Staff Integration

Ref: JP 3-33, Joint Task Force Headquarters (Jan '18), pp. II-18 to II-21.

Effective joint operations require close coordination, synchronization, and information sharing across the staff directorates. There are clear benefits of the J-code structure in terms of effectiveness, efficiency, administration, accountability, and "plug and play" functionality. However, there is a common tendency for knowledge and expertise to "stovepipe" within the J-code directorates due to the sheer number of ongoing staff actions. Effective knowledge sharing and IM plans will increase collaboration and sharing, which can mitigate this risk. The most common technique for promoting cross-functional collaboration is the formation of an organizational structure that blends J-code functional management with task accomplishment by cross-functional teams of subject matter experts from multiple J-codes (Figure II-3).

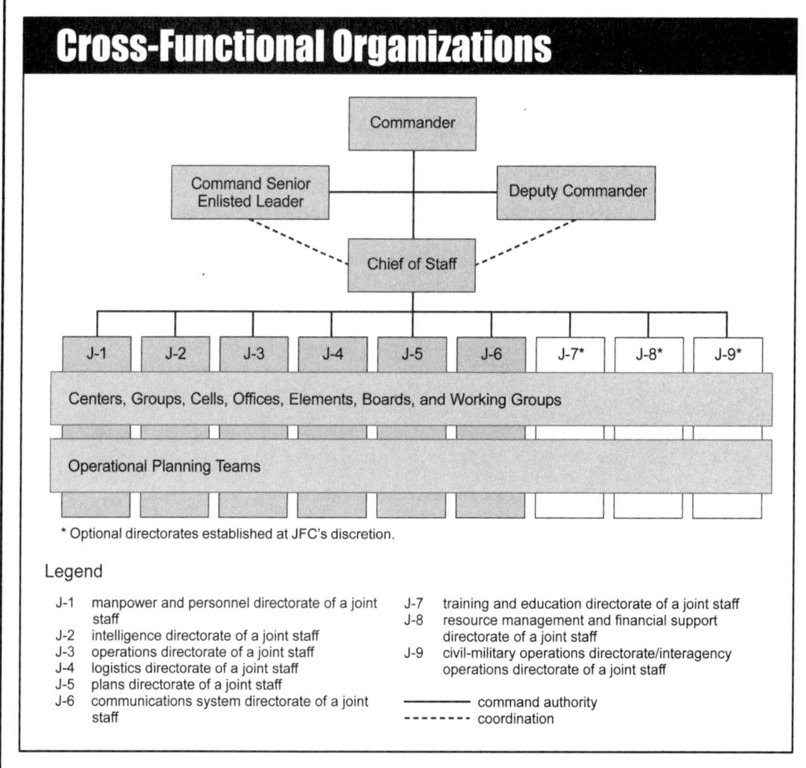

Ref: JP 3-33, fig. II-3. Cross-Functional Organizations and Staff Integration.

Although cross-functional in their membership, most of these teams fall under the principal oversight of the staff directorates or their functional chiefs. This arrangement strengthens the staff effort in ways that benefit the JTF and its commander in mission execution. These organizations are venues through which the cross-functional expertise of the staff is brought to bear on the planning and execution problems being addressed by the commander. Horizontal, cross-functional organizations overlaid on the vertical J-code structure provide a powerful method of staff integration.

Continued on next page

Continued on next page

Continued from previous page

Center

A center is an enduring functional organization, with supporting staff, designed to perform a joint function within a JFC's HQ. Often, these organizations have designated locations or facilities. Examples of centers include the JOC, joint personnel processing center (JPPC), and the CMOC.

Group

A group is an enduring functional organization formed to support a broad HQ function within a JFC's HQ. Normally, groups within a JTF HQ include a JPG that manages JTF HQ planning. JPG functions include leading designated planning efforts, resourcing and managing subordinate planning teams, and coordinating planning activities with other staff directorates.

Cell

A cell is a subordinate organization formed around a specific process, capability, or activity within a designated larger organization of a JFC's HQ. A cell usually is part of both functional and traditional staff structures. An example of a cell within the traditional staff structure could be a joint electronic spectrum operations cell subordinate to the operations branch within the J-3. An example of a cell within a functional staff structure could be a current operations cell within the JOC.

Working Group

A Working Group (fig. II-4) is an enduring or ad hoc organization within a JFC's HQ formed around a specific function whose purpose is to provide analysis to users. The WG consists of a core functional group and other staff and component representatives.

Basic Working Group Model

Core Element

Functional Chief, Deputy and Administrative Section

Future Plans (Planning and Assessment)

Future Operations (Planning and Assessment)

Current Operations (Direct and Monitor)

Other functional members, other stakeholders, and component representatives

Working Group Agenda

- Current operations update
- Future operations estimate
- Future plans estimate
- Functional assessment of the campaign

Expected Outcomes

- Broad understanding of functional priorities in support of planning and decision making
- Approved future operations staff estimate
- Approved future plans staff estimate
- Approved functional estimate in support of campaign assessment

Planning Team
Planning Team
Working Group
Planning Team
Planning Team

Support multiple planning teams on multiple event horizons

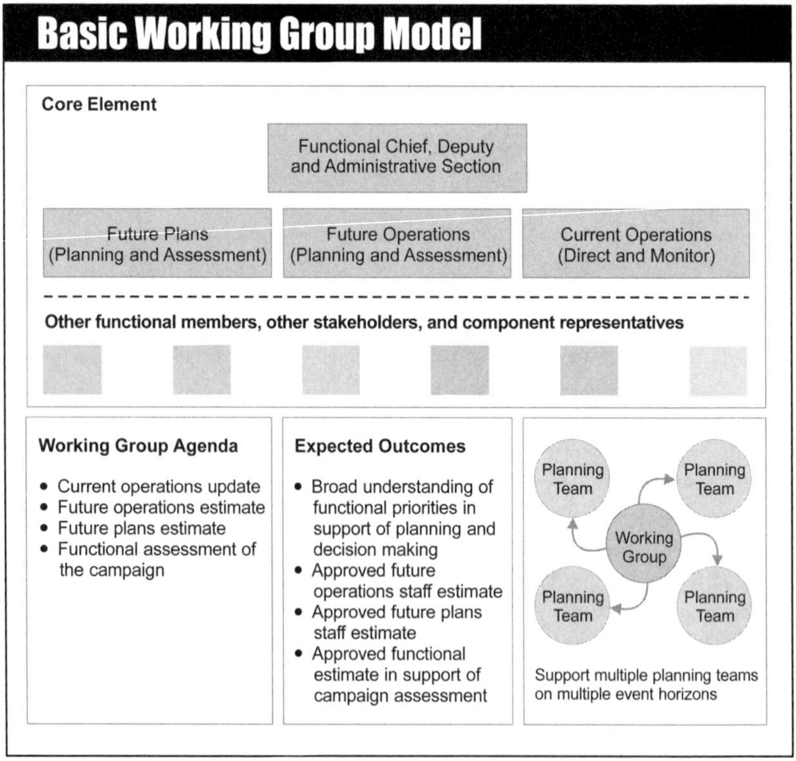

Ref: JP 3-33, fig. II-4. Basic Working Group Model.

Continued from previous page

Office
An office is an enduring organization that is formed around a specific function within a JFC's HQ to coordinate and manage support requirements. An example of an office is the joint mortuary affairs office (JMAO).

Element
An element is an organization formed around a specific function within a designated directorate of a JFC's HQ. The subordinate components of an element usually are functional cells. An example of an element is the joint fires element (JFE).

Board
A board is an organized group of individuals within a JFC's HQ, appointed by the commander (or other authority) that meets with the purpose of gaining guidance or decision. Its responsibilities and authority are governed by the authority that established the board. Boards are chaired by a senior leader with members representing major staff elements, subordinate commands, LNOs, and other organizations as required. There are two different types of boards: command board and functional board.

Operational Planning Team (OPT)
OPTs (fig. II-5) are established to solve a single problem related to a specific task or requirement on a single event horizon. In most cases, OPTs are not enduring and will dissolve upon completion of the assigned task. OPT membership is typically determined by the staff officer responsible for the event horizon in which the OPT is working (i.e., the J-5 [future plans], J-35 [future operations], and J-33 [current operations]).

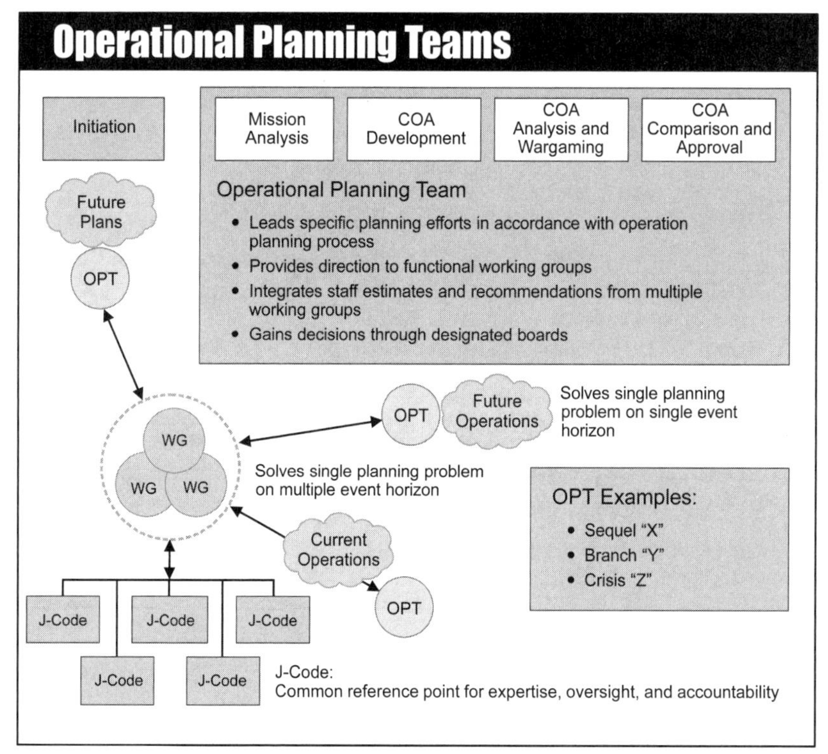

Ref: JP 3-33, fig. II-5. Operational Planning Teams.

VI. Liaison Officers (LNOs)

Ref: JP 3-33, Joint Task Force Headquarters (Jan '18), pp. II-24 to II-25.

When forming the JTF staff, it is important to have LNO representation from all organizations with which the JTF will conduct a significant level of coordination. LNOs facilitate the communication maintained between elements of a JTF to ensure mutual understanding and unity of purpose and action. Liaison is the most commonly employed technique for establishing and maintaining close, continuous, physical communication between commands.

The CJTF must identify the requirement for liaison personnel based on command relationships and mission support requirements. LNOs must be requested at the earliest opportunity. Per this request, any specific qualifications and functions for these personnel should be noted by the CJTF. LNOs to the JTF HQ should be of sufficient rank (recommend equal rank to JTF primary staff officers) to influence the decision-making process. Ideally, LNOs should possess the requisite skill sets (technical training or language) to liaise and communicate effectively with receiving organizations.

Liaison should be established between the JTF HQ and higher commands; between adjacent units; and between supporting, attached, and assigned forces and the JTF HQ. Additionally, the JTF may also exchange LNOs with other interagency partners, MNFs, HNs, and other significant entities.

LNO Guidelines

LNOs provide an essential C2 bridge between the JTF HQ, its parent organizations, and its subordinate organizations. To help ensure LNOs are properly employed and not misused, the JTF should follow certain basic guidelines:

- Liaison officers (LNOs) are personal and official representatives of the sending organizations and should be treated accordingly.
- LNOs support the gaining organizations and serve as critical conduits between organizations.
- LNOs remain in their parent organizations' chain of command.
- LNOs perform four basic functions: monitor, coordinate, advise, and assist.
- LNOs (to include multinational LNOs) must have sufficient access to information to be effective.
- LNOs are not full-time planners.
- LNOs are not watch officers.
- LNOs are not substitutes for delivering critical information through normal command and control channels or a conduit for general information sharing.
- LNOs are not replacements for proper staff-to-staff coordination.
- LNOs are not replacements for augmentees or representatives.
- LNOs do not have the authority to make decisions for their commander without coordination and approval.

For additional information, refer to JP 3-33, Annex C, "Liaison Officers."

II. Joint Task Force Subordinate Commands

Ref: JP 3-33, Joint Task Force Headquarters (Jan '18), chap. III.

A JTF is composed of significant elements, assigned or attached, of two or more Military Departments operating under a single CJTF. The subordinate Service components' HQ and their forces provide the basic building blocks for the JTF's component structure. The CJTF can organize the JTF with Service components, functional components, or a combination based on the nature of the mission and the operational environment. All joint forces include Service component commands because administrative and logistic support for joint forces is provided through Service component commands. Typical JTFs have a combination of Service and functional components. A CJTF also can establish one or more subordinate JTFs if necessary.

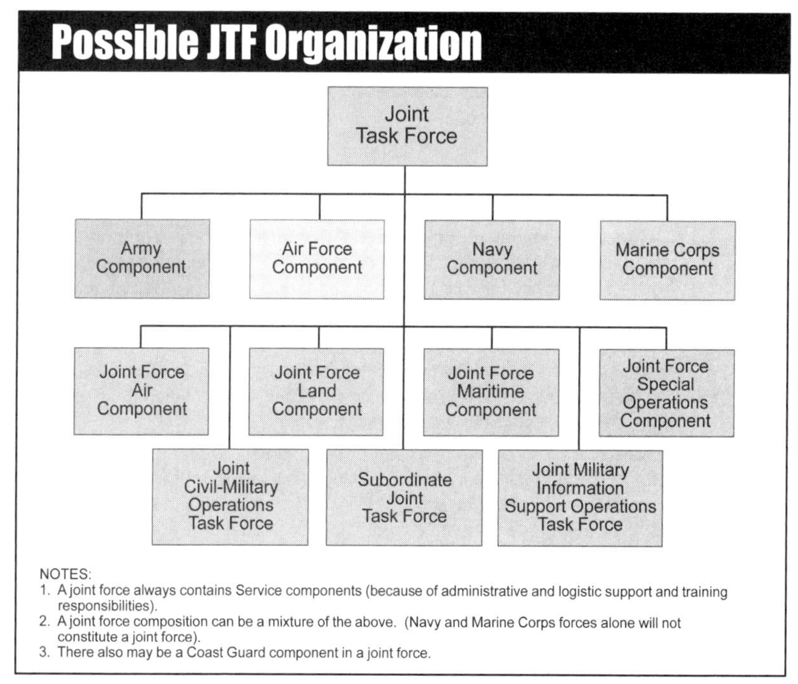

Ref: JP 3-33, Joint Task Force Headquarters, fig. III-1, p. III-2.

In addition to other responsibilities that change according to circumstances, following are typical general responsibilities common to both Service and functional component commanders.

- Plan and execute operations to accomplish missions or tasks assigned by the higher commander's plans or orders.
- Advise the CJTF on employment of the component command's capabilities, progress toward mission accomplishment, and other matters of common concern.
- Assess the progress of operations; integrate, manage, and share information as part of the JTF's assessment activity; and provide timely updates to the

higher commander relating to the progress or regression of tasks, objectives, and/or effects.

- Coordinate with other JTF component commanders to ensure effective and efficient conduct of operations. In addition, coordinate with supporting agencies, supporting commanders, and friendly forces and governments as authorized and as necessary to fulfill assigned responsibilities.
- Provide liaison personnel to other commands and organizations as appropriate.

I. Service Component Commands

A JTF-level Service component command consists of the Service component commander and all Service forces that have been assigned or attached to the JTF. In general, a Service component commander is responsible for all aspects of planning and executing operations as tasked by the next higher commander.

Service component commanders have responsibilities that derive from their roles in fulfilling the Services support function. The CJTF may conduct operations through the Service component commander or, at lower echelons, other Service force commanders. A CJTF can delegate to subordinate commanders no more than the level of authority the establishing authority has given the CJTF. When a Service component commander exercises OPCON of forces, and the CJTF delegates that OPCON or TACON over those forces to another component commander or other subordinate JFC, the Service component commander retains the following responsibilities for certain Service-specific functions:

- Advise the CJTF on the proper employment, task organization, and command relationship of the Service component's forces.
- Accomplish such operational missions as may be assigned.
- Select and nominate specific units of the parent Service component for attachment to other subordinate commands. Unless otherwise directed, these units revert to the Service component commanders' control when such subordinate commands are dissolved.
- Conduct joint training, including the training, as directed, of components of other Services in joint operations for which the Service component commander has or may be assigned primary responsibility or for which the Service component's facilities and capabilities are suitable.
- Inform the CJTF, other component or supporting commanders, and the CCDR, if affected, of planning for changes in logistic support that would significantly affect operational capability or sustainability sufficiently early in the planning process for the JFC to evaluate the proposals prior to final decision or implementation.

Service component commanders or other Service force commanders assigned to a CCDR are responsible through the chain of command, extending to the Service Chief, for the following:

- Internal Service administration and discipline (administrative control [ADCON]).
- Training in joint doctrine and their own Service doctrine, tactics, techniques, and procedures.
- Logistic functions normal to the command, except as otherwise directed by higher authority. The operating details of any Service logistic support system are retained and exercised by the Service component commanders in accordance with instructions of their Military Departments, subject to the directive authority of the CCDR.

II. Functional Component Commands

CJTFs normally establish functional component commands to control military operations. A functional component command typically consists of forces of two or more Military Departments established to perform designated missions.

Functional Component Commands

Ref: JP 3-33, Joint Task Force Headquarters (Jan '18), pp. III-4 to III-6.

CJTFs may normally establish functional component commands to control military operations. A functional component command typically consists of forces of two or more Military Departments established to perform designated missions.

Example functional component commands include the following:

A. Joint Force Air Component Commander (JFACC)

The CJTF usually designates a JFACC to establish unity of command and unity of effort for joint air operations. A CJTF will typically assign JFACC responsibilities to the component commander having the preponderance of forces tasked and the ability to effectively plan, task, and control joint air.

See pp. 6-2 to 6-3 for further discussion.

B. Joint Force Land Component Commander (JFLCC)

The CJTF should designate a JFLCC and establish the commander's authority and responsibilities to exercise C2 over land operations, when forces of significant size and capability of more than one Service component participate in a land operation.

See pp. 6-8 to 6-9 for further discussion.

C. Joint Force Maritime Component Commander (JFMCC)

The CJTF may designate a JFMCC to C2 joint maritime operations. As a functional component commander, the JFMCC has authority over assigned and attached forces and forces or assets made available for tasking to perform operational missions. Generally, maritime assets may include navies, marines, SOF, coast guards and similar border patrol and revenue services, nonmilitary shipping managed by the government, civil merchant marines, army/ground forces (normally when embarked), and air and air defense forces operating in the maritime domain.

See pp. 6-18 to 6-20 for further discussion.

D. Special Operations Joint Task Force (SOJTF)

A SOJTF is a modular, tailorable, and scalable SOF organization that allows the Commander, United States Special Operations Command (CDRUSSOCOM), to more efficiently provide integrated, fully capable, and enabled joint SOF to GCCs and subordinate JFCs based on the strategic-, operational-, and tactical-level context.

See p. 6-30 for further discussion.

E. Joint Special Operations Task Force (JSOTF)

A JSOTF is generally composed of units of two or more SOF Service components formed to unilaterally carry out specific special operations or activities or to support a JFC conducting joint operations. A JSOTF may have CF supporting it for specific missions. A JSOTF is normally established by a JFC. For example, a GCC could establish a JTF to conduct operations in a specific JOA of the theater; then, either the GCC or the CJTF could designate a JSOTF commander and establish a JSOTF, subordinate to that CJTF, to plan and execute special operations.

F. Joint Force Special Operations Component Commander (JFSOCC)

The CJTF may designate a JFSOCC, SOJTF commander, or JSOTF commander to accomplish a specific mission or control SOF in the JOA. The JFSOCC will generally be the commander with the preponderance of SOF and the requisite C2. The commander of the TSOC may function as the SOJTF commander, JSOTF commander, or JFSOCC. In certain situations, the SOJTF commander or JSOTF commander may be appointed by CDRUSSOCOM.

Functional component commanders exercise command authority (e.g., OPCON, TACON) as delegated, over forces or military capabilities made available to them consistent with JP 1, Doctrine for the Armed Forces of the United States. The CJTF designates the military capability that is made available for tasking by the functional component commander, as well as that commander's authority and responsibilities. Establishment of a functional component commander must not affect the command relationships between Service component commanders and the CJTF.

The CJTF establishing a functional component command has the authority to designate its commander. Normally, the Service component commander, with the preponderance of forces to be tasked and the ability to C2 those forces, will be designated as the functional component commander. However, the JFC will always consider the mission, nature, and duration of the operation, force capabilities, and the C2 capabilities in selecting a commander. A Service component commander who is also the functional component commander retains the responsibilities associated with Service component command for assigned Service forces.

Commanders of functional component commands advise the CJTF on the proper employment of the military capability made available to accomplish the assigned responsibilities. They are also responsible for meeting the reporting criteria for entities and events in the JOA as outlined in the CCDR's directives and CJTF's amplifying instructions.

Since a functional component commander will employ forces from more than one Service, the component commander's staff should reflect the composition of the functional component command to provide the JFC with the expertise needed to effectively employ the forces made available. The functional component commander should identify staff billets for the needed expertise and individuals to fill those billets and use the individuals when the functional component staffs are formed for exercises and actual operations. The structure of the staff should be flexible enough to expand or contract under changing conditions without loss in coordination or capability. The commander must also be aware of the constraints imposed by logistic factors on the capability of the assigned and attached forces and the responsibilities retained by the Services.

III. Subordinate Joint Task Forces

A CJTF is authorized to establish subordinate JTFs as circumstances require. For example, a CJTF assigned a large JOA, characterized by difficult terrain that restricts movement and maneuver, might determine C2 could be more effective with a subordinate JTF responsible for operations in a remote portion of the JOA. The decision to do so requires careful consideration because of the many factors that will complicate forming a new JTF during ongoing operations.

It is more common to form special-purpose JTFs that focus on specific functional aspects of the CJTF's operations. Because of the nature and visibility of CMO, counter-improvised explosive device (C-IED) operations, and military information support operations (MISO), the CJTF may establish separate task forces for these activities. As a general rule, CMO, C-IED, and military information support task force commanders work directly for the CJTF. However, in certain circumstances (e.g., crisis response and limited contingency operations), these task forces may also be attached to the JSOTF.

- **Joint Civil-Military Operations Task Force (JCMOTF).** A CJTF may establish a JCMOTF when the scope of CMO requires coordination and activities beyond that which a CA representative on the staff could accomplish. The JCMOTF must be resourced to meet specific CMO requirements.
- **Joint Military Information Support Task Force (JMISTF).** A CJTF may establish a JMISTF when the scope of MISO requires coordination and activities exceed the capability of MISO support element to advise and assist.
- **Counter-Improvised Explosive Device Task Force (C-IEDTF).** A CJTF may establish a C-IEDTF when the scope of C-IED operations exceed the capabilities of the JTF staff to plan and direct.

Chap 5

III. Joint Task Force Command & Control

Ref: JP 3-33, Joint Task Force Headquarters (Jan '18), chap. IV.

Chapter 4 of JP 3-33 describes C2 factors and management processes that influence JTF C2. The C2 factors are: the role of the commander in the JTF C2, command relationships, understanding the OE, OA management, operational limitations, interorganizational coordination considerations, multinational considerations, and CMO considerations. The management processes are: JTF IM, the commander's decision cycle, and the HQ battle rhythm.

I. JTF Headquarters Command & Control Factors

A. Role of the Commander in JTF C2

The CJTF has the authority and responsibility to effectively organize, direct, coordinate, and control military forces to accomplish assigned missions. The CJTF's actions associated with these responsibilities are central to JTF C2. The CJTF leverages the full range of skill, knowledge, experience, and judgment to guide the command through the fog and friction of operations towards mission accomplishment.

See following page (p. 5-19) for discussion of the commander's role in planning.

B. Command Relationships

C2 Functions of the JTF Establishing Authority

The JTF establishing authority exercises either combatant command (command authority) (COCOM) or OPCON of the JTF. The JTF establishing authority either transfers forces from subordinate commands and attaches them to the JTF or transfers forces allocated to the CCMD by SecDef via the GFM process, as appropriate. The JTF establishing authority also establishes the command relationships between the CJTF and other subordinate commanders to ensure the success of the JTF.

C2 Functions of the CJTF

The CJTF normally exercises OPCON or TACON over attached forces through designated component, major subordinate command, or subordinate task force commanders. The CJTF may also be a supported or supporting commander. Further, the CJTF may delegate OPCON or TACON of, or establish support relationships for, specific JTF forces or military capabilities to or between subordinate commanders to accomplish specified tasks or missions

For more details concerning command relationships, see pp. 1-31 to 1-40.

C. Understanding the Operational Environment (OE)

The JFC's operational environment is the composite of the conditions, circumstances, and influences that affect the employment of capabilities and bear on the decisions of the commander. It encompasses physical areas and factors (of the air, land, maritime, and space domains) and the information environment (which includes cyberspace). Included within these are the adversary, friendly, and neutral systems that are relevant to a specific joint operation. Understanding the operational environment helps commanders understand the results of various friendly, adversary, and neutral actions and how these affect the JTF mission accomplishment.

See pp. 2-38 to 2-39 and p. 3-50 for further discussion of the operational environment.

D. Operational Area (OA) Management

A critical function of the CJTF is to organize the OA to assist in the integration, coordination, and deconfliction of joint actions. The CJTF can employ areas of operations (AOs), joint special operations areas (JSOAs), amphibious objective areas (AOAs), and joint security areas (JSAs) to support the organization of the OA within the assigned JOA.

See pp. 2-40 to 2-42 or further discussion of operational areas from JP 3-0.

E. Operational Limitations

Operational limitations include actions required or prohibited by higher authority (a constraint or restraint) and other restrictions that limit the commander's freedom of action (such as diplomatic agreements, ROE, diplomatic/political considerations, economic conditions in affected countries, and HN issues). Authorities, in the form of international and domestic law, national policy, and higher HQ guidance and intent, determine the commander's freedom of action. Authorities can be both permissive and restrictive, at times permitting a wide range of options available to the commander, while at other times restricting the actions that may be taken. An operational constraint is a requirement placed on the command by a higher command that directs an action, thus restricting freedom of action. An operational restraint is a requirement placed on the command that prohibits an action, thus restricting freedom of action. Authorities approved for an operation play an integral role in planning, while operational limitations may restrict or bind COA selection or may even impede implementation of the chosen COA. Commanders must identify the approved authorities and operational limitations, understand their impacts, and develop options that maximize approved authorities. This must be done while minimizing the impact of operational limitations to promote maximum freedom of action during execution. A common area of concern for every commander with regard to authorities and limitations is the use of force in mission accomplishment and self-defense.

F. Interorganizational Cooperation Considerations

Relationships between the JTF and USG departments and agencies, international organizations, NGOs, and the private sector should not be equated to the C2 relationships and authorities of a military operation. Whether supported or supporting, close coordination between the military and other non-DOD agencies is a key to successful interagency coordination. Successful interorganizational and private sector coordination enables the JTF to build support, conserve resources, and conduct coherent operations that efficiently achieve shared objectives through unity of effort.

See chap. 8, Interorganizational Cooperation, for further discussion.

G. Multinational Considerations

The President retains and does not relinquish command authority over US forces. On a case-by-case basis, the President may consider placing appropriate US forces under the OPCON of a competent UN, North Atlantic Treaty Organization (NATO), or multinational commander for specific operations authorized by the UN Security Council, or approved by the North Atlantic Council, or other authorized regional organization.

See chap. 7, Multinational Operations, for further discussion.

H. Considerations for DSCA Operations

The President retains and will never relinquish command authority over federal (Title 10, USC) military forces. On a case-by-case basis, the President may consider placing appropriate Title 10, USC, forces under the OPCON or TACON of a dual-status (Title 10/Title 32, USC) commander in support of the governor of a US state or territory. Title 10, USC, forces under the command (assigned or attached OPCON or TACON) of a dual-status commander will follow the Title 10, USC, RUF unless directed otherwise by the President or SecDef.

II. Commander's Role in Planning Operations

Ref: JP 3-33, Joint Task Force Headquarters (Jan '18), fig. IV-1, p. IV-4.

The commander provides guidance that drives JPP and supervises execution of the products. Early in the JPP, the CJTF's vision of an operation is translated into a broad operational approach that guides subsequent detailed planning and produces plans and orders for execution. The CJTF communicates the operational approach through three important mechanisms: commander's intent, commander's planning guidance, and CCIRs. These mechanisms assist the commander and JTF staff in establishing an effective dialogue to enable efficient planning. Later, during preparation activities and operations, these mechanisms assist the entire JTF in remaining focused on the commander's original vision and desired outcome of the operation.

Commander's Intent

The commander's intent is a clear and concise expression of the purpose of the operation and the desired military end state. The CJTF uses the intent statement to help communicate the operational approach to both the JTF staff and subordinate and supporting commands during planning and execution. It provides focus to the staff and helps subordinate and supporting commanders act to achieve the commander's desired results without further orders, even when operations do not unfold as planned. It also includes where the commander will accept risk during the operation.

Commander's Guidance

The commander's guidance communicates the commander's initial thoughts for a given operation to the staff, which enhances effective planning. This guidance may be as broad or detailed as circumstances require. Although commanders provide guidance to their staffs throughout the planning process, there are two opportunities to provide early guidance to the staff to focus their efforts:

Commander's Initial Guidance

Upon receipt of mission, the commander and staff conduct an analysis of the initiating directive to determine time available to mission execution, the current status of intelligence products and staff estimates and other factors relevant to the specific planning situation. The commander will provide initial guidance to the staff, which could specify time constraints, outline initial coordination requirements, authorize movement of key capabilities within the CJTF's authority, and direct other actions as necessary.

Commander's Planning Guidance

Planning guidance is an important input to subsequent mission analysis, but the completion of mission analysis is another point at which the CJTF may provide updated planning guidance that affects course of action (COA) development. The CJTF may have been able to apply operational design to think through the operation before the staff begins JOPP. In this case, the CJTF provides initial planning guidance to help focus the staff in mission analysis. Otherwise, the CJTF and staff will develop their understanding of the operational environment and problem to be solved during mission analysis. Then the commander will issue planning guidance, as he sees appropriate, to help focus the staff's efforts. At a minimum, the CJTF issues planning guidance, either initial or refined, at the conclusion of mission analysis and provides planning guidance as the operational approach matures. The format for the commanders planning guidance varies based on the personality of the commander and the level of command, but should adequately describe the logic to the commanders understanding of the operational environment and of the problem and the description of the operational approach.

It may include the following elements: a description of the operational environment, a statement of the problem that military operations must solve, a description of the operational approach, the CJTF's initial intent, operational limitations, and other factors as desired.

III. Sample JTF HQ Battle Rhythm

Ref: JP 3-33, Joint Task Force Headquarters (Jan '18), pp. IV-17 to IV-18.

Battle rhythm is the sequencing and execution of actions and events within a joint force HQ that are regulated by the flow and sharing of information that support all decision cycles.

Sample Joint Task Force HQ Battle Rhythm

Time	Event	Location	Participants
Note: Event Time is Situationally Dependent	Shift Change	JOC	Battle Staff/others as required
	Targeting Meeting	Briefing Room	As Required
	Situation Update to CJTF	Briefing Room	CJTF, DCJTF, COS, J-1, J-2, J-3, J-4, J-5, J-6, CJTF's Personal and Special Staffs, Component Liaison, others as required
	Plans Update to CJTF	Briefing Room	CJTF, DCJTF, COS, J-1, J-2, J-3, J-4, J-5, J-6, CJTF's Personal and Special Staffs, Component Liaison, others as required
	CJTF's VTC Call to Components	CJTF Conference Room	CJTF, Component Commanders
	JPG	J-5 Plans Conference Room	J-1, J-2, J-3, J-4, J-5, J-6, Core Planners, Component Liaison, others as required
	JTCB Meeting	Briefing Room	DCJTF, J-2, J-3, JFACC, Component Liaison, others as required
	Joint Information Management Board	Briefing Room	COS, J-3, J-6, Staff Information Management Representatives, Component Liaison, others as required
	IO Working Group CCS Working Group	Briefing Room	IO Staff, CA, PA, DSPD, J-1, J-2, J-3, J-4, J-5, J-6, Component Liaison, JMISTF, others as required
	Battle Update Assessment	Briefing Room	CJTF, DCJTF, COS, J-1, J-2, J-3, J-4, J-5, J-6, CJTF's Personal and Special Staffs, Component Liaison, others as required
	Protection Working Group	JOC	FP Officer, J-1, J-2, J-3, J-4, J-5, J-6, Component Liaison, others as required
	Shift Change	JOC	Battle Staff/others as required
	ROE/RUF Working Group	Briefing Room	J-1, J-2, J-3, J-4, J-5, J-6, SJA Component Liaison, others as required
	Combat Assessment Board	Briefing Room	CJTF, DCJTF, COS, J-1, J-2, J-3, J-4, J-5, J-6, CJTF's Personal and Special Staffs, Component Liaison, others as required

Legend

CA	civil affairs
CCS	commander's communication synchronization
CJTF	commander, joint task force
COS	chief of staff
DCJTF	deputy commander, joint task force
DSPD	defense support to public diplomacy
FP	force protection
IO	information operations
J-1	manpower and personnel directorate of a joint staff
J-2	intelligence directorate of a joint staff
J-3	operations directorate of a joint staff
J-4	logistics directorate of a joint staff
J-5	plans directorate of a joint staff
J-6	communications system directorate of a joint staff
JFACC	joint force air component commander
JMISTF	joint military information support task force
JOC	joint operations center
JPG	joint planning group
JTCB	joint targeting coordination board
PA	public affairs
ROE	rules of engagement
RUF	rules for the use of force
SJA	staff judge advocate
VTC	video teleconferencing

Ref: JP 3-33, fig. IV-3. Sample Joint Task Force Headquarters Battle Rhythm.

A battle rhythm is a routine cycle of command and staff activities intended to synchronize current and future operations. As a practical matter, the HQ battle rhythm consists of a series of meetings, report requirements, and other activities. These activities may be daily, weekly, monthly, or quarterly requirements. Typically, the JTF HQ's battle rhythm is managed by the JTF chief of staff.

IV. Joint Task Force Planning

Ref: JP 3-33, Joint Task Force Headquarters (Jan '18), pp. IX-1 to IX-12.

The CJTF and staff develop plans and orders through the application of operational art, operational design, and JPP. They combine art and science to develop products that describe how (the ways) the joint force will employ its capabilities (means) to attain the military end state (ends).

I. Planning Horizons

Like most complex organizations, JTFs have long-, mid-, and near-term objectives. JTFs organize to conduct future planning, future operations planning, and current operations planning. The division of labor between these planning efforts is linked to time or events and is situation, as well as level of command, dependent. Using time horizons to delineate responsibilities, a JTF HQ may focus current operations on activities inside of 24 hours, focus future operations on activities between 24 and 96 hours, and focus future plans on activities beyond 96 hours and possibly up to and over six months. Using these event horizons to delineate responsibilities, a JTF HQ may focus current operations on activities associated with ongoing operations, focus future operations on branch planning, and focus future plans on sequel planning.

See p. 5-24, JTF Plans and Operations Synchronization, for further discussion.

A. Future Plans Planning

The focus of the JTF's future planning is development of plans and OPORDs. Future planning processes and products generally require significant coordination with entities both internal and external to the JTF staff. They also generally require adequate time to integrate the work of this broader planning audience. During operations, the focus of the J-5 and JPG is on the development of sequel planning. Depending on the situation, the JPG may also be tasked to conduct branch planning. The future planning function usually takes place in the J-5 or JPG. This allows it to leverage the functional expertise that resides there.

B. Future Operations Planning

The focus of the JTF's future operations is the development of orders and FRAGORDs that are beneath the threshold of the long-term efforts of the future plans, but not directly related to the management of current operations. Future planning processes and products generally require significant coordination with elements internal to the JTF. They may also require coordination with entities external to the JTF staff. The future operations function typically takes place in the JOC. Future operations planning also develop the branch plans in support of current and ongoing operations. This allows it to leverage the functional expertise that resides there. The planning for future operations often is accomplished under the cognizance of the J-3 (future operations cell).

C. Current Operations Planning

The focus of the JTF's current operations planning is the management of the current operation(s). This activity often includes the development of FRAGORDs to adjust or sustain these ongoing operations. The current operations planning function normally takes place within the JOC or J-3. This allows it to leverage the functional expertise that resides there.

II. Adaptive Planning and Execution System (APEX)

Joint operation planning is accomplished through the Adaptive Planning and Execution (APEX) system. The joint planning and execution community (JPEC) uses APEX to monitor, plan, and execute mobilization, deployment, employment, sustainment, redeployment, and demobilization activities associated with joint operations. APEX's focus is on developing plans that contain a variety of feasible, embedded options for the President and SecDef to leverage as they seek to shape the situation and respond to contingencies. CCMDs participate routinely in both deliberate planning and CAP. Due to the nature of the organization, a JTF HQ typically participates primarily in CAP. However, Service component HQ designated in peacetime as prospective JTF HQ for specific plans usually participate in the CCMD's deliberate planning effort.

See pp. 3-12 to 3-13 for further discussion.

III. Operational Art and Operational Design

Operational art is the cognitive approach by commanders and staffs—supported by their skill, knowledge, experience, creativity, and judgment—to develop strategies, campaigns, and operations to organize and employ military forces by integrating ends, ways, and means. It is a thought process to mitigate the ambiguity and uncertainty of a complex OE and develop insight into the problems at hand. Operational art also promotes unified action by enabling JFCs and staffs to consider the capabilities, actions, goals, priorities, and operating processes of interagency partners and other interorganizational participants, when they determine objectives, establish priorities, and assign tasks to subordinate forces. It facilitates the coordination, synchronization, and, where appropriate, the integration of military operations with activities of other participants, thereby promoting unity of effort.

Operational design supports commanders and staff in their application of operational art with tools and a methodology to conceive of and construct operations and campaigns. Operational design assists commanders in developing the operational approach, which broadly describes the actions the joint force needs to take to attain their end state.

See pp. 3-43 to 3-68 for further discussion.

IV. Joint Planning Process (JPP)

The JPP is an orderly, analytical process through which the JFC and staff translate the broad operational approach into detailed plans and orders. The process occurs across a collaborative network, which requires dialogue among senior leaders, concurrent and parallel plan development across staff echelons, and extensive coordination and synchronization across multiple planning groups. Clear strategic-level guidance and frequent interaction between senior leaders and planners promote early understanding of, and agreement on, planning assumptions, considerations, risks, and other key factors. The focus is on developing plans that contain a variety of viable, embedded options for the commander to consider as the situation develops. This facilitates responsive plan development and modification, resulting in plans that are continually updated. Key to the JPP is the interaction of CJTF, JTF staff, and the commanders and staffs of the next higher, lower, and supporting commands. Although an ultimate product is the plan or OPORD for a specific mission, the JPP is continuous throughout an operation. Even during execution, it produces plans and orders for future operations as well as FRAGORDs that drive immediate adjustments to the current operation.

See facing page for an overview of JTF plans and operations synchronization. See pp. 3-69 to 3-118 for discussion of JPP from JP 5-0.

V. Joint Planning Group (JPG) Composition

Ref: JP 3-33, Joint Task Force Headquarters (Jan '18), fig. IX-7, p. IX-11.

The JPG is a group of staff and command representatives formed by the J-3 and charged with developing JTF plans and orders. The JPG typically forms for crisis planning, but the J-3 can also use it for more deliberate future operation planning requirements.

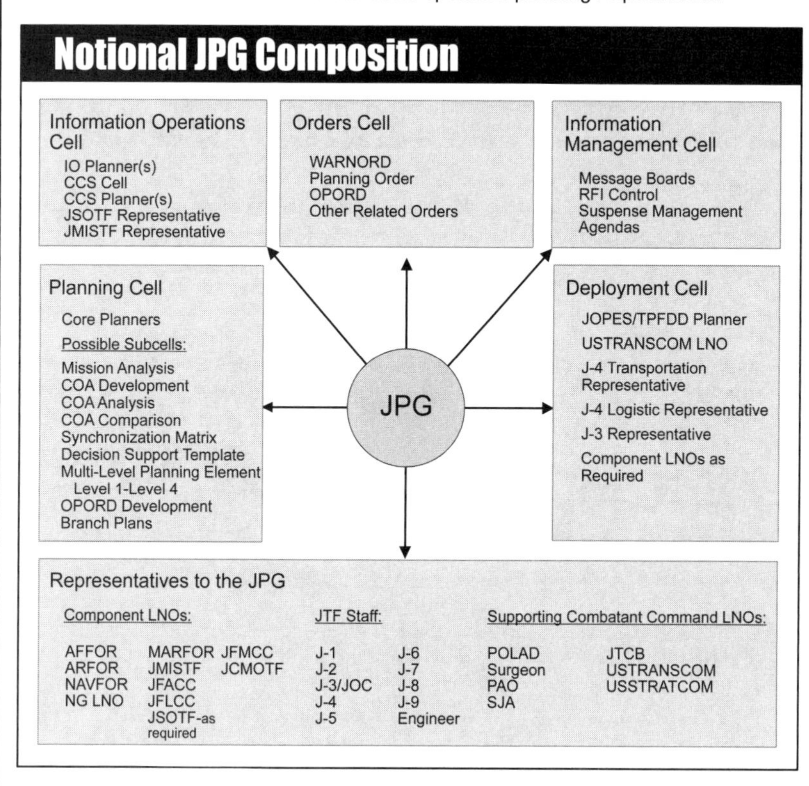

Ref: JP 3-33, fig, IX-4. Notional Joint Planning Group Composition.

Composition of a JPG varies depending on the planning activities being conducted. Normally, all supporting components will have permanent representation in the JPG. There are no mandatory rules to determine the precise number of personnel to staff the JPG. Representation to the JPG should be a long-term assignment to provide continuity of focus and consistency of procedure. Often representatives from the supported and supporting CCMDs and multinational representatives or LNOs will augment the JPG.

JTF Plans and Operations Synchronization

Ref: JP 3-33, Joint Task Force Headquarters (Jan '18), pp. IX-7 to IX-8.

The JPG is often the focal point for OPORD development. The JTF OPORD will typically be based on the establishing authority's OPORD (if available). Upon completion of the plan or OPORD and based on CJTF guidance, designated planning teams focus on execution phase planning. A core JPG should be expanded for select planning functions. Typically, these additional planners will be needed when specific subject matter expertise and staff or component planning input is required.

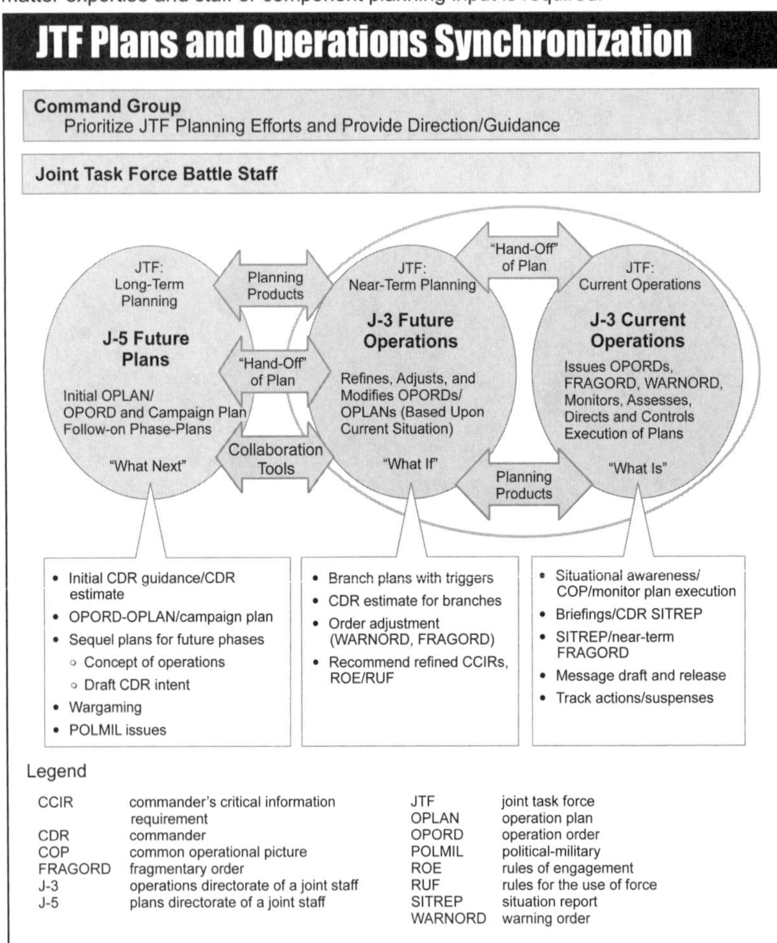

Ref: JP 3-33, fig. IX-3. Joint Task Force Plans and Operations Synchronization.

Upon completion of the plan or OPORD and based on CJTF guidance, designated planning teams focus on execution phase planning. Figure IX-3 (above) represents one organizational option to synchronize long- and short-term planning, assessment, and guidance for commanders.

I. Joint Air Operations

Ref: JP 3-30, Joint Air Operations (Jul '19).

Air Domain

The air domain is the atmosphere, beginning at the Earth's surface, extending to the altitude where its effects upon operations become negligible. While domains are useful constructs for visualizing and characterizing the physical environment in which operations are conducted (the operational area), the use of the term "domain" is not meant to imply or mandate exclusivity, primacy, or C2 of any domain. Specific authorities and responsibilities within an operational area are as specified by the appropriate JFC.

Control of the Air

Historically, control of the air has been a prerequisite to success for modern operations or campaigns because it prevents enemy air and missile threats from effectively interfering with operations of friendly air, land, maritime, space, cyberspace, and special operations forces (SOF), facilitating freedom of action and movement. Dominance of the air cannot be assumed. In the air, the degree of control can range from no control, to a parity (or neutral) situation wherein neither adversary can claim any level of control over the other, to local air superiority in a specific area, to air supremacy over the entire operational area. Control may vary over time. It is important to remember, the degree of control of the air lies within a spectrum that can be enjoyed by any combatant. Likewise, that degree of control can be localized geographically (horizontally and vertically) or defined in the context of an entire theater. The desired degree of control will be at the direction of the JFC and based on the JFC's concept of operations (CONOPS) and will typically be an initial priority objective of joint air operations.

Organization of Forces

A JFC has three basic organizational options for C2 of joint air operations: designate a JFACC, designate a Service component commander, or retain C2 within the JFC's headquarters. In each case, effectively and efficiently organizing the staff, C2 systems, and subordinate forces that will plan, execute, and assess joint air operations is key. Factors impacting selection of each option may include the overall mission, forces available, the ability to C2, and the desired span of control.

When designated, the JFACC is the commander within a combatant command (CCMD), subordinate unified command, or joint task force (JTF) responsible for tasking joint air forces, planning and coordinating joint air operations, or accomplishing such operational missions as may be assigned. The JFACC is given the authority necessary to accomplish missions and tasks assigned by the establishing commander.

Joint Air Operations

Joint air operations are performed by forces made available for joint air tasking. Joint air operations do not include those air operations that a component conducts as an integral and organic part of its own operations. Though missions vary widely within the operational environment and across the range of military operations, the framework and process for the conduct of joint air operations should be consistent.

Joint air operations are normally conducted using centralized control and decentralized execution to achieve effective control and foster initiative, responsiveness, and flexibility. Centralized control is giving one commander the responsibility and authority for planning, directing, and coordinating a military operation or group/category of operations. Because of air power's speed, range, flexibility, and generally limited assets, centralized control best facilitates the integration of forces for the joint air effort and maintains the ability to focus the impact of joint air forces as needed throughout the operational area. Decentralized execution is the delegation of execution authority to subordinate commanders. Decentralized execution is essential to generate the required tempo of operations and to cope with the uncertainty, disorder, and fluidity of combat in air operations.

Factors such as mission requirements, air capabilities, tempo, and scale of operations drive the level of centralized control and decentralized execution required for joint air operations. Close air support (CAS), personnel recovery (PR), and support to special operations missions typically have a high degree of uncertainty and complexity and, therefore, require greater latitude in mission tasking and a higher degree of decentralized execution. Large-scale combat operations will also require greater decentralized execution to meet the wide range of mission tasking. Conversely, highly sensitive strike missions against long-range strategic targets will generally require a higher level of detailed planning and more centralized control. Low-intensity and small-scale operations, which may include very restrictive rules of engagement (ROE), may also require a higher level of centralized control.

I. Joint Force Air Component Commander (JFACC)

The JFC normally designates a JFACC to establish unity of command and unity of effort for joint air operations. The JFC will normally assign JFACC responsibilities to the component commander having the preponderance of forces to be tasked and the ability to effectively plan, task, and control joint air operations. However, the JFC will always consider the mission, nature, and duration of the operation, force capabilities, and the C2 capabilities in selecting a commander.

Authority

The JFC delegates the JFACC the authority necessary to accomplish assigned missions and tasks. The JFACC will normally exercise tactical control (TACON) over forces made available for tasking. Service component commanders will normally retain operational control (OPCON) over their assigned and attached Service forces.

The JFC may designate the JFACC as the supported commander for strategic attack; air interdiction; PR; and airborne intelligence, surveillance, and reconnaissance (ISR) (among other missions). As such, the JFACC is responsible to the JFC for planning, coordinating, executing, and assessing these missions. Other component commanders may support the JFACC in accomplishing these missions, subject to the demands of their own JFC-assigned missions or as explicitly directed by the JFC. Normally, the JFACC is the supported commander for the JFC's overall air interdiction effort and the land and maritime forces commanders are supported commanders for interdiction in their designated area of operations (AO) and have the authority to designate target priority, effects, and timing of fires within their AOs.

JFACC Responsibilities

Ref: JP 3-30, Joint Air Operations (Jul '19), chap. II.

The responsibilities of the JFACC are assigned by the JFC. These include, but are not limited to:

- Develop a JAOP in coordination with the other Service and functional components to best support the JFC's CONOPS or operation plan (OPLAN) (see example in Appendix C, "Joint Air Operations Plan Template").
- Recommend air apportionment priorities to the JFC that should be devoted to the various air operations for a given period of time, after considering objective, priority, or other criteria and consulting with other component commanders.
- Allocate and task the joint air capabilities and forces made available by the Service components based on the JFC's air apportionment decision.
- Provide the JFACC's guidance in the air operations directive (AOD) for the use of joint air capabilities for a specified period that is used throughout the planning stages of the joint air tasking cycle and the execution of the ATO. The AOD may include the JFC's apportionment decision, the JFACC's intent, objectives, weight of effort, and other detailed planning guidance that includes priority of joint air support to JFC and other component operations
- Provide oversight and guidance during execution of joint air operations, to include making timely adjustments to taskings of available joint air forces. The JFACC coordinates with the JFC and affected component commanders, as appropriate, or when the situation requires changes to planned joint air operations.
- Assess the results of joint air operations and forward assessments to the JFC to support the overall assessment effort.
- Perform the duties of the airspace control authority (ACA), if designated.
- Perform the duties of the area air defense commander (AADC), if designated.
- Perform the duties of the space coordinating authority (SCA), if designated. The SCA is responsible for planning, integrating, and coordinating space operations support in the operational area, to include ascertaining space requirements within the joint force. If the individual designated to be the JFACC is also designated the SCA, the individual will normally designate a senior subject matter expert, typically the director of space forces (if the commander, Air Force forces [COMAFFOR], is the JFACC) to execute the daily actions associated with SCA. SCAs for a joint force will normally coordinate their requirements for space effects with the CCMD SCA (if established) or United States Strategic Command's (USSTRATCOM's) Joint Force Space Component Command.
- Perform the duties of the Personnel Recovery (PR) coordinator, as required.
- In concert with the above responsibilities, perform tasks within various mission areas to include, but not limited to:
  - Defensive counterair (DCA) and offensive counterair (OCA)
  - Close air support (CAS)
  - Airborne ISR and incident awareness and assessment
  - Air mobility operations
  - Strategic attack
  - Air interdiction

II. Joint Air Operations Planning

Ref: JP 3-30, Joint Air Operations (Jul '19), chap. III.

Planning for joint air operations begins with understanding the JFC's mission and intent. The JFC's estimate of the operational environment and articulation of the objectives needed to accomplish the mission form the basis for determining components' objectives. The JFACC uses the JFC's mission, commander's estimate and objectives, commander's intent, CONOPS, and the components' objectives to develop a course of action (COA). When the JFC approves the JFACC's COA, it becomes the basis for more detailed joint air operations planning— expressing what, where, and how joint air operations will affect the adversary or current situation. The JFACC's daily guidance ensures joint air operations effectively support the joint force objectives while retaining enough flexibility in execution to adjust to the dynamics of military operations.

Joint Air Operations Planning

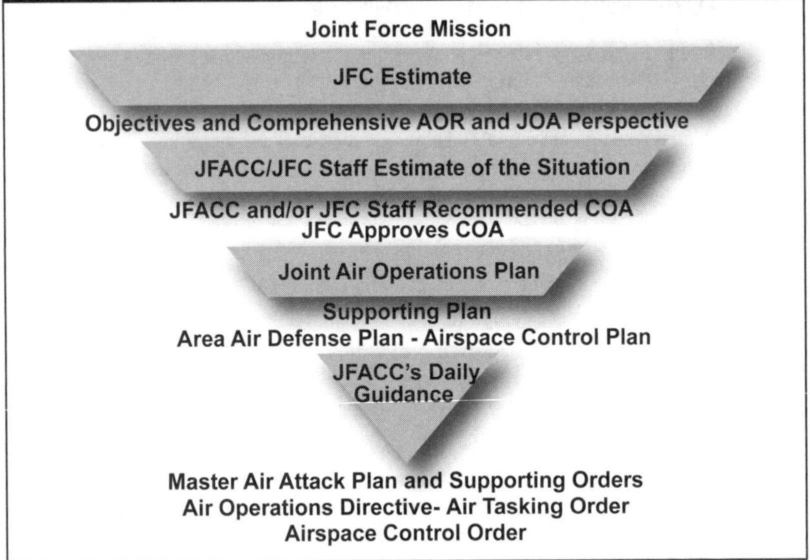

Ref: JP 3-30 (Jul '19), fig. III-1. Joint Air Operations Planning.

The JFACC's role is to plan joint air operations. In doing so, the JFACC provides focus and guidance to the JAOC staff. The amount of direct involvement depends on the time available, preferences, and the experience and accessibility of the staff. The JFACC uses the entire staff during planning to explore the full range of adversary and friendly COAs and to analyze and compare friendly air capabilities with the threat. The JFACC should ensure planning occurs in a collaborative manner with other components. Joint air planners should meet on a regular basis with the JFC's planners and with planners from other joint force components

The Joint Air Operations Plan (JAOP)

The JFACC uses the joint planning process for air (JPPA) to develop a JAOP, which guides employment of air capabilities and forces made available to accomplish missions assigned by the JFC. The JAOP is the JFC's plan to integrate and coordinate joint air operations and encompasses air capabilities and forces supported by, and in support of, other joint force components. The JFACC's planners should anticipate the need to make changes to plans (e.g., sequels or branches) in a dynamic and time-constrained environment.

JPPA Inputs, Steps, Outputs

Key Inputs	JPPA Steps	Key Outputs
Tasking from JFC Guidance from JFACC	Initiation	Initial planning time line JFACC's initial guidance
JFC mission and intent Friendly situation JIPOE Facts and assumptions JFACC tasks/guidance	Mission Analysis	Enemy COAs Mission analysis brief Essential tasks JFACC mission statement JFACC initial operational approach, planning guidance and intent
JFACC operational approach and guidance Enemy COAs Staff estimates supporting JFACC's COAs	COA Development	Support concepts to friendly air COAs JFACC objectives Narratives and graphics
Support concepts to friendly air COAs Enemy most likely/dangerous COAs Coordinated wargame method Coordinated evaluation criteria Coordinated critical events/actions	COA Analysis and Wargaming	Refined, valid air COAs with support concepts Strengths and weaknesses Branch/sequel requirements JFACC decision points and CCIRs
Coordinated evaluation criteria Wargame results Coordinated comparison method	COA Comparison	Decision matrix Preferred COAs
Decision briefing	COA Approval	Selected air COA Summary of operational design/approach JFACC refinement JFC-approved air COA
Approved air COA Staff estimates	Plan/Order Development	Refined and approved JFACC OPLAN/OPORD with appropriate annexes

Legend

CCIR	commander's critical information requirement
COA	course of action
JFACC	joint force air component commander
JFC	joint force commander
JIPOE	joint intelligence preparation of the operational environment
JPPA	joint planning process for air
OPLAN	operation plan
OPORD	operation order

Ref: JP 3-30 (Jul '19), fig. III-4. Joint Planning Process for Air Inputs, Steps, Outputs.

Refer to AFOPS2: The Air Force Operations & Planning SMARTbook, 2nd Ed. (Guide to Curtis E. LeMay Center & Joint Air Operations Doctrine). Topics and references of the 376-pg AFOPS2 include airpower fundamentals and principles (Vol 1), command and organizing (Vol 3); command and control (Annex 3-30/3-52), airpower (doctrine annexes), operations and planning (Annex 3-0), planning for joint air operations (JP 3-30/3-60), targeting (Annex 3-60), and combat support (Annex 4-0, 4-02, 3-10, and 3-34).

III. Airspace Control Authority (ACA)

The ACA is a commander designated by the JFC to assume overall responsibility for the operation of the airspace control system (ACS) in the airspace control area. Developed by the ACA and approved by the JFC, the ACP establishes general guidance for the control of airspace and procedures for the ACS for the joint force operational area. The ACO implements specific control procedures for established time periods. It defines and establishes airspace for military operations as coordinated by the ACA and notifies all agencies of the effective time of activation and the structure of the airspace. The ACO is normally published either as part of the ATO or as a separate document and provides the details of the approved requests for coordination measures such as airspace coordinating measures, air defense measures, and fire support coordination measures (FSCMs). All air missions are subject to the ACO and the ACP. The ACO and ACP provide direction to integrate, coordinate, and deconflict the use of airspace within the operational area. (Note: This does not imply any level of command authority over any air assets.) Methods of airspace control vary by military operation and level of conflict from positive control of all air assets in an airspace control area to procedural control of all such assets or any effective combination.

IV. Area Air Defense Commander (AADC)

The AADC is responsible for DCA operations, which include the integrated air defense system for the JOA. DCA and OCA operations combine as the counterair mission, which is designed to attain and maintain the degree of control of the air and protection desired by the JFC. In coordination with the component commanders, the AADC develops, integrates, and distributes a JFC-approved joint AADP. Typically, for forces made available for DCA, the AADC retains TACON of air sorties, while surface-based air and missile defense forces (e.g., Patriot missile systems) may be provided in support from another component commander. As such, the Army air and missile defense command (AAMDC) should be collocated with the joint air operations center (JAOC), if established, and conduct collaborative intelligence preparation of the battlespace (IPB), planning, and execution control. In distributed operations, the AAMDC may not be in the JAOC but is still functionally tied to it. The Navy component commander (NCC) (or joint force maritime component commander [JFMCC], if designated) exercises OPCON of maritime multi-mission and missile defense ships. When designated, these air and missile defense capabilities are in direct support of the AADC for C2 and execution of air defense.

V. Joint Air Operations Command and Control

Normally, the joint air operation C2 system will be built around the C2 system of the Service component commander designated as the JFACC. Each Service component has an organic system designed for C2 of their air operations. Whether it is the Air Force's theater air control system (TACS), the Army air-ground system (AAGS), the Navy's composite warfare commander (CWC)/Navy tactical air control system (NTACS), Marine air command and control system (MACCS), or the special operations air-ground system (SOAGS) that serves as the nucleus for C2 of joint air operations, the remainder will be integrated to best support the JFC's CONOPS.

Theater Air-Ground System (TAGS)

When all elements of the TACS, AAGS, CWC/NTACS, MACCS with fire support coordination center hierarchy, and SOAGS integrate, the entire system is labeled the TAGS.

Chap 6

II. Joint Land Operations

Ref: JP 3-31, Joint Land Operations (Feb '14).

In the 20th century, joint and multinational operations have encompassed the full diversity of air, land, maritime, and space forces operating throughout the operational area. Advances in capabilities among all forces and the ability to communicate over great distances have made the application of military power in the 21st century more dependent on the ability of commanders to synchronize and integrate joint land operations with other components' operations. Many of these advances have been realized through the use of cyberspace and the electromagnetic spectrum (EMS), which has enabled the US military and allies to communicate and reach across geographic and geopolitical boundaries. However, these advances have also led to increased vulnerabilities and a critical dependence on cyberspace and the EMS for the US and its allies.

Joint land operations include any type of joint military operations, singly or in combination, performed across the range of military operations with joint land forces (Army, Marine, or special operations) made available by Service components in support of the joint force commander's (JFC's) operation or campaign objectives, or in support of other components of the joint force. Joint land operations require synchronization and integration of all instruments of national power to achieve strategic and operational objectives. Normally, joint land operations will also involve multinational land forces.

Joint land operations includes land control operations. These are described as the employment of land forces, supported by maritime and air forces (as appropriate) to control vital land areas. Such operations are conducted to establish local military superiority in land operational areas. Land control operations may also be required to isolate, seize, or secure weapons of mass destruction (WMD) to prevent use, proliferation, or loss.

Joint Publication (JP) 3-0, Joint Operations, establishes the JFC's operational environment as composed of the air, land, maritime, and space domains as well as the information environment (which includes cyberspace). Domains are useful constructs to aid in visualizing and characterizing the physical environment in which operations are conducted. Nothing in the definitions of, or the use of the term domain, implies or mandates exclusivity, primacy, or C2 of that domain. C2 is established by the JFC based upon the most effective use of available resources to accomplish assigned missions. The land domain is the land area of the Earth's surface ending at the high water mark and overlapping with the maritime domain in the landward segment of the littorals. The land domain shares the Earth's surface with the maritime domain.

See chap. 2, Joint Operations.

Land operations are conducted within a complex operational environment. Numbers of civilians, amount of valuable infrastructure, avenues of approach, freedom of vehicular movement, and communications functionality vary considerably among land environments, creating challenges for the JFLCC. In addition, urban or emerging subterranean environments require special consideration for the conduct of joint land operations. As a result, joint land operations require an effective and efficient C2 structure to achieve success.

It is important to understand that in today's complex operational environment, adversary actions can be delivered on, from, within, and outside of the operational area, all with potentially global impacts and influence. To negate those threats, commanders at all levels should consider how space, cyberspace, and EMS capabilities enhance the effectiveness and execution of joint land operations.

Organizing the Joint Land Force

Joint land operations include any type of joint military operations, singly or in combination, performed across the range of military operations with joint land forces (Army, Marine, or special operations) made available by Service components in support of the joint force commander's (JFC's) operation or campaign objectives, or in support of other components of the joint force.

If the JFC does not choose to retain control at the JFC level, there are four primary options available to the JFC for employing land forces from two or more components:

- Subordinate unified command for land operations (available only to a combatant commander)
- Subordinate joint task forces
- Service components
- Functional land component with joint force land component commander (JFLCC)

Not only can the GCC designate a JFLCC, but each subordinate JFC may also designate their own JFLCC. Consequently, there may be multiple LCCs, each with an organization, duties, and responsibilities tailored to the requirements of their specific JFC, within a single AOR. Where multiple JOAs each have land operations being conducted, the JFLCC designated directly by the GCC may also be designated the theater JFLCC. The primary responsibilities of the theater JFLCC may be to provide coordination with other theater-level functional components, to provide general support to the multiple JFLCCs within the AOR, to conduct theater-level planning, or to conduct joint reception, staging, onward movement, and integration (JRSOI) for the entire joint land force. The most likely candidate for a theater JFLCC is the Army Service component command (ASCC)/theater army. Within a JOA or when there is only one JFLCC in an AOR, the JFC forms a functional land component to improve combat efficiency, unity of effort, weapons system management, component interaction, or control the land scheme of maneuver. Forming a functional land component is a key organizational decision, which will significantly influence the conduct of land operations.

I. The Joint Force Land Component Command (JFLCC)

The JFC has the authority to organize forces to best accomplish the assigned mission based on the CONOPS. The JFC establishes subordinate commands, assigns responsibilities, establishes or delegates appropriate command relationships, and establishes coordinating instructions for the component commanders. Sound organization provides for unity of command, centralized planning and direction, and decentralized execution. Unity of command is necessary for effectiveness and efficiency. Centralized planning and direction is essential for controlling and coordinating the efforts of the forces. Decentralized execution is essential because no one commander can control the detailed actions of a large number of units or individuals. When organizing joint forces, simplicity and clarity are critical; by making the JFLCC the single commander for joint land operations, the JFC has the ability to enhance synchronization of operations not only between US ground and component forces, but also with multinational land forces.

The JFC defines the authority and responsibilities of the functional component commanders based upon the CONOPS, and may alter this authority during the course of an operation.

The designation of a JFLCC normally occurs when forces of significant size and capability of more than one Service component participate in a land operation and the JFC determines that doing this will achieve unity of command and effort among land forces.

JFLCC Responsibilities

Ref: JP 3-31, Joint Land Operations (Feb '14), chap. II.

The JFLCC's overall responsibilities and roles are to plan, coordinate, and employ forces made available for tasking in support of the JFC's CONOPS. The responsibilities of the JFLCC include, but are not limited to, the following:

- Advising the JFC on the proper employment of forces made available for tasking. Developing, integrating, maintaining, and sharing with the JFC an accurate representation of the land common operational picture (COP) (objects and events) within the JFLCC's operational area, as an input to the JFC's COP.
- Developing the joint land operation plan (OPLAN)/operation order (OPORD) in support of the JFC's CONOPS and optimizing the operations of task-organized land forces.
- Directing the execution of land operations as specified by the JFC, which includes making timely adjustments to the tasking of forces and capabilities made available.
- Coordinating the planning and execution of joint land operations with the other components and supporting agencies.
- Evaluating the results of land operations to include the effectiveness of interdiction operations and forwarding these results to the JFC to support the combat assessment effort.
- Synchronizing and integrating movement and maneuver, fires, and interdiction in support of land operations.
- Designating the target priorities, effects, and timing for joint land operations.
- Planning and conducting personnel recovery (PR) in support of joint land operations and for isolating events occurring within assigned operational area or as tasked by the JFC.
- Providing mutual support to other components by conducting operations such as suppression of enemy air defenses and suppression of threats to maritime operations.
- Coordinating with other functional and Service components' sustainment support in accomplishment of JFC objectives.
- Providing an assistant or deputy to the area air defense commander (AADC) for land-based joint theater integrated air and missile defense (AMD) operations and coordination as determined by the JFC.
- Supporting the JFCs information operations (IO) by developing the IO requirements that support land operations and synchronizing the land force information-related capabilities (IRCs) when directed.
- Establishing standard operating procedures (SOPs) and other directives based on JFC guidance.
- Providing inputs into the JFC-approved joint operational area air defense plan (AADP) and the airspace control plan (ACP).
- Planning and determining requirements for, and coordinating implementation of, the JFLCC's communications systems, integrating them into the theater's Department of Defense information networks (DODIN) architecture.
- Integrating cyberspace operations (CO) into plans.
- Integrating special operations, as required, into overall land operations.
- Performing joint security functions, such as serving as the joint security coordinator (JSC), as designated by the JFC.
- Supervising detainee operations as designated by the JFC.
- Facilitating interorganizational coordination, as required.

II. Support to Joint Operation Planning

Ref: JP 3-31, Joint Land Operations (Feb '14), chap. III.

JFLCC planning tasks are to:

- Prepare and coordinate required land component OPLANs or OPORDs in support of assigned JFC missions.
- Coordinate land component planning efforts with higher, lower, adjacent, and multinational headquarters as required.
- Develop land component COAs within the framework of the JFC-assigned objectives or missions, forces available, and the commander's intent.
- Determine land component force requirements and coordinate land force planning in support of the selected COAs. The JFLCC conducts planning using the planning processes of the command that forms the core of the headquarters. While almost all headquarters use the planning cycle described in joint planning publications, the specific steps in the process may have different names and somewhat different activities. The JFLCC's staff, provided by Services other than the core of the headquarters and integrated into the core staff, must quickly adapt to the planning processes and battle rhythm of the staff they are joining.

Joint Planning Process (JPP)

The joint planning process (JPP) is a proven analytical process that provides an orderly approach to planning at any point of joint operations. JPP may be used by a JFLCC's staff during deliberate and crisis action planning. The focus of JPP is the interaction for planning between commanders, staffs, and echelons. JPP is also linked with the joint intelligence preparation of the operational environment (JIPOE). JIPOE is the analytical process used by joint intelligence organizations to produce intelligence assessments, estimates, and other intelligence products in support of the JFC's decision¬making process. The process is used to analyze the physical domains; the information environment; political, military, economic, social, information, and infrastructure systems; and all other relevant aspects of the operational environment, and to determine an adversary's capabilities to operate within that environment.

See pp. 3-69 to 3-118 for discussion of the joint planning process (JPP) from JP 5-0. See pp. 3-72 to 3-73 for an overview and discussion of joint intelligence preparation of the operational environment (JIPOE).

Commander's Operational Approach

The JFLCC's planners must first frame the strategic and operational problem by developing an understanding of the situation before addressing operational design and ultimately OPLANs. Several cognitive models exist to assist JFLCC's and their staffs as they plan and execute joint land operations. The operational approach is the commander's visualization of how the operations should transform current conditions at end state.

The operational approach is based largely on an understanding of the operational environment and the problem facing the JFLCC.

Developing a commander's operational approach provides for problem framing as one method for establishing the context of a situation within which a commander and staff must act to achieve the strategic objectives. The essence of problem framing is to examine the problem from multiple perspectives and set conditions for learning about the problem throughout the planning and execution of military operations. Framing can also support the commander's discourse with superiors regarding the nature of the problem the commander has been asked to solve. It also assists in developing a mutual understanding of the operational environment.

See p. 3-53 and 3-54 for related discussion from JP 5-0.

Joint Planning Group (JPG)

The primary planning element for the JFLCC to support the JFC's planning or to perform component planning is the JPG. Planners from the JFLCC's core headquarters staff element are the nucleus around which the JPG is normally built. It includes personnel from each of the primary coordinating, functional, and special staff elements, LNOs, and when necessary, planners from the JFLCC's subordinate commands or multinational land forces (see Figure III-1). The JPG develops and disseminates staff planning guidance and schedules. It confirms the process and products to be developed and delivered to support the JFLCC's planning effort. The JPG is the planning hub and synchronization center for future plans. The JPG develops the CONOPS for each plan.

See p. 5-23 for related discussion of the JPG from JP 3-33.

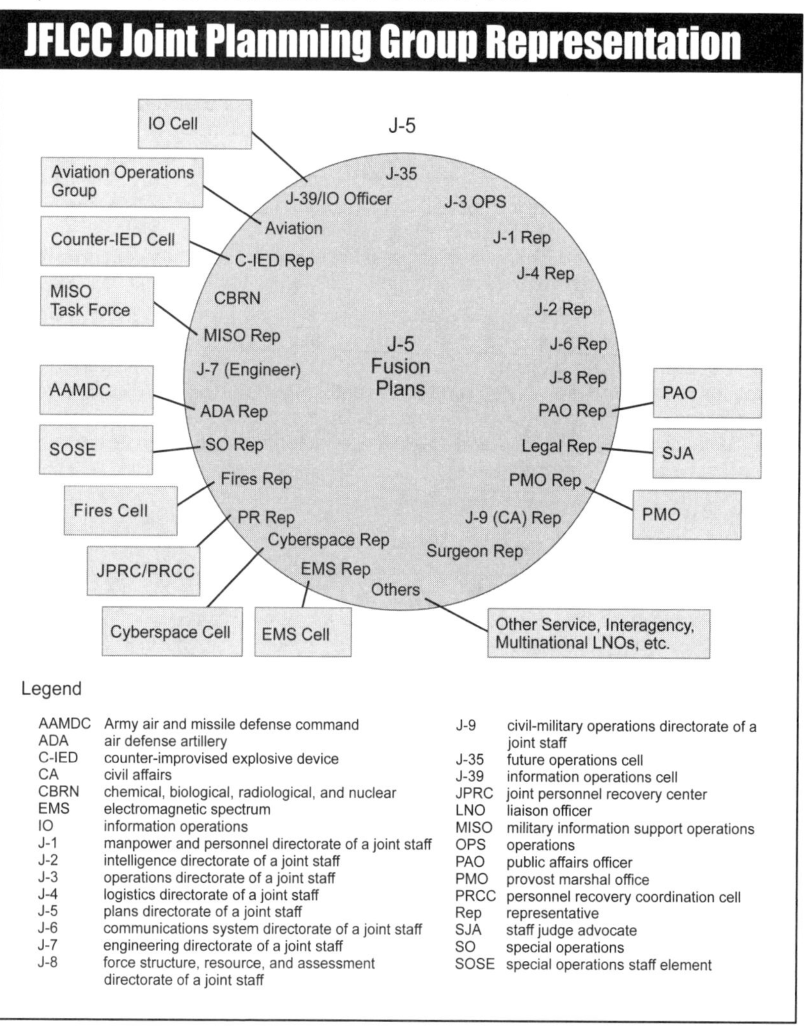

Ref: JP 3-31 (Feb '14), fig. III-1. Joint Force Land Component Commander Joint Planning Group Representation.

Joint Force Operations

III. The Army in Joint Operations

Ref: ADP 3-0, Operations (Jul '19).

The Army's primary mission is to organize, train, and equip its forces to conduct prompt and sustained land combat to defeat enemy ground forces and seize, occupy, and defend land areas. The Army accomplishes its mission by supporting the joint force and unified action partners in four strategic roles: shape operational environments, prevent conflict, prevail in large-scale ground combat, and consolidate gains. The strategic roles clarify the enduring reasons for which the Army is organized, trained, and equipped. Strategic roles are not tasks assigned to subordinate units.

Unified Land Operations

Unified land operations is the Army's warfighting doctrine, and it is the Army's operational concept and contribution to unified action. Unified land operations is an intellectual outgrowth of both previous operations doctrine and recent combat experience. It recognizes the nature of modern warfare in multiple domains and the need to conduct a fluid mix of offensive, defensive, and stability operations or DSCA simultaneously. Unified land operations acknowledges that strategic success requires fully integrating U.S. military operations with the efforts of interagency and multinational partners. Army forces, as part of the joint force, contribute to joint operations through the conduct of unified land operations. Unified land operations is the simultaneous execution of offense, defense, stability, and defense support of civil authorities across multiple domains to shape operational environments, prevent conflict, prevail in large-scale ground combat, and consolidate gains as part of unified action.

The goal of unified land operations is to establish conditions that achieve the JFC's end state by applying landpower as part of a unified action to defeat the enemy. Unified land operations is how the Army applies combat power through 1) simultaneous offensive, defensive, and stability, or DSCA, to 2) seize, retain, and exploit the initiative, and 3) consolidate gains. Military forces seek to prevent or deter threats through unified action, and, when necessary, defeat aggression.

Decisive Action

Decisive action is the continuous, simultaneous execution of offensive, defensive, and stability operations or defense support of civil authority tasks. Army forces conduct decisive action. Commanders seize, retain, and exploit the initiative while synchronizing their actions to achieve the best effects possible. Operations conducted outside the United States and its territories simultaneously combine three elements of decisive action—offense, defense, and stability. Within the United States and its territories, decisive action combines elements of DSCA and, as required, offense and defense to support homeland defense.

Army forces create depth in time and space through combined arms, economy of force, continuous reconnaissance, and joint capabilities. Conducting operations across large areas forces an adversary or enemy to react in multiple directions and opens up opportunities that can be further exploited to create additional dilemmas.

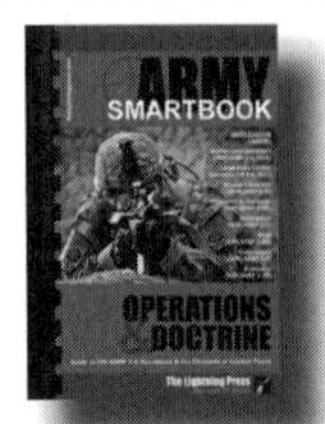

The Army's Operational Construct

The Army's operational concept of unified land operations—including its principles, tenets, and operational structure—serves as the basic framework for all operations across the conflict continuum. It is the core of Army doctrine and guides how Army forces contribute to unified action.

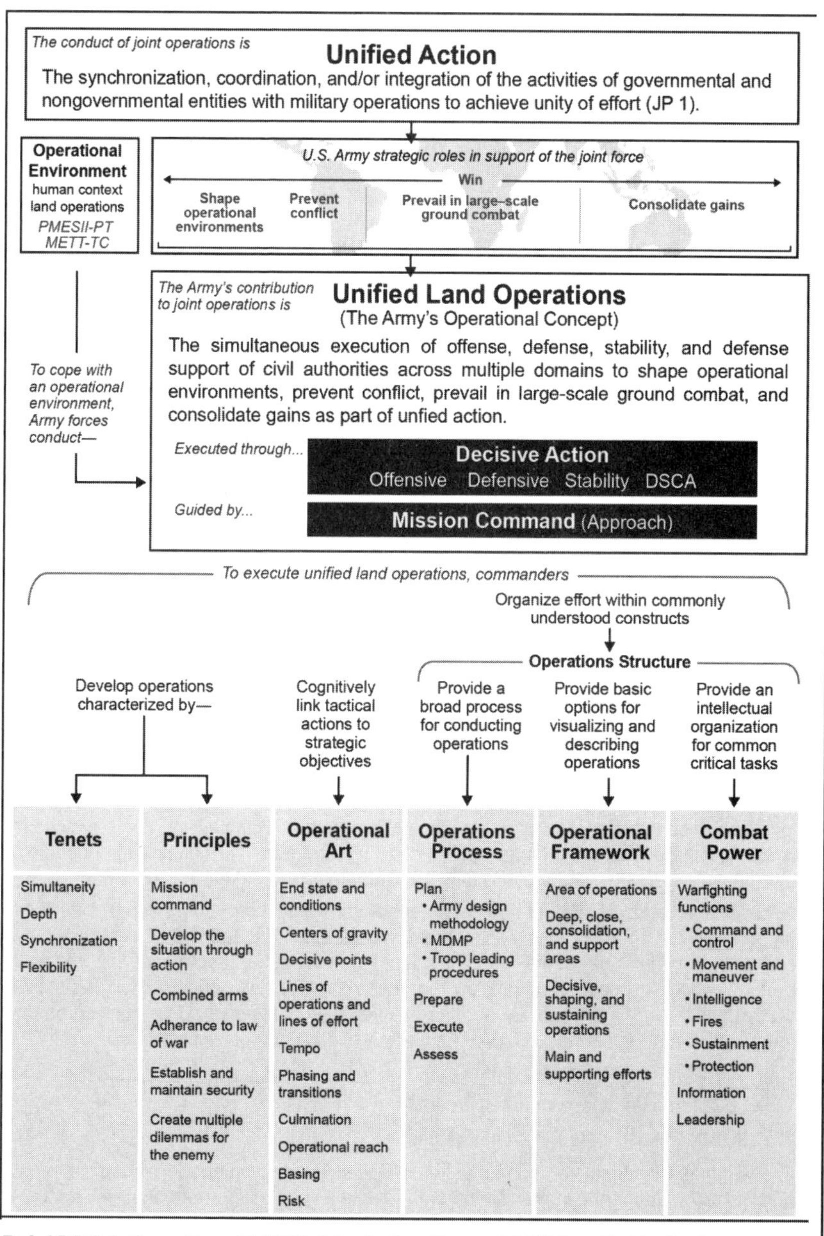

Ref: ADP 3-0, Operations (Jul '19), Introductory figure. ADP 3-0 unified logic chart.

Joint Force Operations

IV. Marine Corps Forces and Expeditionary Operations

Ref: MCDP 1-0, Marine Corps Operations (Aug '11).

The nation requires an expeditionary force-in-readiness capable of responding to a crisis anywhere in the world. The Marine Corps provides self-sustainable, task organized combined arms forces capable of conducting a full spectrum of operations in support of the joint force commander. These missions might include forcible entry operations, peace enforcement, evacuation of American citizens and embassies, humanitarian assistance/disaster relief, or operations to reinforce or complement the capabilities of other Services to provide balanced military forces to the joint force commander. The unique capabilities of the Marine Corps as a sea service and partner with the U.S. Navy allow the use of the sea as both a maneuver space and a secure base of operations from which to conduct operations in the littoral areas of the world. The ability to remain at sea for long periods of time without the requirement of third nation basing rights makes the Marine Corps the force of choice in emerging crises.

Expeditionary Maneuver Warfare Concept

Expeditionary maneuver warfare is the Marine Corps capstone operational concept. It applies the doctrine of maneuver warfare to Marine Corps expeditionary operations to achieve desired effects across the spectrum of conflict. Supporting operational concepts such as Operational Maneuver From The Sea (OMFTS), Ship-To-Objective Maneuver (STOM), MPF 2010, and Expeditionary Bases and Sites are all elements of expeditionary maneuver warfare.

Operations as Part of the Joint Force

Marine Corps forces normally conduct operations as part of a joint force. Regardless of the level of the joint force or how a joint force commander organizes his force, if Marine Corps forces are assigned, there is always a Marine Corps Service component. There are two levels of Marine Corps components-a Marine Corps component under a unified command and a Marine Corps component under a subordinate unified command or a joint task force.

Forward-deployed naval forces, including Marine Corps forces, are usually the first conventional forces to arrive in an austere theater or AO during expeditionary operations. The Marine Corps component commander's inherent capability to command and control Marine Corps forces—and attached or assigned forces of other Services or nations—allows him to serve as a functional component commander. Such assignments may be for limited contingencies or for some phases of a major operation or campaign, depending upon the size, scope, nature of the mission, and the functional area assigned.

If the Marine Corps component commander has functional component commander responsibilities, he normally executes them with his subordinate MAGTF. A Marine Corps component commander can also act as a functional component commander. This may be for a particular phase of an operation or for its full duration, depending upon the size, scope, and nature of the mission and the functional area assigned. The most common functional components the joint force commander may establish include:

- Joint force maritime component commander
- Joint force land component commander
- Joint force air component commander

In addition to functional component duties, the joint force commander can assign the Marine Corps component commander other joint duties such as the area air defense commander or airspace control authority. Again, these functions are normally accomplished by the assigned MAGTF.

The Marine Air-Ground Task Force (MAGTF)

The MAGTF is an air-ground combined arms task organization of Marine Corps forces under a single commander, structured to accomplish a specific mission. It is the Marine Corps' principal organization for all missions across the range of military operations. It is designed to fight, while having the ability to prevent conflicts and control crises. All MAGTFs are task-organized and vary in size and capability according to the assigned mission, threat, and battlespace environment. They are specifically tailored for rapid deployment by air or sea and ideally suited for a forward presence role. A MAGTF provides the naval, joint, or combined commander with a readily available force capable of operating as—

- The landing force of an amphibious task organization
- A land force in sustained operations ashore
- A land force or the landward portion of a naval force conducting operations such as non-combatant evacuations, humanitarian assistance, disaster relief, or the tactical recovery of an aircraft or aircrew
- A forward deployed force providing a strong deterrence in a crisis area
- A force conducting training with allied forces as part of a theater engagement plan

All MAGTFs are, by design, expeditionary, and comprised of four core elements: a command element (CE), a ground combat element (GCE), an aviation combat element (ACE), and a Logistics Combat Element (LCE). MAGTF combat forces reside within these four elements.

A MAGTF is an integrated combined arms forces structured to accomplish a specific mission. To provide a frame of reference for general sizing, a given MAGTF may be categorized in one of the following four types:

- Marine Expeditionary Force (MEF)
- Marine Expeditionary Brigade (MEB)
- Marine Expeditionary Unit (MEU)
- Special Purpose MAGTF

The MAGTF is the Marine Corps' principal organization for the conduct of all missions across the range of military operations. MAGTFs are integrated, combined-arms forces with organic ground, aviation, and sustainment elements. They are flexible, task-organized forces that can respond rapidly to a contingency anywhere in the world and are able to conduct a variety of missions. Although organized and equipped to participate as part of naval expeditionary forces, MAGTFs also have the capability to conduct sustained operations ashore. The MAGTF provides a combatant commander or other operational commander with a versatile expeditionary force that is capable of responding to a broad range of crisis and conflict situations. MAGTFs are organized, trained, and equipped to perform missions ranging from humanitarian assistance to peacekeeping to intense combat and can operate in permissive, uncertain, and hostile environments. They may be shore- or sea-based in support of joint and multinational major operations and/or campaigns. MAGTFs deploy as amphibious, air-contingency, or maritime prepositioned forces (MPF), either as part of a naval expeditionary force or via strategic lift.

See p. 6-21, Operational Employment for Amphibious Ready Groups with Embarked MEUs, and p. 6-26, Amphibious Operations, for related discussions.

Refer to MEU2: The Marine Expeditionary Unit (MEU) SMARTbook, 2nd Ed. (Guide to Battle Staff Operations & the Rapid Response Planning Process) for further discussion. The MEU SMARTbook is designed to be a reference for MEU and PHIBRON Commanders, MEU and PHIBRON staffs and the commanders and staffs of the Major Subordinate Elements (MSE) and Naval Support Elements (NSE) of the ARG-MEU team.

V. Forms of Operations

JFCs strive to apply the many dimensions of military power to address both traditional warfare and irregular warfare (IW) simultaneously across the depth, breadth, and height of the operational area. Consequently, JFCs normally achieve concentration in some operations or in specific functions and require economy of force in others. All joint OPLANs must feature an appropriate combination and balance between offensive and defensive operations and stability operations in all phases. Planning for stability operations should begin when joint operation planning is initiated. Planning for the transition from sustained land combat operations to the termination of joint operations and then a complete handover to civil authority and redeployment must commence during plan development and be ongoing during all phases of a joint campaign or major operation. Even while sustained land operations are ongoing, there will be a need to establish or restore security and provide humanitarian relief as succeeding areas are occupied or bypassed.

Types of Military Operations

The US employs its military capabilities at home and abroad in support of its national security goals in a variety of operations. Some operations conducted by a JFLCC may involve only steady-state, routine, recurring military activities that do not relate directly to either traditional war or IW. Other operations, such as COIN, support to insurgency, and combating terrorism, primarily involve IW. Operations such as nation assistance, foreign internal defense, peace operations, FHA, combating terrorism, counterdrug operations, show of force operations, and arms control are applied to meet military engagement, security cooperation, and deterrence objectives. Major operations and campaigns are typically characterized by large-scale combat operations associated with traditional war. All of these circumstances—each potentially with different root causes and objectives—can exist concurrently within a single operational area and may require consideration by a JFLCC.

See p. 2-44 for related discussion from JP 3-0, Joint Operations.

Major operations and campaigns, whether or not they involve large-scale ground combat, normally will include some level of both offense and defense.

Offensive land operations are combat operations conducted to defeat and destroy enemy land forces and seize terrain, resources, and population centers. Offensive land operations impose the commander's will on the enemy.

Defensive operations are combat operations conducted to defeat an enemy attack, gain time, economize forces, and develop conditions favorable for offensive or stability operations. Defensive land control operations retain terrain, guard populations, and protect critical capabilities against enemy attacks and are used to gain time and economize forces so offensive tasks can be executed elsewhere.

Stability operations encompass various military missions, tasks, and activities conducted outside the US in coordination with other instruments of national power to maintain or reestablish a safe and secure environment, provide essential governmental services, emergency infrastructure reconstruction, and humanitarian relief. Stability operations will not only include stability tasks, but will often have elements of offense and defense.

Military operations inside the US and its territories, though limited in many respects, are conducted to accomplish two missions: homeland defense and defense support of civil authorities (DSCA). A JFLCC is often used to provide C2 for land operations for DSCA. DSCA consists of Department of Defense support to US civil authorities for domestic emergencies, both man-made and natural, and for designated law enforcement and other activities, such as national special security events.

See p. 2-44 for related discussion from JP 3-0, Joint Operations.

III. Joint Maritime Operations

Ref: JP 3-32, Joint Maritime Operations (Jun '18).

Maritime forces operate on (surface), under (subsurface), or above (air) the sea and/or above and on the land in support of amphibious operations, port security, infrastructure protection, strike, and integrated air and missile defense (IAMD) operations and other types of operations across the range of military operations.

- Movement and maneuver of forces within international waters can take place without prior diplomatic agreement.
- Maritime forces are mostly a self-deploying, self-sustaining, sea-based expeditionary force and a combined-arms team. Maritime forces are manned, trained, and equipped to operate with limited reliance on ports or airfields.

Maritime Domain

The maritime domain is the oceans, seas, bays, estuaries, islands, coastal areas, and the airspace above these, including the littorals. Nothing in the definition of, or the use of the term domain, implies or mandates exclusivity, primacy, or C2 of that domain. Per JP 2-01.3, Joint Intelligence Preparation of the Operational Environment, the littoral comprises two segments of the OE. First, "seaward: the area from the open ocean to the shore, which must be controlled to support operations ashore." Second, "landward: the area inland from the shore that can be supported and defended directly from the sea."

The maritime domain also has unique economic, diplomatic, military, and legal aspects. US naval forces operate in the deep waters of the open ocean and other maritime environments, including coastal areas, rivers, estuaries, and landward portions of the littorals, including associated airspace. In many regions of the world, rivers mark and define international borders and facilitate intracontinental trade. Ensuring access and securing these waterways are often priorities of state governments seeking to maintain stability and sovereignty.

Maritime forces can participate in multiple operations ashore. They can execute, support, or enable missions ashore by conducting forcible entry operations (such as an amphibious assault), seizing/establishing expeditionary advance bases, seabasing of assets, moving land forces into the operational area via sealift, providing fire and air support, and influencing operations through deterrence. Maritime forces may be employed in littoral waters for the conduct of sea control or denial, ballistic missile defense (BMD), and to support joint force or component C2 platforms. Joint forces can support maritime operations with surveillance, logistics, fires, air support, and military engineering.

Joint Maritime Operations

Joint maritime operations are performed with maritime forces, and other forces assigned, attached, or made available, in support of the JFC's operation or campaign objectives or in support of other components of the joint force. The JFC may designate a JFMCC to C2 a joint maritime operation. As a functional component commander, the JFMCC has authority over assigned and attached forces and forces made available for tasking.

> *The Sea Services have historically organized, trained, and equipped to perform four essential functions: deterrence, sea control, power projection, and maritime security. Because access to the global commons is critical, this strategy introduces a fifth function: all domain access. This function assures appropriate freedom of action in any domain—the sea, air, land, space, and cyberspace, as well as in the electromagnetic (EM) spectrum.*
>
> *- A Cooperative Strategy for 21st Century Seapower (Mar 2015)*

The degree of integration and coordination between joint force component commanders varies depending on the situation. For some joint maritime operations, the JFMCC may operate without the support of other Service component forces (e.g., open ocean submarine operations); whereas for others, there may be detailed integration between components (e.g., attack of enemy submarines in port or their supporting critical infrastructures ashore). In other cases, tactical control (TACON) of maritime forces may be delegated to other joint force components (e.g., close air support [CAS] and strategic attack). For sea control operations, TACON of another joint force component's forces may be delegated to the JFMCC (e.g., air operations in maritime surface warfare [SUW]). In certain situations, specification of operational control (OPCON) or TACON of forces may not be practical. In these cases, the JFC should establish a support relationship, as required. All major operations generally necessitate some degree of maritime support to deploy, sustain, withdraw, and redeploy forces.

Joint maritime operations occur in blue water, green water, brown water environments, and in the landward areas in the littorals, each with its own challenges. Operations in blue water (high seas and open oceans) require forces capable of remaining on station for extended periods, largely unrestricted by sea state, and with logistics capability to sustain these forces indefinitely. Operations in green water (coastal waters, ports, and harbors) stretching seaward require ships, amphibious warfare ships and landing craft, and patrol craft with the stability and agility to operate effectively in surf, in shallows, and the near-shore areas of the littorals. Operations in brown water (navigable rivers, lakes, bays, and their estuaries) involve shallows and congested areas that constrain maneuver but do not subject maritime forces to extreme surf conditions. Operations on land in the littorals may involve landing forces going ashore by embarked aircraft, landing craft, and amphibious vehicles from amphibious warfare ships.

I. Joint Force Maritime Component Commander (JFMCC)

JFCs at the combatant command (CCMD), subordinate unified command, and/or joint task force (TF) command levels all organize staffs and forces to accomplish the mission based on their vision and concept of operations (CONOPS). Organizing the maritime force should take into account the nature of today's complex OE; technological advances in communications; intelligence collection systems; improved weapons capabilities; and how multinational forces organize, train, equip, and conduct operations. Equally important in determining how a JFC organizes joint

JFMCC Roles and Responsibilities

Ref: JP 3-32, Joint Maritime Operations (Jun '18), fig. II-1, p. II-2.

JFMCC responsibilities are to plan, coordinate, allocate, and task joint maritime operations based on the JFC's CONOPS and apportionment decisions. Specific responsibilities that are normally assigned to the JFMCC include:

- Develop a joint maritime operations plan to best support joint force objectives.
- Provide centralized direction for the allocation and tasking of forces/capabilities made available.
- Request forces of other component commanders when necessary for the accomplishment of the maritime mission.
- Make maritime apportionment recommendations to the joint force commander (JFC).
- Provide maritime forces to other component commanders in accordance with JFC maritime apportionment decisions.
- Control the operational level synchronization and execution of joint maritime operations, as specified by the JFC, to include adjusting targets and tasks for available joint capabilities/forces. The JFC and affected component commanders will be notified, as appropriate, if the joint force maritime component commander (JFMCC) changes the planned joint maritime operations during execution.
- Act as supported commander within the assigned area of operations (AO).
- Assign and coordinate target priorities within the assigned AO by synchronizing and integrating maneuver, mobility and movement, fires, and interdiction. The JFMCC nominates targets located within the maritime AO to the joint targeting process that may potentially require action by another component commander's assigned forces.
- Evaluate results of maritime operations and forward assessments to the JFC in support of the overall effort.
- Support JFC information operations with assigned assets, when directed.
- Function as a supported and supporting commander, as directed by the JFC.
- Perform other functions, as directed by the JFC.
- Establish a personnel recovery coordination cell to account for and report the status of isolated personnel and to coordinate and control maritime component personnel recovery events; and, if directed by the JFC, establish a separate joint personnel recovery center for the same purpose in support of a joint recovery event.
- Coordinate the planning and execution of joint maritime operations with the other components and supporting agencies.
- Integrate the JFMCC's communications systems and resources into the theater's networked communications system architecture, or common operational picture, and synchronize JFMCC's critical voice and data requirements. Ensure these communications systems requirements, coordination issues, and capabilities are integrated in the joint planning and execution process

The JFC establishes the authority and command relationships of the JFMCC. The JFMCC normally exercises OPCON over their own Service forces and TACON over other Service forces made available for tasking. Regardless of organizational and command arrangements within joint commands, Service component commanders are responsible for certain Service-specific functions and other matters affecting their forces: internal administration, training, logistics, and Service-unique intelligence operations. The JFMCC should be aware of all such Service-specific responsibilities forces are the threat's nature, capabilities, and the OE (e.g., geography, accessibility, climate, and infrastructure).

The JFC establishes subordinate commands, assigns responsibilities, establishes appropriate command relationships, provides coordinating instructions to optimize the capabilities of each subordinate, and gains synergistic effects for the joint force as a whole. The JFC may designate a JFMCC to facilitate unity of effort, focus, and synchronize efforts while providing subordinate commanders flexibility and opportunity to exercise initiative and maintain the joint forces' operational tempo.

Normally, joint forces are organized with a combination of Service and functional component commands with operational responsibilities. The JFC normally designates the forces and maritime assets that will be made available for tasking by the JFMCC and delegates the appropriate command authority the JFMCC will exercise over assigned and attached forces and maritime assets made available for tasking. Generally, these forces and maritime assets include navies, marines, special operations forces (SOF), coast guards and similar border patrol and revenue services, nonmilitary shipping managed by the government, civil merchant marines, army/ground forces (normally when embarked), and air and air defense (AD) forces. Establishment of a JFMCC must not affect the command relationships between Service component commanders and the JFC.

In cases where the JFC does not designate a JFMCC, the JFC may elect to directly task maritime forces. Typically, this would occur when an operation is of limited duration, scope, or complexity. If this option is exercised, the JFC's staff assists in planning and coordinating maritime operations for JFC approval. The JFC may elect to centralize selected functions (plan, coordinate, and task) within the staff to provide direction, control, and coordination of the joint force.

Naval command relationships are based on a philosophy of mission command involving centralized guidance, collaborative planning, and decentralized control and execution. With a long-standing practice of using mission-type orders, naval C2 practices are intended to achieve relative advantage through organizational ability to rapidly observe, orient, decide, and act. Mission-type orders enable continued operations allowing subordinates to exercise initiative consistent with the higher commander's intent and act independently to accomplish the mission in conditions where communications are restricted, compromised, or denied.

Since the JFC normally designates a Service component commander to also serve as a functional component commander, the dual-designated Service/functional component commander will normally exercise OPCON as a Service component commander over their own Service forces and TACON as a functional component commander over other Services forces made available for tasking. USN multi-mission ships are rarely made available for tasking outside the maritime component, because their multi-mission capabilities will require them to fulfill JFMCC operational requirements. However, some capabilities of multi-mission ships and other maritime forces may be made available to other components in direct support.

Forward-deployed maritime force packages, commonly called adaptive force packages, are normally comprised of units that train together prior to deploying. These tailored force packages may include carrier strike groups (CSGs), expeditionary strike groups, and amphibious ready groups (ARGs) with an embarked Marine expeditionary unit (MEU). Force packages can be scaled up by adding ships and capabilities or scaled down into smaller surface action groups (SAGs), individual ships, or special purpose forces designed to conduct numerous types of military operations.

II. Operational Employment for Amphibious Ready Groups with Embarked MEUs

Ref: JP 3-32, Joint Maritime Operations (Jun '18), fig. II-2, p. II-3 to II-4.

The ARG/MEU is a forward-deployed, flexible, sea-based force that provides the President and the geographic combatant commander (GCC) with credible deterrence and decision time across the range of military operations. The ARG and MEU affords the GCC a responsive, flexible, and versatile capability to shape the OE, respond to crises, and protect US and allied interests in permissive and select uncertain and hostile environments. ARG and MEU capabilities support initial crisis response, introduce follow-on forces, support designated SOF, and other missions in permissive and select uncertain and hostile environments, which include, but are not limited to: amphibious assaults, amphibious raids, amphibious demonstrations, amphibious withdrawals, and amphibious force support to crisis response and other operations (e.g., noncombatant evacuation operations, humanitarian assistance, or MSO). The ARG and Navy detachments are organized under the command of a Navy O-6, while the MEU, with its embarked Marine air-ground task force (MAGTF), is under the command of a Marine Corps O-6.

Aggregated

The most common form where the amphibious ready group (ARG)with embarked Marine expeditionary unit (MEU) is employed under a single geographic combatant commander (GCC) who maintains operational control (OPCON) or tactical control (TACON) of the ARG/MEU. "Split" is a subset of aggregated, where the ARG and MEU remains employed within a single GCC's area of responsibility(AOR), but the units are separated by time, distance, or task while operating beyond the reach of tilt-rotor aircraft or landing craft. Aggregated is the preferred employment construct.

Disaggregated

This construct is driven by emergent requirements wherein the ARG and MEU is divided into parts to support multiple GCCs. The ARG and MEU elements operate within the distinct OPCON/TACON chains oft he respective GCCs. Disaggregation comes with a corresponding degradation of ARG and MEU operational readiness, training, and maintenance. This is the least preferred employment construct.

Distributed

The ARG and MEU is partitioned for emergent requirements for multiple GCCs. However, the original GCC to whom it was allocated retains OPCON while another exercises TACON of elements that are distributed for a specific mission or duration mission. The ARG and MEU is able to sustain its elements, facilitate planning, and conduct military engagement and joint/combined training across AOR boundaries, and is supported throughout operations. ARG and MEU communication and computers systems are critical for supporting distributed operations. The GCC that has OPCON may request re-aggregation at any time, and the ARG and MEU commanders cannot make changes to capabilities allocated OPCON or TACON without approval. Distributed is the preferred employment construct to support multiple GCCs.

See pp. 6-14 to 6-15 for an overview and related discussion of Marine Corps Forces and Expeditionary Operations.

III. Organizing and Manning the Maritime Component Headquarters

Ref: JP 3-32 (Chg 1), Cmd & Control for Joint Maritime Operations (May '08), app. G.

The component HQ organization and staffing will differ depending upon the mission, OE, existing and potential adversaries, nature of the crisis (e.g., tsunami, cyclone, earthquake), time available, and desired end state. The JFMCC's staff is typically built from an existing Service component, numbered fleet, MAGTF, or subordinate Service force staff and then augmented as required. A joint air component coordination element is often included to coordinate JFACC missions.

Maritime Operations Center (MOC) and N-Code Structure

In a maritime HQ, two complementary methods of organizing people and processes exist. The first is the doctrinal N-code structure, which organizes people by the function they perform (i.e., intelligence, logistics). The second is a cross-functional staff that organizes the staff into boards, centers, cells, and working groups that manage specific processes or tasks that do not fit well under the N-code structure and require cross-functional participation, such as targeting and assessment. The fast pace of military operations and cross-talk needed to support an operational-level command has made the cross-functional approach the preferred manner of organization, while maintaining the doctrinal roles of the N-code structure. The maritime operations center (MOC) can be thought of as a loosely-bound network of staff entities overlaying the N-code structure. If a Navy component or numbered fleet commander is designated as the JFMCC, their existing staff or MOC will normally form the nucleus of the JFMCC staff or MOC. The formalized addition of this cross-functional network and process to the doctrinal N-code organizational structure is what constitutes the MOC. The MOC's focus is on operational tasks and activities (versus fleet management or support). It must be recognized, however, that when a commander establishes a MOC, the traditional staff code organization does not disappear. Indeed, the doctrinal N-code directorates are the foundation of the MOC. They supply the manpower, expertise, and facilities needed by the MOC to function. As a practical matter, the commander establishes and maintains only those boards, centers, cells, and working groups that enhance planning and decision making within the HQ. A fires cell, for example, is likely not required during a disaster relief operation. The commander establishes, modifies, and dissolves these functional entities as the needs of the command evolve. MOCs provide an organizational framework through which maritime commanders may exercise operational-level C2.

For more information on Navy MOCs, see Navy Warfare Publication (NWP) 3-32, Maritime Operations at the Operational Level of War, and Navy Tactics, Techniques, and Procedures (NTTP) 3-32.1, Maritime Operations Center.

Liaison elements from and to other joint force and Service components are also considerations in composition and required infrastructure. Joint force command relationships, the nature of the mission, and standing Service agreements help determine liaison manning requirements. The naval and amphibious liaison element is the primary coordination element at the joint air operations center (AOC).

Task Organization of Subordinate Forces

The JFMCC normally delegates the authority to plan and execute tactical missions to subordinate CTF or task group (TG) commanders. This enables the JFMCC to focus attention on the operational level and empowers subordinate commanders to employ their forces to support the commander's intent. Individual platforms are assigned or attached to these subordinate CTFs. Each CTF is assigned a commander, and only the commander reports to the JFMCC. The CTF may further subdivide the TF into TGs, task units, and task elements to exercise control at the tactical level. These subdivisions may be organized based on capabilities, missions, geography, or a hybrid of all three.

The JFMCC establishes the support relationships between the subordinate CTFs for various lines of operation. Further, given the nature of maritime operations and tasks assigned to a CTF, each CTF will likely be both a supported and supporting commander for a number of missions. As the common superior, the JFMCC organizes the TFs structure, delegates appropriate authorities, and establishes supporting relationships across the CTFs for the planned operation. These relationships may change by phase of an operation.

For more information on task organization with respect to amphibious forces, refer to JP 3-02, Amphibious Operations.

Navy Composite Warfare Doctrine

USN tactical commanders typically exercise decentralized control over assigned forces through use of composite warfare doctrine. This doctrine establishes a composite warfare organization within the task organization by assigning the commander's warfare command functions to subordinates. The composite warfare construct allows the officer in tactical command (OTC) to assign some or all of the command functions associated with mission areas to warfare commanders, functional group commanders, and coordinators, thus supporting decentralized execution. The composite warfare organization enables offensive and defensive combat operations against multiple targets and threats simultaneously. Flexibility of implementation, reinforced by clear guidance to subordinates and use of command by negation, are keys to decentralized control of the tactical force. The OTC may implement a composite warfare organization whenever and to whatever extent required, depending upon the composition and mission of the force and the capabilities of the threat. Within the composite warfare construct, the OTC may establish a subordinate composite warfare commander (CWC) who in turn may establish subordinate warfare commanders and/or functional warfare commanders. The warfare commanders that may be established include the air and missile defense commander (AMDC), the antisubmarine warfare commander (ASWC), the information operations warfare commander (IWC), the strike warfare commander (STWC), and the surface warfare commander (SUWC). The functional group commanders that may be established include the BMD commander, the maritime interception operations commander (MIOC), the mine warfare commander (MIWC), the screen commander, and the underway replenishment group commander. When the levels of activity and complexity in the mission areas involved are considered manageable, the tasks of ASWC and SUWC can be assigned to one commander, titled the sea combat commander.

In maritime usage, the OTC is the senior officer present eligible to assume command or the officer to whom the senior officer has delegated tactical command. If only one task organization (e.g., TF, TG) is operating independently in a portion of the maritime operational area, the commander of that task organization is the OTC. However, when multiple task organizations are operating together in the maritime operational area, the OTC is either the common superior or the commander to whom the common superior has assigned OTC command functions. In a maritime operational area that has multiple TFs operating within it, the common superior will be the NCC/JFMCC. Unless this commander assigns OTC command functions to one of the CTFs, the command will simultaneously be an operational- and tactical-level command. Care has to be exercised to ensure cross-functional working groups within these commands have clear charters and understandings on which level they are supporting and how their products support the commander's decision making associated with that level. When warfare functions are assigned to subordinate commanders, it is assumed the necessary authority for command, control, direction, and coordination required for the execution of those functions are delegated with it.

While acknowledged in joint doctrine, the OTC and CWC are maritime, unique constructs. Joint community understanding of these C2 constructs is important when coordinating or working with maritime forces. The OTC controls CWC and subordinate warfare commander's actions through command by negation. Allied and multinational maritime procedures and instructions use the term command by veto to mean the same thing.

IV. Operational-Level Maritime Operations

Ref: JP 3-32, Joint Maritime Operations (Jun '18), chap. 4.

The USN's traditional and doctrinal warfighting configuration is the fleet, commanded by a numbered fleet commander. Typically, the fleet commander task-organizes assigned and attached forces using the Navy's administrative organization as its foundation. This is a historical organizational framework from which extensive warfare doctrine flows.

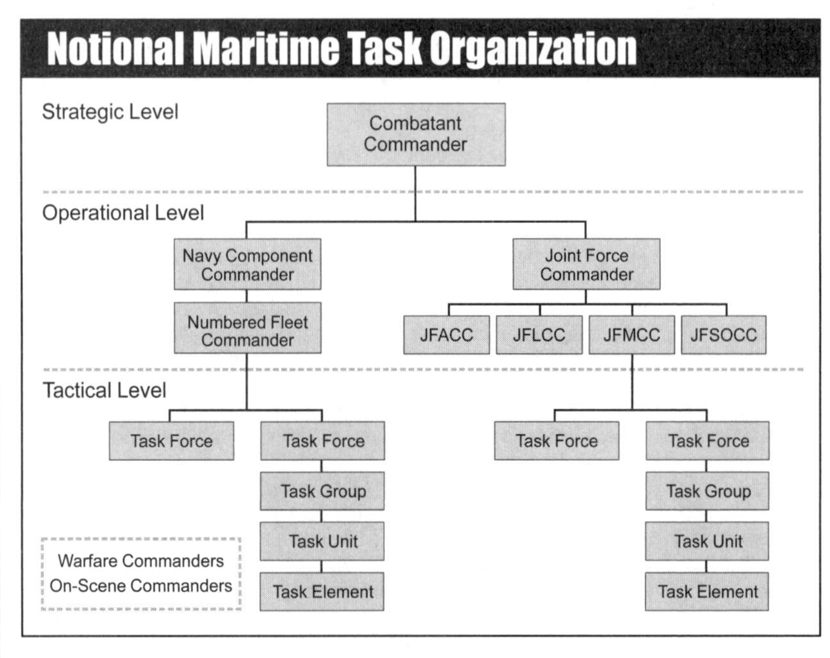

Ref: JP 3-32 (Jun '18), fig. IV-1. Notional Maritime Task Organization.

The JFMCC may create subordinate TFs, who may in turn create further subordinate organizations. In each case, the establishing authority must designate the command authorities for each subordinate organization, to include support relationships as required. Although the CTF is normally the CWC, the CTF can designate a subordinate commander to be the CWC. CTFs will typically assign forces under TACON to subordinate commanders. A CTF who has OPCON can designate a support command authority between two or more subordinate force commanders.

Surface Warfare

Surface Warfare encompasses operations conducted to destroy or neutralize enemy naval surface forces and merchant vessels. These operations typically include the planning and directing of surveillance of the maritime domain, interdiction, and strikes by aircraft and missiles.

Air and Missile Defense

Countering air and missile threats consists of a combination of theater counterair and integrated air and missile defense (IAMD). Counterair is the foundational framework at the theater level. IAMD synchronizes aspects of counterair with global missile defense, homeland defense, and global strike.

Undersea Warfare (USW)

Undersea warfare (USW) operations are conducted to establish dominance in the undersea portion of the maritime operational area, which permits friendly forces to operate throughout the maritime operational area and denies an opposing force the effective use of underwater systems and weapons. USW includes offensive and defensive submarine, antisubmarine warfare, and MIW operations.

Mine Warfare (MIW)

Maritime MIW is divided into two basic subdivisions: the laying of mines to degrade the enemy's capabilities to wage warfare and the countering of enemy-laid mines to permit friendly maneuver.

Strike Warfare

Strike warfare operations are naval operations to destroy or neutralize targets ashore, including attacks against strategic or tactical targets, such as manufacturing facilities and operating bases, from which the enemy is capable of conducting or supporting air, surface, or subsurface operations against friendly forces.

Amphibious Operations

Amphibious operations are complex and may involve all components of the joint force. They are typified by close integration of forces trained, organized, and equipped for different combat functions. The JFC and JFMCC should shape the amphibious objective area or operational area by employing carrier strike groups and other maritime and joint assets prior to the commencement of the amphibious operation.

Naval Surface Fire Support (NSFS)

Naval surface fire support (NSFS) units provide direct or general support to other joint force components or subordinate forces of the JFMCC. When supporting a landing force or other ground forces, an NSFS spotting team is usually attached to the maneuvering forces for fire support coordination purposes.

Other Maritime Operations

- Maritime intercept operations (MIO)
- Maritime security operations (MSO)
- Maritime homeland defense and defense support of civil authorities
- Global maritime partnerships and security cooperation
- Sea-based operations
- Counterdrug operations
- Noncombatant evacuation operations (NEO)
- Protection of shipping
- Maritime pre-positioning force operations
- Foreign humanitarian assistance (FHA)
- Maritime operational threat response (MOTR)
- Riverine operation

Refer to The Naval Operations & Planning SMARTbook (Guide to Designing, Planning & Conducting Maritime Operations) for complete discussion of essential Navy keystone warfighting doctrine and maritime operations at the JFMCC/CFMCC, Fleet and JTF levels. Topics include maritime forces, organization and capabilities; maritime operations; maritime headquarters (MHQ) and the maritime operations center (MOC); the maritime operations process; naval planning; naval logistics; and naval theater security cooperation.

V. Amphibious Operations

Ref: JP 3-02, Amphibious Operations (Jan '19).

An amphibious force is an amphibious task force and landing force together with other forces that are trained, organized, and equipped for amphibious operations.

An amphibious operation is a military operation launched from the sea by an amphibious force (AF) to conduct landing force (LF) operations within the littorals. The littorals include those land areas (and their adjacent sea and associated air space) that are predominantly susceptible to engagement and influence from the sea and may reach far inland.

The AF executes rapid, focused operations to accomplish the joint force commander's (JFC's) objectives. The commander, amphibious task force (CATF), and commander, landing force (CLF), plan and execute operations based on maneuver warfare philosophy. Operations should create freedom of action for the AF, while controlling the tempo better than the enemy can.

Types of Amphibious Operations

- Amphibious raids are conducted as independent operations or in support of other operations.
- Amphibious demonstrations are a show of force intended to influence or deter an enemy's decision.
- Amphibious assaults are launched from the sea by the amphibious force on a hostile or potentially hostile shore.
- Amphibious withdrawals are operations conducted to extract forces in ships or craft from a hostile or potentially hostile shore.

Command and Control

The command relationships established between the CATF and CLF and other designated commanders are important decisions. The AF commanders are coequal in planning matters. Planning decisions should be reached on a basis of common understanding of the mission; objectives; and tactics, techniques, and procedures and on a free exchange of information. Typically, a support relationship is established between the commanders and is based on the complementary rather than similar nature and capabilities of the amphibious task force (ATF) and LF.

Amphibious Force (AF) Organization

No standard organization is applicable to all situations that may be encountered in an amphibious operation. Depending on the amphibious operation, an ATF may vary in size and can be task-organized. Forward-deployed ATFs are normally organized into amphibious ready groups with three amphibious warfare ships (an amphibious assault ship [general purpose]/amphibious assault ship [multipurpose], amphibious transport dock, and dock landing ship).

Landing Force (LF) Organization

The LF may be composed of United States Marine Corps (USMC) and United States Army (USA) forces, other US forces, and multinational forces (MNFs). Organizations that can be assigned as LF include the Marine expeditionary force, Marine expeditionary brigade, Marine expeditionary unit (MEU), and, in some cases, a special purpose Marine air ground task force. If USA forces comprise part of the LF, they will be task-organized with appropriate combat and sustainment capabilities to support the LF.

See pp. 6-14 to 6-15 for an overview and related discussion of Marine Corps Forces and Expeditionary Operations.

IV. Special Operations

Ref: JP 3-05, Special Operations (Jul '14).

Special operations require unique modes of employment, tactics, techniques, procedures, and equipment. They are often conducted in hostile, denied, or politically and/or diplomatically sensitive environments, and are characterized by one or more of the following: time-sensitivity, clandestine or covert nature, low visibility, work with or through indigenous forces, greater requirements for regional orientation and cultural expertise, and a higher degree of risk. Special operations provide JFCs and chiefs of mission (COMs) with discrete, precise, and scalable options that can be synchronized with activities of other interagency partners to achieve United States Government (USG) objectives. These operations are designed in a culturally attuned manner to create both immediate and enduring effects to help prevent and deter conflict or prevail in war. They assess and shape foreign political and military environments unilaterally, or with host nations (HNs), multinational partners, and indigenous populations. Although special operations can be conducted independently, most are coordinated with conventional forces (CF), interagency partners, and multinational partners, and may include work with indigenous, insurgent, or irregular forces. Special operations may differ from conventional operations in degree of strategic, physical, and political and/or diplomatic risk; operational techniques; modes of employment; and dependence on intelligence and indigenous assets.

Special operations can be a single engagement, such as direct action (DA) against a critical target; as a protracted operation or series of activities such as support to insurgent forces through unconventional warfare (UW); or support to a HN force through foreign internal defense (FID) or security force assistance (SFA). Military information support operations (MISO) can be used during special operations to influence selected target audiences' behavior and actions. Civil affairs operations (CAO) also provide essential support to a JFC or country team. Special operations, synchronized with MISO and CAO, can create effects disproportionate to the size of the units involved.

I. Designated Special Operations Forces

United States Special Operations Command (USSOCOM) is a unified combatant command (CCMD). It is unique among the CCMDs in that it performs Service-like functions and has Military Department-like responsibilities and authorities. A theater special operations command (TSOC) is a subordinate unified command of USSOCOM. TSOCs perform broad, continuous missions uniquely suited to special operations forces (SOF) capabilities. Secretary of Defense (SecDef) has assigned operational control (OPCON) of the TSOCs and attached SOF tactical units to their respective geographic combatant commander (GCC) via the Global Force Management Implementation Guidance.

United States Army Special Operations Command is the designated Army component command for USSOCOM and provides manned, trained, and equipped Army special operations forces. Naval Special Warfare Command is designated the Navy component command of USSOCOM and mans, trains, equips, and provides SEALs. US Air Force Special Operations Command is designated the Air Force component of USSOCOM and organizes, trains, equips, and provides trained Air Force special operations forces. US Marine Corps Forces, Special Operations Command is designated the Marine Corps component of USSOCOM and trains, equips, and provides Marine Corps special operations forces.

II. Special Operations Core Activities

Ref: JP 3-05, Special Operations (Jul '14), chap. II.

USSOCOM organizes, trains, and equips SOF for special operations core activities, and other such activities as may be specified by the President and/or SecDef. Special operations missions may include more than one core activity. The special operations core activities are: direct action, special reconnaissance, countering weapons of mass destruction, counterterrorism, unconventional warfare (UW), foreign internal defense, security force assistance, hostage rescue and recovery, counterinsurgency, foreign humanitarian assistance, military information support operations, and civil affairs operations.

Special operations are inherently joint because of the integration and interdependency that is established among ARSOF, NAVSOF, AFSOF, and MARSOF to accomplish their missions. SOF conduct joint and combined training both within the SOF community, with CF, and with interagency and multinational partners. When employed, SOF deploy with its C2 structure intact, which facilitates integration into the joint force, retains SOF cohesion, and provides a supported JFC with the control mechanism to address specific special operations concerns and coordinates its activities with other components and supporting commands.

Direct Action

Direct action entails short-duration strikes and other small-scale offensive actions conducted with specialized military capabilities to seize, destroy, capture, exploit, recover, or damage designated targets in hostile, denied, or diplomatically and/or politically sensitive environments.

Special Reconnaissance

Special reconnaissance entails reconnaissance and surveillance actions normally conducted in a clandestine or covert manner to collect or verify information of strategic or operational significance, employing military capabilities not normally found in conventional forces (CF).

Countering Weapons of Mass Destruction

SOF support USG efforts to curtail the development, possession, proliferation, use, and effects of weapons of mass destruction, related expertise, materials, technologies, and means of delivery by state and non-state actors

Counterterrorism

UW consists of operations and activities that are conducted to enable a resistance movement or insurgency to coerce, disrupt, or overthrow a government or occupying power by operating through or with an underground, auxiliary, and guerrilla force in a denied area.

Unconventional Warfare

Foreign internal defense refers to US activities that support a host nation's (HN's) internal defense and development strategy and program designed to protect against subversion, lawlessness, insurgency, terrorism, and other threats to their internal security, and stability.

Foreign Internal Defense

USG security sector reform (SSR) focuses on the way a HN provides safety, security, and justice with civilian government oversight. The Department of Defense's (DOD's)

primary role in SSR is to support the reform, restructure, or reestablishment of the HN armed forces and the defense aspect of the security sector, which is accomplished through security force assistance.

Security Force Assistance

Hostage rescue and recovery operations are sensitive crisis response missions in response to terrorist threats and incidents. Offensive operations in support of hostage rescue and recovery can include the recapture of US facilities, installations, and sensitive material overseas.

Hostage Rescue and Recovery

Hostage rescue and recovery operations are sensitive crisis response missions in response to terrorist threats and incidents. Offensive operations in support of hostage rescue and recovery can include the recapture of US facilities, installations, and sensitive material overseas.

Counterinsurgency

Counterinsurgency is a comprehensive civilian and military effort designed to simultaneously defeat and contain insurgency and address its root causes.

Foreign Humanitarian Assistance (FHA)

Foreign humanitarian assistance is a range of DOD humanitarian activities conducted outside the US and its territories to relieve or reduce human suffering, disease, hunger, or privation.

Military Information Operations (MISO)

Military information support operations (MISO) are planned to convey selected information and indicators to foreign audiences to influence their emotions, motives, objective reasoning, and ultimately the behavior of foreign governments, organizations, groups, and individuals in a manner favorable to the originator's objectives.

Civil Affairs Operations

Civil affairs operations are actions planned, executed, and assessed by civil affairs that enhance the operational environment; identify and mitigate underlying causes of instability within civil society; or involve the application of functional specialty skills normally the responsibility of civil government

Refer to these SMARTbooks for related material:

Refer to TAA2: Military Engagement, Security Cooperation & Stability SMARTbook (Foreign Train, Advise, & Assist) for further discussion. Topics include the Range of Military Operations (JP 3-0), Security Cooperation & Security Assistance (Train, Advise, & Assist), Stability Operations (ADRP 3-07), Peace Operations (JP 3-07.3), Counterinsurgency Operations (JP & FM 3-24), Civil-Military Operations (JP 3-57), Multinational Operations (JP 3-16), Interorganizational Cooperation (JP 3-08), and more.

Refer to CTS1: The Counterterrorism, WMD & Hybrid Threat SMARTbook for further discussion. CTS1 topics and chapters include: the terrorist threat (characteristics, goals & objectives, organization, state-sponsored, international, and domestic), hybrid and future threats, forms of terrorism (tactics, techniques, & procedures), counterterrorism, critical infrastructure, protection planning and preparation, countering WMD, and consequence management (all hazards response).

III. Command and Control Considerations

SOF units based in the US are generally assigned to and under combatant command (COCOM) of Commander, United States Special Operations Command (CDRUSSOCOM), with OPCON exercised through the USSOCOM Service component commands. SecDef assigns the TSOCs to USSOCOM under CDRUSSOCOM's COCOM, and assigns OPCON of the TSOCs to the GCCs. SecDef also authorizes CDRUSSOCOM/GCCs to establish support relationships when SOF commanders are required to simultaneously support multiple operations or commanders.

Special Operations Forces Joint Task Force (SOJTF)

The special operations joint task force (SOJTF) is the principal joint SOF organization tasked to meet all special operations requirements in major operations, campaigns, or a contingency. A SOJTF is a modular, tailorable, and scalable SOF organization that allows USSOCOM to more efficiently provide integrated, fully capable, and enabled joint SOF to GCCs and subordinate JFCs based on the strategic, operational, and tactical context. Depending on circumstances, the SOJTF may be directed to serve as the joint task force (JTF), or a joint force special operations component commander (JFSOCC).

Command and Control of Special Operations Forces in Theater

The TSOC plans and conducts operations in support of the GCC. The GCC normally exercises OPCON of attached SOF through the commander, theater special operations command (CDRTSOC), who may exercise OPCON of subordinate forces directly from the TSOC location, or through a smaller special operations command-forward, located elsewhere in the theater of operations.

Special Operations Coordination, Liaison, and Distributed Command Elements

SOF commanders have elements to liaise with various organizations. The special operations command and control element (SOCCE) is the focal point for SOFCF coordination, and the synchronization of special operations activities with other joint operations. The SOCCE is normally employed when SOF conducts operations in support of a CF. It performs command and control (C2) or liaison functions according to mission requirements and as directed by the establishing SOF commander (JFSOCC, commander, SOJTF, or commander, joint special operations task force [CDRJSOTF]). A special operations liaison element (SOLE) is typically a joint team provided by the JFSOCC/CDRJSOTF to the joint force air component commander (if designated) at the joint air operations center, or appropriate Service air component C2 center to coordinate, deconflict, and synchronize special operations air, surface, and subsurface activities with joint air operations. The special operations liaison officer (SOLO) is a SOF officer with language, cultural, military, and civilian training in addition to SOF staff experience. SOLOs are assigned to a HN's national SOF headquarters as part of a recurring and permanent US SOF presence in select HNs. The SOF representative is an experienced SOF officer proficient n the language most commonly used for partner nation government business. SOF representatives are assigned to the US embassies in selected partner nations as part of a recurring and persistent US SOF presence.

Interdependence of Conventional Forces and Special Operations Forces

SOF and CF often share the same operational areas for extended periods when they are mutually reliant on each other's capabilities. SOF-CF synchronization facilitates unity of effort; maximizes the capability of the joint force; and allows the JFC to optimize the principles of joint operations in planning and execution.

Chap 7 Multinational Operations

Ref: JP 3-16, Multinational Operations (Mar '19), chap. I and executive summary.

Multinational operations are operations conducted by forces of two or more nations, usually undertaken within the structure of a coalition or alliance. Other possible arrangements include supervision by an intergovernmental organization (IGO) such as the United Nations (UN), the North Atlantic Treaty Organization (NATO), or the Organization for Security and Cooperation in Europe. Two primary forms of multinational partnership that the joint force commander (JFC) will encounter are an alliance or a coalition.

Alliance

An alliance is the relationship that results from a formal agreement between two or more nations for broad, long-term objectives that further the common interests of the members.

Coalition

A coalition is an arrangement between two or more nations for common action. Coalitions are typically ad hoc; formed by different nations, often with different objectives; usually for a single problem or issue, while addressing a narrow sector of common interest. Operations conducted with units from two or more coalition members are referred to as coalition operations.

I. Strategic Context

Nations form regional and global geopolitical and economic relationships to promote their mutual national interests, ensure mutual security against real and perceived threats, conduct foreign humanitarian assistance (FHA), conduct peace operations (PO), and promote their ideals. Cultural, diplomatic, psychological, economic, technological, and informational factors all influence multinational operations and participation. However, a nation's decision to employ military capabilities is always a political decision.

Since Operation DESERT STORM in 1991, the trend has been to conduct US military operations as part of a multinational force (MNF). Therefore, US commanders should be prepared to perform either supported or supporting roles in military operations as part of an MNF. These operations could span the range of military operations and require coordination with a variety of United States Government (USG) departments and agencies, foreign military forces, local authorities, international organizations, and nongovernmental organizations (NGOs). The move to a more comprehensive approach toward problem solving, particularly in regard to counterinsurgency operations, other counter threat network activities, or stability activities, increases the need for coordination and synchronization among military and nonmilitary entities.

Refer to TAA2: Military Engagement, Security Cooperation & Stability SMARTbook (Foreign Train, Advise, & Assist) for further discussion. Topics include the Range of Military Operations (JP 3-0), Security Cooperation & Security Assistance (Train, Advise, & Assist), Stability Operations (ADRP 3-07), Peace Operations (JP 3-07.3), Counterinsurgency Operations (JP & FM 3-24), Civil-Military Operations (JP 3-57), Multinational Operations (JP 3-16), Interorganizational Cooperation (JP 3-08), and more.

II. Security Cooperation (SC)

Security cooperation (SC) provides ways and means to help achieve national security and foreign policy objectives. US national and Department of Defense (DOD) strategic guidance emphasizes the importance of defense relationships with allies and PNs to advance national security objectives, promote stability, prevent conflicts, and reduce the risk of having to employ US military forces in a conflict. SC activities are likely to be conducted in a combatant command's (CCMD's) daily operations. SC advances progress toward cooperation within the competition continuum by strengthening and expanding the existing network of US allies and partners, which improves the overall warfighting effectiveness of the joint force and enables more effective multinational operations. SC activities, many of which are shaping activities within the geographic combatant commander (GCC) campaign plans—the centerpiece of the planning construct from which OPLANs/concept plans (CONPLANs) are now branches—are deemed essential to achieving national security and foreign policy objectives. SC activities also build interoperability with NATO Allies and other partners in peacetime, thereby speeding the establishment of effective coalitions—a key factor in potential major combat operations with near-peer competitors.

The Guidance for Employment of the Force (GEF) provides the foundation for all DOD interactions with foreign defense establishments and supports the President's National Security Strategy. With respect to SC, the GEF provides guidance on building partner capacity and capability, relationships, and facilitating access (under the premise that the primary entity of military engagement is the nation state and the means which GCCs influence nation states is through their defense establishments). The GEF outlines the following SC activities: defense contacts and familiarization, personnel exchange, combined exercises and training, train and equip/provide defense articles, defense institution building, operational support, education, and international armaments cooperation.

GCC theater strategies, as reflected in their combatant command campaign plans (CCPs), typically emphasize military engagement, SC, and deterrence activities as daily operations. GCCs shape their areas of responsibility through SC activities by continually employing military forces to complement and reinforce other instruments of national power. The GCC's CCP provides a framework within which CCMDs conduct cooperative military activities and development. Ideally, SC activities lessen the causes of a potential crisis before a situation deteriorates and requires substantial US military intervention.

The CCP is the primary document that focuses on each command's activities designed to attain theater strategic end states. The GEF and Chairman of the Joint Chiefs of Staff Instruction (CJCSI) 3110.01, (U) Joint Strategic Campaign Plan (JSCP) (referred to as the JSCP), provide regional focus and SC priorities.

DOD components may develop supporting plans that focus on activities conducted to support the execution of the CCPs and on their own SC activities that directly contribute to the campaign end states and/or DOD component programs in support of broader Title 10, US Code, responsibilities. The Services conduct much of the detailed work to build interoperability and capacity with NATO Allies and mission partners.

The DOD State Partnership Program establishes enduring relationships between emerging PNs of strategic value and individual US states and territories. The DOD State Partnership Program is an important contribution to the DOD SC programs conducted by the GCCs in conjunction with the National Defense Strategy, National Security Strategy, National Military Strategy, Department of State (DOS), campaign plans, and theater SC guidance to promote national and combatant commander (CCDR) objectives, stability, and partner capacity.

See pp. 2-57 to 2-62 for related discussion of engagement, security cooperation, and deterrence missions, tasks and actions from JP 3-0, Joint Operations.

III. Nature of Multinational Operations

Ref: JP 3-16, Multinational Operations (Mar '19), pp. I-2 to I-5.

After World War II, General Dwight D. Eisenhower noted that "mutual confidence" is the "one basic thing that will make allied commands work." While the tenets discussed below cannot guarantee success, ignoring them may lead to mission failure due to a lack of unity of effort.

1. Respect

In assigning missions and tasks, the commander should consider that national honor and prestige may be as important to a contributing nation as combat capability. All partners must be included in the planning process, and their opinions must be sought in mission assignment. Understanding, discussion, and consideration of partner ideas are essential to building effective relationships, as are respect for each partner's culture, customs, history, and values.

2. Rapport

US commanders and staffs should establish rapport with their counterparts from partner countries, as well as the multinational force commander (MNFC). This requires personal, direct relationships that only they can develop. Good rapport between leaders will improve teamwork among their staffs and subordinate commanders and overall unity of effort. The use of liaisons can facilitate the development of rapport by assisting in the staffing of issues to the correct group and in monitoring responses.

3. Knowledge of Partners

US commanders and their staffs should have an understanding of each member of the MNF. Much time and effort is spent learning about the enemy; a similar effort is required to understand the doctrine, capabilities, strategic goals, culture, customs, history, and values of each partner. This will facilitate the effective integration of multinational partners into the operation and enhance the synergistic effect of their forces.

4. Patience

Effective partnerships take time and attention to develop. Diligent pursuit of a trusting, mutually beneficial relationship with multinational partners requires untiring, evenhanded patience. This is more difficult to accomplish within coalitions than within alliances; however, it is just as necessary. It is therefore imperative that US commanders and their staffs apply appropriate resources, travel, staffing, and time not only to maintain, but also to expand and cultivate multinational relationships. Without patience and continued engagement, established partnerships can easily dissolve.

5. Mission Focus

When dealing with other nations, US forces should temper the need for respect, rapport, knowledge, and patience with the requirement to ensure that the necessary tasks are accomplished by those with the capabilities and authorities to accomplish those tasks. This is especially critical in the security line of operation, where failure could prove to have catastrophic results. If operational necessity requires tasks being assigned to personnel who are not proficient in accomplishing those tasks, then the MNF commander must recognize the risks and apply appropriate mitigating measures.

6. Trust and Coordination

Commanders should engage other leaders of the MNF to build personal relationships and develop trust and confidence. Developing these relationships is a conscious collaborative act rather than something that just happens. Commanders build trust through words and actions. Trust and confidence are essential to synergy and harmony, both within the joint force and also with our multinational partners. Coordination and cooperation among organizations are based on trust.

IV. Rationalization, Standardization, and Interoperability (RSI)

Ref: JP 3-16, Multinational Operations (Mar '19), pp. I-8 to I-11.

International rationalization, standardization, and interoperability (RSI) with PNs is important for achieving practical cooperation; efficient use of research, development, procurement, support, and production resources; and effective multinational capability without sacrificing US capabilities.

Rationalization

In the RSI construct, rationalization refers to any action that increases the effectiveness of MNFs through more efficient or effective use of defense resources committed to the MNF. Rationalization includes consolidation, reassignment of national priorities to higher multinational needs, standardization, specialization, mutual support or improved interoperability, and greater cooperation. Rationalization applies to both weapons and materiel resources (the processes to loan and/or transfer equipment to another nation participating in an MNF operation) and non-weapons military matters.

Standardization

Unity of effort is greatly enhanced through standardization. The basic purpose of standardization programs is to achieve the closest practical cooperation among multinational partners through the efficient use of resources and the reduction of operational, logistic, communications, technical, and procedural obstacles in multinational military operations.

Standardization agreements like AJPs, MPs, STANAGs, and ABCANZ standards provide a baseline for cooperation within a coalition. In many parts of the world, these multilateral and other bilateral agreements for standardization between potential coalition members may be in place prior to the formation of the coalition. However, participants may not be immediately familiar with such agreements.

Interoperability

Interoperability greatly enhances multinational operations through the ability to operate in the execution of assigned tasks. Nations whose forces are interoperable across materiel and nonmateriel capabilities can operate together effectively in numerous ways. For example, as part of developing PN security forces, the extent of interoperability can be used to gauge the effectiveness of SC/SFA activities. Although frequently identified with technology, important areas of interoperability may include doctrine, procedures, communications, and training.

Factors that enhance interoperability start with understanding the nature of multinational operations as described in paragraph 3, "Nature of Multinational Operations." Additional factors include planning for interoperability and sharing information, the personalities of the commander and staff, visits to assess multinational capabilities, a command atmosphere permitting positive criticism and rewarding the sharing of information, liaison teams, multinational training exercises, and a constant effort to eliminate sources of confusion and misunderstanding. The establishment of standards for assessing the logistic capability of expected participants in a multinational operation should be the first step in achieving logistic interoperability among participants. Such standards should already be established for alliance members when the preponderance of NATO nations are representative of a particular alliance.

Factors that inhibit interoperability include restricted access to national proprietary defense information; time available; any refusal to cooperate with partners; differences in military organization, security, language, doctrine, and equipment; level of experience; and conflicting personalities.

I. Multinational Command & Coordination

Ref: JP 3-16, Multinational Operations (Mar '19), chap. II.

I. Command and Control of U.S. Forces in Multinational Operations

Although nations will often participate in multinational operations, they rarely, if ever, relinquish national command of their forces. As such, forces participating in a multinational operation will always have at least two distinct chains of command: a national chain of command and a multinational chain of command.

National Command

As Commander in Chief, the President always retains and cannot relinquish national command authority over US forces. National command includes the authority and responsibility for organizing, directing, coordinating, controlling, planning employment of, and protecting military forces. The President also has the authority to terminate US participation in multinational operations at any time. All nations participating in a multinational operation will have their own form of national command. NATO and the European Union (EU) use the term "full command" to describe national command by their member states.

Multinational Command

Command authority for an MNFC is normally negotiated between the participating nations and can vary from nation to nation. In making a decision regarding an appropriate command relationship for a multinational military operation, the President carefully considers such factors as mission, size of the proposed US force, risks involved, anticipated duration, and rules of engagement (ROE). Command authority will be specified in the implementing agreements that provide a clear and common understanding of what authorities are specified over which forces.

II. Unified Action

Unified action during multinational operations involves the synergistic application of all instruments of national power as provided by each participating nation; it includes the actions of nonmilitary organizations as well as military forces. This construct is applicable at all levels of command. In a multinational environment, unified action synchronizes, coordinates, and/or integrates multinational operations with the operations of other HN and national government agencies, international organizations (e.g., UN), NGOs, and the private sector to achieve unity of effort in the operational area (OA). When working with NATO forces, it can also be referred to as a comprehensive approach.

III. Multinational Force (MNF)

MNFC is a generic term applied to a commander who exercises command authority over a military force composed of elements from two or more nations. The extent of the MNFC's command authority is determined by the participating nations or elements. This authority can vary widely and may be limited by national caveats of those nations participating in the operation. The MNFC's primary duty is to unify the efforts of the MNF toward common objectives. An operation could have numerous MNFCs.

IV. Command Structures of Forces in Multinational Operations

Ref: JP 3-16, Multinational Operations (Mar '19), pp. II-4 to II-8.

No single command structure meets the needs of every multinational command but one absolute remains constant; political considerations will heavily influence the ultimate shape of the command structure. Organizational structures include the following:

A. Integrated Command Structure

Multinational commands organized under an integrated command structure provide unity of effort in a multinational setting. A good example of this command structure is found in the North Atlantic Treaty Organization where a strategic commander is designated from a member nation, but the strategic command staff and the commanders and staffs of subordinate commands are of multinational makeup.

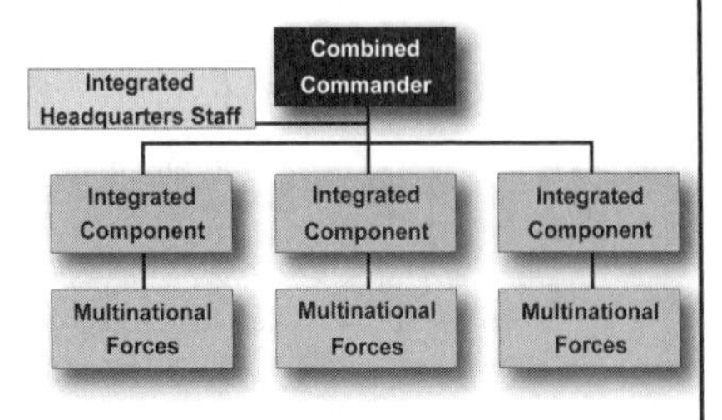

B. Lead Nation Command Structure

A lead nation structure exists when all member nations place their forces under the control of one nation. The lead nation command can be distinguished by a dominant lead nation command and staff arrangement with subordinate elements retaining strict national integrity. A good example of the lead nation structure is Combined Forces Command-Afghanistan wherein a US-led headquarters provides the overall military C2 over the two main subordinate commands: one predominately US forces and the other predominately Afghan forces.

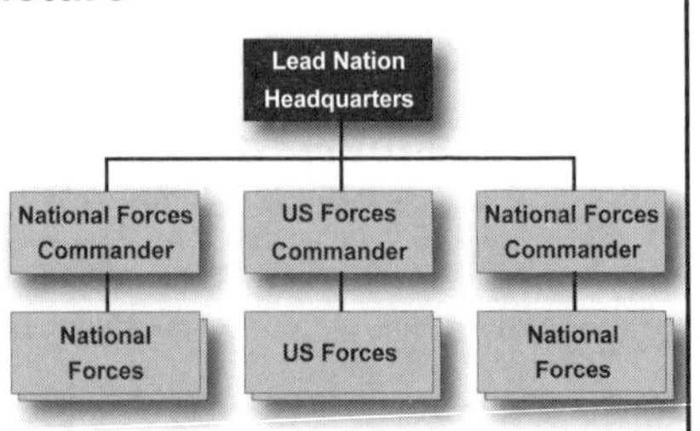

C. Parallel Command Structures

Under a parallel command structure, no single force commander is designated. The coalition leadership must develop a means for coordination among the participants to attain unity of effort. This can be accomplished through the use of coordination centers. Nonetheless, because of the absence of a single commander, the use of a parallel command structure should be avoided if at all possible.

See pp. 8-16 to 8-17 for discussion and listing of US organizational structures in foreign countries: Ambassadors, Chiefs of Mission, Country Team, Defense Attaches, etc.

Multinational Operations

II. Multinational Planning & Execution

Ref: JP 3-16, Multinational Operations (Mar '19), chap. III and executive summary.

I. Diplomatic and Military Considerations

Any number of different situations could generate the need for a multinational response, from man-made actions (such as interstate aggression) to natural disasters (like an earthquake). In responding to such situations, nations weigh their national interests and then determine if, when, and where they will expend their nation's resources. Nations also choose the manner and extent of their foreign involvement for reasons both known and unknown to other nations. The composition of an MNF may change as partners enter and leave when their respective national objectives change or force contributions reach the limits of their nation's ability to sustain them.

Factors Affecting Military Capabilities of Nations

- National Interests
- Domestic Politics
- Objectives
- Arms Control Limitations
- Doctrine
- Organization
- Training
- Leader Development
- Equipment
- History
- Defense Budget
- Domestic Law
- Treaties

Ref: JP 3-16, fig. III-1. Factors Affecting the Military Capabilities of Nations.

Some nations may even be asked to integrate their forces with those of another, so that a contribution may, for example, consist of an infantry company containing platoons from different countries. The only constant is that a decision to "join in" is, in every case, a calculated diplomatic decision by each potential member of a coalition or alliance. The nature of their national decisions, in turn, influences the MNTF's command structure. In a parallel command structure, national forces essentially operate under their own doctrine and procedures within the guidelines determined by the strategic national guidance and are not significantly impacted by multinational influences. Under the integrated and LN command structures, more multinational involvement and interaction occurs.

As shown in Figure III-1 above, numerous factors influence the military capabilities of nations. The operational-level commander must be aware of the specific operational limitations and capabilities of the forces of participating nations and consider these differences when assigning missions and conducting operations. MNTF commanders at all levels may be required to spend considerable time consulting and negotiating with diplomats, HN officials, local leaders, and others; their role as diplomats should not be underestimated.

II. Factors in Multinational Participation

Ref: Adapted from JP 3-0, Joint Operations (Aug '11), pp. II-4 to II-8.

Editor's Note: The following overview of factors in multinational participation is from the previous edition of JP 3-0, Joint Operations (Aug '11). Refer to JP 3-16, Multinational Operations (Mar '19) for expanded discussion of general and operational considerations.

Joint forces should be prepared for combat and noncombat operations with forces from other nations within the framework of an alliance or coalition under US or other-than-US leadership. Following, contributing, and supporting are important roles in multinational operations — often as important as leading.

1. National Goals

No two nations share exactly the same reasons for entering a coalition or alliance. To some degree, participation within an alliance or coalition requires the subordination of national autonomy by member nations. The glue that binds the multinational force is trust and agreement, however tenuous, on common goals and objectives. However, different national goals, often unstated, cause each nation to measure progress in its own way. Consequently, perceptions of progress may vary among the participants. JFCs should strive to understand each nation's goals and how those goals can affect conflict termination and the national strategic end state. Maintaining cohesion and unity of effort requires understanding and adjusting to the perceptions and needs of member nations.

2. Cultural and Language Differences

Each partner in multinational operations possesses a unique cultural identity — the result of language, values, religion, and economic and social outlooks. Language differences often present the most immediate challenge. Information lost during translation can be substantial, and misunderstandings and miscommunications can have disastrous effects. To assist with cultural and language challenges, JFCs should employ linguists and area experts, often available within or through the Service components or from other US agencies. Linguists must be capable of translating warfighting-unique language to military forces of diverse cultures.

3. Command and Control

By law, the President retains command authority over US forces. This includes the authority and responsibility for effectively using available resources and for planning employment, organizing, directing, coordinating, controlling, and protecting military forces for the achievement of assigned missions. JFCs should have a responsive and reliable link to appropriate US agencies and political leadership. In all multinational operations, even when operating under the operational control (OPCON) of a foreign commander, US commanders will maintain the capability to report separately to higher US military authorities in addition to foreign commanders. Further, the President may deem it prudent or advantageous (for reasons such as maximizing military effectiveness and ensuring unified action) to place appropriate US forces under the control of a foreign commander to achieve specified military objectives. In making this determination, the President carefully considers such factors as the mission, size of the proposed US force, risks involved, anticipated duration, and ROE. Coordinated policy, particularly on such matters as alliance or coalition commanders' authority over national logistics (including infrastructure) and theater intelligence, is required.

4. C2 Structures

Alliances typically have developed C2 structures, systems, and procedures. Allied forces typically mirror their alliance composition, with the predominant nation providing the allied force commander. Staffs are integrated, and subordinate commands often are led by senior representatives from member nations. Doctrine, standardization agreements, close military cooperation, and robust diplomatic relations characterize alliances. Coalitions may adopt a parallel or lead nation C2 structure or a combination of the two.

See pp. 7-7 to 7-8 for discussion of multinational command and coordination.

Multinational Operations

5. Liaison

Coordination and liaison are important considerations. Regardless of the command structure, coalitions and alliances require a significant liaison structure. Differences in language, equipment, capabilities, doctrine, and procedures are some of the interoperability challenges that mandate close cooperation through, among other things, liaisons.

JFCs often deploy robust liaison teams with sufficient communications equipment to permit instantaneous communication between national force commanders during the early stages of coalition formation and planning.

6. Information and Intelligence Sharing

The success of a multinational operation hinges upon timely and accurate information and intelligence sharing. As DOD moves toward a net-centric environment, it faces new challenges validating intelligence information and information sources, as well as sharing of information required to integrate participating multinational partners. This information sharing can only occur within a culture of trust, based upon an effective information-sharing environment, that uses the lowest classification level possible, must support multilateral or bilateral information exchanges between the multinational staff and forces, as well as the military staffs and governments for each participating nation. Actions to improve the ability to share information such as establishing metadata or tagging standards, agreeing to information exchange standards, and using unclassified information (e.g., commercial imagery) need to be addressed early (as early as the development of military systems for formal alliances).

7. Logistics

Multinational logistics is a challenge; however many issues can be resolved or mitigated by a thorough understanding of capabilities and procedures before operations begin. Potential problem areas include differences in logistic doctrine, stockage levels, logistic mobility, interoperability, infrastructure, competition between the Services and multinational partners for common support, and national resource limitations.

Nonetheless, JFCs need to coordinate for the effective and efficient use of all logistic support to include lift, distribution, and sustainment assets as well as the use of infrastructure such as highways, rail lines, seaports, and airfields in a manner that supports mission accomplishment. The notion that logistics is primarily a national responsibility cannot supplant detailed logistic planning in seeking multinational solutions. Multinational force commanders (MNFCs) typically form multinational logistic staff sections early to facilitate logistic coordination and support multinational operations. Careful consideration should be given to the broad range of multinational logistic support options; from lead nation and role specialization nations, to the formation of multinational integrated logistic units to deliver effective support while achieving greater efficiency. Standardization of logistic systems and procedures is an ongoing, iterative process and MNFCs should ensure that the latest techniques, procedures, and arrangements are understood for the current operation. Interoperability of equipment, especially in adjacent or subordinate multinational units, is desirable and should be considered during concept development. The acquisition and cross-servicing agreement (ACSA) is a tool for mutual exchange of logistic support and services. ACSA is a reimbursable, bilateral support program that allows reimbursable logistics-exchanges between US and foreign military forces. An ACSA provides the necessary legal authority to allow mutual logistic support between the US and multinational partners.

For further guidance on multinational logistics, refer to JP 4-08, Joint Doctrine for Logistic Support of Multinational Operations. There are numerous other important multinational considerations relating to mission assignments, organization of the operational area, intelligence, planning, ROE, doctrine and procedures, and PA. Expanded discussions on these and the previously discussed considerations are provided in JP 3-16, Multinational Operations.

III. Building and Maintaining a Multinational Force

Ref: JP 3-16, Multinational Operations (Mar '19), pp. III-3 to III-5.

Building an MNF starts with the political decisions and diplomatic efforts to create a coalition or spur an alliance into action. Discussion and coordination between potential participants will initially seek to sort out basic questions at the national strategic level. The result of these discussions should determine the nature and limits of the response; the command structure of the response force; and the essential strategic guidance for the response force to include military objectives and the desired end states.

Partner Nation Contributions

	Contribution	Examples
Combat Forces	Troops Ships Aircraft Staff Officers Trainers	
Noncombat Forces	Diplomatic Support	• Diplomatic recognition • Opening embassy or mission • Supporting United Nations security resolution
	Financial Support	• Debt forgiveness • Unfreezing assets • Direct financial assistance
	Logistics, Lift, and Sustainment OCS	• Logistic infrastructure • Strategic air and sea lift • Intratheater lift • Operational contract support (OCS)
	Basing, Access, and Overflight Support	• Basing rights • Access to facilities • Overflight rights
	Stabilization and Reconstruction Support	• Humanitarian assistance • Public infrastructure
	Governance and Ministerial Support	• Constitutional support • Ministerial mentoring • Civil service training support

Ref: JP 3-16, fig. III-2. Partner Nation Contributions.

Refer to TAA2: Military Engagement, Security Cooperation & Stability SMARTbook (Foreign Train, Advise, & Assist) for further discussion. Topics include the Range of Military Operations (JP 3-0), Security Cooperation & Security Assistance (Train, Advise, & Assist), Stability Operations (ADRP 3-07), Peace Operations (JP 3-07.3), Counterinsurgency Operations (JP & FM 3-24), Civil-Military Operations (JP 3-57), Multinational Operations (JP 3-16), Interorganizational Cooperation (JP 3-08), and more.

Chap 8

Interorganizational Cooperation

Ref: JP 3-08, Interorganizational Cooperation (val. Oct '17), chap. I.

JP 3-08 describes the joint force commander's (JFC's) coordination with various external organizations that may be involved with, or operate simultaneously with, joint operations. This coordination includes the Armed Forces of the United States; United States Government (USG) departments and agencies; state, territorial, local, and tribal government agencies; foreign military forces and government agencies; international organizations; nongovernmental organizations (NGOs); and the private sector. Interagency coordination describes the interaction between USG departments and agencies and is a subset of interorganizational cooperation.

Interorganizational Cooperation Relationships

Ref: JP 3-08, fig. I-1, Interorganizational Cooperation Relationships.

The Department of Defense (DOD) conducts interorganizational cooperation across a range of operations, with each type of operation involving different communities of interest, structures, and authorities. The terms "interagency" and "interorganizational" do not define structures or organizations, but rather describe processes occurring among various separate entities.

I. Foundations of Cooperation

Commitment to interorganizational cooperation can facilitate cooperation in areas of common interest, promote a common operational picture (COP), and enable sharing of critical information and resources. This commitment is based on recognition that external organizations will affect the JFC's mission, and vice versa. Interorganizational cooperation may enable:

Unity of Effort

The translation of national objectives into unified action is essential to unity of effort and ultimately mission success.

Common Objectives

Joint and multinational operations are integrated at the strategic level and coordinated at the operational and tactical level with the activities of participating United States Government (USG) departments and agencies, relevant international organizations, nongovernmental organizations (NGOs), host nation (HN) agencies, and elements of the private sector to achieve common objectives.

Common Understanding

Common understanding can enable the joint force commander (JFC) to identify opportunities for cooperation, assist in mitigating unnecessary conflict or unintended consequences, and operate effectively in the same space as external organizations.

II. Unity of Effort

Within the USG, unity of effort can be diminished by organizational stovepiping, crisis-driven planning, different core missions, and divergent organizational processes and cultures. When USG departments and agencies do not coordinate sufficiently, they may interpret national policy guidance differently, develop different objectives and strategies, and set different priorities, and therefore, not act in concert toward national objectives. In a coalition, the interests and practices of participating foreign governments and military forces, international organizations, NGOs, and private sector entities are distinct from, and at times can compete with, USG interests, further exacerbating these issues. Interorganizational cooperation can build mutual understanding of respective goals.

Interorganizational cooperation seeks to find commons goals, objectives, or principles between different organizations; set the conditions for unified action through planning and preparation; and leverage cross-organizational capabilities for unity of effort during execution.

In military operations, unity of effort is facilitated by first establishing unity of command. Unity of command is based on the designation of a single commander with the authority to direct and coordinate the efforts of all assigned and attached forces in pursuit of a common objective. In operations involving interagency partners and other stakeholders where the commander may not control all elements, commanders should seek cooperation and build consensus to achieve unity of effort toward mission objectives.

While unity of command and the exercise of C2 apply strictly to military forces and operations, unified action among all stakeholders is necessary to achieve unity of effort in military operations involving civilian organizations and foreign military forces or military participation in civilian-led operations. Unified action is the DOD doctrinal term that represents a comprehensive approach. Unified action is promoted through close, continuous coordination and cooperation, which seeks to minimize confusion over objectives, inadequate structure or procedures, and bureaucratic or personnel limitations.

III. US Government Department and Agency Coordination

Ref: JP 3-08, Interorganizational Cooperation (val. Oct '17), pp. I-12 to I-14.

One difficulty of coordinating operations among USG departments and agencies is determining appropriate counterparts and exchanging information among them when habitual relationships are not established. Organizational differences exist between the US military and USG departments' and agencies' hierarchies, particularly at the operational level. In defense support of civil authorities (DSCA), military forces may not be familiar with existing structures for disaster response. In foreign operations, existing structures may be limited or not exist at all. Further, overall lead authority in a crisis response and limited contingency operation is likely to be exercised not by the GCC, but by a US ambassador, COM (usually, but not always, the ambassador), or other senior civilian who will provide policy and goals for all USG departments and agencies. When a disaster is declared, the DOS geographic bureau of the affected area becomes the key participating bureau.

Relative Organizational Structures (Levels)

Levels of Decision Making	United States Executive Departments and Agencies
National Strategic	National Security Council Secretary of Defense Joint Chiefs of Staff Secretaries of State, Homeland Security, etc. United States Agency for International Development administrator
Theater Strategic	Combatant Commander Ambassador Chief of Mission Federal Emergency Management Agency (FEMA) administrator
Operational	Joint Task Force Defense Coordinating Officer Embassy Country Team staffs FEMA region directors Federal coordinating officers
Tactical	Task Force or Service Component Commander Agency Field Representatives (e.g., Office of United States Foreign Disaster Assistance disaster assistance response team) Domestic response teams

Ref: JP 3-08, fig. I-2. Relevant United States Agency Organizational Structures Levels.

Decision making at the lowest levels is frequently thwarted because field coordinators may not be vested with the authority to speak for parent organizations. Physical or virtual interagency teaming initiatives, such as joint interagency task forces (JIATFs), joint interagency coordination groups (JIACGs), or routine interagency video teleconferences, improve reachback and expedite decision making. Figure I-2 depicts comparative organizational structures in the context of four levels of decision making.

IV. Coordinating Efforts

Ref: JP 3-08, Interorganizational Cooperation (val. Oct '17), pp. I-6 to I-12; I-14 to I-18.

Strategic Direction

Strategic direction is the common thread that integrates and synchronizes the activities of the Joint Staff (JS), combatant commands (CCMDs), Services, combat support agencies, and other USG departments and agencies.

Applying the Military Component

JFCs have long coordinated with USG departments and agencies such as DOS, Department of Justice (DOJ), Department of Transportation (DOT), United States Agency for International Development (USAID), and the intelligence community (IC). JFCs preparing for domestic operations maintain relationships with the Department of Homeland Security (DHS) and its component agencies; DOJ, particularly the Federal Bureau of Investigation (FBI); and the National Guard Bureau (NGB), which provides coordination with the adjutants general (TAGs) of the 50 states, District of Columbia (through its commanding general), and the three territories. Other JFCs preparing for foreign operations coordinate selectively with multinational partners, international organizations (e.g., the UN and NATO), NGOs, and HN agencies.

Solutions to complex problems seldom reside exclusively in one department or agency. Joint campaign and operation plans should be developed to optimize the core authorities, competencies, and operational capabilities of other departments and agencies, particularly when DOD provides support to those departments and agencies.

In many situations, DOD serves in a supporting role. Thus, commanders and their staffs may need to adjust or adopt procedures, especially planning and reporting, to coordinate military operations with the activities of other organizations. DOD interaction with international organizations; NGOs; state, local, and tribal authorities; and the private sector is conducted with the knowledge, approval, assistance, and cooperation of the chief of mission (COM) in foreign countries and DHS (e.g., Federal Emergency Management Agency [FEMA]) in the US. USAID and DHS may also facilitate DOD interaction with these entities. DOD may also develop direct relations with these organizations to prepare for potential coordinated response operations.

Capitalizing on Organizational Diversity

Each stakeholder brings its own culture, authorities, missions, philosophy, goals, practices, expertise, and skills to the mission. This diversity is an asset when viewed as an opportunity to see the problem from multiple perspectives, for generating innovative solutions, and for bringing various capabilities to the fight.

Gathering the Right Resources

Commanders should identify available resources and capabilities, and determine how to work with mission partners to apply them. Despite potential philosophical, cultural, and operational differences, JFCs should foster an atmosphere of cooperation that contributes to unity of effort.

Identifying Authorities

Each USG department and agency derives authorities from the US Constitution, federal law, federal charters, Presidential directives, congressional mandates, and strategic direction. These authorities should be identified and documented early in the joint planning process (JPP). International organization authorities are based on their formal agreements with member governments and rely largely on consensus among their members. NGOs are independent of the USG, although they may have certain contractual obligations when they receive USG funding. Private sector organizations are bound by the laws of various jurisdictions, including their home jurisdiction and any location where they do business.

Key Coordination Terms

The following terms are a range of interactions that occur among stakeholders. The following descriptions provide a baseline for common understanding.

- **Collaboration** is a process where organizations work together to attain common goals by sharing knowledge, learning, and building consensus.
- **Compromise** is a settlement of differences by mutual concessions without violation of core values; an agreement reached by adjustment of conflicting or opposing positions, by reciprocal modification of an original position. Compromise should not be regarded in the context of win/lose.
- **Consensus** is a general or collective agreement, accord, or position reached by a group as a whole. It implies a serious consideration of every group member's position and results in a mutually acceptable outcome even if there are differences among parties.
- **Cooperation** is the process of acting together for a common purpose or mutual benefit. It involves working in harmony, side by side and implies an association between organizations. It is the alternative to working separately in competition. Cooperation with other departments and agencies does not require giving up authority, autonomy, or becoming subordinated to the direction of others.
- **Coordination** is the process of organizing a complex enterprise in which numerous organizations are involved, and bringing their contributions together to form a coherent or efficient whole.
- **Synchronization** is the process of planning when and how—across time and space—stakeholders will apply their resources in a sequenced fashion.

Key Considerations

Joint planning should include key external stakeholders, ideally starting with mission analysis. Within the area of responsibility (AOR) and the joint operations area (JOA), structures are established at the CCMD, subordinate joint task force (JTF) headquarters (HQ), task force, and Service component levels to coordinate and resolve military, political, humanitarian, and other issues. The crux of interorganizational cooperation is understanding the civil-military relationship as collaborative rather than competitive.

Organizational Environments

Sharing information among department and agency participants is critical to ensure no participant is handicapped by a lack of situational awareness, uncertainties are reduced as much as possible, and interagency decision making is empowered by a common operational picture. Common unifying goals should be clarified with a discussion on the way to achieve them based upon the roles and responsibilities of each organization with their assigned resources.

Commander's Communication Synchronization (CCS)

The USG uses strategic guidance and direction to coordinate use of the informational instrument of national power in specific situations. Commander's communication guidance is a fundamental component of national security direction.

See p. 3-10 for further discussion.

Cyberspace Considerations

Access to the Internet provides adversaries the capability to compromise the integrity of US critical infrastructures/key resources in direct and indirect ways. Threats to all interorganizational networks present a significant risk to national security and global military missions.

See pp 2-26 to 2-27 for further discussion of cyberspace considerations.

V. Whole-of-Government Approach

For domestic operations, USG departments and agencies aspire to a whole-of-government approach. This approach involves the integration of USG efforts through interagency planning that set forth detailed concepts of operations; descriptions of critical tasks and responsibilities; detailed resource, personnel, and sourcing requirements; and specific provisions for the rapid integration of resources and personnel directed in Presidential Policy Directive (PPD)-8, National Preparedness. The National Preparedness System contains five national planning frameworks (NPFs) (https://www.fema.gov/national-planning-frameworks) that spell out USG departments and agencies' responsibilities to prevent, protect against, mitigate, respond to, and recover from the threats and hazards that pose the greatest risk, as called for in *PPD-8, National Preparedness*-National Prevention Framework, National Protection Framework, National Mitigation Framework, National Response Framework (NRF), National Disaster Recovery Framework.

For international operations, there is no similarly robust interagency framework with equivalent statutory authorities or designated interagency roles and responsibilities. Some policies, processes, and organizations facilitate whole-of-government efforts (for example, National Security Presidential Directive (NSPD)-44, Management of Interagency Efforts Concerning Reconstruction and Stabilization; Department of Defense Instruction [DODI] 3000.05, Stability Operations; DOS's Bureau of Conflict and Stabilization Operations [CSO]; and the Office of the Director of National Intelligence [ODNI] to coordinate the activities of all USG intelligence agencies). Security cooperation programs under Title 10, United States Code (USC), Section 2282 authority.

See pp. 8-7 to 8-8 for further discussion.

VI. Working Relationships and Practices

In an interagency sense, the concept of supported/supporting is less about command relationships and more about the methods used to obtain and provide support. Civilian departments and agencies tend to operate via coordination and communication structures, rather than C2 structures.

The most common technique for collaboration is the identification or formation of boards, centers, cells, working groups, offices, elements, planning teams, and other enduring or temporary cross-functional staff organizations that manage specific processes and accomplish tasks in support of mission accomplishment.

Direct, early liaison can facilitate the flow of accurate and timely information about the crisis area, especially when civilian department, agency, and organizational activities precede military operations. Early liaison can also build working relationships based upon trust and open communications among all organizations.

VII. Considerations for Effective Cooperation

Military policies, processes, and procedures can be different from those of civilian organizations. These differences may present significant challenges to interorganizational cooperation. The various USG departments and agencies often have different, and sometimes conflicting, legal authority, policies, procedures, and decision¬making techniques, which make unified action a challenge. In addition, some international organizations, NGOs, and private sector entities may have policies that are antithetical to those of the USG, particularly the US military. US membership in an international organization does not guarantee that the international organization will act in a manner that supports or is consistent with US policy or objectives. While many NGOs might not be hostile to DOD goals, they may choose to not cooperate with DOD or USG efforts to maintain their neutrality. Private sector entities are largely motivated by business or other institutional interests. USG contractors are legally constrained by the language of their contract and generally report to, and are accountable to, only the contracting officer of the contracting agency.

I. Joint Planning Considerations

Ref: JP 3-08, Interorganizational Cooperation (val. Oct '17), chap. II.

USG organizations working to achieve national security objectives require increased and improved communications and coordination. This section provides a frame of reference that reflects all levels of interorganizational involvement.

Joint Planning

Joint planning should include key external stakeholders, ideally starting with mission analysis. Where direct participation is not feasible, joint planners should consider the activities and interests of external stakeholders that affect the command's mission. The CCDR, through the campaign plan, works with civilian organizations to build annex V (Interagency Coordination) of the joint plan. Emphasis should be placed on operationalizing the theater TCP or functional campaign plan (FCP) to facilitate cooperation among all partners, awareness of non-partners, and collective problem framing and synchronization of the CCDR's campaign plan with other interagency planning products. Subordinate JFCs leverage the planning efforts of the CCMD while also building civilian organization participation into their plan and participate in integrated planning with the embassies. Within the area of responsibility (AOR) and the joint operations area (JOA), structures are established at the CCMD, subordinate joint task force (JTF) headquarters (HQ), task force, and Service component levels to coordinate and resolve military, political, humanitarian, and other issues. This section identifies tools for the commander to facilitate interorganizational cooperation in domestic or foreign operations.

I. Whole-of-Government Approach

A whole-of-government approach integrates the collaborative efforts of USG departments and agencies to achieve unity of effort. Under unified action, a whole-of¬government approach identifies combinations of USG capabilities and resources that could be directed toward the strategic objectives in support of US regional goals as they align with global security priorities. Commanders integrate the expertise and capabilities of participating USG departments and agencies, within the context of their authorities, to accomplish their missions.

Whole-of-government planning refers to NSC-sponsored processes by which multiple USG departments and agencies come together to develop plans that address challenges to national interests. Whole-of-government planning is distinct from the contributions of USG departments and agencies to DOD planning, which remains a DOD responsibility.

Planning and consulting with stakeholders optimizes the instruments of national power to achieve operational objectives and attain strategic end states in support of US regional goals in support of global security priorities.

USG civilian departments and agencies have different cultures and capacities, and understand planning differently. Many organizations do not conduct operational planning. To compensate for these differences, commanders should ensure joint force organization initiatives and broader interagency processes help sustain civilian presence in military planning.

Hallmarks of successful whole-of-government planning and operations include:

- A designated lead or primary agency.
- All USG instruments of national power are integrated into the process.
- Agency core missions are related to mission goals.
- Participants forge a common understanding of the operational environment and the problem USG activities are intended to solve.
- Active lines of communications and pre-established relationships to allow for the ease of information sharing.
- A shared USG goal and clearly stated objectives to achieve results through comprehensive integration and synchronization of activities at the implementation level.
- A common determination of what resources and capabilities are to be aligned to achieve the planning objectives.
- A defined strategic objective.

Guidelines to operationalize a whole-of-government approach require that:

- Commanders and civilian decision makers consider all USG capabilities to achieve objectives.
- Planning groups include personnel from all sectors and organizations.
- Ongoing or existing policies and programs are reassessed, modified where necessary, and integrated into the objectives and desired outcomes defined for the mission and strategic end state.
- Planners consider and incorporate interagency capabilities, resources, activities, and comparative advantages in the application of the instruments of national power.

II. Joint Planning and Interorganizational Cooperation

CCMD campaign plans, also known as TCPs and FCPs, implement the military portion of national policy and defense strategy as identified in the Guidance for Employment of the Force (GEF) or other issuances, and implement the military portion of national policy and defense strategy. Designated CCMD campaign plans direct the activities the command will do to shape the operational environment and deter crises on a daily basis. The commander identifies the resources assigned and allocated to the CCMD, prioritizes objectives (to include the contingencies the command is directed to prepare for), and commits those resources to shape the operational environment and support the national strategic objectives. The commander assesses the commitment of resources and makes recommendations to civilian leadership on future resources and national efforts in the region. CCMD campaign plans direct military activities (including ongoing operations, security cooperation activities, intelligence collection, exercises, and other shaping or preventive activities) that shape the operational environment to prevent, prepare for, or mitigate contingencies.

Strategic Guidance

CCDRs develop objectives based on strategic guidance provided by the President, SecDef, and CJCS. CCDRs coordinate planning for operations, actions, and activities at the theater, strategic, and operational levels to achieve strategic objectives.

Adaptive Planning and Execution (APEX)

Once approval has been provided within the proper chains of command, the CCDRs coordinate with affected USG entities throughout the Adaptive Planning and Execution (APEX) enterprise to align the instruments of national power. The CCDR is guid-

The National Security Council (NSC)

Ref: JP 3-08, Interorganizational Cooperation (val. Oct '17), pp. II-1 to II-2.

The NSC is the President's principal forum to consider and decide national security policy. The NSC is the President's principal arm to coordinate these policies among various USG departments and agencies.

DOD Role in the National Security Council System (NSCS)

Key DOD players in the NSCS come from the OSD and JS. SecDef is a regular member of the NSC and the National Security Council/Principals Committee (NSC/PC). The Deputy Secretary of Defense is a member of the National Security Council/Deputies Committee (NSC/DC). If appointed, an Under Secretary of Defense may chair a National Security Council/interagency policy committee (NSC/IPC).

A primary statutory responsibility assigned to the CJCS in Title 10, USC, is to act as the principal military advisor to the President, SecDef, and the NSC. The CJCS does this through the NSCS. CJCS regularly attends NSC meetings and provides advice and views in this capacity. The other members of the JCS may submit advice or an opinion in disagreement with, or in addition to, the advice provided by the CJCS.

The Military Departments implement, but do not participate directly in, national security policy-making activities of the interagency process. They are represented by the CJCS.

Joint Staff Role in the NSCS

The CJCS acts as spokesperson for CCDRs operational requirements and represents CCMD interests in the NSCS through direct communication with the CCDRs and their staffs.

The JS provides operational input and staff support through the CJCS (or designee) for policy decisions made by the OSD. It coordinates with the CCMDs, Services, and other USG departments and agencies, and prepares directives (e.g., warning, alert, and execute orders) for SecDef approval. These orders include definitions of command and interagency relationships.

Within the JS, the offices of the CJCS, Secretary of the JS, Joint Staff J-2 [Intelligence Directorate], Joint Staff J-3 [Operations Directorate], Joint Staff J-4 [Logistics Directorate], Joint Staff J-5 [Plans Directorate], and Joint Staff J-7 [Joint Force Development Directorate] are focal points for NSC-related actions. The JS J-3 provides advice on execution of military operations, the JS J-4 assesses logistic implications of contemplated operations, and the JS J-5 often focuses on a particular NSC matter for policy and planning purposes. Each JS directorate solicits Service input through the Military Departments. SecDef may also designate one of the Service Chiefs or functional CCDRs as the executive agent for direction and coordination of DOD activities for specific mission areas.

For more information on the NSC and its membership, refer to JP 1, Doctrine for the Armed Forces of the United States, and Chairman of the Joint Chiefs of Staff Instruction (CJCSI) 5715.01, Joint Staff Participation in Interagency Affairs.

Combatant Commander (CCDRs') Role in the NSCS

Although the CJCS presents the views of the CCDRs at the NSC, the CJCS may request and leverage CCMD participation at key NSC forums, including IPCs, NSC/DCs, NSC/PCs, and other events (e.g., Cabinet-level exercises). Execution of CCMD campaign plans by CCDRs is enhanced by robust interaction with interagency partners based on standing authorities. JS and OSD will coordinate authorizations through the NSCS.

ed by USG strategic guidance and planning, with respect to USG departments and agencies, and disseminates that guidance to the joint force in annex V (Interagency Coordination) of the CCMD's campaign plans. Considering how best to integrate civilian and military efforts in a mutually supportive way is essential throughout the JPP. Developing a well-crafted and articulated annex V to joint plans is the primary method within POLMIL planning, from the strategic through operational levels, for explaining the linkages and tasks necessary for mission success across all phases of integrated civil-military efforts. Interagency partners should participate at the earliest phases of the operation or campaign. Linking agency actions to phases of the operation enables scheduling and coordination. The development of annex V should enhance early operational coordination with planners from the other USG departments and agencies that will be involved in the operation's planning and execution.

When developing joint plans, planners should identify opportunities to support and promote a unified USG approach to achieve national security objectives.

CCMD campaign plans direct military activities (including ongoing operations, security cooperation activities, intelligence collection, exercises, and other shaping or preventive activities) that shape the operational environment to prevent, prepare for, or mitigate contingencies.

During plan development, planners should identify decision points and desired preparatory activities to be performed by external stakeholders that enable transition from a DOD-supported to DOD-supporting role. Concurrently, planners should analyze and plan for reverting back to a DOD-supported role in the operation.

Flexible Deterrent Options (FDOs)

Flexible deterrent options (FDOs) are pre-planned, deterrence-oriented, diplomatic, informational, military, and economic actions that are carefully tailored to send a signal to influence an adversary's actions.

See pp. 2-74 to 2-75 for further discussion of FDOs.

A. Plan Development and Coordination

Although planning is conducted in anticipation of future events, there may be crisis situations that call for an immediate US military response (e.g., noncombatant evacuation operation or FHA). CCDRs frequently develop courses of action (COAs) based on recommendations and considerations originating in DOS joint/regional bureaus or in one or more US embassies. The country team provides resident agency experience and links through the CCMD and by extension to agency HQ in Washington, DC. Emergency action plans at every US embassy cover a wide range of contingencies and crises and can assist the commanders in identifying COAs, options, and constraints to military actions and support activities. The GCC's staff also consults with JS and other organizations to coordinate military operations and synchronize actions at the national strategic and theater strategic levels. Under the promote cooperation program, the JS J-5 facilitates periodic interagency working groups that include CCMD planning staffs, other DOD offices, and agency partners for collaboration on planning (e.g., campaign and other contingency plans). Promote cooperation events enable interagency partners' insights on environmental changes to be shared with the CCMDs.

B. Guidance for Employment of the Force (GEF)

The GEF translates national security objectives and high-level strategy into DOD priorities and comprehensive planning direction. The GEF identifies SecDef's strategic priorities and policy, and conveys his guidance for near-term plans and defense posture. The JSCP is promulgated by the CJCS. It implements the strategic policy guidance provided in the GEF and initiates the planning process for the development of plans. Together, the GEF and JSCP provide guidance and task CCDRs and staffs to develop TCPs and FCPs that integrate security cooperation (shaping and

III. Stakeholders

International Organizations

An international organization is an organization created by a formal agreement (e.g., a treaty) between two or more governments on a global, regional, or functional basis to protect and promote national interests shared by member states. International organizations may be established for wide-ranging or narrowly defined purposes. Examples include the UN, NATO, Organization of American States (OAS), and the African Union (AU). NATO and the Organization for Security and Cooperation in Europe are regional security organizations, while the European Union (EU), the AU, and the OAS are general regional organizations. However, some general regional organizations and sub-regional organizations conduct security-related activities. For example, the AU conducted peacekeeping operations in the Sudan through the AU Mission in Sudan and operations in Somalia. These organizations have defined structures, roles, and responsibilities, and may be equipped with the resources and expertise to participate in complex interorganizational cooperation.

Nongovernmental Organizations (NGOs)

NGOs are private, self-governing, not-for-profit organizations dedicated to alleviating human suffering; and/or promoting education, health care, economic development, environmental protection, human rights, and conflict resolution; and/or encouraging the establishment of democratic institutions and civil society. Where long¬term problems precede a deepening crisis, NGOs are frequently on scene before the US military and may have an established presence in the crisis area. NGOs frequently work in areas where military forces conduct military engagement, security cooperation, and deterrence activities. They will most likely remain long after military forces have departed.

The Private Sector

The private sector is an umbrella term that may be applied to any or all of the nonpublic or commercial individuals and businesses, specified nonprofit organizations, most of academia and other scholastic institutions, and selected NGOs. Private sector organizations range from large and multinational to small with limited resources and focused on one country. The private sector also includes contractors. There may be a plethora of small private sector entities and NGOs in a country. The private sector can help the USG obtain information, identify risks, conduct vulnerability assessments, and provide other assistance. Private organizations' assistance to the USG is most prominent during security cooperation, combat support, and reconstruction.

The Civil-Military Relationship

The crux of interorganizational cooperation is understanding the civil-military relationship as collaborative rather than competitive. While the military normally focuses on achieving clearly defined and measurable objectives within given timelines under a C2 structure, civilian organizations are concerned with fulfilling shifting political, economic, social, and humanitarian interests using negotiation, dialogue, bargaining, and consensus building. Civilian organizations may have a better appreciation of the political-social¬cultural situation, and have better relief, development, and public administration experience, thus potentially acting as agents of change within that society. They work at the local and national, government level, focusing on state-to-state and ministry-to¬ministry relations, and in community development activities where they have substantial insight into local conditions and local operational requirements. At the same time, civilian agencies generally work in permissive environments and may not understand military goals and operations.

IV. Joint Task Forces (JTFs) in the Interagency Process

Ref: JP 3-08, Interorganizational Cooperation (val. Oct '17), pp. II-21 to II-28.

When it is necessary to establish a Joint Task Force (JTF), the establishing authority is normally a CCDR. The CCDR develops the mission statement and CONOPS based upon direction from SecDef, as communicated through the CJCS. The CCDR appoints a commander, joint task force (CJTF), and, in conjunction with the CJTF, determines the capabilities required to achieve military objectives. The CJTF has the authority to organize forces and the JTF HQ, to accomplish the objectives.

The mission assigned to a JTF will require not only the execution of responsibilities involving two or more military departments but, increasingly, the mutual support of numerous USG departments and agencies, and collaboration with international organizations, NGOs, and the private sector. Normally, a JTF is dissolved when the purpose for which it was created has been achieved. The JTF HQ commands and controls the joint force and coordinates military operations with the activities of other USG departments and agencies, MNFs, international organizations, NGOs, the private sector, and HN forces and agencies. A principal distinction between a JTF and a Service command is the JTF's greater emphasis on interorganizational cooperation to achieve unity of effort.

JTFs in the Interagency Process

Unlike the military, most USG departments and agencies are not equipped and organized to create separate staffs at the strategic, operational, and tactical levels. Therefore, JTF personnel interface with individuals who are coordinating their organization's activities at more than one level. The USG interagency process requires the JTF HQ to be especially flexible, responsive, and cognizant of the capabilities of USG departments and agencies, international organizations, HN forces and agencies, NGOs, and the private sector. The JTF HQ provides an important basis for a unified effort, centralized planning and direction, and decentralized execution. Depending on the type of operation; the extent of military operations; and the degree of agency, international organization, NGO, and private sector involvement, the focal point for operational- and tactical-level coordination with civilian departments and agencies may occur at the JTF HQ, the country team, the joint field office (JFO), the CMOC, or the humanitarian operations center (HOC). JTF personnel may also participate actively, or as observers, in a civilian-led functional coordinating group concentrating on a specific issue or project.

- Upon activation of a JTF outside the US, the CCDR determinates whether the JTF has direct liaison authority with the affected COM and with the senior defense official/defense attaché (SDO/DATT) to provide consistent, efficient communication with the COM and the country team.
- JTFs should channel most communications through the LNO team to avoid overwhelming interagency partners with JTF coordination and planning requests.
- JTFs should designate the staff office responsible for interorganizational cooperation. Many JTFs designate their plans directorate of a joint staff or CMOC, while others may form a separate directorate.
- When a large country team or JFO and a military JTF exist side-by-side, detailed procedures should be developed for staff coordination.
- For DSCA operations, the JTF HQ is ideally colocated with the JFO per NRF guidance. All ESFs are represented in the JFO.
- JTFs should consider how to integrate military elements that may not be part of the core JTF (e.g., special operations forces).

Joint Interagency Task Force (JIATF)

Ref: JP 3-08, Interorganizational Cooperation (Oct '16),app. E.

A JIATF may be formed when the mission requires close integration of two or more USG departments and agencies.

The joint interagency task force (JIATF) is a force multiplier that uses a unique organizational structure to focus on a single mission. A JIATF is typically formed for a specific task and purpose as are most task forces. JIATFs are formal organizations usually chartered by the DOD and one or more civilian agencies and guided by a MOA or other founding legal documents that define the roles, responsibilities, and relationships of the JIATF's members. The JIATF is staffed and led by personnel from multiple agencies under a single commander or director.

Forming a national level JIATF takes a national charter that lays out authorities and mandates membership and resourcing. An executive order, national level directive, or mandate from the NSC/HSC that directs all agencies involved to support the JIATF with actual resources may be required. SecDef may, in cooperation with other Cabinet members, form a JIATF through the establishment of detailed memoranda of agreement. JFCs can form JIATFs with one or more USG agencies based on mutual cooperation and agreement.

The establishment of functional and enduring JIATFs transcends the internal capabilities and authorities of combatant commands and JTFs. Based upon the analysis and the desire to establish JIATFs, the JIACG (or equivalent organization) or another designated staff entity should document the requirements for formal submission through command channels to JS and OSD for approval and pursuit through the NSC or HSC system. Success would be manifest in interagency consensus, commitment, and MOAs or MOUs that infuse JIATFs with supporting policy, legitimacy, defined purpose, authorities, leadership parameters, functional protocols, and resources.

Coordinating authorities, channels, and terms of reference must be carefully established and documented for JIATFs, with the aim of facilitating their missions and flexibility while not promoting duplication of effort and confusion. Such authorities constitute the rules of the road for JIATFs, and they must contribute to unity of effort and common situational awareness.

Increasingly, JIATFs are being formed to achieve unity of effort and bring all instruments of national power to bear on asymmetric threats. JIATFs are often created to address problems such as militias, "bad neighbors," and foreign fighters, all of which complicate the security environment. JIATFs may be separate elements under the JFC, or they may be subordinate to a functional component command, a joint special operations task force, or a staff section such as the J-3. JIATF members can coordinate with the country team, their home agencies, JIACGs (or equivalent organization) in the area of interest, and other JIATFs in order to defeat complex hostile networks. Because they use more than the military instrument of national power, JIATFs are generally not a lethal asset, but rather develop and drive creative nonlethal solutions and policy actions to accomplish their mission.

Joint Support Force (JSF)

A joint support force may be formed when the mission is a DSCA operation and DOD is operating in support of one or more USG departments and agencies. Although organized similar to a JTF (i.e., with a commander, command element, and forces), the title indicates a more cooperation-focused organization to the interorganizational community.

other military engagement activities) planning, contingency plans as campaign plan branches, and subordinate campaign plans as CCDRs consider necessary.

C. Annex V (Interagency Coordination)

The efficiency and effectiveness of USG departments and agencies is diminished if they are excluded from joint planning. The quality of military planning and the subsequent effectiveness of operations and achievement of US strategic objectives can be enhanced by early and sustained participation of relevant USG stakeholders during strategic assessment, policy formulation, and planning.

Annex V (Interagency Coordination) should be consistent with the planning guidance contained in CJCSM 3130.03, Adaptive Planning and Execution System (APEX) Planning Formats and Guidance. A CCDR is responsible for developing annex V for TCPs, FCPs, OPLANs, and usually for Level 3 CONPLANs. A supported JFC, not a CCDR, is responsible for developing annex V for OPLANs and usually for Level 3 CONPLANs. Annex V should specify, for participating USG departments and agencies, the following: the capabilities desired by the military, the shared understanding of the situation, and the common objectives required to accomplish the mission. Annex V also provides a single location in a plan to capture potential contributions of USG departments and agencies; identify potential DOD supporting roles to other USG departments and agencies; and frame mutually agreeable, integrating relationships (coordination and collaboration processes), linkages, and methods. This enables agency planners to plan in concert with the military, to better determine their support requirements, and to suggest other USG activities or organizations that could contribute to the operation.

DOD plans are approved by SecDef (or designee). While they are not "cosigned" by other USG departments and agencies, DOD typically seeks input from them to gain broad USG consensus. The military plan may be in support of a wider USG effort. In this case, annex V may approach the same level of effort and importance as annex C (Operations).

D. Joint Interagency Coordination Group (JIACG)

The JIACG is an interagency staff group that establishes regular, timely, and collaborative working relationships between civilian and military operational planners. Composed of USG civilian and military experts accredited to the CCDR and tailored to meet a supported CCDR's requirements, the JIACG (or equivalent organization) provides the CCDR with the capability to collaborate at the operational level with other USG civilian departments and agencies. JIACGs (or equivalent organizations) complement the interagency coordination that takes place at the national strategic level through the DOD and the NSC. Members participate in planning and provide links back to their parent civilian departments and agencies to help synchronize JTF operations with their efforts.

JIACG is a common DOD term across CCMDs. Each CCMD has formed unique organizations with similar functions to respond to a wide range of missions across operational environments. If augmented with other partners, such as international organizations, NGOs, and multinational representatives, the JIACG (or equivalent organization) enhances the capability to conduct interorganizational cooperation.

II. Domestic Considerations

Ref: JP 3-08, Interorganizational Cooperation (val. Oct '17), chap. III.

I. Key Government Stakeholders

The Department of Homeland Security (DHS) leads the unified national effort to secure America by preventing terrorism and enhancing security, securing and managing our borders, enforcing and administering immigration laws, safeguarding and securing cyberspace, and ensuring resilience to disasters. Within DOD, the Secretary of Defense (SecDef) has overall authority and is the President's principal advisor on military matters concerning use of federal forces in homeland defense (HD) and defense support to civil authority (DSCA). CNGB is SecDef's principal advisor, through the CJCS, for non-federalized NG forces. The Assistant Secretary of Defense (Homeland Defense and Global Security) (ASD[HD&GS]) serves as the principal staff assistant delegated the authority to manage and coordinate HD and DSCA functions at the SecDef level.

The two CCMDs with major HD and DSCA missions are United States Northern Command (USNORTHCOM) and United States Pacific Command (USPACOM), as their AORs include the US and its territories. USNORTHCOM and USPACOM HD missions include conducting operations to deter, prevent, and defeat threats and aggression aimed at the US, its territories, and interests within the assigned AOR; and, as directed by the President or SecDef, provide DSCA.

These geographic CCMDs may also have senior DHS representatives and a NG representative assigned as advisors. The senior DHS representative advises the commander and staff on HS and DSCA issues and requirements, and facilitates information sharing, coordination, and collaboration between the command and the operational agencies of DHS (e.g., FEMA, US Customs and Border Protection [CBP], and United States Coast Guard [USCG]).

II. State, Local, Territorial, and Tribal Considerations

When a disaster threatens or occurs, a governor may request federal assistance. If DOD support is required and approved as part of that federal assistance, then DOD may execute mission assignments in support of the primary federal agency that often result in a wide range of assistance to local, tribal, territorial, and state authorities. Incidents can have a mix of public health, economic, social, environmental, criminal, and political implications with potentially serious long-term effects. Significant incidents require a coordinated response across organizations and jurisdictions, political boundaries, sectors of society, and multiple organizations.

Federal law, as codified in Title 10 and Title 32, USC, creates distinct mechanisms for local and state authorities to request NG forces or resources. Local and state authorities may also request federal forces (active and reserve) under Title 10, USC, authority for a contingency response. The NG of the US is administered by the NGB, which is a joint activity under DOD and provides communication for NG to DOD to support unified action. The NG active, reserve framework is built on mechanisms that coordinate among federal, state, territorial, tribal, and local governments to prevent, protect against, and respond to threats and natural disasters. NG forces operate under state active duty, Title 32, USC, or federal active duty, Title 10, USC, depending on activation status.

III. DOMESTIC Considerations for Interorganizational Cooperation

Ref: JP 3-08, Interorganizational Cooperation (val. Oct '17), pp. III-6 to III-15.

DOD works closely with other USG departments and agencies when planning. The supported GCCs are DOD principal planning agents and provide joint planning directives for peacetime assistance rendered by DOD within their assigned AORs. Upon issuance of an execute order by the CJCS, at the direction of the President or SecDef, to initiate or conduct military operations, the supported commander implements and relays the authority of the order with their own orders directing action to subordinate commanders, supporting commanders, and directors of supporting agencies. Thorough joint planning requires that a GCC's operations and activities align with national security objectives contained in strategic guidance. The GEF prioritizes these objectives and DOD priorities for each CCMD, which then develop a FCP or TCP, as required. In addition to participating in interagency steering groups and councils, DOD has responsibilities under the NRF. The salient frameworks and directives that will guide DSCA operations are the following:

Robert T. Stafford Disaster Relief & Emergency Assistance Act

This Act provides the authority for the USG to assist with state and local government response to a major disaster or emergency. The act gives the President the authority to declare an area a major disaster, declare an area an emergency, exercise 10-day emergency authority, and send in federal assets when an emergency occurs in an area over which the federal government exercises primary responsibility.

National Response Framework (NRF)

The NRF is a guide that details how the nation conducts all-hazards response—from the smallest incident to the largest catastrophe. The NRF identifies the response principles, to include DOD, as well as the roles and structures that organize national response. It describes how communities, states, the USG, private sector, and US NGO partners apply these principles to coordinate a national response.

National Incident Management System (NIMS) / Incident Command System (ICS)

The NIMS, and its associated ICS, provides a systematic, proactive approach to guide departments and agencies at all levels of government, NGOs, and the private sector to work seamlessly to prevent, protect against, respond to, recover from, and mitigate the effects of incidents, regardless of cause, size, location, or complexity, to reduce the loss of life and property and harm to the environment. NIMS is integrated with the NRF. The NIMS provides the template for the management of incidents, while the NRF provides the structure and mechanisms for national-level policy for incident management. The NIMS is a tested system that interagency partners utilize and practice regularly. Leaders with NRF responsibilities should have an understanding of its principles, structures, and techniques.

To align DOD planning with the needs of those requiring DSCA, DOD coordinates with interagency partners and with the NGB. Coordination should align national frameworks, NIMS, and interagency guidelines provided in the JSCP. The standing CJCS DSCA Execute Order delegates limited approval authority to supported CCDRs to respond to domestic emergencies and/or disasters and aligns with the NRF to provide a unified national response.

Domestic Operating Environment

The domestic operating environment for DSCA presents unique challenges to the JFC. When executing DSCA, the US military is normally in support of another USG department or agency that is coordinating the federal response. The President can direct DOD to be the lead for the federal response; however, this would only happen in extraordinary situations and would involve other DOD core mission areas. US federal and NG forces may also support state, territorial, local, or tribal activities. Commanders and staffs at all levels must understand the relationships, both statutory and operational, among all USG departments and agencies involved in the operation. It is equally important to understand DOD's role in supporting other USG departments and agencies. DOD can provide assistance to the primary agency as authorized by SecDef or the President.

Military commanders are authorized to take action under immediate response authority in certain circumstances. In response to a RFA from a civil authority, under imminently serious conditions and if time does not permit approval from higher authority, DOD officials (i.e., military commanders, heads of DOD components, and responsible DOD civilian officials) may provide an immediate response by temporarily employing the resources under their control—subject to any supplemental direction provided by higher HQ—to save lives, prevent human suffering, or mitigate great property damage within the US.

Military forces may also help DOJ or other federal, state, or local law enforcement agencies (LEAs) when requirements are met. This includes military assistance in response to civil disturbances. In addition to emergency or disaster assistance, other USG departments and agencies may request DOD assistance as part of HS. Military commanders should review, with legal counsel, each request for domestic aid for statutory compliance, especially for law enforcement assistance to civil authorities. SecDef must personally approve any request to assist LEAs in preplanned national events. Requests for DOD assets in support of law enforcement require careful review during planning to ensure DOD support conforms to law and policy and does not degrade the mission capability of CCDRs. The US Constitution, federal laws, and USG policies and regulations restrict domestic military operations. Requests for DOD assistance should be coordinated with the supporting organization's legal counsel or SJA

CBRN Response

Supporting CBRN missions requires a number of specialized capabilities. These capabilities may be required to support civil authorities as part of efforts ranging from the prevention of an attack to technical nuclear forensics to support attribution. For example, the 2011 Interagency Domestic Radiological/Nuclear Search Plan specifies that DOD maintains an operational radiological/nuclear search capability. Additionally, managing the consequences of a CBRN incident is a USG effort.

Refer to our series of related ***Homeland Defense & DSCA, Counterterrorism, and Disaster Response SMARTbooks*** *for further discussion. The US Armed Forces have a historic precedent and enduring role in supporting civil authorities during times of emergency, and this role is codified in national defense strategy as a primary mission of DOD. In the past decade alone, natural disasters of considerable severity resulted in 699 Presidential Disaster Declarations, an average of nearly six per month. Disaster management (or emergency management) is the term used to designate the efforts of communities or businesses to plan for and coordinate all the personnel and materials required to mitigate or recover from* ***natural or man-made disasters, or acts of terrorism****.*

IV. Homeland Defense & Defense Support of Civil Authorities (DSCA)

Ref: JP 3-08, Interorganizational Cooperation (val. Oct '17), pp. III-4 to III-6.

The use of the Armed Forces inside the US and its territories, though limited in some respects, falls into two mission areas: HD—for which DOD is lead agency and employs military forces to conduct military operations in defense of the homeland; and DSCA—for which DOD supports other USG departments and agencies by providing military resources in support of civil authorities. DSCA is consistent with the national frameworks in that it supplements the efforts and resources of other USG departments and agencies in support of state, local, territorial, and tribal governments, as well as NGOs and volunteer organizations. In most cases, the President and SecDef determine when DOD will be involved in HD and DSCA missions. Interorganizational cooperation for HD and DSCA is particularly sensitive when joint forces conduct operations in proximity to our domestic population and critical infrastructure. While the HD and DSCA missions are distinct, some department roles and responsibilities overlap, and operations require extensive coordination between lead and supporting agencies. HD and DSCA operations may occur concurrently and require extensive integration and synchronization. DOD may conduct HD operations in a lead agency role, while at the same time providing DSCA in response to the consequences of an attack or natural disaster. In addition, operations may also transition from HD to DSCA to HS or vice versa (e.g., the USCG in maritime security) with the lead shifting depending on the situation and USG's desired outcome. However, the designation of the federal department or agency with lead responsibility is not always predetermined. In certain time-critical situations, on-scene leaders are empowered to conduct operations in response to a particular threat.

Homeland Defense (HD)

HD is the protection of US sovereignty, territory, domestic population, and critical infrastructure against external threats and aggression, or other threats, as directed by the President. DOD is responsible for the HD mission and leads HD responses, with other USG departments and agencies in support. DOD's capability to respond quickly to multiple threats in a variety of situations can strain limited resources. For example, the same force constituted to deploy on a contingency operation overseas may also be the most qualified force for a HD mission. For HD missions, the President authorizes military action to counter threats to and within the US.

Defense Support of Civil Authorities (DSCA)

DOD DSCA activities must be specifically authorized and are generally conducted in support of a primary civilian agency. The exceptions are those noted in the NRF (ESF #3 and ESF # 9 where DOD/US Army Corps of Engineers and DOD have primary or shared primary responsibilities. RFAs from another department or agency may be predicated on mutual agreements between organizations or stem from a presidential designation of a federal disaster area or a federal state of emergency. DOD support is typically requested only when the resources of state, local, and tribal governments or other USG departments and agencies prove insufficient to provide critical support in a timely manner, or when specialized military assets are required.

Refer to The Homeland Defense & DSCA SMARTbook (Protecting the Homeland / Defense Support to Civil Authority) for complete discussion. Topics and references include homeland defense (JP 3-28), defense support of civil authorities (JP 3-28), Army support of civil authorities (ADRP 3-28), multi-service DSCA TTPs (ATP 3-28.1/MCWP 3-36.2), DSCA liaison officer toolkit (GTA 90-01-020), key legal and policy documents, and specific hazard and planning guidance.

Chap 8

III. Foreign Considerations

Ref: JP 3-08, Interorganizational Cooperation (val. Oct '17), chap. IV.

Within the executive branch, DOS is the lead foreign affairs agency, assisting the President in foreign policy formulation and execution. DOS oversees the coordination of DOD external political-military relationships with overall US foreign policy. USAID is the lead agency for overseas development and disaster response and carries out programs that complement DOD efforts in stabilization, foreign internal defense, and security force assistance.

I. USG Structure in Foreign Countries

A. The Diplomatic Mission

The United States Government (USG) has bilateral diplomatic relations with almost all of the world's independent states. The US bilateral representation in the foreign country, known as the diplomatic mission, is established IAW the Vienna Convention on Diplomatic Relations, of which the US is a party.

Missions are organized under DOS regional and functional bureaus. The boundaries for the DOS regions roughly approximate those of the CCMDs and therefore geographic and functional seams must be addressed and managed. DOS provides the core staff of a diplomatic mission and administers the presence of representatives of other USG departments and agencies in the country. A diplomatic mission is led by a COM, usually the ambassador, but at times another person designated by the President, or the chargé d'affaires (the chargé) when no US ambassador is accredited to the country or the ambassador is absent from the country. The deputy chief of mission (DCM) is second in charge of the mission and usually assumes the role of chargé in the absence of the COM.

For countries with which the US has no diplomatic relations, the embassy of another country represents US interests and at times houses an interests section staffed with USG employees. In countries where an international organization is headquartered, the US may have a multilateral mission to the international organization in addition to the bilateral mission to the foreign country.

See following pages (pp. 8-20 to 8-21) for an overview and further discussion

B. Combatant Commands (CCMDs)

USG departments and agencies augment CCMDs to help integrate the instruments of national power in plans. GCCs, functional CCDRs, and, increasingly, JTF commanders are assigned a POLAD by DOS. POLADs are senior DOS officers (often flag-rank equivalent) detailed as personal advisors to senior US military leaders and commanders, and they provide policy analysis and insight regarding the diplomatic and political aspects of the commanders' duties. The POLAD is directly responsible to the CCDR or CJTF. They do not serve as DOS representatives.

See following pages (pp. 8-20 to 8-21) for an overview and further discussion.

USG Structure in Foreign Countries

Ref: JP 3-08, Interorganizational Coordination (val. Oct '17), pp. IV-2 to IV-7.

The Diplomatic Mission

The US has bilateral diplomatic relations with 190 of the world's other 193 independent states. The US bilateral representation in the foreign country, known as the diplomatic mission, is established in accordance with the Vienna Convention on Diplomatic Relations, of which the US is a signatory. DOS provides the core staff of a diplomatic mission and administers the presence of representatives of other USG agencies in the country. A diplomatic mission is led by a COM, usually the ambassador, but at times the chargé d'affaires (the chargé), when no US ambassador is accredited to the country or the ambassador is absent from the country. The deputy chief of mission (DCM) is second in charge of the mission and usually assumes the role of chargé in the absence of the COM. For countries with which the US has no diplomatic relations, the embassy of another country represents US interest and at times houses an interests section staffed with USG employees. In countries where an IGO is headquartered, the US may have a multilateral mission to the IGO in addition to the bilateral mission to the foreign country.

- **The Ambassador.** The President, with the advice and consent of the Senate, appoints the ambassador. The ambassador is the personal representative of the President to the government of the foreign country or to the IGO to which accredited and, as such, is the COM, responsible for recommending and implementing national policy regarding the foreign country or IGO and for overseeing the activities of USG employees in the mission. While the majority of ambassadors are career members of the Foreign Service, many are appointed from outside the Foreign Service. The ambassador has extraordinary decision-making authority as the senior USG official on the ground during crises.
- **Chief of Mission (COM).** The bilateral COM has authority over all USG personnel in country, except for those assigned to a combatant command, a USG multilateral mission, or an IGO. The COM may be accredited to more than one country. The COM interacts daily with DOS's strategic-level planners and decision makers. The COM provides recommendations and considerations for CAP directly to the GCC and commander of a JTF. While forces in the field under a GCC are exempt from the COM's statutory authority, the COM confers with the GCC regularly to coordinate US military activities with the foreign policy direction being taken by the USG toward the HN. The COM's political role is important to the success of military operations involving the Armed Forces of the United States. Generally, each COM has a formal agreement with the GCC as to which DOD personnel fall under the security responsibility of each.
- **Deputy Chief of Mission (DCM).** The DCM is chosen from the ranks of career foreign service officers through a rigorous selection process to be the principal deputy to the ambassador. Although not appointed by the President, the DCM wields considerable power, especially when acting as the COM while in chargé. The DCM is usually responsible for the day-to-day activities of the embassy.
- **The Embassy.** The HQ of the mission is the embassy, located in the political capital city of the HN. Although the various USG agencies that make up the mission may have individual HQ elsewhere in the country, the embassy is the focal point for interagency coordination. The main building of the embassy is termed the chancery; the ambassador's house is known as the residence.
- **Consulates.** The size or principal location of commercial activity in some countries necessitates the establishment of one or more consulates—branch offices of the mission located in key cities, often at a distance from the embassy. A consulate is headed by a principal officer.

Department of State (DOS) Plans

The overall global plan is the DOS/USAID Joint Strategic Plan. The Quadrennial Diplomacy and Development Review is a study completed by DOS every four years that analyzes the short-, medium-, and long-term blueprint for the US diplomatic and development efforts abroad. In addition, DOS regional bureaus influence specific geographic areas and functional bureaus focus on specific interests such as terrorism or arms control. US missions prepare an integrated country strategy every three years that sets country-level US foreign policy goals and objectives, and establishes an action plan to achieve those objectives. The integrated country strategy is a concise, streamlined document that facilitates long-term diplomatic and assistance planning. They are coordinated among the departments and agencies represented on the country team, both in their embassy and in the Washington, DC, interagency community. DOS regional bureaus in Washington, DC, and their joint regional strategies cover geographic regions that are not identical to the GCCs' AORs.

The Country Team

The country team, headed by the COM, is the senior in-country interagency coordinating body. It is composed of the COM, DCM, section heads, the senior member of each US department or agency in country, and other USG personnel as determined by the COM. Each member presents the position of his or her parent organization to the country team and conveys country team considerations back to the parent organization. The COM confers with the country team to develop and implement foreign policy toward the HN and to disseminate decisions to the members of the mission.

Combatant Commands (CCMDs)

USG departments and agencies augment CCMDs to help integrate the instruments of national power in plans.

- **Policy Advisor (POLAD).** GCCs, functional CCDRs, and, increasingly, JTF commanders are assigned a POLAD by DOS. POLADs are senior DOS officers (often flag-rank equivalent) detailed as personal advisors to senior US military leaders and commanders, and they provide policy analysis and insight regarding the diplomatic and political aspects of the commanders' duties.
- **USAID.** USAID also places senior development advisors (SDAs) at most geographic CCMDs, USSOCOM, and Pentagon (JS) to coordinate GCC relations with USAID HQ and field missions. These advisors are senior USAID Foreign Service officers. They inform GCC planning and operations concerning USAID programs and processes and serve as the CCDR's principal advisor on all development matters in the AOR. OFDA places HA advisors at the CCMDs and Pentagon to coordinate responses involving DOD assistance, provide training, and participate in planning.
- **Joint Interagency Coordination Group (JIACG).** The JIACG (or equivalent organization), when formed, participates in planning efforts. Each JIACG (or equivalent organization) is a multifunctional, advisory element that facilitates information sharing. It provides regular, timely, and collaborative day-to-day support to plan, coordinate, prepare, and implement agency activities.
- **Other USG agencies may detail liaison personnel** to combatant command staffs to improve interagency coordination.
- **DOD Regional Centers** are aligned with the geographic CCMD's programs and objectives and are DOD's primary instruments for regional outreach and alumni network-building efforts among US and foreign military, civilian, and non-government actors.
- DOS and USAID assess conflict prevention, mitigation, and stabilization activities with Interagency Conflict Assessment Framework (ICAF) and Conflict Assessment Framework (CAF), respectively.

II. Foreign Operations

The POLMIL Dimension

Within the executive branch, DOS is the lead foreign affairs agency, assisting the President in foreign policy formulation and execution. DOS oversees the coordination of DOD external POLMIL relationships with overall US foreign policy. USAID is the lead agency for overseas development and disaster response and carries out programs that complement DOD efforts in stabilization, foreign internal defense, and security force assistance. USG policy, treaties, and agreements bring DOD into a wide range of external POLMIL relationships that include:

- Bilateral military relationships
- Multinational military forces
- Multilateral mutual defense alliances
- Treaties and agreements involving DOD activities or interests
- Use of US military assets for FHA or PO (or conducted under UN auspices)

A. Theater or Regional Focus

CCDRs implement DOD external POLMIL relations within their campaign plans. The geographic and functional CCMD's operations and activities align with the DSR and the GEF. The GEF prioritizes campaign objectives for each CCMD, which then develops a TCP or FCP. The CCMD campaign plan and nested country plans should complement the current DOS joint regional strategy, the integrated country strategy, and if applicable, USAID's country development cooperation strategy. The geographic CCMD's regional focus is similar to the regional focus of DOS's regional bureaus; however, the geographic boundaries differ. Most other USG foreign affairs agencies are regionally organized as well, again with varying geographic boundaries. The CCMDs include security cooperation activities requiring interorganizational cooperation in their campaign plans, which include posture and country-specific security cooperation sections. In contrast, the DOS focal point for formulation and implementation of regional foreign policy strategies requiring interorganizational cooperation is the regional bureau headed by an assistant secretary at DOS in Washington, DC. USAID has a similar structure, with geographic bureaus headed by assistant administrators in Washington, DC. Although the CCDRs will often find it more expeditious to approach the COMs for approval of an activity in HNs, the political effect of the proposed US military activity usually goes beyond the boundaries of the individual HN. In such cases, the CCDR should not assume that the position of the COM corresponds to the region-wide position of DOS. The CCDR's POLAD can assist in ascertaining whether the activity has regional bureau approval.

B. CCMD Campaign Plans, Crisis Response, and Limited Contingency Operations

The CCMD's campaign plan and nested country plans should complement DOS integrated country strategies and other plans developed by the country teams and USG interagency partners. In a crisis response and limited contingency operation, coordination between DOD and other USG departments and agencies normally occurs within the NSC/IPC and, if directed, during development of the USG strategic plan. During lesser operations and operations not involving armed conflict, the CCDR's staff may deal directly with a COM or members of the country team regarding issues that do not transcend the boundaries of the HN. In some operations, a special envoy of the President may be involved.

See following pages (pp. 8-24 to 8-25) for further discussion of crisis action organizational considerations.

III. Stakeholders

Ref: JP 3-08, Interorganizational Cooperation (val. Oct '17), pp. IV-6 to IV-13.

International Organizations

United Nations (UN). Coordination with the UN begins at the national level with DOS, through the US ambassador to the UN, officially titled the US Permanent Representative. The ambassador typically has the status of Cabinet rank and is assisted at the US mission to the UN by a military assistant who coordinates military interests primarily with the United Nations Department of Peacekeeping Operations (UNDPKO) and the UNOCHA. USG coordination with UN PO missions or agencies in-theater is through the US country team, which includes DOS's refugee coordinators focused on humanitarian response through UN agencies and the International Committee of the Red Cross (ICRC).

North Atlantic Treaty Organization (NATO). NATO is an alliance of 28 countries from North America and Europe committed to fulfilling the goals of the North Atlantic Treaty. IAW the treaty, the fundamental role of NATO is to safeguard the freedom and security of its member countries by political and military means. NATO has no operational forces of its own other than those assigned to it by member countries or contributed by partner countries to carry out a specific mission. It has a number of mechanisms available for the defense planning and resource planning that form the basis of cooperation within the Alliance.

Nongovernmental Organizations (NGOs)

NGOs typically operate under approval of the HN and provide humanitarian or other assistance in many of the world's trouble spots. NGOs range in size and experience from those with multimillion dollar budgets and decades of global experience in developmental and humanitarian relief to newly created, small organizations dedicated to a particular emergency or disaster. The capability, equipment and other resources, and expertise vary greatly from one NGO to another. NGOs are involved in such diverse activities as education, technical projects, relief activities, refugee assistance, public policy, development programs, human rights, and conflict resolution. The number of lives they affect, the resources they provide, and the moral authority conferred by their humanitarian focus sometimes enable NGOs to wield a great deal of influence within the interagency and international communities.

In a hostile or uncertain environment, the military's initial objective is stabilization and security for its own forces. NGOs normally seek to address humanitarian needs first and are often unwilling to subordinate their objectives to military missions, which they had no part in determining. Many NGOs view their relationship with the military under the UNOCHA Guidelines on the Use of Foreign Military and Civil Defence Assets in Disaster Relief, commonly referred to as the Oslo Guidelines, and UNOCHA Guidelines on the Use of Military and Civil Defence Assets to Support United Nations Humanitarian Activities in Complex Emergencies that define the humanitarian principles and the importance of distinction and last resort. The Guidelines for Relations Between US Armed Forces and Non-Governmental Humanitarian Organizations in Hostile or Potentially Hostile Environments, agreed to by the DOD, InterAction, and USIP, should facilitate interaction between the US military forces and NGOs.

For more information, refer to "Civil-Military Guidelines and Reference for Complex Emergencies" and JP 3-08, Appendix C, "Nongovernmental Organizations."

The Private Sector

The private sector possesses the skills and expertise to contribute to US objectives. These capabilities can be used to reduce operational requirements and maximize use of finite resources. A number of DODIs regulate the conduct of private military and security companies operating with DOD.

Crisis Action Organization

Ref: JP 3-08, Interorganizational Cooperation (val. Oct '17), pp. IV-15 to IV-19.

The combatant command crisis action organization is activated upon receipt of the CJCS warning or alert order or at the direction of the CCDR. Activation of other crisis action cells to administer the specific requirements of task force operations may be directed shortly thereafter. These cells support not only functional requirements of the JTF such as logistics, but also coordination of military and nonmilitary activities and the establishment of a temporary framework for interagency coordination. Liaison and coordinating mechanisms that the CCDR may elect to establish to facilitate the synchronization of military and nonmilitary activities include:

Humanitarian Assistance Survey Team (HAST)

Early in crisis response planning, an assessment can help identify resources to immediately mitigate a humanitarian crisis. The supported CCDR may organize and deploy a HAST to acquire information for planning. This assessment should analyze existing conditions and recommend FHA force structure. Before deploying, the HAST should review the current threat assessment; current intelligence; geospatial information and services support; and embassy, DOS, and USAID POCs. The disaster assistance response team (DART) and USAID mission can provide some of this information to the HAST. Once deployed, the HAST can assess the HN government's capabilities, identify primary POCs, determine the threat, survey facilities that may be used for FP purposes, and coordinate support arrangements for the delivery of food and medical supplies. If dislocated civilians are an element of the crisis, the DOS Bureau of Population, Refugees, and Migration (PRM), International Organization for Migration (IOM), or the United Nations High Commissioner for Refugees (UNHCR) can also be resources. The HAST works closely with the DART to prevent duplication of effort. Unlike the DART, which assesses overall humanitarian conditions and requirements in the affected country, the HAST focuses its efforts to assess the opportunities and conditions to provide specific military support to civilian agencies.

Disaster Assistance Response Team (DART)

USAID, through its OFDA DART, is the lead agency for foreign disaster response. USAID/OFDA may deploy a DART into the crisis area to coordinate the FHA effort and activate an on-call, Washington, DC-based response management team. The DART links the geographic CCMD and USG departments and agencies, international organizations, and NGOs that participate in FHA operations. The DART team leader represents the USG response and may be supported by DOD. In addition to personnel from OFDA and other parts of USAID, the DART may include liaisons from DOS, parts of DOD (e.g., US Army Corps of Engineers), or other USG departments and agencies (e.g., Centers for Disease Control and Prevention [CDC]), depending on the nature of the response. DARTs provide specialists in a variety of DR skills to help US embassies and USAID missions manage the USG response to foreign disasters. DARTs assess the disaster situation and recommend follow-up actions.

Humanitarian Assistance Coordination Center (HACC)

The supported GCC may establish a HACC to plan and coordinate with interagency partners. Normally, the HACC is a temporary body that operates during the early stages of the operation. Once a CMOC or civilian HOC has been established, the role of the HACC diminishes, and its functions transition to one or both of these organizations. Staffing for the HACC should include a director appointed by the supported GCC, a CMO planner, a USAID/OFDA advisor or liaison, a PA officer, an NGO advisor, and other augmentation (e.g., preventive medicine physician, veterinarian).

Joint Logistics Operations Center (JLOC)

The JLOC supports the geographic CCMD's joint operations center and the operations planning teams. The GCC reviews the requirements and establishes priorities to use supplies, facilities, mobility assets, and personnel effectively. The geographic CCMD may also be responsible for provision of supplies for certain interagency personnel. Formed at the discretion of the GCC and operated by the logistics directorate of a joint staff at the geographic CCMD, a JLOC functions as the single POC to coordinate logistic response into the AOR, relieving the JTF of as much of this function as possible. The JLOC may also coordinate with strategic-level providers (e.g., the Defense Logistics Agency, USTRANSCOM, the Services, and the geographic CCMD's staff) to meet JTF support requirements.

Liaison Section

The liaison section in foreign operations coordinates with USG departments and agencies, NGOs, international organizations, and private sector entities. A liaison section coordinates military activities among MNFs, other USG departments and agencies, participating international organizations and NGOs, the private sector, HNs, and indigenous populations. Military forces, participating agencies, and HNs should consider exchanging liaison personnel to maximize information flow. Information should flow between all parties. NGO liaisons should have access to the military. The CMOC can facilitate coordination. Alternatively, the HN may establish a coordination center.

The Multinational Planning Augmentation Team (MPAT)

The Multinational Planning Augmentation Team (MPAT) program is a cooperative multinational effort to facilitate establishment and augmentation of a multinational task force HQ. The MPAT provides multinational expertise in planning and integrates other nations' militaries, international organizations, and NGOs in the planning process. The MPAT uses a trained cadre that has worked with international organizations and NGOs prior to a crisis and deploys to the task force HQ once a crisis occurs. The MPAT program develops techniques and exercises multinational planning activities for operational-level task forces. This includes coordination, collaboration, and cooperation with USG organizations, international organizations, NGOs, private sector entities, and HN government agencies.

Interagency Management System (IMS)

When IMS is activated, an integration planning cell may deploy from the CRSG and colocate with the designated GCC's HQ. The integration planning cell should be established in conjunction with the development of a US strategic plan. It supports the GCC in integrating the military plan with the civilian components of the US strategic and implementation plans and serves as the representation of all participating agencies and the CRSG to the GCC.

PA and Media Support

The proactive release of accurate information to domestic and international audiences puts joint operations in context; facilitates informed perceptions about military operations; undermines adversarial propaganda; and helps achieve national, strategic, and operational objectives. By conveying the facts about joint force activities in a proactive manner, PA helps the JFC to impact the information environment, particularly as it relates to public support. The JFC's PA officer helps inform USG departments and agencies and NGOs concerning joint force operations. The PA officer also coordinates public information activities to align messages.

At the national level, the ASD(PA) interfaces with USG departments and agencies in the NSC/DC and issues PA guidance; advises on public information, command information, and community relations; and provides DOD information to the public, Congress, and the media. At the theater level, PA planning includes coordination with USG departments and agencies, the ambassador and country team (particularly the embassy PA section), the HN, national and international media, media elements of member forces, and other external stakeholders.

IV. Joint Task Force (JTF) Considerations

JFCs are responsible to conduct civil-military operations (CMO). They may establish a JCMOTF when the scope of CMO requires coordination and activities beyond the organic CMO capability. The US Army CA command and brigade, or the United States Marine Corps (USMC) CA group, are staffed to provide the operational core of a JCMOTF. NGOs in the operational area may not have a similarly defined structure.

Operations by USG departments and agencies, the equivalent agencies of other national governments, international organizations, NGOs, and private sector entities, in concert with or supplementing those of HN entities, may be in progress when US forces arrive in a JOA.

See chapter 5, Joint Task Forces (JTFs) for further discussion.

Regional Strategy

In further analyzing the mission, consider how the theater or functional strategy will affect joint force planning and operations in the projected JOA. The NSC, DOS, COM, and the supported CCDR will provide the regional strategy and an appreciation for how the regional strategy affects the countries involved in projected operations. This may affect COA development, themes and messages, and planning and execution activities. A well-defined regional strategy will delineate the military mission and assist in determining force requirements and defining the theater objectives.

JTF Assessment Team

A valuable tool in the mission analysis process is the deployment of a JTF assessment team to the projected JOA. The purpose of the assessment team is to establish liaison with the ambassador or COM, country team, HN, and, if present, multinational members, UN representatives, and IGO and NGO representatives. USAID, because of the extensive contacts it develops in carrying out development work at the community level, can provide key situational awareness for JTF assessments. The JTF assessment team is similar in composition to the HAST and, if provided early warning of pending operations, may be able to conduct assessment in association with the HAST.

Organizational Tools for the JTF

The CJTF should establish structures to coordinate all activities in the JOA. In addition to military operations, these structures should include political, civil, administrative, legal, and humanitarian elements, as well as international organizations, NGOs, private sector entities, and the media. The CJTF should consider how joint force actions and those of other organizations contribute to the desired objectives. This consideration requires liaison and routine contact with all parties, as well as reliable communications. An assessment team can develop recommendations for the CJTF concerning formation of an executive steering group (ESG), CMOC, and liaison teams.

V. Civil-Military Teams (JIATFs/PRCs)

A civil-military team combines diplomatic, informational, military, and economic capabilities to enhance the legitimacy and the effectiveness of the HN government. A civil-military team can combine military and civil efforts to diminish the means and motivations of conflict, while developing provincial, district, state, or local institutions so they can lead in governance, provide basic services and economic development, and enforce the rule of law. Civil-military teams of interagency experts can be formed to conduct specific missions (e.g., agricultural, economic, and CT). Examples of civil-military teams include JIATFs, and provincial reconstruction teams (PRTs) in Iraq and Afghanistan.

For more information on JIATFs and civil-military teams, refer to JP 3-08 Appendix E, "Joint Interagency Task Force," and JP 3-08 Appendix F, "Civil-Military Teaming," respectively.

VI. Civil-Military Operations Center (CMOC)

Ref: JP 3-08, Interorganizational Cooperation (val. Oct '17), pp. IV-22 to IV-27.

Military forces should normally develop relations with USG departments and agencies, civilian authorities, international organizations, NGOs, private sector entities, and the population during contingency operations. The CMOC is a mechanism to coordinate CMO that can also provide operational and tactical level coordination between the JFC and other stakeholders. The CMOC generally does not set policy or direct operations, but rather coordinates and facilitates. The CMOC is the meeting place of stakeholders. It may be physical or virtual, and conducted collaboratively through online networks, as NGOs may be reluctant to conduct coordination meetings in settings managed by the military. The organization of the CMOC is theater- and mission-dependent. A commander at any echelon may establish a CMOC. In fact, more than one CMOC may be established in an operational area and each is task-organized based on the mission. Horizontal and vertical synchronization among multiple CMOCs assists in unity of effort.

During large-scale FHA operations, US forces may organize using the CMOC. If both are established, the CMOC should colocate with the HOC to facilitate operations and assist in later transition of any CMOC operations to the HOC.

In FHA operations, the UN organizes along key clusters. Coordination meetings hosted by UN elements may supplant the need for a US-military run CMOC. Commanders should complement, rather than compete with, the UN cluster meetings. NGOs are far more likely to participate in UN-sponsored meetings than US- (especially US military) sponsored coordination and deconfliction meetings.

The CJTF must carefully consider where to locate the CMOC. Security, FP, and easy access for external stakeholders are all valid considerations. The location should be distinct and separate from the joint force operations center, even if geographically colocated, and should be segregated from any nation's classified information. If security conditions permit, every effort should be made to locate the CMOC "outside the wire" to maximize participation by organizations that want to minimize the appearance of close association with military operations.

Political representatives in the CMOC may provide the CJTF with avenues to align operational considerations and concerns with political actions. Additionally, the CMOC provides stakeholders a single point to coordinate with the military, which facilitates the efforts of a joint force and the relief community.

- The military should not attempt to dictate USG civilian counterpart or international organization, NGO, and private sector partner activities, but to coordinate a team approach to problem resolution.
- JFCs cannot direct organizations or people not under their command to cooperate. However, a JFC can work with these entities to forge unity of effort on issues like security, logistic support, information sharing, communications, Periodic meetings can be scheduled in the CMOC to match civil-sector needs to organizations capable of meeting them. USG validated RFAs go to the appropriate JTF or agency representative for action.

Refer to TAA2: Military Engagement, Security Cooperation & Stability SMARTbook (Foreign Train, Advise, & Assist) for further discussion. Topics include the Range of Military Operations (JP 3-0), Security Cooperation & Security Assistance (Train, Advise, & Assist), Stability Operations (ADRP 3-07), Peace Operations (JP 3-07.3), Counterinsurgency Operations (JP & FM 3-24), Civil-Military Operations (JP 3-57), Multinational Operations (JP 3-16), Interorganizational Cooperation (JP 3-08), and more.

VII. Aligning Words with Deeds

Ref: JP 3-08, Interorganizational Cooperation (val. Oct '17), pp. IV-27 to IV-30.

The USG builds on coordinated actions and information to maintain credibility and trust with foreign populaces, governments, adversaries, and US citizens alike. This is done through accuracy, consistency, timeliness, and transparency in words and deeds. Credibility is important to build relationships that advance our national interests.

All USG departments and agencies share responsibility to use information as an instrument of national power. This includes developing processes to access and analyze communication and to deliver information to key audiences, both US and foreign. DOD synchronizes, aligns, and coordinates communication to facilitate understanding by key audiences. This is done to create, strengthen, or preserve conditions favorable to USG objectives. National strategic direction provides the building blocks for the JFC's communication guidance. It is also essential to DOD initiatives to achieve unity of effort through unified action with our interagency partners and the broader interorganizational community. Key audience beliefs, perceptions, and behavior are essential to develop any strategy, plan, or operation. PA, IO, and defense support to public diplomacy (DSPD) are supporting capabilities. While CCDRs directly control assigned PA and IRC assets, they do not direct those assets conducting public diplomacy, which are the responsibility of DOS or the local US embassy.

Media reports influence public attitudes about operations, which in turn can affect policy decisions. Most USG departments and agencies have representatives dedicated to reporting their activities, each with multiple sources in the respective organization. DOD's primary media representatives are PA personnel. Potential operations draw intense media scrutiny. This scrutiny can influence USG departments and agencies, international organizations, and NGOs from the strategic level of the NSC, to the field, as international organizations and NGOs vie for public attention and charitable contributions. Responding to competing or contradictory news reports can divert personnel from planning and execution. Commanders and their staffs evaluate the impact of information on the operation and the interagency coordination process, to integrate PA expertise in crisis planning for operations. The White House Office of Global Communications facilitates USG communication with foreign audiences. The DOS Bureau of International Information Programs is the communications service for the US foreign affairs community.

Whole-of-government themes and an overarching narrative provide a foundation to build unity of effort. Subordinate themes, messages, or stories tailored to specific audiences and built on cultural understanding and knowledge of the key communicators in the operational area can help implement unified action. What works in one environment will not necessarily work in another due to variables such as the sophistication of the local populations, differences in government, value systems, media, and communication systems. JFCs synchronize themes, messages, images and actions, selecting the delivery vehicle, optimizing types of media, and infusing messages with beliefs and attitudes to influence the audience.

CCDRs and staffs should evaluate communication considerations with the interagency partners when planning joint operations. Joint operations can influence and inform key foreign audiences, foster understanding of US policy, and advance US interests. Words, images, and actions can shape the operational environment. CCDRs plan, execute, and assess activities to implement security cooperation plans in support of US embassies' information programs, public diplomacy, and PA programs directly supporting DOD missions.

Defense support to public diplomacy (DSPD) are DOD activities and measures to support and facilitate USG public diplomacy efforts. DSPD helps align DOD activities with a coherent and compelling DOS diplomacy of deeds in concert with other USG departments and agencies. DOS leads public diplomacy, with the DOD in a supporting role.

(JFODS5-1) Index

L

M

N

O

Purchase/Order

SMARTsavings on SMARTbooks! Save big when you order our titles together in a SMARTset bundle. It's the most popular & least expensive way to buy, and a great way to build your professional library. If you need a quote or have special requests, please contact us by one of the methods below!

View, download FREE samples and purchase online:

www.TheLightningPress.com

Order SECURE Online

Web: www.TheLightningPress.com

Email: SMARTbooks@TheLightningPress.com

24-hour Order & Customer Service Line

Place your order (or leave a voicemail) at 1-800-997-8827

Phone Orders, Customer Service & Quotes

Live customer service and phone orders available Mon - Fri 0900-1800 EST at (863) 409-8084

Mail, Check & Money Order

2227 Arrowhead Blvd., Lakeland, FL 33813

Government/Unit/Bulk Sales

The Lightning Press is a **service-disabled, veteran-owned small business**, DOD-approved vendor and federally registered—to include the SAM, WAWF, FBO, and FEDPAY.

We accept and process both **Government Purchase Cards** (GCPC/GPC) and **Purchase Orders** (PO/PR&Cs).

15% OFF
RETAIL EVERYDAY

Buy direct from our website and always get the latest editions and the best pricing. Join our SMARTnews email list for free notification of changes and new editions.

www.TheLightningPress.com